Suspension System

Michael Trzesniowski

Suspension System

Springer Vieweg

Michael Trzesniowski
Drivetrain and Suspension Systems
Pankl Racing Systems AG
Kapfenberg, Austria

ISBN 978-3-658-39846-0 ISBN 978-3-658-39847-7 (eBook)
https://doi.org/10.1007/978-3-658-39847-7

This Springer Vieweg imprint is published by the registered company Springer Fachmedien Wiesbaden GmbH, part of Springer Nature.
The registered company address is: Abraham-Lincoln-Str. 46, 65189 Wiesbaden, Germany

Preface

Now the manual series is already in its second edition, and the family of these books is growing. While there were already five volumes at the beginning, another volume has been added for this edition. The authoritative idea that individual special volumes can go into depth without space problems has thus proven itself. Since the publication of the first edition, further findings have been added, which have found their way into the corresponding chapters, or new chapters have been added.

That the contents nevertheless fit together and complement each other as if in a single book – one of the great strengths of the original book *Race Car Technology* – is ensured by the editor in a comparable way to how the project manager keeps an eye on the overall function in a major design project.

The racing car technology handbook series is dedicated to the racing vehicle from conception, design and calculation to operation and its (further) development.

The first volume, "Basic Course in Race Car Technology", thus offers not only current considerations but also a historical overview of motorsport, racing operations, such as the rescue chain, and a comprehensive overview of the technology used in racing cars as a general introduction to the subject. For more than fifteen years, the author has been concerned with the driving dynamics and chassis tuning of production passenger cars.

Volume two, "Complete Vehicle", starts with the chronological design process and therefore begins with concept considerations, takes into account safety aspects and the design of the driver's environment, describes aerodynamic influences and then looks at the frame and bodywork design.

Volume three, "Powertrain", deals with all forms of drive systems and their energy storage and continues in the sense of load flow via starting elements and transmission up to the side shafts. Electrical systems and electronic driving aids have also found their appropriate place in this volume.

Volume four, "Chassis", is devoted exclusively to the decisive subsystem and its components that determine driving behaviour. Tyres and wheels, wheel-guiding parts, springs and dampers, steering and brakes are covered.

Volume five "Data Analysis, Tuning and Development" deals with the phase that follows once the vehicle has been designed and built. The development and tuning of a racing vehicle require a much different approach than its construction and key tools – such as data acquisition and analysis, simulation and testing – are therefore presented. The subject of data acquisition and analysis is profoundly presented by an author who is confronted with this activity on a daily basis.

For volume six, "Practical Course in Vehicle Dynamics", authors have been recruited who have decades of experience as race engineers and race drivers on the race track. In their work, they describe the practical tuning of racing vehicles, underpin what they present with examples of calculations and thus also build a bridge to the theoretical considerations in the other volumes.

I wish all readers that they will find "their" volume in the abundance offered and that they will get essential impulses for their studies, profession and/or leisure time from reading it, be it because they are designing a vehicle, building one, operating and improving one or because they are analysing one with a thirst for knowledge.

Kapfenberg, Austria
Spring 2019

Michael Trzesniowski

Greeting

27.05.2019
EC Todsen

Second Edition: Racing Car Technology Handbook: Six Volumes

Motorsport continues to inspire. For as long as there have been cars, drivers have been pushing their racing cars to their technical and physical limits, engaging in gripping and exciting competitions. But the competition doesn't just take place on the race track. The foundation for success is laid in the development departments and design offices. In-depth knowledge of vehicle technology and development methodologies, along with thorough and timely project management, creative problem-solving skills and unconditional team play, determine victory or defeat.

Motorsport continues to be a model and guide for technological progress – be it in lightweight design, material selection or aerodynamics. Chassis and tire technology also benefit immensely, and new safety concepts are often based on experience from the racetrack. However, the influence of motorsport is particularly evident in the powertrain: In addition to the impressive increase in performance and efficiency of the classic internal combustion engine drive system, the key future technologies of hybrid and purely electric drive have also successfully arrived in motorsport competition and are continuing this successfully and with public appeal in a partly completely new setting. It remains very exciting to observe which attractive innovations digital networking solutions and autonomous driving systems will generate in motorsport. I recommend that young engineers in particular acquire the tools for their future careers in motorsport. What you learn in motorsport sticks. Formula Student already offers an ideal environment to start with.

I am very pleased that the book series *Handbuch Rennwagentechnik* has been so well received and that the second edition has been published within two years. This shows that the competencies addressed are clearly presented in this work and conveyed in an understandable way.

This book series has deservedly become a well-known and valued reference work among experts. The work brings students closer to the fascination of motorsport and racing enthusiast laymen to a deeper technical understanding.

I wish you much success on and off the race track!

Prof. Dr.-Ing. Peter Gutzmer

Deputy Chairman of the Executive Board and Chief Technology Officer, Schaeffler AG, Herzogenaurach, Germany

Abbreviations, Formula Symbols and Units

Equations given in the text are generally quantity equations. The quantities can be used in any units, preferably in the SI units (meter-kilogram-second system). The unit of the quantity to be calculated then results from the selected units of the variables. Sometimes the numerical value equations commonly used in practice are also given. With these, the equation is only correct if it is calculated with the specified units. The unit of the result variable is therefore also given in the text.

Geometric Points

Bo	*Body centre of gravity*
C to *G*	*Reference points, in general*
M	*Centre point*
O	*Pitch centre*
P	*Instantaneous centre*
Ro	*Roll centre*
T and *U*	*Tie rod or linkage pivot point*
U_f or U_r	*Wheel centre point, front or rear*
V	*Vehicle centre of gravity*
W	*Centre of tyre resp. wheel contact*

Indices

If more than one index occurs, they are separated by a comma. The order of indices is this:

For forces, the first index indicates the location or point at which the force is applied and the second index indicates the direction of the force, e.g. $F_{W,\,Z}$ . . . wheel contact force (high force at the tyre contact point). The vehicle fixed coordinate system used is defined in the glossary.

Additional specifications, such as front, rear, driven, etc., follow as further indices.

0	*Zero-point position or starting point. Ambient*
1	*To the top/ in jounce/ in compression/ in*
2	*To the bottom/ in rebound/ out*
A	*Ackermann steering angle*
a	*Driven, accelerating (one wheel only)*
ax	*Axial*
b	*Bending*
b	*Braking (one wheel only)*
B	*Braking (one axle)*
Ba	*Stabiliser bar, anti-roll bar*
Vol.	*Brake disc*
Bl	*Blade of anti-roll bar*
Bo	*Body*
D	*Damping*
dyn	*Dynamic*
e	*Due to the elasticity (compliances)*
Ex	*Excess*
f	*Front, forward*
fr	*Friction*
H	*Steering wheel*
h	*Master-*
hyd	*Hydraulic*
i	*Inner wheel, inner*
k	*Kinematic (due to wheel travel)*
critical	*Critical*
L	*Aerodynamic*
l	*Left, left-hand side*
lo	*Slipping, lock resp.*
ls	*Loss*
m or med	*Middle, mean*
max	*Maximum permissible*
min	*Minimum*
n	*Rated value*
o	*Outer wheel, outer*
p	*Driver*
Pi	*Piston*
pl	*Design position (partial loaded)*
Pr	*Piston rod*

(continued)

r	*Rear, aft*
R	*Rolling (wheel)*
rad	*Radial*
Rd	*Rod, linkage resp.*
Re	*Residual*
ref	*Reference ~*
Ro	*Roll*
rs	*Right, right-hand side*
rsl	*Resulting*
S	*Steering*
S	*Anti-roll bar, stabilizer*
Sg	*Steering gear*
Sp	*Spring*
t	*Total, nominal value resp.*
T	*Tyre*
tan	*Tangent*
ts	*Torsional*
U	*Unsprung weight or axle*
V	*Overall vehicle*
W	*Wheel*
X or x	*Longitudinal direction in general*
Y or y	*Lateral direction*
Z or z	*Vertical direction*

Distances **in mm**

Δb	*Track change*
a to p	*Distances and length (in general)*
b_f or b_r	*Track width, front or rear*
d or D	*Diameter, in general*
D_S	*Track circle diameter (front)*
e	*Wheel offset*
e_R	*Offset of vertical force at rolling tyre*
f	*Diagonal spring travel*
h or H	*Height, in general*
h_{Bo}	*Height of body centre of gravity*
h_{ref}	*Ride height*
$h_{Ro,f}$ or $h_{Ro,r}$	*Height of roll centre at front or rear axle*

(continued)

h_{Sg}	*Travel of steering gear (rack and pinion)*
h_V	*Height of the vehicle centre of gravity*
j	*Distance between the two steering axis at the ground*
l	*Wheelbase*
$l_{Bo,f}$ or $l_{Bo,r}$	*Distance of body centre of gravity to the middle of the front or rear axle*
l_f or l_r	*Distance of vehicle centre of gravity to middle of front or rear axle*
n_τ	*Caster offset at wheel centre*
$n_{\tau,k}$	*Kinematic lateral force lever arm due to caster*
$n_{\tau,t}$	*Lateral force arm, in total*
OD_T	*Outer diameter of the tyre*
q	*Force lever of vertical force*
r	*Effective control arm length or force lever in general*
R	*Path radius, corner radius*
r_Δ	*Static toe-in (one wheel only)*
r_σ	*Scrub radius, static*
$r_{\tau,e}$	*Elastokinematic caster offset at ground*
$r_{\tau,k}$	*Kinematic caster offset at ground*
$r_{\Delta,t}$	*Total static toe-in (both wheels of one axis)*
$r_{\sigma,t}$	*Total transverse offset at ground*
$r_{\tau,t}$	*Total caster offset at ground*
$r_{\tau,T}$	*Pneumatic trail*
r_U	*Deflection force lever arm*
r_a	*Kingpin offset at hub*
r_{dyn}	*Force dynamic rolling radius of the tyre at 60 kph*
r_{stat}	*Static loaded radius of the tyre*
r_T	*Force offset in the centre of tyre contact (+) inside or (-) outside of curve*
s	*Travel or stroke, in general*
s_1	*Wheel travel in jounce*
s_2	*Wheel travel in rebound*
s_{Re}	*Residual wheel travel*
s_t	*Total wheel travel*
s_T	*Static tyre deflection*
t	*(Wall) thickness*

Angle in ° or rad

α_f or α_r	*Slip angle of front or rear tyre*
β	*Angle, in general*

(continued)

β	*Sideslip angle (attitude angle)*
χ	*Acceleration reaction support resp. diagonal springing angle*
δ	*steer angle*
$\delta_{A,o}$	*Ackermann steer angle, nominal value outer wheel*
δ_H	*Steering wheel angle*
δ_m	*Mean steer angle*
δ_i or δ_o	*Actual steer angle, inner or outer wheel*
δ_{Sg}	*Turn angle of steering-column stalk*
$\delta_{V,0,t}$	*Total static toe angle*
$\Delta\delta$	*Differential steer angle (actual value)*
$\Delta\delta_A$	*Differential steer angle according to Ackermann (nominal value)*
$\Delta\delta_F$	*Steering flaw*
$\Delta\delta_{H,e}$	*Part of steering-wheel angle due to compliances*
$\Delta\delta_H$	*Range of steering wheel angle*
ε or ε_W	*Camber angle*
ε	*Brake reaction support angle*
φ	*Body roll angle*
λ	*Steering arm angle*
σ	*Kingpin inclination angle*
τ	*Caster angle of the steered front wheels*

Masses, Weights in kg

m	*Mass, weight or load in general*
$m_{1,Bo,f}$ or $m_{1,Bo,r}$	*Part of body mass on one side of the front or rear axle*
m_{Bo}	*Vehicle body weight*
$m_{Bo,f}$ or $m_{Bo,r}$	*Part of body mass on front or rear*
$m_{U,f}$ or $m_{U,r}$	*Unsprung axle mass, front or rear*
m_V	*Weight of vehicle in running condition*
$m_{V,f}$ or $m_{V,r}$	*Axle load, front or rear*
$m_{V,t}$	*Gross vehicle weight*
m_W	*Weight of one wheel*

Forces *in* N

ΔF	*Change of force*
$\Delta F_{V,Z}$	*Axle load transfer*
$\Delta F_{W,Z}$	*Change of vertical force at one wheel*
F_D	*Damping force*
F_H	*Steering-wheel force*
$F_{L,Z}$	*Aerodynamic downforce*
F_O	*Force at pitch centre*
F_{Pi}	*Piston rod force*
F_R	*Rolling resistance of the tyre*
F_{rsl}	*Resulting force*
F_{Sp}	*Spring force, one side of the axle*
F_T	*Tie rod or push rod force*
$F_{T,Y,\varepsilon}$	*Lateral force due to camber*
$F_{V,X,B}$	*Brake force at the centre of gravity of the vehicle*
$F_{V,Y}$	*Lateral force at vehicle*
$F_{V,Z,f}$ or $F_{V,Z,r}$	*Axle load front or rear*
$F_{V,Z,t}$	*Cross vehicle weight*
$F_{W,X,a}$ or $F_{W,X,A}$	*Accelerating force in the centre of tyre contact of one wheel (a) or both wheels (A)*
$F_{W,X,b}$	*Brake force in the centre of tyre contact of one wheel*
$F_{W,X,B,f}$ resp. $F_{W,X,B,r}$	*Brake reaction force to the front or rear axle*
$F_{W,Y}$	*Lateral force at wheel*
$F_{W,Z}$	*Vertical force at the centre of tyre contact*

Torques and Moments in Nm

M_a or M_A	*Driving torque to one wheel (a) or axle (A)*
M_b	*Bending moment*
M_b or M_B	*Braking torque applied to one wheel (b) or axle (B)*
M_{fr}	*Moment of friction*
M_H	*Steering wheel torque*
M_{Sg}	*Torque at steering gear*
M_W	*Self-aligning torque*
T	*Torsional moment in general*

Spring Rates (in N/mm)

c_f or c_r	*Rate of the body supporting spring at parallel springing, related to the center of tyre contact of one axle side, front or rear*
c_S	*Rate of the anti-roll bar at reciprocal springing*
c_{SP}	*Static rate of the spring*
c_T	*Vertical tyre (spring) rate*

(in Nm/°)

c_α	*Rate of torsional bar*
$c_{Ro,S,f}$	*Roll stiffness of anti-roll bar acting on front axle*
$c_{Ro,Sp,f}$	*Roll stiffness of suspension springs acting on front axle*
$c_{Ro,f}$	*Roll stiffness front axle*

Dimensionless Key Figures

η_G	*Efficiency of lateral force proportioning*
Φ_f or Φ_r	*Brake force fraction front or rear*
$\mu_{W,X}$	*Coefficient of friction in longitudinal direction*
$\mu_{W,Y}$	*Coefficient of friction in lateral direction*
c_A	*Downforce coefficient*
c_W	*Drag coefficient*
D	*Damping ratio*
i_m	*Axle load ratio front/rear*
i_S	*Overall kinematic steering ratio*
i_{Sg}	*Steering gear ratio*
i_{Sp}	*Ratio of spring to wheel (motion ratio, installation ratio)*
i_T	*Ratio of steering linkage*
j	*Number in general*
k	*Correction factor for tyre stiffness*
k_B	*brake coefficient*
k_p	*Correction factor for tyre pressure because of camber*
k_Q	*Correction factor for load capacity because of camber*
k_R	*Rolling resistance coefficient*
$k_{R,0}$	*Rolling resistance coefficient measured on a tyre test rig*
k_χ	*Anti-squat coefficient*

(continued)

$k_{\varepsilon,f\ \text{resp.}\ r}$	*Anti-dive coefficient*
$S_{W,X,a}$ or $S_{W,X,b}$	*Longitudinal slip under accelerating or braking*
z	*Braking factor, normalized deceleration*

Other Sizes

θ	*Celsius temperature*	°C
ρ	*Density*	kg/m³
ρ_L	*Density of air*	kg/m³
σ	*Stress*	N/m²
τ	*Shear stress*	N/m²
Δ	*Difference*	–
ω	*Circular frequency*	s^{-1}
A	*Area, cross-section area*	m²
a_x	*Longitudinal acceleration in general*	m/s²
a_y	*Lateral acceleration*	m/s²
$c_{\alpha,T}$	*Cornering stiffness of a tyre*	N/°
c_α	*Cornering stiffness of an axle*	N/°
E	*Modulus of elasticity, Young's modulus*	N/mm²
f	*Frequency*	Hz
g	*Acceleration due to gravity*	m/s²
G	*Shear modulus*	N/m²
I	*Area moment of inertia*	mm⁴
J	*Dynamic moment of inertia*	kgm²
k_{ap}	*Critical damping*	Ns/m
k_D	*Damping value*	Ns/m
N	*Specific stopping power*	kW/cm²
n	*Revolutions per minute or vibration frequency*	min^{-1}
P	*Power*	W
p	*Pressure*	N/cm²
p_0	*Ambient pressure*	bar[1]
P_{ls}	*Power loss*	W
p_T	*Tyre (inflation) pressure*	bar[1]
R_e	*Yield strength*	N/m²

(continued)

[1] 1 bar = 100 kPa. Although the valid SI unit for pressure is Pascal (Pa), the book uses the unit bar, which is more "handy" in practice.

R_m	*Ultimate tensile strength*	N/m^2
$R_{p0.2}$	*0.2% yield strength*	N/m^2
T	*Thermodynamic temperature*	K
t	*Time*	s
V	*Volume*	$L = dm^3$
v_D	*Piston velocity in shock absorber*	m/s
v_V or v_X	*Longitudinal velocity*	m/s or km/h
v_W	*Circumferential tyre velocity*	m/s
W	*Work*	J

Contents

Tyres and Wheels

1

While wheels are still in the general field of vision because of their aesthetic effect, the spectators at the race track hardly pay any attention to the tires. The situation is no different for production vehicles. Studies have shown that hardly any drivers regularly check the inflation pressure of their tyres. Yet both parts count as unsprung masses, are relevant to safety and the tyre is by far the most important single (mechanical) component on a vehicle and especially on a competition vehicle. The tyres transmit all the forces to the car which ensure its desired position in relation to the road surface (the aerodynamic forces merely

M. Trzesniowski, *Suspension System*,
https://doi.org/10.1007/978-3-658-39847-7_1

support it in this respect or even have a disruptive effect in the event of lift at the axles or in the event of crosswinds), and are therefore the limiting link in the overall racing vehicle system. Apart from this, the tyres are also the most difficult component to control, because they are subject to large deformations and also change their properties strongly with the temperature and with the state of wear, i.e. already during a race.

1.1 Terms

Tyres are characterised by the outer diameter, section width, rim diameter and in many cases section height, Fig. 1.1. The operating width $b_{T,\max}$ can be up to 6% greater than the section width. Under load, the tyre compresses (static tyre deflection). Therefore, the static radius is slightly less than half the outer diameter. When the tyre rotates rapidly in operation, the radius increases, see Fig. 1.11. At 60 km/h, this value is used as a reference and is called the dynamic rolling radius. It is determined indirectly from the number of wheel revolutions and the distance travelled.[1]

See also	DIN 70020 Part 5:	Tire terms
	DIN 7803 Part 5:	Radial tyre marking
	DIN 74361 Part:	Disc wheels
	DIN 74361 Part 2:	Wheels mounting
	DIN 7829:	Rims marking

[1] For more details, see Racing Car Technology Manual Vol. 3 Powertrain, Chap. 4 Powertrain Design.

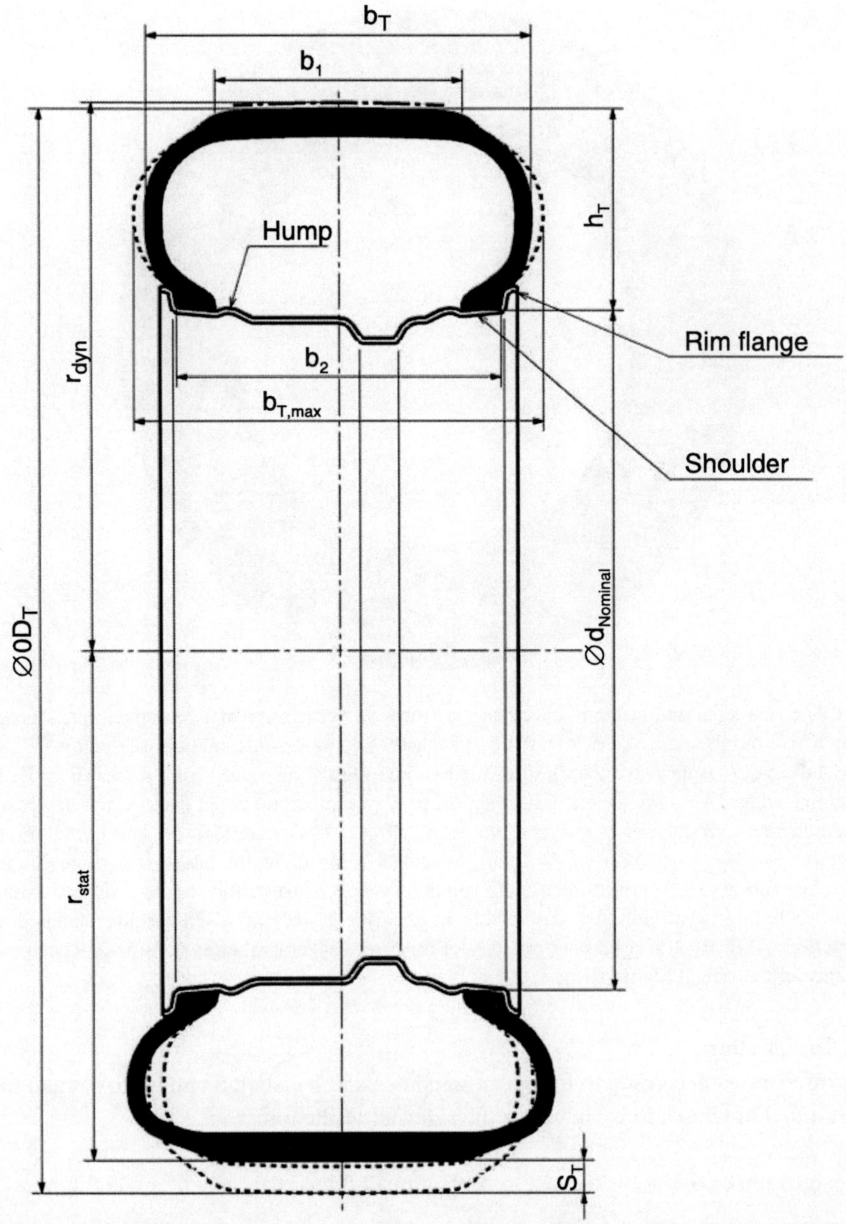

Fig. 1.1 Main wheel and tyre dimensions according to DIN 70020 T5. Rim: $d_{nominala}$ *Rim* diameter, b_2 *rim width*; *Tyre, AE: tire*: Nominal diameter, b_T Section width determined on measuring rim at 1.8 bar (or 2.3 bar for V, W and ZR tyres) inflation pressure, b_1 *Tread* width, $b_{T,max}$ Operating width (overall width*)*, OD_T Outside diameter determined on measuring rim at 1.8 bar (or 2.3 bar for V, W and ZR tyres) inflation pressure, h_T Cross section height, s_T Static tyre deflection, r_{stat} *Static loaded radius*, r_{dyn} dynamic rolling radius

Fig. 1.2 Some legal and standard specifications for road vehicle tyres. 1 Manufacturer, 2 Product name, 3 Size designation, see below, 4 Load capacity and speed information, see below, 5 Tubeless = tubeless, 6 Approval according to international regulations with test number (E = ECE R 30 nominal values. 4 = country that issued approval), 7 Manufacturer code: Factory, tire design, date of manufacture (production week/year: 43 week 2006), 8 Department of Transportation (U.-S. Department of Transportation), 9 T.W.I. = Tread Wear Indicator, transverse ridges in tread grooves appearing at 1.6 mm residual tread height), 10 Tire construction details and max. Inflation pressure, 11 Treadwear: relative service life of the tyre based on a US-specific standard test, 12 Traction: A, B or C. Wet braking capacity of the tyre, 13 Temperature: A, B or C. Temperature resistance of the tyre at higher speeds

Tyre Designation

The tyre marking according to European standards and legislation applies to car and truck tyres up to 270 km/h and is shown on the sidewall of the tyre, Fig. 1.2.

Designation example for the size specification and load capacity of a radial tyre:

245/40 R 18 88 Y with:	245	=	Width b_T of the new tyre in mm on the measuring rim at an inflation pressure of 1.8 bar
	40	=	Cross-section ratio in %
	R	=	Tyre type: R = radial tyre/2 = diagonal Tyre/ D = emergency wheel
	18	=	Rim diameter $d_{nominal}$ in inches
	88	=	Load index, determines the load capacity, see Table 1.2: Max. 560 kg at 2.5 bar inflation pressure
	Y	=	Speed symbol, see Table 1.1: Up to 300 km/h

Designation example for the size specification of a diagonal tyre:

5.60–15/4 PR with:	5.60	=	Tyre width in inches
	15	=	Rim diameter in inches
	4 PR	=	Number of plies (in this case 4). Indication of the carcass strength, PR stands for *ply rating*

The minimum inflation pressure of passenger car tyres is determined by the wheel load and the maximum speed of the vehicle, see Tables 1.2 and 1.1.

Above 160 km/h, the inflation pressures must be increased linearly by Δp_T or the load-bearing capacity is reduced at the same pressure corresponding to a reduction by Δp_T:

Speed, km/h	160	170	180	190	200	210
Δp_T, bar	0	0.06	0.12	0.18	0.24	0.3

The maximum pressures according to Table 1.3 must not be exceeded

For camber angles ε above 2°, the inflation pressures must be corrected i.e. multiplied by the following factor k_p:

Camber angle ε, °	≤ 2	2.5	3	3.5	4
Correction factor k_p, –	1	1.03	1.07	1.1	1.14

(Intermediate values are to be interpolated linearly)

or the load-bearing capacity must be corrected with the following factor k_Q:

Camber angle ε, °	≤ 2	3	4
Correction factor k_Q, –	1	0.95	0.90

(Intermediate values are to be interpolated linearly)

At higher speeds, the specified minimum inflation pressures (Table 1.2) must be increased by the stated values [bar] for safety reasons:

Speed, km/h		160	170	180	190	200	210	220	230	240	250	260	270	300
Speed symbol	H	0	0.06	0.12	0.18	0.24	0.3	–	–	–	–	–	–	–
	V	0	0	0	0	0	0	0.1	0.2	0.3	–	–	–	–
	W	0	0	0	0	0.1	0.2	0.3	0.4	0.5	0.5	0.5	0.5	–
	Y	0	0	0	0	0	0	0	0.1	0.2	0.3	0.4	0.5	0.5

The maximum pressures according to Table 1.3 must not be exceeded.

The inflation pressure values listed in Tables 1.2 and 1.1 are minimum values. The actual pressures for optimum driving behaviour may be higher. In addition, the pressures must be increased, e.g. due to higher driving speed or deviating camber angles. However, the maximum values listed in Table 1.3 must not be exceeded under any circumstances.

Table 1.1 Speed symbol and load capacity

Maximum speed of the vehicle, km/h	Speed symbol	Maximum speed of the vehicle, km/h	Speed symbol and correction factor, –		
			V	W	Y
120	L				
130	M	210	1	1	1
140	N	220	0.97	1	1
150	P	230	0.94	1	1
160	Q	240	0.91	1	1
170	R	250	–	0.95	1
180	S	260	–	0.90	1
190	T	270	–	0.85	1
200	U	280	–	–	0.95
210	H	290	–	–	0.90
Over 240	ZR	300	–	–	0.85

The following applies to V, W, and Y: Lower load capacities must be accepted at higher speeds. The permissible load capacities are obtained by multiplying the values in Table 1.2 by the correction factors given. In the case of ZR tyres which are no longer manufactured, these values were determined jointly by the vehicle manufacturer and the tyre manufacturer

The designation for racing tires differs from that of street tires. It is structured as follows: Tyre width/outer diameter – Rim diameter.

Designation examples racing tires:

250/530–13	250	=	Tyre width in mm
	530	=	Outer diameter in mm
	13	=	Rim diameter in inches
10.0/20.0–13	10.0	=	Tyre width in inches
	20.0	=	Outer diameter in inch
	13	=	Rim diameter in inches

Height-to-Width Ratio (Aspect Ratio)

The ratio, which is influential for tyre properties, is defined as:

$$Sectionratio = \frac{\text{Height}}{\text{Wide}} = \frac{h_T}{b_T} \cdot 100\%$$

A height/width ratio of, for example, 50% is referred to as a 50 series tyre or a 50 series tyre.

The development history of passenger car tires, which goes back about a hundred years, shows a clear decrease in the aspect ratio from 113% to up to 25% [14], i.e. the tires became

Table 1.2 Load capacity indices (load index)

	Tyre load capacity[a] [kg] at inflation pressure[b] [bar]										
Load index	1.5	1.6	1.7	1.8	1.9	2.0	2.1	2.2	2.3	2.4	2.5
69	215	225	240	250	260	270	285	295	305	315	325
70	225	235	245	260	270	280	290	300	315	325	335
71	230	240	255	265	275	290	300	310	325	335	345
72	235	250	260	275	285	295	310	320	330	345	355
73	245	255	270	280	295	305	315	330	340	355	365
74	250	260	275	290	300	315	325	340	350	365	375
75	255	270	285	300	310	325	335	350	360	375	387
76	265	280	295	310	320	335	350	360	375	385	400
77	275	290	305	315	330	345	360	370	385	400	412
78	280	295	310	325	340	355	370	385	400	410	425
79	290	305	320	335	350	365	380	395	410	425	437
80	300	315	330	345	360	375	390	405	420	435	450
81	305	325	340	355	370	385	400	415	430	445	462
82	315	330	350	365	380	395	415	430	445	460	475
83	325	340	360	375	390	405	425	440	455	470	487
84	330	350	365	385	400	420	435	450	470	485	500
85	340	360	380	395	415	430	450	465	480	500	515
86	350	370	390	410	425	445	460	480	495	515	530
87	360	380	400	420	440	455	475	490	510	525	545
88	370	390	410	430	450	470	485	505	525	540	560
89	385	405	425	445	465	485	505	525	545	560	580
90	400	420	440	460	480	500	520	540	560	580	600

[a]The load indices are valid for all passenger car tyres up to speed symbol "W"

[b]The inflation pressure refers to a speed of 160 km/h and camber angles up to 2°. For the actual pressures to be applied to the vehicle, other criteria are important, such as maximum speed, driving behaviour, etc.

Table 1.3 Limit values for the inflation pressure

Speed symbol	Max. inflation pressure[a], bar
Up to and including T	3.2
H, V, W, Y, ZR	3.5

[a]When the tyre is cold at the start of operation. The pressure increase caused by flexing must not be compensated by releasing air

wider and wider while the sidewall height decreased at the same time. This development can also be observed in racing tires [9]. Wider tyres transmit (lateral) forces better and short sidewalls lead to stiffer tyres (albeit with a loss of comfort, which must be compensated for by the chassis). A tyre with a small aspect ratio also allows a larger rim to be used with the

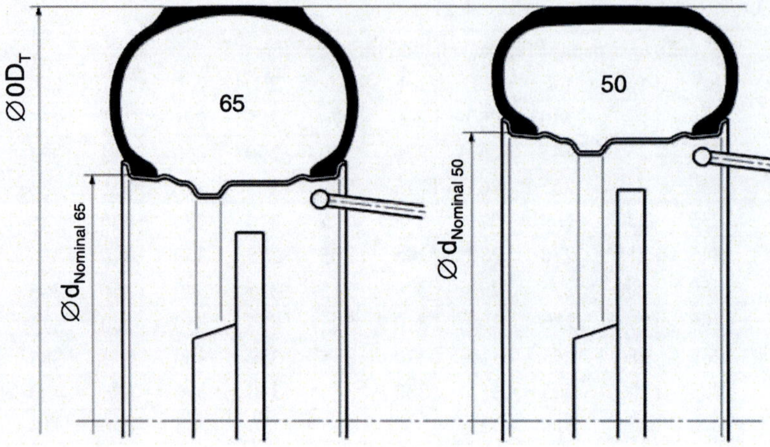

Fig. 1.3 Effect of a smaller aspect ratio with constant outer diameter. The outer diameter OD_T is specified. The 50 series tyre results in a larger rim diameter $d_{nominal\ 50}$ and allows a larger diameter brake disc to be fitted and the joints to be fitted further out on the wheel carrier

same outer diameter. This in turn offers more space for larger brake discs and further effective distances of suspension joints, Fig. 1.3.

Formula 1 tires are not typical low profile tires, quite the opposite balloon tires. But this is hardly due to technical reasons. It rather comes from the regulations (max. 13″ rims are allowed and with the allowed outer diameters the high cross-section results) and the spectators have been used to these voluminous tyres for decades. This familiar picture will change with the 18″ wheels from the 2021 season.

The section height can be calculated from the information given in the tyre designation:

$$h_T = 0.5 \cdot (OD_T - 25.4 \cdot d_{nenn}) \tag{1.1}$$

With:	h_T and OD_T in mm
	$d_{nominal}$ in inches (1 inch = 25.4 mm)

Designation of Rims and *Wheels*

Wheels are defined by the rim diameter and the rim width. Further important sizes result from the design of the area that accommodates the tire bead.

Rims, if supplied loose without wheel discs, are marked near the valve hole, Fig. 1.4.

Wheels are marked on their outside between the fixing holes, Fig. 1.5. for details see Sect. 1.3.2.

Fig. 1.4 Marking of rims according to DIN 7829. **1** Manufacturer, **2** Rim number, **3** Size designation, see Sect. 1.3.2, **4** Date of manufacture: production week/year

Fig. 1.5 Marking of wheels according to DIN 7829. **1** Rim size (width x diameter), **2** Hump and rim base design, **3** Offset 45 mm, **4** Manufacturer, part number and date of manufacture (week/year stamp)

1.2 Tyres

Tyres are the single most important part of a vehicle in terms of impact on performance. They transmit all forces and moments to the road surface via four relatively small surfaces – called *contact patches* – and other assemblies, such as suspension and aerodynamic aids, merely support the tyres in this function. They also cushion the wheels and the chassis parts that move with them. Thus, in addition to comfort (mechanical and acoustic) and economy (mileage, fuel consumption), the tyres primarily influence a vehicle's handling and driving safety (self-steering behaviour, steering precision, driving stability, adhesion). The following evaluation criteria for the assessment of tyres, which are important for racing vehicles, result from this [14]:

- based on traction, braking distance, lap time, aquaplaning.
- on driving stability: straight line stability, cornering stability, braking in corners.
- the steering behaviour in the 0° range and in the limit range, as well as the steering precision.
- on durability: structural durability, high speed capability, dielectric strength.

1.2.1 Requirements

The requirements placed on a racing tyre can be succinctly formulated: Maximum grip values with minimum wear and sufficient shape stability. The ideal tire also does not change its behavior during the race, unlike its real-life counterpart.

The tyres should have as low a mass as possible so that they can easily follow the course of the road and a low mass moment of inertia so that they do not require unnecessary acceleration forces during braking and propulsion.

1.2.2 Basics

Tyre Construction

Both radial and bias-ply tyres are used, Fig. 1.6. However, bias-ply tyres no longer play a role in the equipment of cars and trucks.

In radial tyres (Fig. 1.10) the carcass (casing substructure) connects the two bead cores (steel cords) with cords running radially (name!). The stiffness required for the tyre to function is ensured by belt plies which lie under the tread on the carcass. Hence the alternative name belted tyre. The sidewalls of this type of tyre are extremely flexible due to the carcass structure described above. This means that the sidewalls have little influence on the contact of the tread. The tire contact area is primarily determined by the stiff belt (usually made of steel strands). Compared to the diagonal design, less heat is released in the contact patch, the rolling resistance and the wear of the tread are lower. With the same service life, a softer rubber compound can thus be used on the tread of a radial tyre, allowing greater frictional forces. There are few disadvantages to the radial tire – one

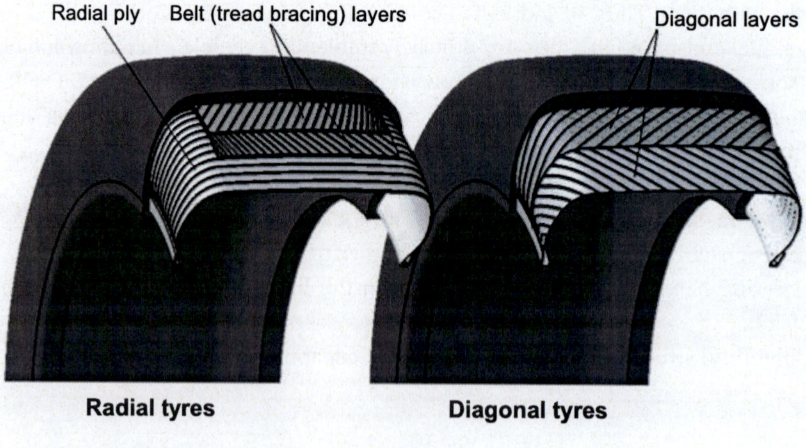

Fig. 1.6 Basic structure of radial and bias-ply tyres

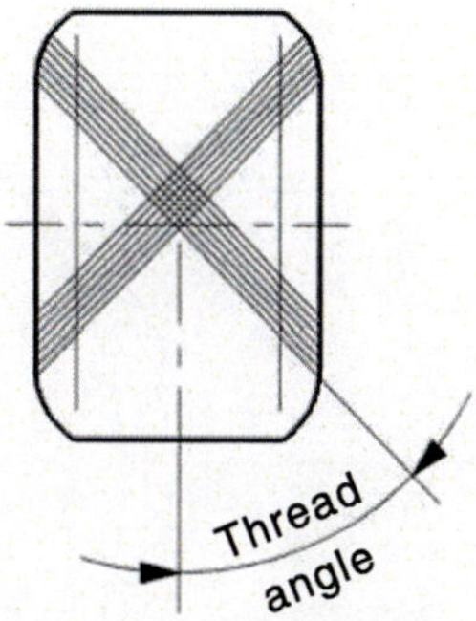

Fig. 1.7 Cord angle (zenith angle, thread angle) of substructure layers. The cord angle is the angle that the cords make with the tire center plane

reason it is so dominant. Its sidewall is relatively vulnerable, and on cobblestones there can be longitudinal vibrations that are perceived as booming noises. The latter is a comfort problem and only relevant for production vehicles. For this reason, elastic mounts decouple the control arms from the vehicle body.

In bias-ply tires, the substructure (carcass) consists of at least two plies of rubberized cords whose cord angle is between 38° and 40° for standard tires, between 30° and 35° for high-speed tires, and 26° for racing tires [18], Fig. 1.7. The cord angle influences essential tire properties such as lateral stiffness, vertical stiffness, and rolling resistance. An obtuse cord angle increases ride comfort but decreases lateral stability. An acute cord angle increases ride stability at the expense of ride comfort. Due to the absence of (steel) belt plies, the mass moment of inertia is lower than with radial tyres. Above all, this characteristic makes this type of construction still interesting for racing tires.

A diagonal carcass has a roughly circular cross-section, see Fig. 1.8. To make the tread wider, correspondingly more rubber must be applied to the sides. This limits the widening of tyre cross-sections or, more precisely, the reduction of the height/width ratio. On radial tyres, the desired flat tread is created by the belt placed on the carcass.

The transmission of a pure circumferential force is not possible with a 90° cord angle. The cords can only transmit tensile forces and, when the rim rotates relative to the tread, they align themselves accordingly at an angle, Fig. 1.9. In contrast to standard radial tyres with one ply at 90° zenith angle, racing tyres in radial construction therefore generally have two-ply carcasses with cord angles between 85° and 70° because of the greater load [15].

In the Anglo-Saxon world, there is a third type of tyre which lies between the two mentioned, the *bias-belted tyre*. Belt plies lie on a diagonal carcass, as in a radial tyre. However, the characteristics of this type of tyre are not as good as those of a radial tyre. The tire rolls softer in the lower speed range, but is not necessarily suitable for higher speeds.

These different types of tyres sometimes require different chassis designs so that the greatest possible forces can be built up.

On radial tyres, the contact patch deforms less than on cross-ply tyres, but the sidewall is more compliant and a slip angle is built up more quickly. However, this leads to a narrower limit range when cornering, which makes it more difficult for the driver to estimate the

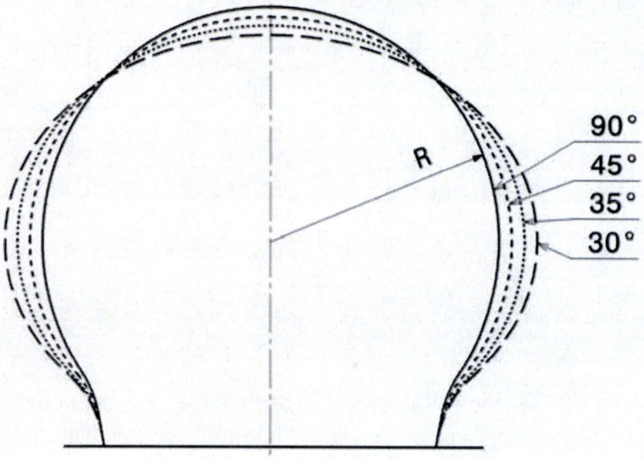

Fig. 1.8 Influence of the cord angle on the carcass cross-section, after [15]. A thread angle of 90° results in a circular cross-section. With smaller angles the cross section becomes more oval

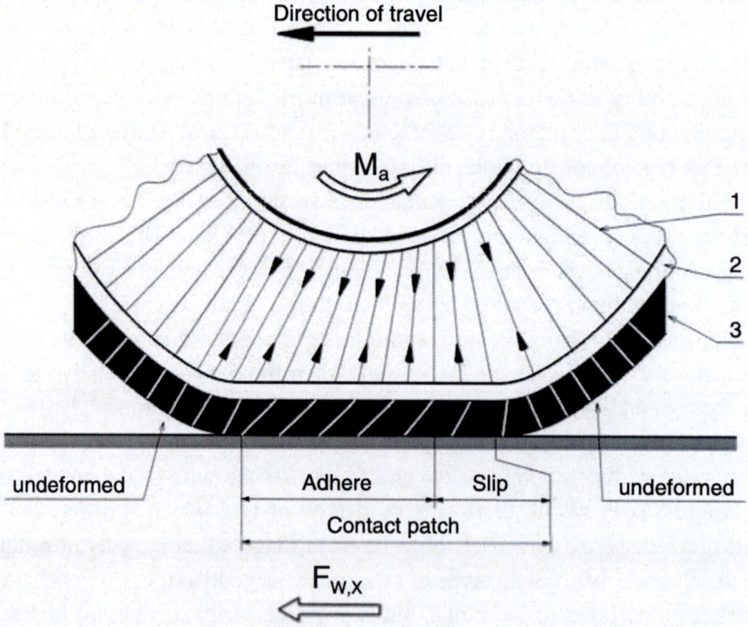

Fig. 1.9 Transmission of a circumferential force from the tread of a radial tyre to the rim. When a moment is transmitted, the cords align themselves obliquely, deviating from the original radial direction, and thus transmit the circumferential force from the tire contact patch. The rubber elements in the tread are deformed by the friction when running into the lathe (area: sticking) and slide back into the normal position when running out (area: sliding). **1** Cords, **2** Carcass, crown plies, **3** Tread. $F_{W,x}$ Force exerted by the road on the tyre, M_a Drive torque on the rim

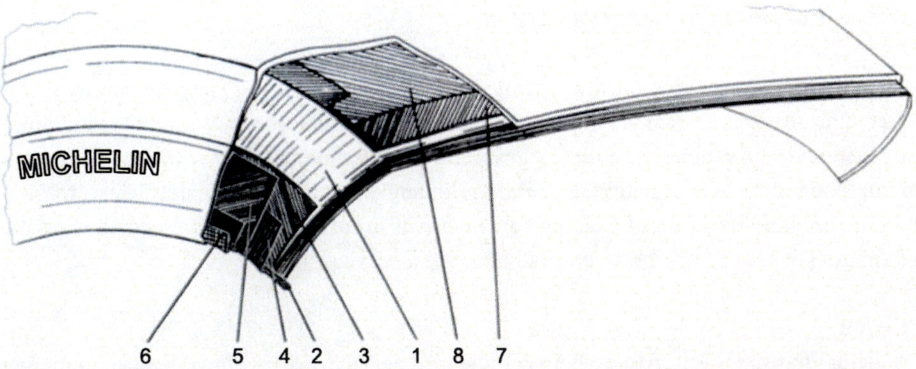

Fig. 1.10 Tyre structure of a racing tyre in radial construction [8]. **1** Radial casing plies, **2** Bead wire, **3** First sidewall protector, **4** Turn-up of one casing ply, **5** Second sidewall protector, **6** Turn-up of another casing ply, **7** First crown ply, **8** Second crown ply

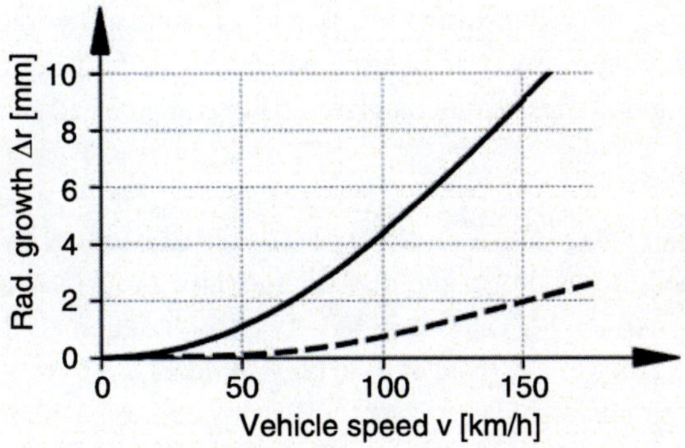

Fig. 1.11 Radial tire growth versus speed measured at the zenith of the tire, after [1]. Tire dimensions: bias-ply tyre 6.00–13, Radial tyre 165 R13

drivable slip angle [8]. Radial tyres require a greater camber change for efficient lateral force build-up (Fig. 1.10).

With radial tyres, the dynamic rolling radius remains approximately constant. The (steel) belt below the tread is deformed oval, but its circumference remains the same, as with a tracked vehicle. This is not the case with bias ply tires. Their cross-section becomes, as it were, triangular with increasing circumferential speed and the outer diameter increases, cf. Figure 1.11. Racing tyres are designed for higher speeds and therefore show a smaller increase in diameter than standard tyres.[2]

[2]In vol. *3 Powertrain of* the series Racing Car Technology Manual there is an example of a racing tyre, Chap. 4

1.2.3 Influence on Driving Behaviour

The influence of the tyres on the driving behaviour is manifold. They determine under which deformations a force is built up between the road and the vehicle and thus influence the position of the vehicle. The most significant variables are the slip, at which the greatest circumferential force is transmitted, and that slip angle at which the greatest lateral force is built up, because the driver of a racing vehicle drives in such a way that (almost) always the maximum possible forces between road and tyre are exhausted.

Friction

The actual transmission of force between the tyre and the road surface depends on many factors, such as wheel load, temperature, sliding speed and surface condition.

First of all, the behaviour of rubber as a material is interesting. It has an enormously high elasticity, i.e. it easily undergoes changes in shape and returns to its original position after being relieved, and it is fascinating because of its extremely high ductility. The molecular chains form a tangle when at rest, which is unravelled into strands when stretched. Depending on the rubber compound, the return to the initial state is damped by internal friction or even a deformation residue remains (visco-elastic behavior). In the latter case, the curves for loading and unloading in a stress-strain diagram do not coincide and enclose an area (hysteresis), Fig. 1.12.

The larger this area is, the more of the deformation energy is converted into heat by this rubber compound and is not returned. If you drop a ball made of rubber without hysteresis, it almost bounces back to its original height. One made of rubber with extremely large hysteresis remains deformed on the ground. The tread of road tyres, unlike racing tyres, has a relatively small hysteresis. To compare rubber compounds, consider some Shore A hardness values at room temperature of some types of tires: Indy Car road course 55–70, Formula 1 50–65, dragster 25–45 [25], passenger car winter tires 60.

In addition, rubber shows a changing behaviour over deformation speed and load.

If this material is pulled over a surface, a force has to be overcome which is composed as follows:

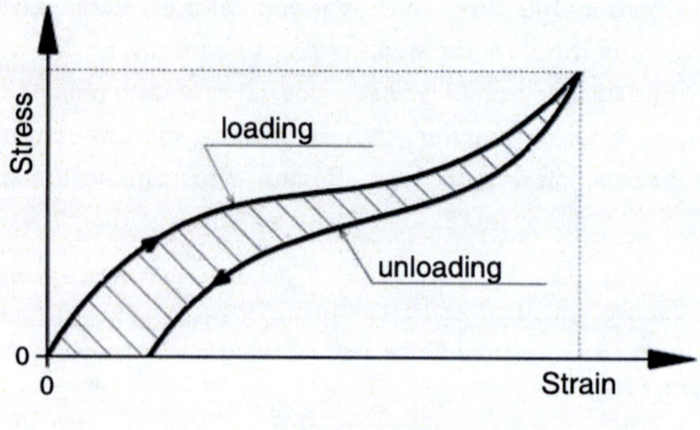

Fig. 1.12 Visco-elastic behaviour of rubber. The area enclosed by the curves corresponds to the work of deformation that is not recovered and converted into heat

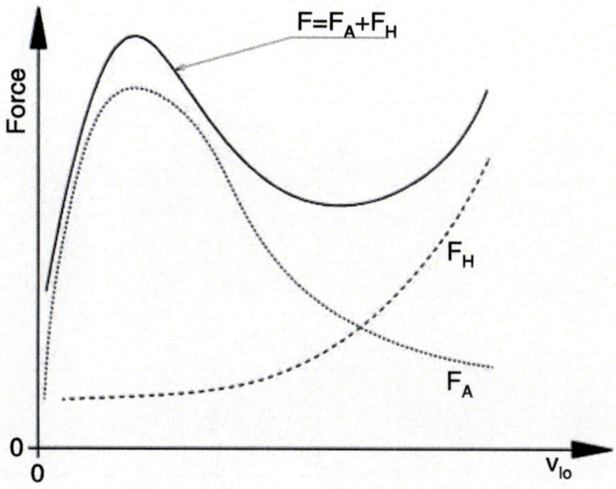

Fig. 1.13 Main composition of rubber friction and influence of sliding speed. F_A Adhesive forces, F_H Forces due to hysteresis, v_{lo} Sliding speed

$$F_{fr} = F_{Adhesion} + F_{Hysteresis} + F_{Cohesion} + F_{Viscose}$$

The proportion of adhesion forces (intermolecular adhesive *forces*) is the dominant component. Cohesive forces cause wear in the case of frictional contact and are of secondary importance. Figure 1.13 shows the progression of the decisive force components over the sliding speed.

It can be seen that adhesion forces are decisive at low sliding speeds (e.g. front slippage area in ABS braking), while hysteresis forces determine the transmissible forces in the area of high sliding speeds (e.g. in locking brakes). Hysteresis forces arise particularly in the case of movements relative to rough surfaces. The frictional force is created by a pressure difference between the pressure side of the rubber running up onto the elevation and the side running down with losses due to hysteresis. The remaining pressure force counteracts the sliding movement, Fig. 1.14.

Hysteresis forces are therefore caused by *mechanical keying* between the road surface and the rubber.

The optimum sliding speed for large frictional forces is approximately between 0.05 and 0.5 m/s. Values around 0.01 m/s are assigned to adhesion and those around 1 m/s to sliding. If there is a separating film between the road surface and the rubber of the tread, the adhesion forces are no longer effective and only the lower parts of the frictional force remain. For this reason, treaded tyres are used in rainy conditions. These displace the water into the tread valleys and thus establish a certain contact between the rubber of the tread blocks and the road surface. On dry roads, non-profiled tyres (slicks) have the highest adhesion values. Similarly, on the road side, a smooth, flat surface is more favorable than rough surfaces that provide less contact area. Following the dry friction of solid bodies, the frictional force of tyres is also expressed as a function of the wheel load:

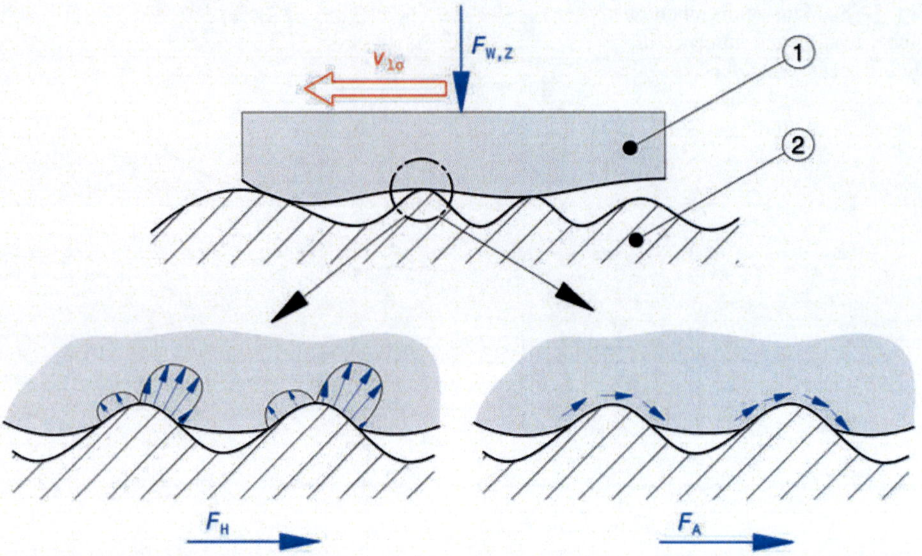

Fig. 1.14 Mechanisms of rubber road friction. 1 Rubber, 2 Roadway. $F_{W,Z}$ Wheel load, F_H Forces due to hysteresis, F_A Adhesion forces

$$F_{W,X} = \mu_{W,X} \cdot F_{W,Z} \,\text{resp.}\, F_{W,Y} = \mu_{W,Y} \cdot F_{W,Z} \tag{1.2}$$

$\mu_{W,X}$, $\mu_{W,Y}$	Coefficient of friction in circumferential or lateral direction, –
$F_{W,X}$ or $_Y$ and $_Z$	Forces acting at the Tyre contact point in circumferential, lateral and vertical directions, N

As explained above for the friction force composition, the coefficients of friction are not constant for a given rubber compound to road surface pairing, but depend, among other things, on the sliding speed, Fig. 1.15. In general, the lower the contact pressure and the more uniform the pressure distribution within the contact patch, the greater the adhesion values that occur.

As soon as a tyre transmits circumferential forces (driving, braking), a relative speed occurs between the tyre contact surface and the road. Figure 1.9 shows the deformation of the tread. The rubber elements in the incoming slippery area adhere to the road surface, the rim continues to rotate and the tread deforms elastically (form slip, shear). This deformation increases towards the run-out area such that elements begin to slide (*slip*) until they eventually oscillate back to the undeformed initial state as the tread section lifts off the road. The circumferential speed of the tyre is greater or less than the vehicle speed, depending on whether traction or braking forces are being transmitted.

The ratio of the two speeds is the (longitudinal) slip S_W:

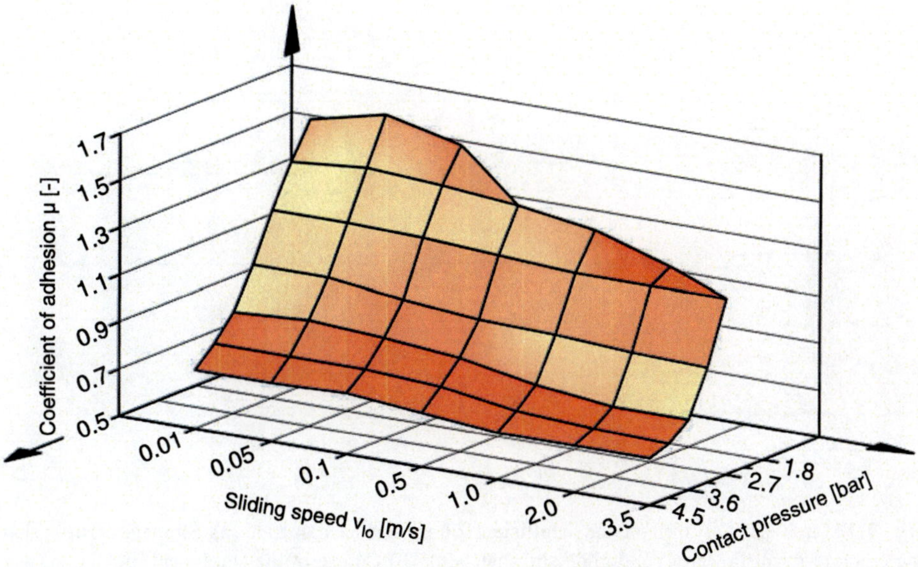

Fig. 1.15 Coefficient of adhesion as a function of sliding speed and contact pressure, according to [27]. A typical tread compound of a passenger car tyre was measured on corundum-180

$$S_{W,X,b} = \frac{v - v_W}{v} \cdot 100\ [\%] \tag{1.3a}$$

$$S_{W,X,a} = \frac{v_W - v}{v_W} \cdot 100\ [\%] \tag{1.3b}$$

	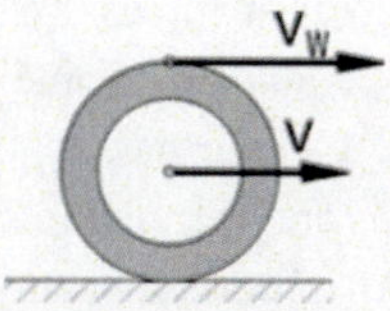
v	Speed of the vehicle, m/s
v_W	Circumferential speed of the tyre, m/s $v_W = r_{dyn} \cdot \omega_W$
r_{dyn}	Dynamic tyre radius, m
ω_W	Wheel angular velocity, s^{-1}
$S_{W,X,b}$	Slip occurring during braking, %
$S_{W,X,a}$	Slip occurring during driving, %

A purely rolling wheel therefore has a slip of 0%, a locking or spinning wheel 100%. The basic influence of slip on the coefficient of friction is shown in Fig. 1.16. In the area dominated by form slip (Fig. 1.16b), the friction increases approximately linearly with the

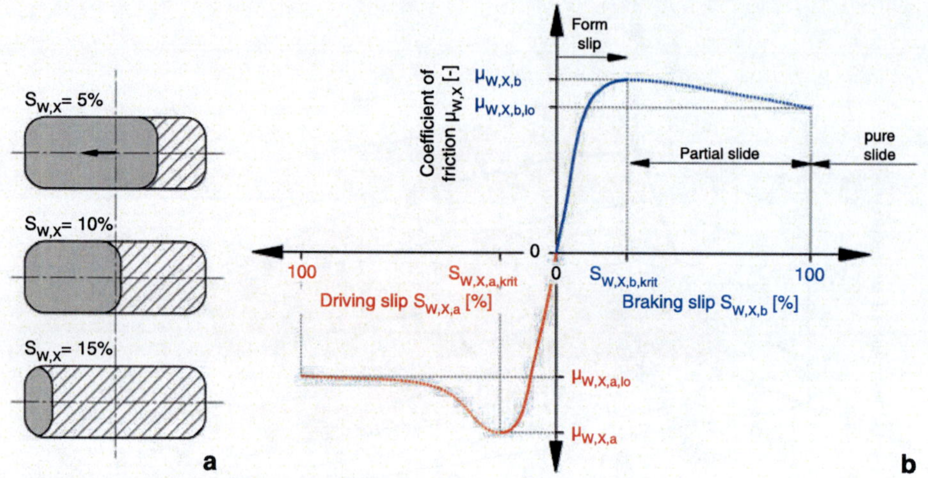

Fig. 1.16 Influence of slip on the coefficient of friction, schematic. (**a**) Slippage distribution according to form slip (grey) and glide slip (hatched), (**b**) Course of the coefficient of friction over the slip

slip until a critical slip value is reached, at which the coefficient of friction reaches its maximum. The following decreasing part is shown dotted, because it is practically not stationary, but changes to pure sliding within fractions of a second. In the tyre contact patch, the proportions (Fig. 1.16a) of moulded slip in the incoming area (grey) and sliding slip in the outgoing area (hatched) change with the slip value $S_{W,X}$. The greater the slip value, the greater the proportion of the sliding area.

The circumferential force is therefore generated by the sliding of the rubber on the road surface and the greatest forces are generated at the optimum sliding speed of the rubber compound, i.e. a certain slip. Some exemplary curves of coefficients of friction versus slip are shown in Fig. 1.17 for passenger car tyres and Fig. 1.18 for racing tyres.

Racing vehicles are operated in such a way that the maximum forces dictated by the tire are transmitted. An important parameter for longitudinal forces is slip. The maximum value of the coefficient of friction is called the static friction coefficient μ_W and occurs at around 10–30% slip, depending on the tyre and road surface. The lowest value is measured when the wheel locks. It is called the coefficient of sliding friction $\mu_{W,lo}$.

The static friction coefficients of racing tires on dry pavement can reach values of 3 and peak values up to 5 for a short period of time (about 500 ms) [13].

This slip-dependent behavior of friction is of central importance for the control logic of traction control and launch control. For the best acceleration, the on-board computer must adjust to the slip of the static friction coefficient and not ensure that the tire does not slide at all.

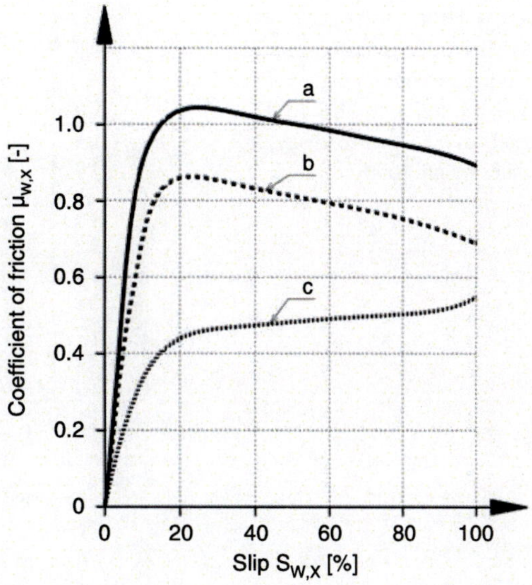

Fig. 1.17 Coefficients of friction of a profiled passenger car tyre, according to [11]. (**a**) Dry asphalt, (**b**) Wet asphalt, (**c**) Loose gravel

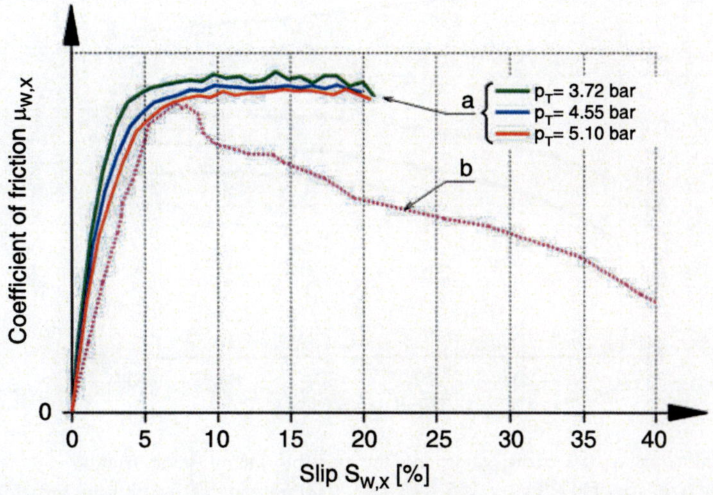

Fig. 1.18 Course of the coefficient of friction versus slip of racing tyres, (**a**) according to [23], (**b**) according to [13]. For measurement series a, the influence of the tyre inflation pressure p_T is also entered. A lower pressure leads to a larger contact area, which can transmit a greater circumferential force

The sliding speed-dependent behaviour of the rubber friction can also be seen when looking at the course of the maximum adhesion coefficients over the driving speed, Fig. 1.19. The coefficient of friction (both sliding and adhesion coefficients) decrease with speed. This effect is even more pronounced on wet roads.

Fig. 1.19 Course of the coefficient of adhesion versus driving speed for a passenger car tyre, according to [26]. (**a**) Dry roadway, (**b**) Wet roadway, h_w water height [mm]

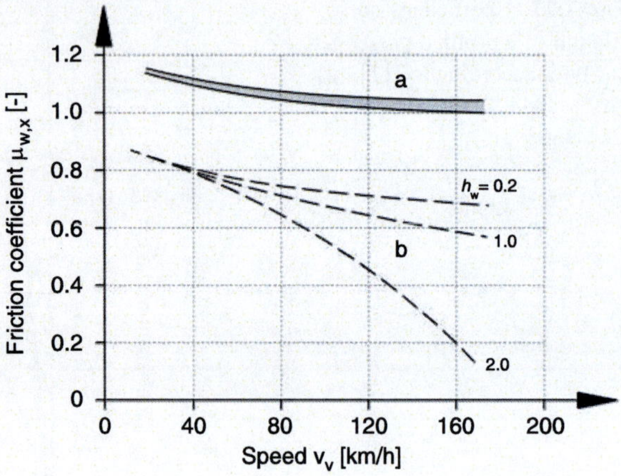

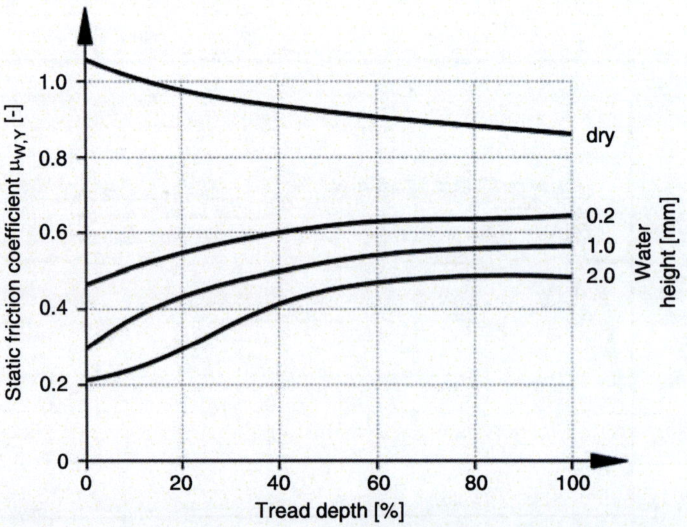

Fig. 1.20 Influence of the tread pattern on the possible lateral static friction $\mu_{W,Y}$ of a radial passenger car tyre, after [11]. Tyres: 155 R13 78 S, 100% tread depth = 8 mm, inflation pressure: 1.8 bar, driving speed: 60 km/h, slip angle: 10

Tread Condition

In some vehicle classes, treaded tyres are mandatory. It is found that tyres with the minimum permitted tread depth give the best lap times. On dry roads, where adhesion forces dictate the frictional behaviour, the tyre with a larger contact area with the road has an advantage, Fig. 1.17. With a treaded tyre, the contact area is interrupted by tread grooves and the transmissible force is reduced. In wet conditions the situation is exactly the opposite, Fig. 1.20. Other phenomena that explain the better performance of worn

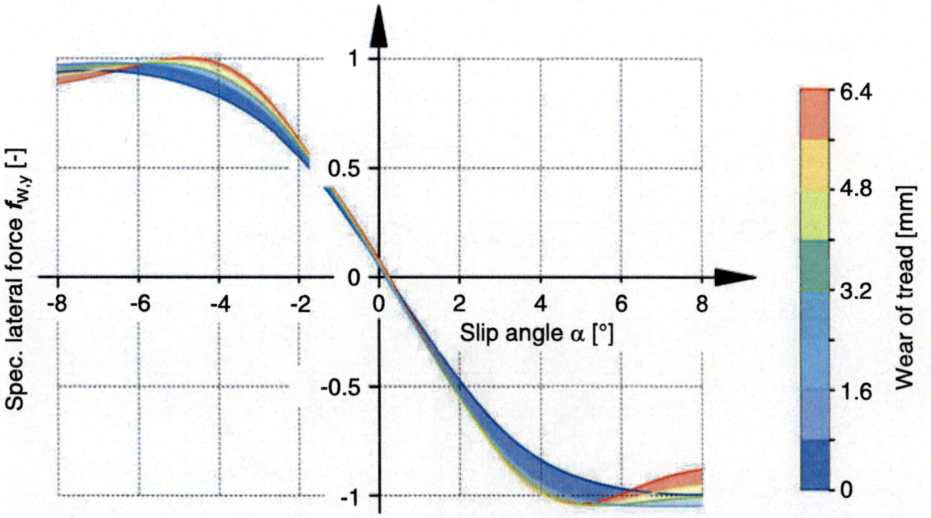

Fig. 1.21 Lateral force over side slip under the influence of tyre wear [33]. For comparison purposes, the absolute value of the lateral force is not plotted, but this is related to its maximum in the original state (wear 0 mm, blue)

tyres are: Firstly, the deformation capacity increases with the tread depth, which the driver perceives as a spongy ride, and secondly, the tread heats up more with greater deformation, which in addition to reducing the transmissible force can lead to overheating of the tire. In addition, high tread blocks make it more difficult to dissipate heat to the tyre's substructure. This also leads to thermal problems. Treaded racing tyres are produced in some racing classes by mechanically working a tyre down to the specified minimum tread depth (approximately 2.4–6 mm) [7]. In this process, the service life of such a tyre need not be at all worse than that of the unmachined tyre. On the contrary, because of the heating phenomena described above, the machined tyres actually last longer in racing than those driven with the full tread depth [15]. However, there are also profiled racing tires that are manufactured with a tread depth of only 2.4–3.2 mm and can be used right away. Summer tyres for passenger cars have comparatively at least 8–10 mm tread when new from the factory.

Some of the phenomena described above also occur in a tyre when its tread wears, i.e. its thickness decreases during operation, Fig. 1.21. At the beginning of an (endurance) race the tyre shows a comparatively "tame" characteristic (blue edge curve, wear 0 mm). At the greatest wear (red edge curve), when the tread has the lowest thickness, the tyre develops its greatest lateral force and also shows greater lateral stiffness (slope of the curve at the origin, *cornering stiffness*). Between these two extremes, the characteristics change accordingly, but not linearly. With the first millimetres of wear, the maximum lateral force increases more rapidly than with the following ones.

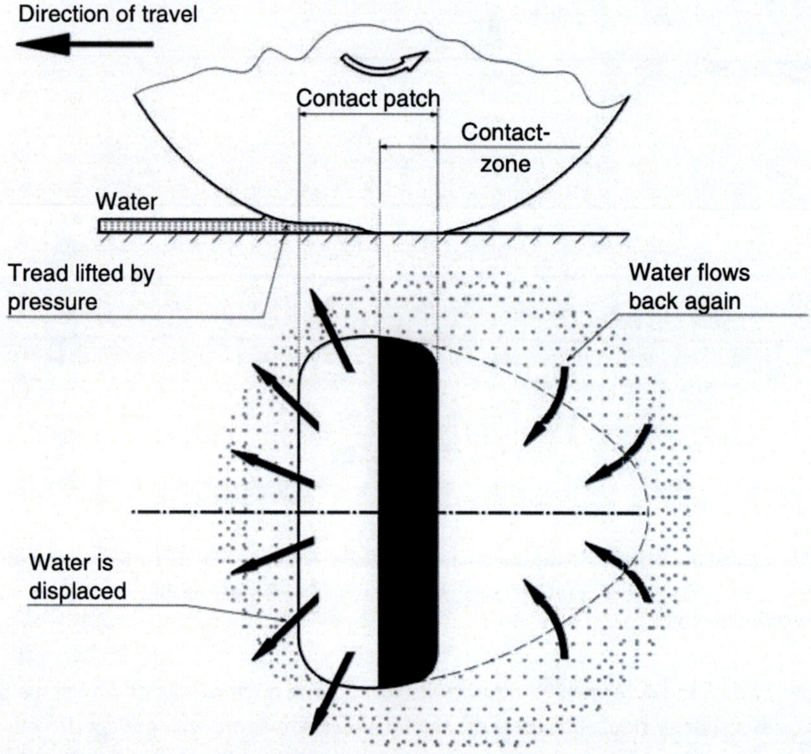

Fig. 1.22 Road contact of a slick in wet conditions, after [20]. In the inlet area of the tyre, the tread lifts off due to the pressure of the water. Real frictional contact only occurs in the rear zone of the contact patch. The water film is thus pressed away from the contact area and the water flows back into this area of the road after the tyre has rolled off with a delay

Rubber friction is mainly caused by adhesion and deformation. If a separating film comes between the two contact partners, tread and road, the adhesion component is drastically reduced or even eliminated and practically only the smaller deformation component remains. At low water depths, even a slick still builds up useful forces, because a certain part of the tyre contact area is still in direct contact with the road surface, Fig. 1.22.

The aim in developing a rain tyre must therefore be to remove the water between the rubber and the road surface. This is done by appropriately designing the channel-like depressions between the tread blocks. If the tread blocks are small, the spaces between them act as an escape space for the water. There are, however, limits to the enlargement of this escape space due to the contact area of the rubber, which is reduced to the same extent. The contact area can overheat. If the spaces between the profiles are designed as channels, the water is drained off to the front, rear and side. However, the water is only forced out of the contact patch through these channels when the wheel is rotating – this effect is lost

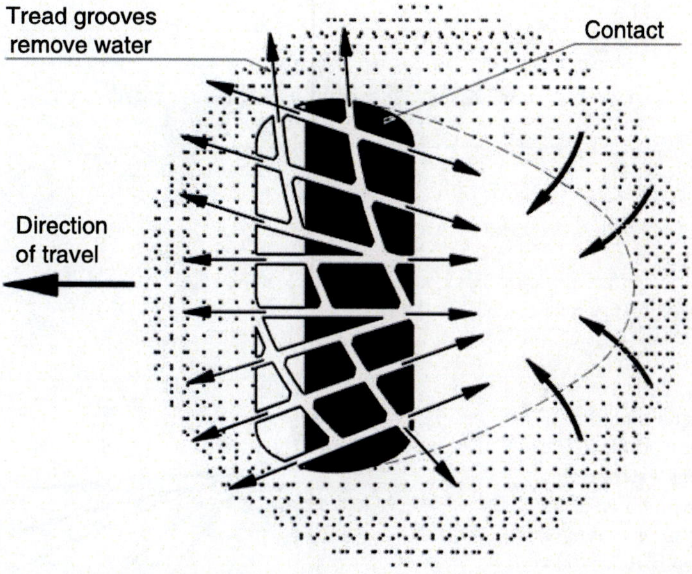

Fig. 1.23 Water displacement effect of a profiled tyre, according to [20]. The water is forced into the gaps by the incoming tread blocks and directed to the front, rear and side. A part of the tyre surface thus finds dry contact with the road surface

when the wheels are locked, Fig. 1.23. The water-displacing effect is also reflected in the rolling resistance. This increases up to ten-fold in wet conditions.

A great deal of calculation[3] and testing is required for the efficient design of the channels. The development is complicated by the fact that a tread must be optimized for different water depths and driving speeds, which can at best be a compromise. Tire slippage is asymmetrical due to camber, toe-in and lateral forces. That is why asymmetric profiles are also offered.

Even when the water from the tread grooves has been removed from the tyre contact patch, a lubricating film of liquid still remains for the incoming tread block. Therefore, to increase friction, this film must also be removed. Figure 1.24 shows the process. The leading edge of a tread block wipes the water film from the pavement, allowing dry contact for the remaining block. Additionally, cuts can be made in profiled cleats. The resulting lamellae (sipe) form further edges and improve the wiping effect.

In addition, a tread pattern causes an increase in pressure on the remaining tread blocks in the contact area, because the contact areas become smaller for the same wheel load. The water cools the tread and so the edges of the tread blocks do not overheat even in the case of rubber compounds with a relatively low glass transition temperature (see appendix).

The effect of a water film on the friction behaviour is shown in Fig. 1.25.

[3]E.g. with CFD (Computational Fluid Dynamics), computer programs for flow simulation.

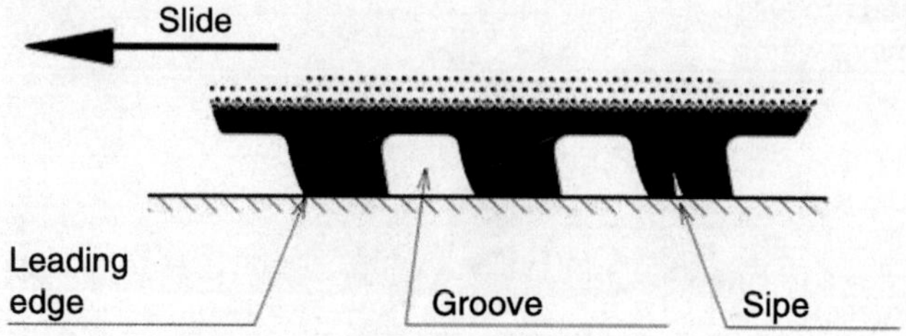

Fig. 1.24 Wiping effect of profile block, after [20]. The leading edge of a tread block wipes away the water film, allowing dry road contact for the block

Fig. 1.25 Static friction coefficients of a profiled series tyre at different water heights, after [11]. Measured on a passenger car summer tyre with 8 mm tread depth. Below 60 km/h hardly any influence can be detected. At large water depths of 3 mm, the tire floats at high speeds on

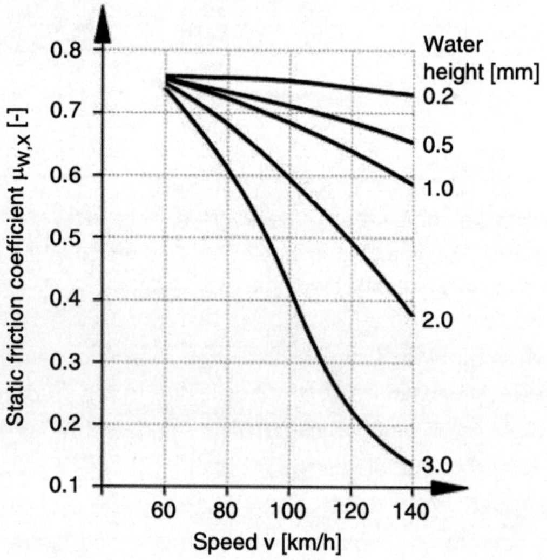

It should be noted that water depths of more than 2 mm are rare on wet roads. Greater values occur during heavy downpours and in ruts and other depressions. At high speeds, small displacement cross-sections and large water depths, the tyre floats up completely, the dreaded aquaplaning. In contrast to dry roads, the transmission behaviour improves in this case with higher inflation pressures and smaller tyre contact areas (as a result, the water film is more likely to be "broken through" by the tyre). Wider tyres are also at a disadvantage in this context due to the relatively short contact patch.

So a rain tyre cannot deliver the performance of a slick, but it does postpone the speed at which aquaplaning occurs. The tyre sidewall can also be designed to be softer than a slick because rain tyres build up lower forces. The tyre inflation pressure is usually chosen lower as well.

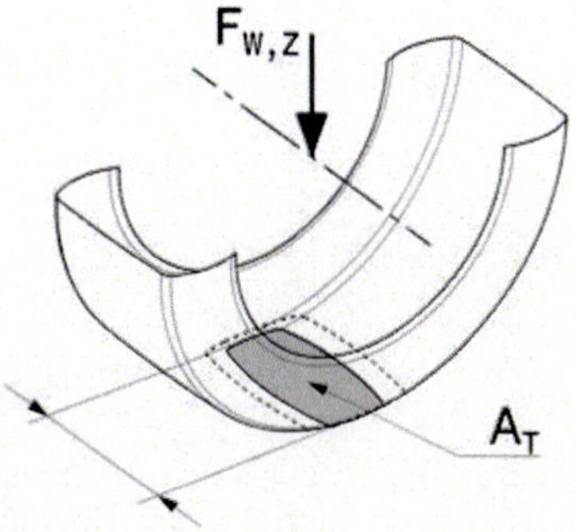

Fig. 1.26 Load-bearing behaviour of an air-filled tyre. The wheel load $F_{W,Z}$ is absorbed by a corresponding flat surface A_T. If the wheel load is greater, the surface will expand to the contour shown in dashed lines

Tyre *Inflation Pressure*

In simplified terms, the tyre can be imagined as a flexible diaphragm which is held in place by the internal pressure p_T. With this model it is easy to calculate the contact area A_T of a tyre for a given wheel load $F_{W,Z}$, cf. also Fig. 1.26:

$$A_T = \frac{k \cdot F_{W,Z}}{p_T} \tag{1.4}$$

k	Correction factor to account for tire stiffness, –. $k \approx 0.9$–0.85 [14]

In fact, the area is somewhat smaller because the tyre sidewalls contribute and transmit a bending moment to the tread, thus relieving the load on the diaphragm. This is taken into account with the factor *k*. With increasing load (or decreasing inflation pressure) the tread becomes longer, its width remains approximately the same [13]. A longer contact area has the disadvantage that the tread has to bend through a greater angle when it is applied to the road and when it leaves it – the so-called flexing (fulling) work increases, the tyre becomes warmer and fails in extreme cases.

Figure 1.27 shows how the actual contact pressure can be distributed in the lathe bed. The pressure is distributed relatively evenly over the contact area. Only in the run-in area is a pressure increase clearly visible. This is a typical feature of radial tyres, where the belt has a balancing effect. In the case of cross-ply tyres, on the other hand, the two shoulder areas are subjected to considerably higher pressure than the centre of the tread [32].

Excessive inflation pressure reduces the contact area on paved surfaces and thus the forces that can be transmitted. On loose ground, traction is better with lower inflation pressure, which is used in raid and rally vehicles. In production vehicles, the inflation pressure is comparatively lower than in racing vehicles for reasons of comfort. In this context, the ideal is a system that can vary the inflation pressure while driving. At low driving speeds, a low inflation pressure is set, which is increased as the speed increases.

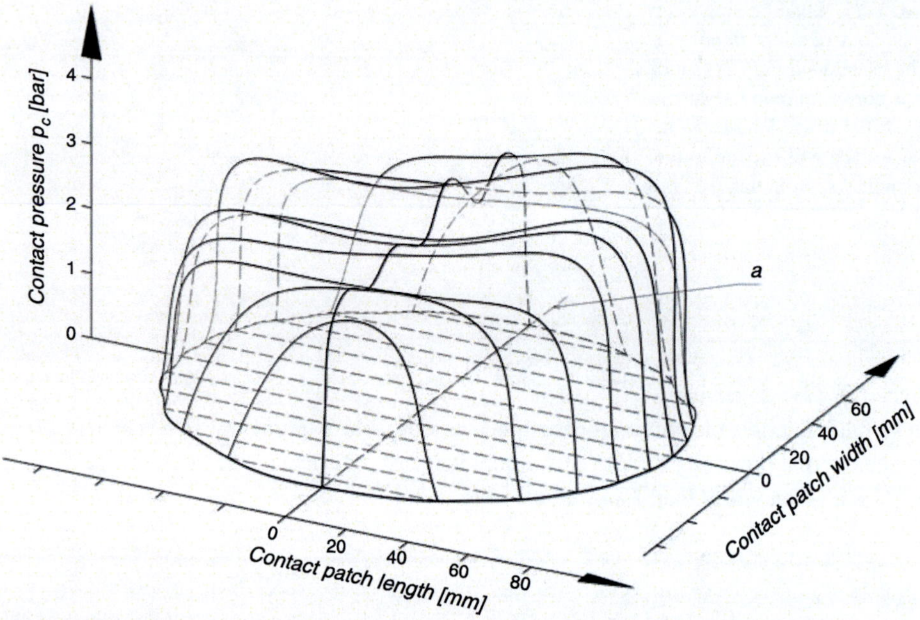

Fig. 1.27 Pressure distribution in the contact patch, according to [31]. Free rolling radial tyre with driving speed = 60 km/h. *a* Projection of the wheel axis. The rolling direction is to the right

Figure 1.28 shows some basic influences of the inflation pressure on various tyre characteristics. The rolling resistance can be reduced by inflating the tyres more strongly (diagram top left: curve of rolling resistance over wheel load). This also makes the spring effect of the tyre stiffer (cf. diagram of wheel load versus static tyre pressure). The lateral stiffness increases with increasing tyre inflation pressure (cf. curve lateral force over slip angle). For this reason, the steering tendency of any vehicle can be corrected in the direction of understeer by increasing the internal tire pressure on the rear axle. With the increased tyre inflation pressure, the quality of the lateral force distribution also improves [1]. However, an excessively high inflation pressure reduces the contact area and thus reduces the transferable frictional forces. This is noticeable to the driver, among other things, through a reduced self-aligning torque.

When the tyre pressure decreases, the wheel speed increases due to the decreasing contact patch diameter (dyn. rolling radius). So-called passive tyre pressure monitoring systems use this effect by calculating the tyre pressure by comparing the wheel speeds.

Figure 1.29 provides an informative overview of the correlation between inflation pressure and tyre behaviour. The target pressure depends on the occurring wheel load and increases linearly with it. If the lateral transfer behaviour of the tyre is in the foreground (e.g. slalom competition, typical circuits), one will tend to orientate oneself towards the upper limit of the target range (green). If a high circumferential force is desired for the

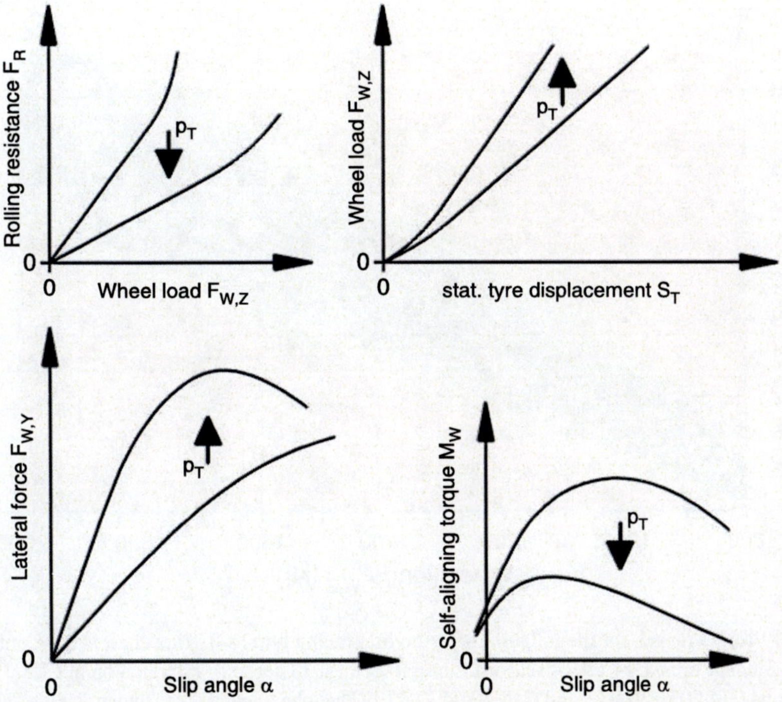

Fig. 1.28 Influence of inflation pressure on various tyre characteristics in principle. p_T Tyre inflation pressure

competition under consideration (e.g. acceleration competition), a somewhat lower inflation pressure is aimed for.

If the volume in the tyre is considered to be invariant (isochoric), the pressure increase in the tyre due to heating can be easily calculated, Fig. 1.30:

$$p_{T,2} = p_{T,1} \frac{273.15 + \vartheta_{T,2}}{273.15 + \vartheta_{T,1}} + p_0 \frac{\vartheta_{T,2} - \vartheta_{T,1}}{273.15 + \vartheta_{T,1}} \qquad (1.5)$$

$p_{T,1}$ or $p_{T,2}$	Overpressure in the tyre at temperature $\vartheta_{T,1}$ or $\vartheta_{T,2}$, bar
$_{T,1}$ ϑor $\vartheta_{T,2}$	Temperature of the air in the tyre, °C
p_0	Ambient pressure, bar

At this point it must be mentioned that professional racing drivers feel a pressure change in the tire of 0.01 (= 1/100!) bar.

The average inflation pressure is 1–1.25 bar in Formula 1 [8]. The width of the target pressure range, however, is only approx. 0.14 bar even for these voluminous tyres [20]. There are race tracks where different inflation pressures are used on the left and

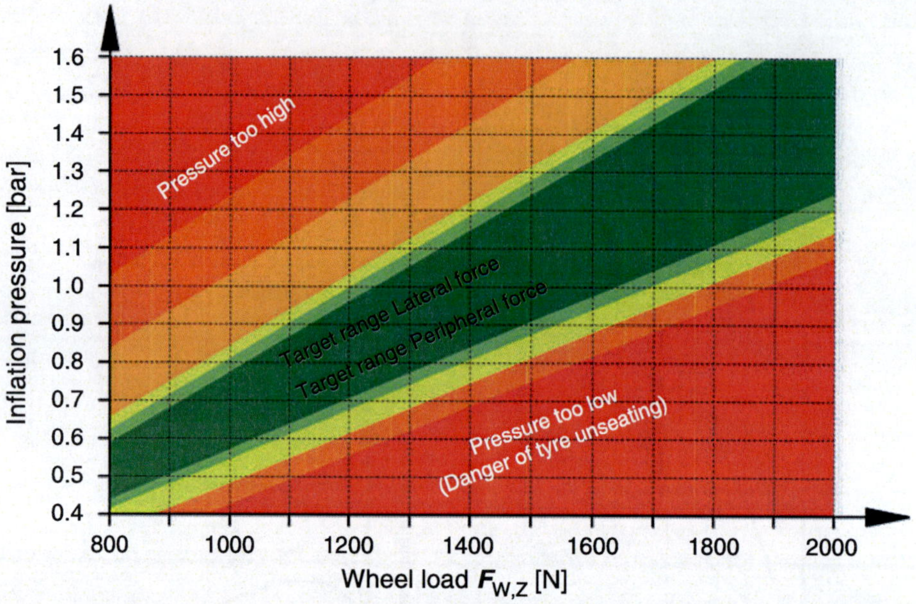

Fig. 1.29 Target ranges for the inflation pressure of a racing tyre [34]. This chart is for Continental Formula Student tires, C14 (2014 season) through C18 (2018 season), mounted on a 7 J × 13″ rim. Dimensions C14 205/510 R13 and C18 205/470 R13. The wheel load is the (dynamically) occurring wheel load during operation

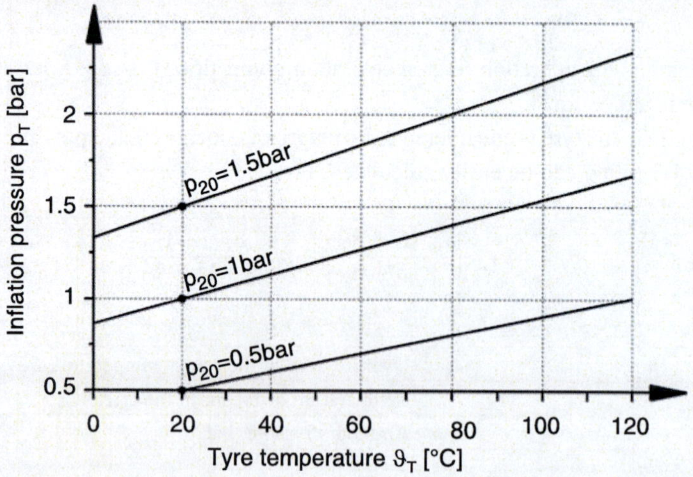

Fig. 1.30 Isochoric pressure rise in the air-filled tyre. Due to heating during driving, the pressure increases at constant tyre volume. The increase is entered for three selected tyre pressures p_T at 20 °C. The ambient pressure is 1 bar

right. In Barcelona, Formula 1 cars start with lower pressure in the left tyres. By driving through the three fast right turns, the temperature of the left tires increases due to the greater contact force and thus the internal pressure increases. Thus, the pressure ratio left to right is equalized in the course of the race [4].

Racing tires are often filled with nitrogen (N_2) instead of air. With a content of at least 95% N_2, the advantages are lower pressure loss (larger N_2 molecules diffuse more heavily through rubber), more stable pressure conditions (different specific heat capacity) and higher temperature resistance (lower pressure increase) [10]. In addition, nitrogen inflation reduces the natural content of water vapor in the air and thus phenomena such as vapor formation (the pressure in the tire then increases disproportionately) and corrosion. However, it must be noted that the tire must be correctly evacuated beforehand or at least well purged with nitrogen. An inexpensive alternative is to pump dried air into the tyre. The N_2 content of natural air is about 70%.

Wheel Load (*Corner Weight*)

The load capacity of a tyre depends mainly on: Inflation pressure, rim diameter, tread width and section height.

As expected, the transferable circumferential force of a tyre increases with wheel load, Fig. 1.31.

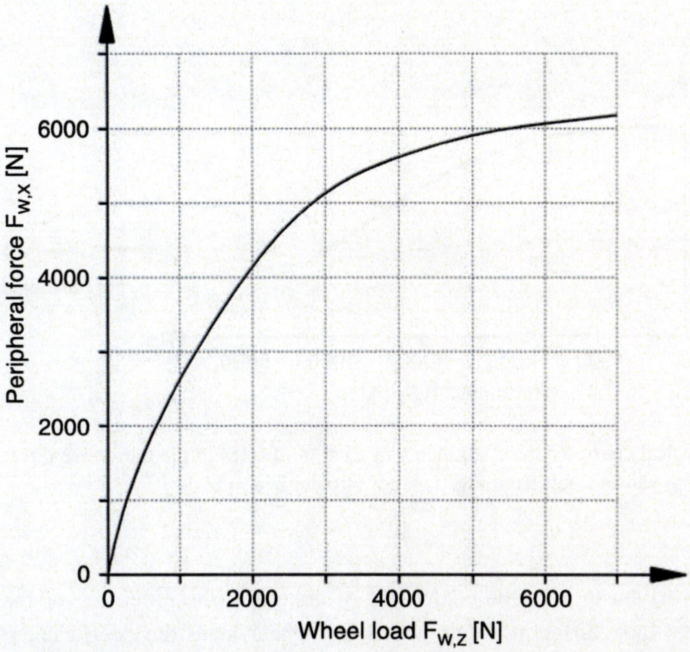

Fig. 1.31 Longitudinal force $F_{W,X}$ of a tyre as a function of the wheel load $F_{W,Z}$, according to [13]. The transmittable circumferential force increases with increasing wheel load

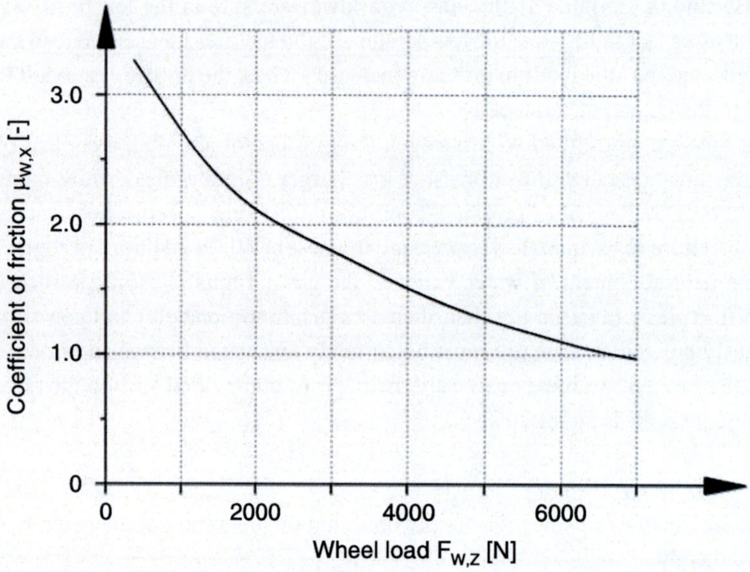

Fig. 1.32 Course of the coefficient of friction of a tyre in circumferential direction over the wheel load. This curve corresponds to the one in Fig. 1.31. (After [13])

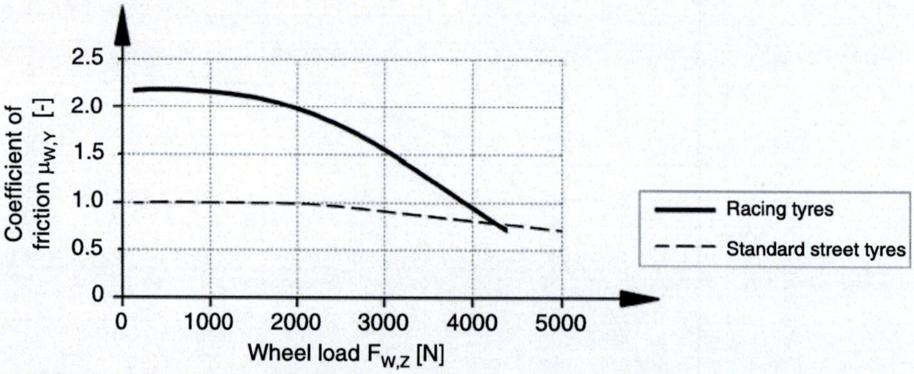

Fig. 1.33 Typical course of the coefficients of friction of tyres in the transverse direction over the wheel load. The values result from Fig. 1.38 for slip angle $\alpha = 8°$

However, if you look at the coefficient of adhesion (the efficiency of the tire, so to speak), things look different. With increasing wheel load, the coefficient of adhesion generally decreases. This is true not only for circumferential forces but also for lateral forces, cf. Figs. 1.32 and 1.33.

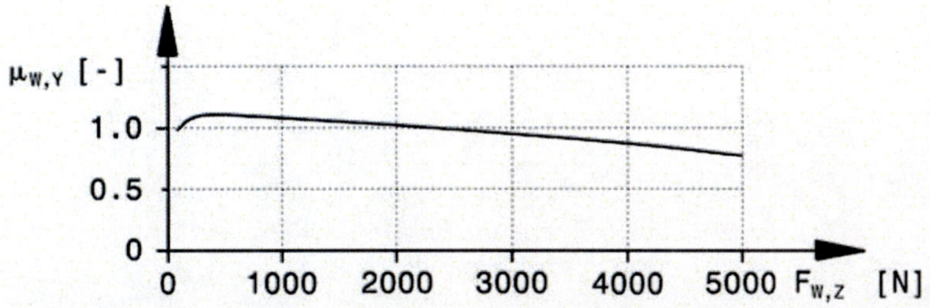

Fig. 1.34 Course of the coefficient of adhesion in the transverse direction over the wheel load. A certain wheel load is required so that a frictional force can be built up at all. However, the maximum friction is already reached at low wheel loads, from which point the frictional force decreases with increasing wheel load

For a given wheel load, wider tyres are therefore better because there is less pressure in the tyre contact area and a greater frictional force is built up (cf. also Fig. 1.44). However, "over-tyring" is still bad: A certain pressure is required in the contact patch for a large frictional connection, which can be seen in Fig. 1.34. If the wheel load is too low, the frictional force decreases again compared to the maximum value.

At extreme angular accelerations of the tire, such as occur in drag racing, a large slip occurs and the contact force in the contact patch inlet increases considerably due to mass effects of the tire. This increases the acceleration capability of such vehicles. In addition, the tire growth, which increases with speed (cf. Fig. 1.11), replaces a gearbox so that high speeds can be achieved without shifting maneuvers [15].

The values shown in the diagrams for circumferential and lateral forces are maximum values that can be achieved for a specific tyre/road pairing. However, the tire needs a certain amount of time (and thus a distance) until it has built up the lateral force. For passenger car tyres, typical values of this so-called *relaxation length* are between 0.2 and 0.7 m [14]. If wheel load fluctuations now occur – as is unavoidable during real driving – this causes a lateral force loss, the magnitude of which depends essentially on shock absorbers, tire inflation pressure and the control arm inboard joints/bearings. However, wheel load variation also causes a loss of lateral force simply due to the degressive behaviour of the tyre, which is explained in the following section, Fig. 1.35. In the example shown, the wheel load $F_{W,Z}$ varies sinusoidally (blue). Due to the degressive tyre identification $F_{W,Y} = f(F_{W,Z})$ (red), this results in an asymmetrical fluctuation of the lateral force (green). Its axis of symmetry – i.e. the average lateral force – is lowered by the amount $\Delta F_{W,Y,m}$. The amount of the average lateral force loss in passenger cars is quantified in [11] as 40 N/°slip angle.

Interestingly, the influence of this vertical force variation is much more pronounced for cornering forces than for circumferential forces. In the case of the braking distance, unevenness of the road surface has only a very slight effect in the road test, but there are significant differences in the cornering capacity [30].

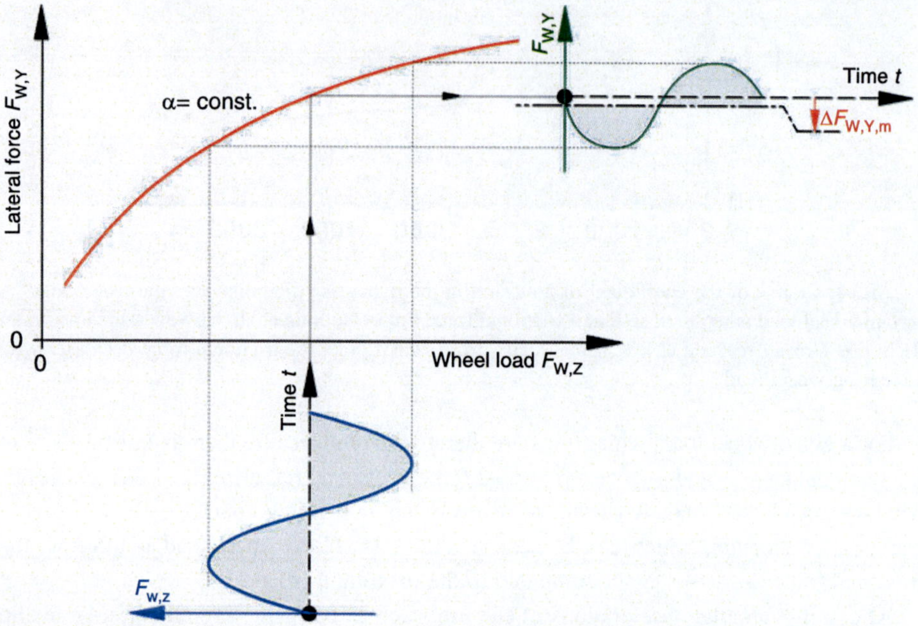

Fig. 1.35 Loss of lateral force due to wheel load sway, after [27]. α slip angle

Tyre Side Slip

The direction of movement deviating from the straight line as a projection of the wheel plane onto the road surface is always referred to as side sliping, irrespective of the road surface contact.

Tyre slip is the dominant parameter determining driving stability. The tyre slip angle as a function of the lateral force is influenced by the tyre type, but also in particular by the tyre inflation pressure and the wheel load. Furthermore, tyre construction and tread condition, as well as the superposition of a circumferential force have an influence.

In the case of a side slipping wheel, the rubber elements are deflected from their normal position by the lateral force acting in the slipper, Fig. 1.36. Initially, the elastic self-aligning force of the tyre acts against the deflection, which increases (approximately linearly) over the running length. If the maximum transmissible frictional force of rubber elements in the contact patch is exceeded (this is the case, for example, with large slip angles), these elements slip and the deflection no longer increases over the running length. When the rubber elements reach the run-out end of the slipper, the contact force decreases and the elements eventually spring back to their initial position. Due to the structural strength of the tire, the tread is deflected even before it enters the contact patch and returns to its initial position only after it has run out. According to the lateral deformation, the course of the cornering force in the contact patch results. It is worth noting here that the resulting total force $F_{W,Y}$ around the tyre trail (*pneumatic trail*) $r_{\tau,T}$ acts behind the geometric wheel contact point. This leads to the so-called self-aligning moment (see below). In a rough

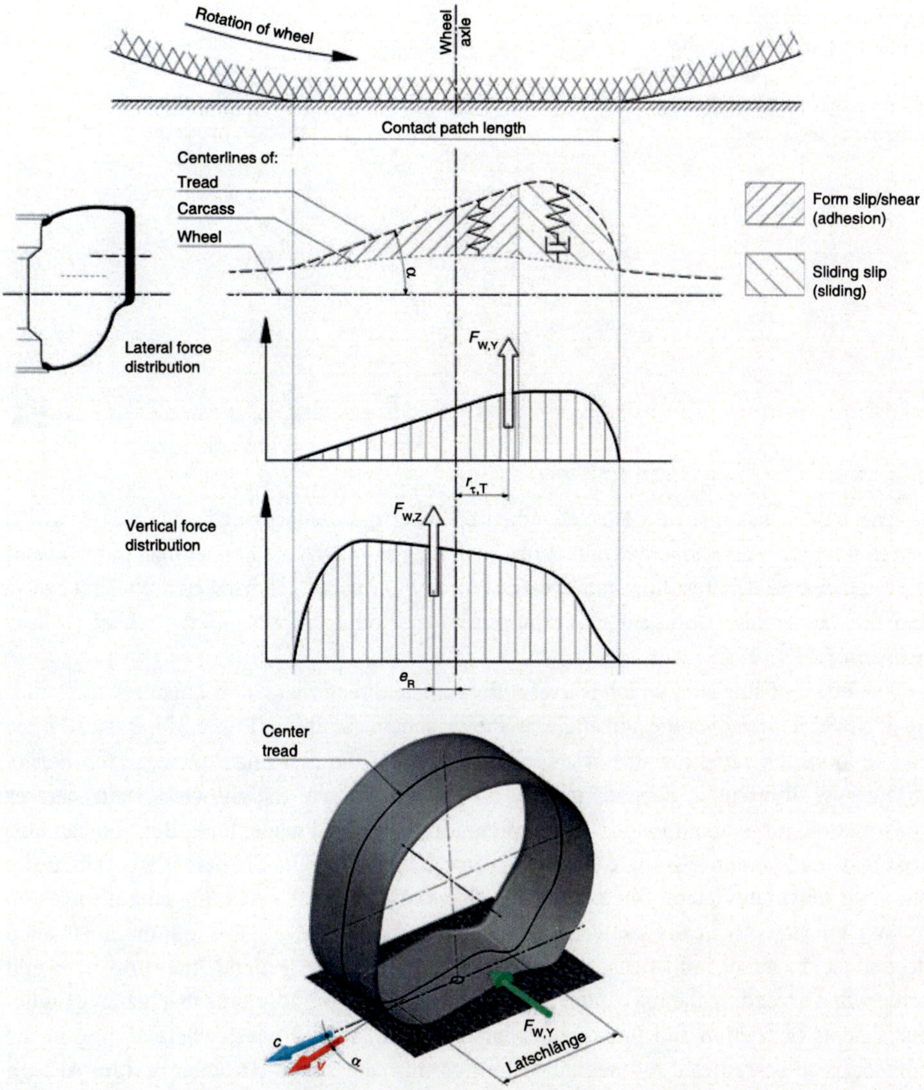

Fig. 1.36 Deformation of a diagonally rolling tyre. Above: schematic courses, below: axonometric representation. v Direction tyre heading, c median plane wheel, $F_{W,Y}$ lateral force, $F_{W,Z}$ vertical force, $r_{\tau,T}$, e_R center offset distance of forces, α slip angle

approximation (exact if the lateral force distribution is triangular), $r_{\tau,T} = 1/6\ l_T$, with load length l_T. If the sliding fraction in the contact patch increases, the lateral force distribution becomes trapezoidal and $r_{\tau,T}$ becomes smaller. With full sliding (rectangular course), $r_{\tau,T} = 0$.

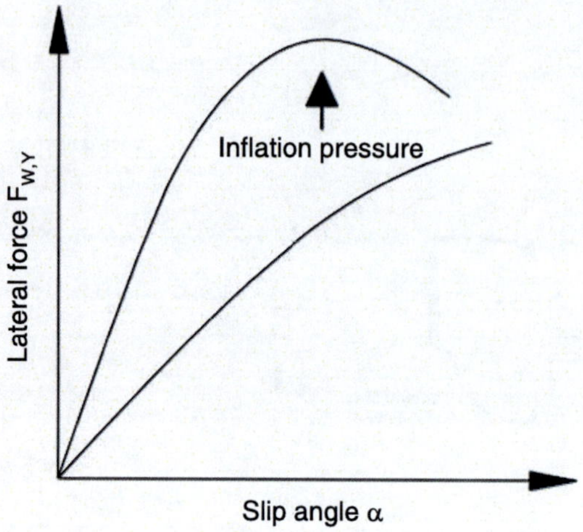

Fig. 1.37 Principle curve of the cornering force over the slip angle. In addition, the influence of an increasing tyre inflation pressure is registered

The tread must first be deflected when entering the contact patch. Mass effects and material rigidity mean that the contact forces in the run-in area are greater than in the rest of the contact area. The resulting total vertical force $F_{W,Z}$ acts e_R in front of the wheel centre and thus generates a moment that counteracts the rolling motion of the wheel (*rolling resistance*).

The effect of side slip, which is useful for vehicle directional control, namely the build-up of a lateral force over the slip angle, is shown schematically in (Fig. 1.37). Basically, the course looks the same for all tyres. Only the slope of the first linear increase (*cornering stiffness*) is different. It depends mainly on the type of tyre (height/width ratio, carcass construction, rubber compounds, . . .), inflation pressure and wheel load. But you can also see that from a certain slip angle the lateral force decreases again. So there is an optimal slip angle for maximum lateral force. For racing tires, the range of maximum lateral forces on dry pavement is typically between 3° and 7° [25]. However, this optimum is again dependent on many influences, including driving speed. In general, the drop in lateral force when exceeding the maximum is more pronounced for those tires that achieve higher coefficients of friction and thus greater lateral forces. Radial tyres, where sliding in the contact patch is restricted by the belt, belong to this category due to their design. At high speeds, the side sliping causes the tyre to heat up to a not inconsiderable extent. For this reason, smaller slip angles are generally used at high speeds so that the tyres are not destroyed.

A higher inflation pressure increases the lateral stiffness of a tyre (Fig. 1.37). Too low an inflation pressure has a negative effect on the maximum lateral steering force. The driving behaviour of a car can thus be influenced by the inflation pressure. If the pressure of the tyres of an axle is increased, a smaller slip angle occurs at this axle with the same cornering force.

For comparison, Fig. 1.38 shows the influence of wheel load and slip angle on the lateral force for two different tyres. With increasing slip angle, the lateral force increases – albeit

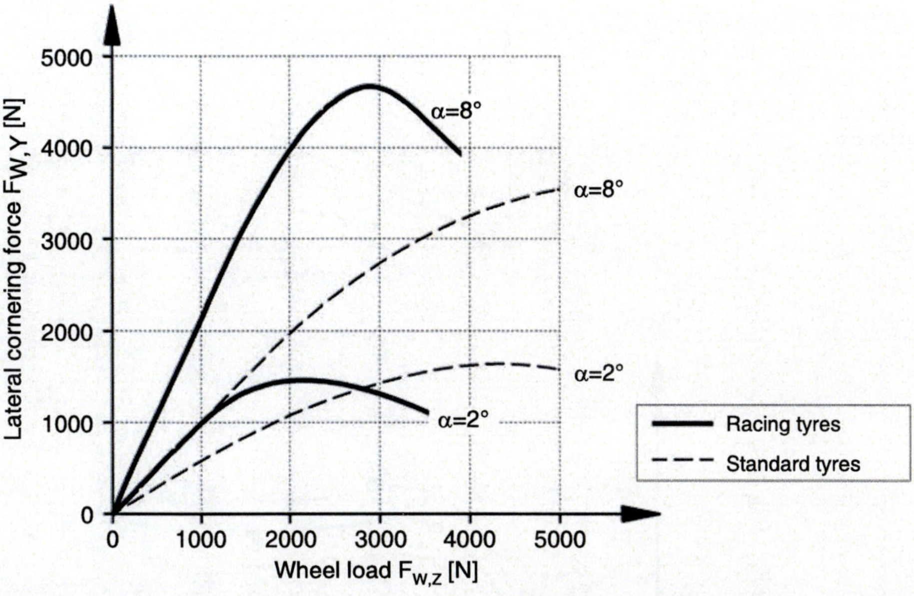

Fig. 1.38 Typical cornering force $F_{W,Y}$ of two tyres as a function of the wheel load $F_{W,Z}$ Two curves are shown for different slip angles α. It can be seen that the cornering force increases with increasing slip angle. Furthermore, the racing tyre shows a considerably steeper lateral force build-up over the wheel load compared to the standard tyre. On the other hand, the lateral force also drops steeply again after exceeding its maximum. The standard tyre, on the other hand, behaves considerably better and is therefore more suitable for the average driver

not linearly, but degressively. The racing tyre achieves higher lateral forces than its counterpart for passenger cars, but these forces fall more sharply with increasing wheel load.

Camber

Another influencing variable is the tyre plane inclined to the road surface (= camber), definition and sign see Fig. 1.41.

The wheel rolling freely under camber behaves like a truncated cone and describes a circular arc on the road surface. If the wheel is prevented from doing so (e.g. by the wheel suspension), it generates a force which points towards the apex of this truncated cone. The magnitude of the lateral forces caused by camber (Fig. 1.39) are, in relation to the relevant angle in degrees, about one fifth to one tenth of those forces generated by side sliping (Fig. 1.40).

If another force is superimposed on this camber side force when cornering, the total side force of the tyre increases or decreases depending on how the wheel has cambered, Fig. 1.41.

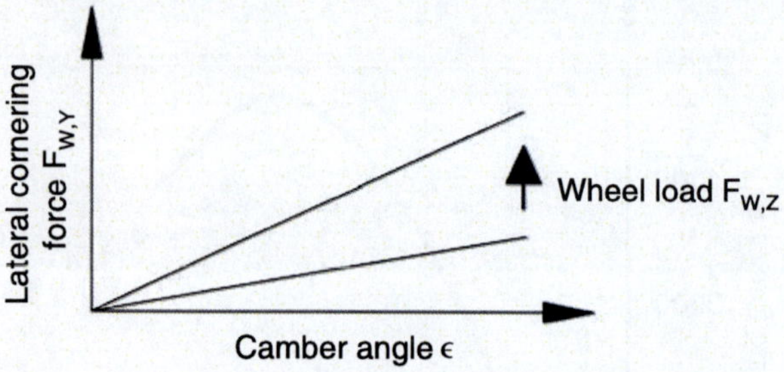

Fig. 1.39 Principle course of the lateral force over the camber angle. In addition, the influence of increasing wheel load is registered

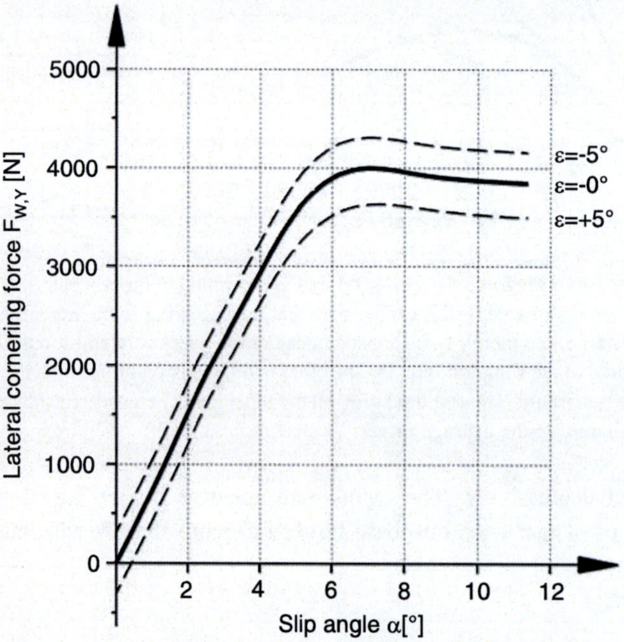

Fig. 1.40 Typical curve of the cornering force $F_{W,Y}$ of a racing tyre over the slip angle α. In addition, the change of the course influenced by the camber angle ε is entered. A camber angle also causes a lateral force when driving straight ahead (α = 0°). With a negative camber, this increases the existing lateral force of the side slip. A positive camber decreases the side slip lateral force

A camber side force can also occur without the wheel having fallen. An asymmetrical tyre structure (e.g. one wall is stiffer than the other) produces a similar effect. The tire basically deforms under the wheel load as seen in (Fig. 1.41) on the left, causing a side force during roll. This possibility of a camber "built into the tire" is used in racing series where only small camber angles are possible due to regulations and where corners are only negotiated in one sense. This applies, for example, to the NASCAR series with the oval stadiums popular in the USA [15]. The "built-in" camber is sometimes so severe that the mechanics have to countersteer if they want to push the car straight into the pits.

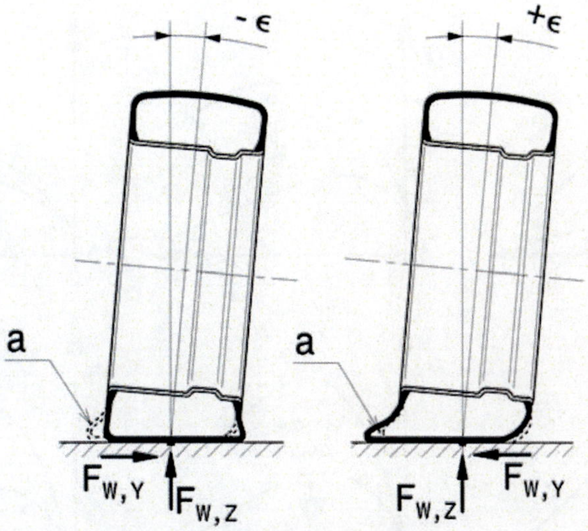

Fig. 1.41 Influence of camber with simultaneous application of lateral forces. **a** Contour of the deformed tyre due to conical rolling. In the left part of the picture, the camber angle is negative: The additional lateral force $F_{W,Y}$ acts in the same direction as the camber lateral force, i.e. it counteracts the tyre deformation. The total transmittable lateral force increases. The occurring slip angle is positive. In the right part of the picture, the additional lateral force $F_{W,Y}$ acts against the camber lateral force and thus reduces the total lateral force. The slip angle that occurs is negative

Self-Aligning Torque

The cornering force of a steered, bias-ply tyre is initially built up steadily in the contact area starting from the run-in and decreasing again as it progresses, depending on the friction conditions (cf. Figs. 1.36, 2.6, and 5.12). The resulting total force acts behind the geometric contact point and thus generates a moment about the vertical axis of the tyre. This moment wants to turn the wheel back to the straight-ahead position and is therefore called self-aligning moment. The magnitude of the self-aligning moment is therefore dependent on the adhesion and sliding processes in the contact patch and provides the driver with a valuable indication of the force conditions in the tyre contact patch via the moment on the steering wheel. It is of particular importance that the self-aligning torque reaches its maximum before the lateral force above the slip angle. This allows the exceptionally sensitive driver to specifically approach the limit of the tires (= maximum lateral force).

The most important influences and their effects are shown in (Fig. 1.42). It can be seen that all measures which increase the transmissible forces in the contact area also increase the self-aligning torque. Thus increasing wheel load, decreasing inflation pressure, large surface (= low to no profile) and increasing adhesion.

The so-called Gough diagram (Fig. 1.43) provides a summary representation of tyre forces. It combines the parameters lateral force, self-aligning torque, pneumatic trial, wheel load and slip angle in one picture.

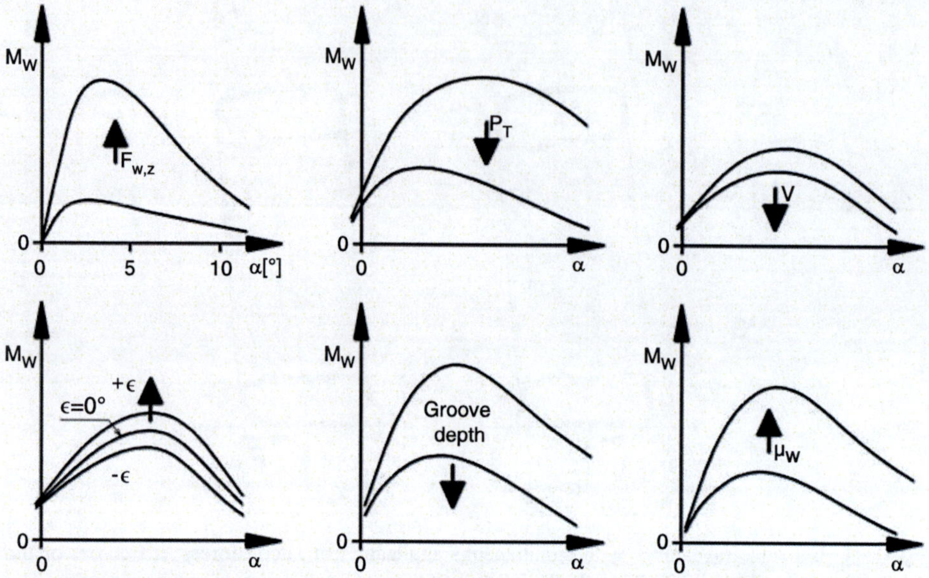

Fig. 1.42 Some influences on the self-aligning torque, M_W self-aligning torque, α slip angle, $F_{W,Z}$ Wheel load, p_T Tyre inflation pressure, v Driving speed, ε camber angle, μ_W friction

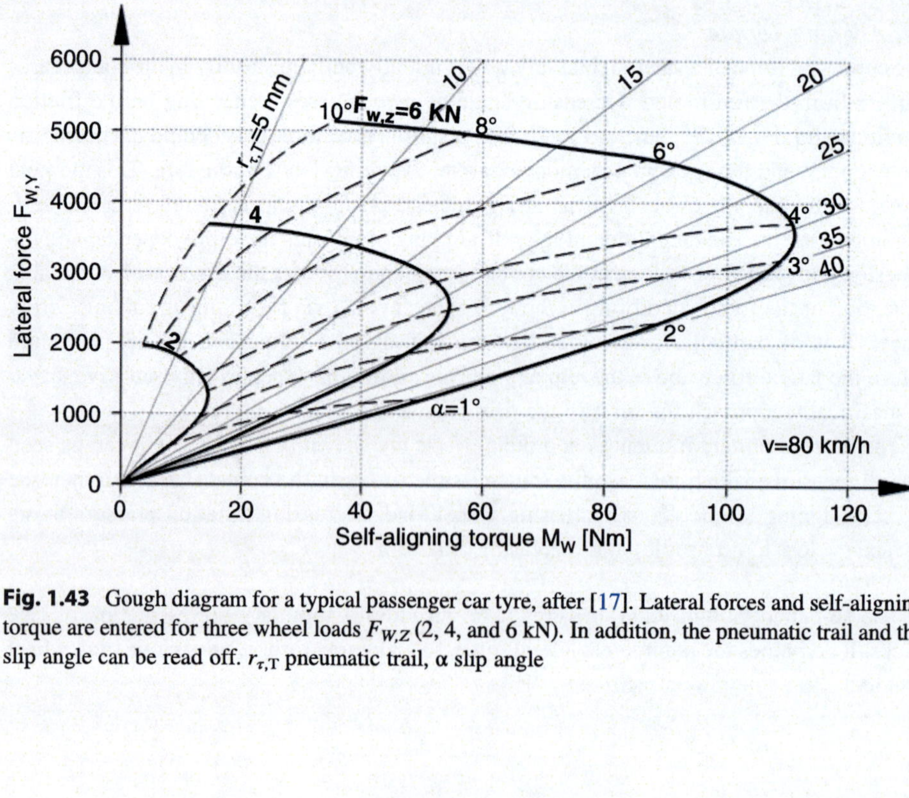

Fig. 1.43 Gough diagram for a typical passenger car tyre, after [17]. Lateral forces and self-aligning torque are entered for three wheel loads $F_{W,Z}$ (2, 4, and 6 kN). In addition, the pneumatic trail and the slip angle can be read off. $r_{\tau,\mathrm{T}}$ pneumatic trail, α slip angle

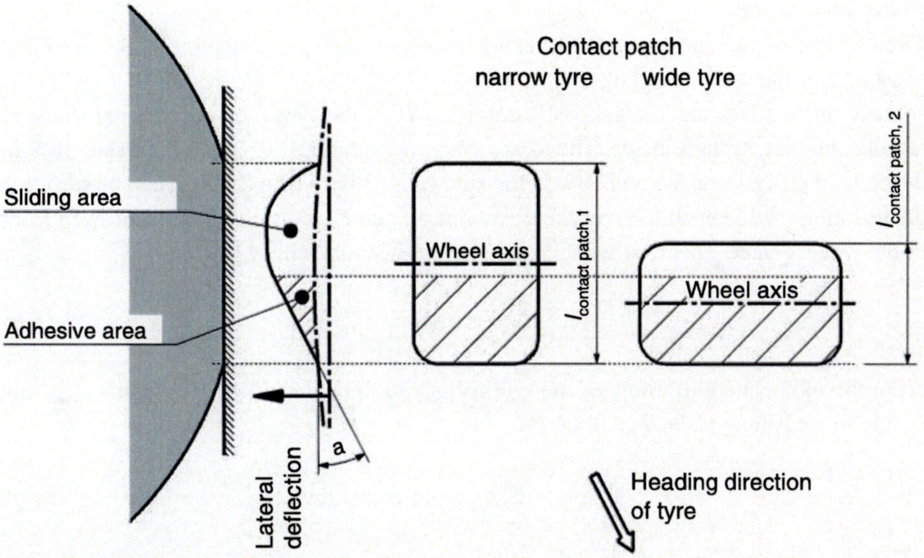

Fig. 1.44 Effect of different tyre widths on the frictional behaviour in the contact area of a side slipping wheel. α slip angle. The lateral displacement of the rubber elements in the tread is shown. The first area in which the displacement increases linearly is the adhesion area (form slip). This is followed by the sliding area. Under the same conditions (wheel load, inflation pressure, slip angle), approximately the same deformation conditions occur. In the case of a wide tyre with the shorter contact patch length $l_{contact\ patch,2}$ this leads to a high proportion of grip area (hatched) in relation to the total contact area

Tyre Width

The contact area depends primarily on the inflation pressure and the wheel load as explained above. With the same wheel load and the same inflation pressure (and therefore the same surface area), a wide tyre has a shorter (but wider) tread than a narrow tyre. If we now consider the deformation behaviour of a tyre under the effect of a lateral force (e.g. Fig. 1.36), it is obvious that, for the same slip angle, the proportion of adhesive surface is greater in the contact patch of the wider tyre than in the narrower tyre. Figure 1.44 represents this consideration schematically. The greater proportion of adhesive in the contact area means greater transmissible lateral forces for the wider tyre under the same conditions.

Wide tyres also transmit circumferential forces better and are therefore superior to narrower ones. A short contact patch requires a smaller bending angle from the tread when entering the flat road surface than a long one. Thus this part of the flexing work is smaller with the wide tyre. A disadvantage of wide tyres results from the short contact patch in wet conditions. The aquaplaning behaviour of short contact patches is considerably worse than that of long ones.

Rolling Resistance

Tire deflection and the associated flexing work is one of the main causes of rolling resistance of the straight rolling tire. The tread is buckled and compressed as it enters the contact patch. Likewise, the sidewalls deform. When the tyre runs out, the deformation recedes, but not without losses. The losses occur approximately 50% in the tread, 20% in the belt, 10% in the carcass and 10% in the sidewalls [26]. In (Fig. 1.36), it can be seen that the resulting total vertical force of the tire contact patch $F_{W,Z}$ attacks an amount e_R in front of the wheel center. This results in a rolling resistance moment M_R:

$$M_R = F_{W,Z} \cdot e_R \tag{1.6}$$

From the moment equilibrium on the tyre rolling straight ahead with the radius r_{dyn}, this results in the rolling resistance force F_R:

$$F_R = \frac{e_R}{r_{dyn}} F_{W,Z} = k_R \cdot F_{W,Z} \tag{1.7}$$

k_R	Rolling resistance coefficient, –

The rolling resistance coefficient depends on the tyre design, the road surface, the wheel load, the driving speed and the inflation pressure (cf. Fig. 1.28). Principle curves can be taken from (Fig. 1.45).

These values are determined on a drum test rig. The wheel is pressed against a drum. Compared to the measurements on the drum, there are sometimes considerable deviations on real road surfaces. On asphalt, rolling friction increases by about 20%, on rough concrete by at least 30%. The values are lower on new road surfaces than on worn surfaces.

The actual value k_R therefore results from the measured value $k_{R,0}$ to:

$$k_R = i_R \cdot k_{R,0}$$

i_R	Ratio, –

Values for i_R:	Surface	i_R
	Asphalt	1.2
	Concrete	1.3–1.4
	Cobblestone	Approx. 1.5
	Compact sand	Approx. 4
	Loose sand	Up to 20

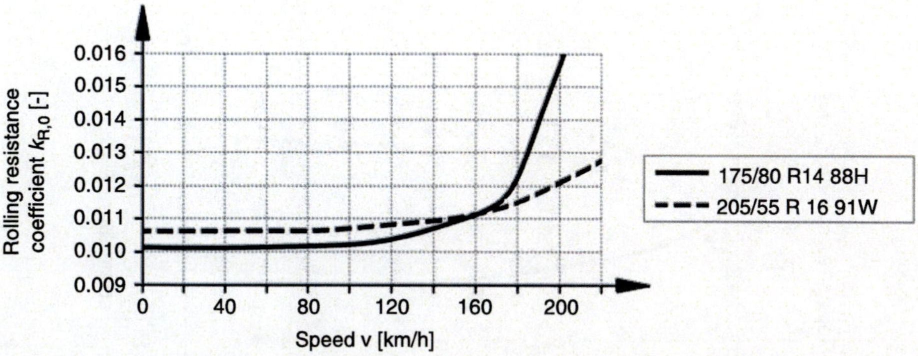

Fig. 1.45 Average values of rolling resistance figures for series radial tyres measured on a drum test rig, according to [11]. Actual values deviate from the drum measured values, see text. H tyres are only permitted up to 210 km/h and have a lower rolling resistance below 160 km/h than e.g. V and W tyres

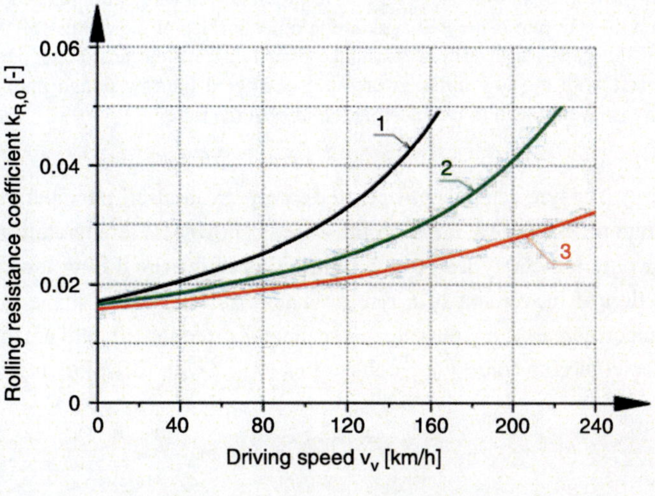

Fig. 1.46 Rolling resistance coefficients of different tyres, after [22]. 1 Passenger car tyre, 2 Sports car tyres, 3 Racing tyres

The influence of tyre design on rolling resistance is shown (Fig. 1.46) by some typical curves of the drag coefficient versus driving speed. A racing tyre (3) is designed for much higher speeds than a passenger car tyre (1) and this is reflected, among other things, in much lower rolling resistance at the same speeds.

Rain tyres displace road water from the tyre contact patch in tread grooves. This work can increase rolling resistance by up to 10 times.

The influence of temperature on rolling resistance is shown in Fig. 1.47. Over the test time, the tyre temperature increases due to flexing work and the rolling resistance decreases. It can also be seen that the increase in temperature over time tends towards a

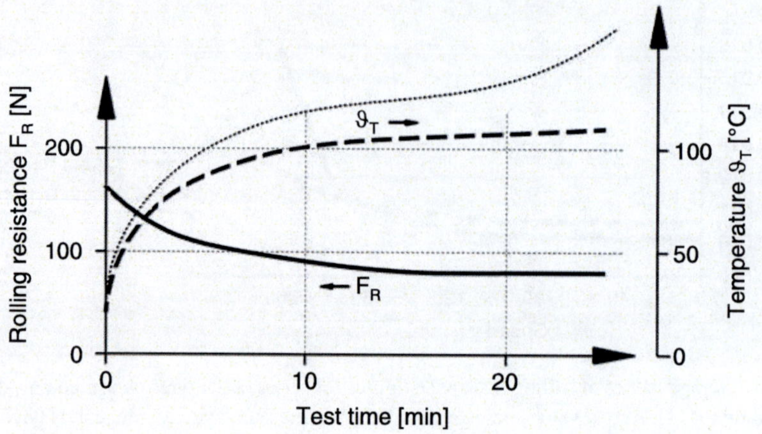

Fig. 1.47 Rolling resistance as a function of tire temperature θ_T, after [1]. Tyres: 7.25–13 nylon; speed: 165 km/h; wheel load: 4000 N; internal tyre pressure: 1.5 bar; Drum diameter: 2.5 m; drum curvature convex, i.e. the tyre is pressed against the outer surface of the drum. The increase in tyre temperature and the decrease in rolling resistance over the test time are plotted. In addition, the behaviour of a tyre with too low inflation pressure is entered (narrow dotted line). Unacceptable temperature increase would lead to tyre destruction after some time

permissible limit. The tyre can thus be operated safely. In addition, a temperature curve of a tyre with an impermissibly low inflation pressure is shown. Its temperature rises continuously, which would inevitably lead to the destruction of the tire during longer operation.

If the direction of movement of a tyre deviates from the wheel plane (side slip), the rolling resistance increases in relation to the actual direction of movement of the tyre (Fig. 1.48). The cornering force $F_{W,Y}$ contributes to this with the following share:

$$F_{W,\bar{x}} = F_{W,Y} \cdot \sin \alpha \tag{1.8}$$

Of the rolling resistance F_R when the wheel is running straight ahead, the following component counteracts the movement:

$$F_{R,\bar{x}} = F_R \cdot \cos \alpha \tag{1.9}$$

Due to this side sliping influence, the following circumstances have a resistance-increasing effect:

- Toe-in/toe-out (about 1% change per $\delta_{V,0} = 10'$) or wheel position error.
- Increasing corner speed (larger slip angles)
- Track width change during compression and extension
- Production run-out of the tyre
- Camber position of the tyre (influence can be neglected up to ±2°).

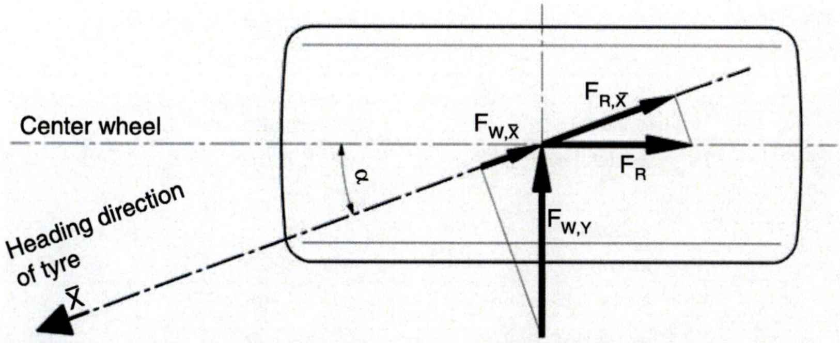

Fig. 1.48 Forces acting against the direction of movement of a slanted tyre (top view). Both the cornering force and the rolling resistance have a contribution which counteracts the direction of movement of the tyre

Due to the slip and the slip angle, the tire absorbs additional energy without converting it into propulsion. Part of the power is therefore provided to deform the tire:

$$P_{ls,\alpha} = F_{W,\bar{x}} \cdot v \tag{1.10a}$$

$$P_{ls,S} = F_{W,X} \cdot v \cdot \frac{S_{W,X,a}}{1 - S_{W,X,a}} \tag{1.10b}$$

$P_{ls,\alpha}$	Power loss due to side slip
$P_{ls,S}$	Power loss due to drive slip

An idea of the magnitude of such losses in Formula 1 cars (total mass 675 kg) is given in Table 1.4. On the oval track at Indianapolis, Indycars decelerate at 3.7 m/s^2 only by steering.

When exceeding a certain speed, more precisely at a certain wheel speed, the tyre is destroyed by increased flexing work and centrifugal acceleration. Each tyre therefore has a maximum design speed which must not be exceeded.

Springing Properties

The tyre acts like a spring due to its structure and the load-bearing medium air. When the wheel load is increased, the centre of the wheel drops proportionally and when the load is reduced, it returns to its original position. This behaviour must be taken into account when designing spring-damper systems for wheel suspension, (see Sect. 3.2.1) *Vibrations*. The vertical stiffness increases (slightly) progressively with the inflation pressure p_T. But also the driving speed and the side sliping influence the suspension behaviour, (Fig. 1.49).

Table 1.4 Power loss of the tyres during acceleration, braking and cornering of a Formula 1 car, according to [20]

Speed km/h	Driving status	Deformation state	Power loss kW
100	1.2 *g* acceleration	3% slip	6.7
100	1.2 *g* acceleration	6% slip	13.4
300	3.5 *g* braking	3% slip	58.2
300	3.5 *g* braking	6% slip	116.3
240	3.5 *g* lateral acceleration	3° side slip	80.5
240	3.5 *g* lateral acceleration	6° side slip	160.3

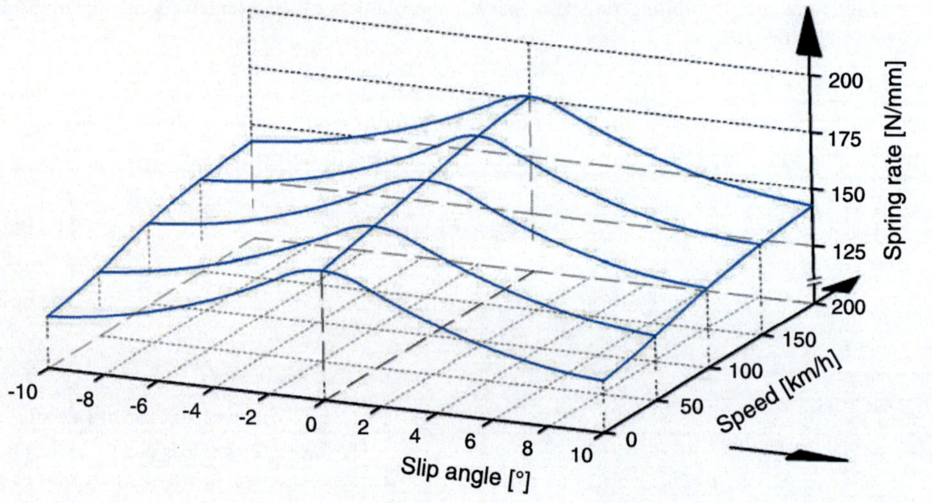

Fig. 1.49 Tire spring rate as a function of side slip and speed, after [11]. Radial tyre 185/70 R13 86 S, $p_T = 2.1$ bar. As driving speed increases, the spring rate increases because the belt wants to increase the tyre radius due to centrifugal force, thus counteracting the indentation. Side sliping, on the other hand, displaces the tread laterally, causing the tire sidewalls to take over some of the deformation and softening the tire

Temperature

Heat is released by friction and deformation (cf. Figure 1.12). Part of this heat is dissipated to the environment by convection. However, rubber is a poor conductor of heat and a relatively high material temperature is reached. On hot days, maximum values around 130 °C [8] can be measured immediately after arrival in the pit. During tyre development, therefore, great attention is paid to the temperature of the rubber compound during operation. Usually temperatures are measured when the tyre is stationary. The needle-like measuring sensor is pressed into the tread up to the substructure. However, there are

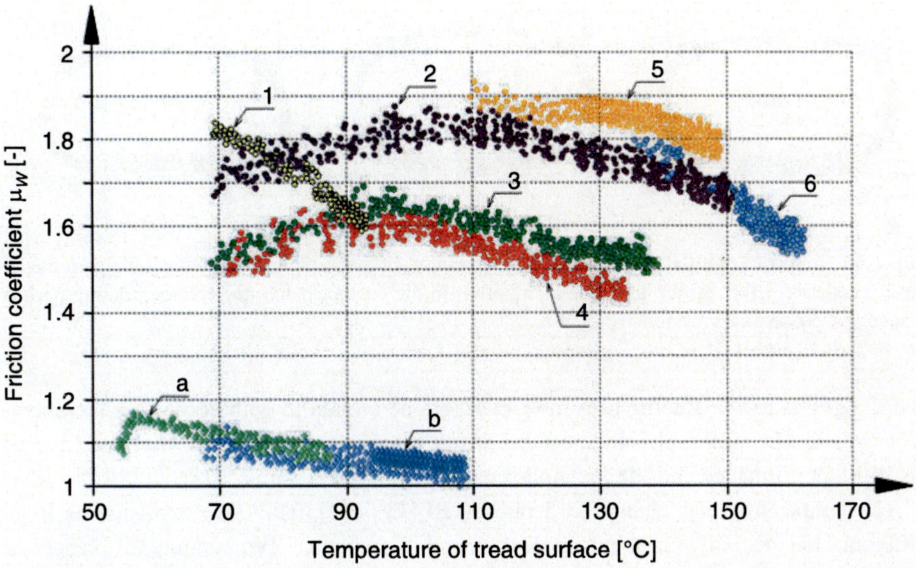

Fig. 1.50 Influence of rubber compound and temperature on the coefficient of friction [24]. Tyre types 1 to 6: racing tyres. Tyre types a and b: passenger car tyres. The temperatures were recorded with optical measuring sensors

also optical measuring methods that record the surface temperature.[4] Such non-contact measurements are also possible while driving and show values of 150 °C [15]. Peak values are around 180–190 °C.

The range of the effect of tread temperature on the coefficient of friction can be seen in Fig. 1.50. Typical passenger car tyres (a and b) develop their maximum friction at relatively low tread temperatures and exhibit an almost constant behaviour. Racing tyres show a stronger influence of temperature and, depending on the rubber compound, a greater variation. Type 1 racing tyres develop high coefficients of friction even at low temperatures, but these fall off rapidly with temperature. Such a tyre could be interesting for hill climbs. Type 2 is much more advantageous in the sense that the usable bandwidth is wider. Type 3 and 4 are similar in this respect, but have a smaller spread at constant temperature, so they are easier for the race engineer and driver to assess. The highest friction coefficients are transmitted by tyre type 5, but only at relatively high temperatures. The highest temperatures of approx. 160 °C are reached by tyre type 6. Racing tyres are supplied only partially vulcanised and only cure in operation after a few temperature cycles.[5] Then they achieve similar properties to passenger car tyres: The hardness is higher

[4] See, e.g., Racing Car Technology Manual Vol. 5 *Data Analysis, Tuning and Development*, Chap. 5. Fig. 5.3.

[5] A complete temperature cycle in this case is the temperature progression over time starting from the ambient temperature up to the maximum value and back to the initial temperature.

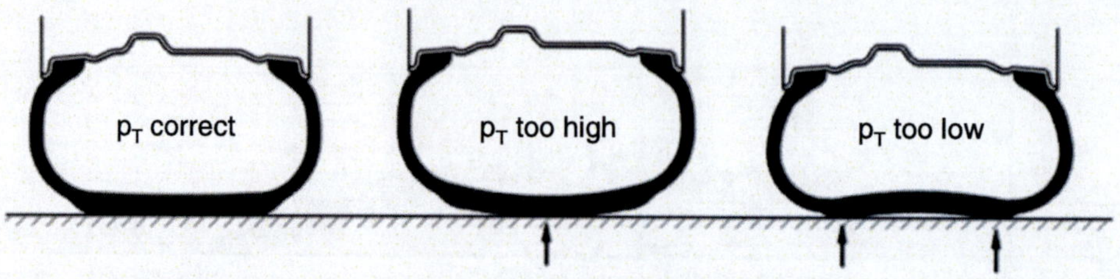

Fig. 1.51 Influence of inflation pressure on tyre heating. Both too high and too low inflation pressure have a negative effect on the pressure distribution in the contact patch and reduce running performance and power transmission

and the grip is lower. Racing tires must therefore be broken in with coordinated temperature cycles. The cycles driven must be documented by the teams so that the time of optimum performance and the remaining service life can be estimated.

The temperature can change at a rate of 10 °C/s [21]. This also explains the large difference between measurements in the box and in operation. Temperature differences of 40–65 °C between the inner and outer tyres are also not uncommon.

The temperature is influenced by: Vehicle weight, vehicle handling, tyre inflation pressure, rubber composition and tyre structure, ambient temperature and road temperature [8].

The vehicle designer can influence the weight and the driving behaviour. The temperature can be influenced within limits by changing the tyre diameter (see Sect. 1.2.4) [8].

The inflation pressure influences the pressure distribution in the contact patch and thus the temperature. If the pressure is too high, the temperature in the middle of the tread is higher than at the edges, (Fig. 1.51). If the inflation pressure is too low, the temperatures at the tyre shoulders will be higher.

For the greatest possible friction force of a tyre, the temperature must be constant across the tread width. The chassis geometry influences the temperature distribution. Higher temperatures on one side can be caused by excessive camber or toe-in. The tyre shoulders, which are in contact with the road surface, exhibit the higher temperatures because they perform the greater deformation work. In this context it must be mentioned that the tread rubber generally releases its stored heat more quickly in contact with the road surface than to the air. In the case of treaded tyres, for example, the temperature in the tread valleys is higher than on the adjacent webs of the treads.

Behaviour under Combined *Forces*

If the curves from the circumferential and lateral forces are combined to form a characteristic diagram, it becomes clear that the envelope of the maximum values lie on an approximate ellipse, curve $F_{W,max}$ in (Fig. 1.52).

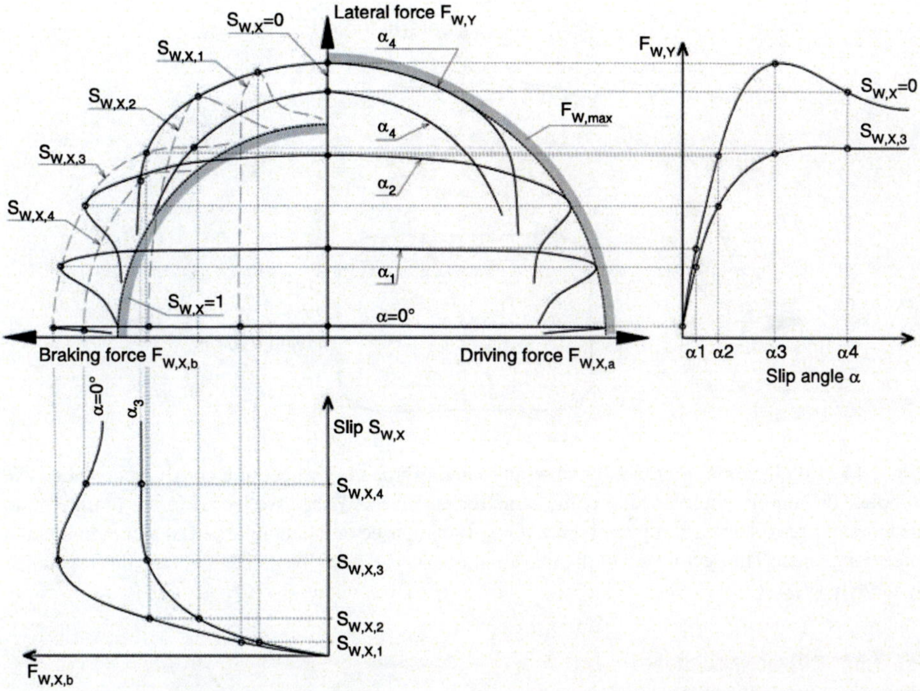

Fig. 1.52 Composition of a tyre map from the curves lateral force over side slip and circumferential force over slip, idealised, according to [16]. In addition to the envelope of the maximum transmittable force $F_{W,max}$, the limit curve for the locked wheel without side slip ($S_{W,X} = 1$) is entered

A tyre can only transmit the maximum possible force in one direction. If a vehicle is to build up maximum lateral acceleration, the driver must neither brake nor accelerate. Conversely, full acceleration is not possible in a corner. If one nevertheless wants to accelerate in a corner, the applied lateral force $F_{W,Y}$ of the outer wheel dictates how large the circumferential force $F_{W,X}$ may be in order to maintain the lateral force, Fig. 1.53. If this maximum transmissible force is exceeded, the transmissible lateral force is reduced accordingly, the wheel begins to slide and the car breaks away at the corresponding axle.

Compared to the purely rolling wheel, the contact area of the accelerated tyre moves forwards or backwards, depending on how the circumferential forces act (Fig. 1.54). It is obvious from this that braking and driving forces do not have the same effect on the tyre.

If we look more closely at the distribution of forces in the contact area, we obtain the idealised curve shown in (Fig. 1.55). The circumferential shear in the slippage is not even constant for the free-rolling wheel and changes its course considerably due to the effect of external forces. In the case of the driven wheel, the emphasis of the longitudinal forces lies in the front slippage area, and in the case of braking, in the outward slippage area. Because

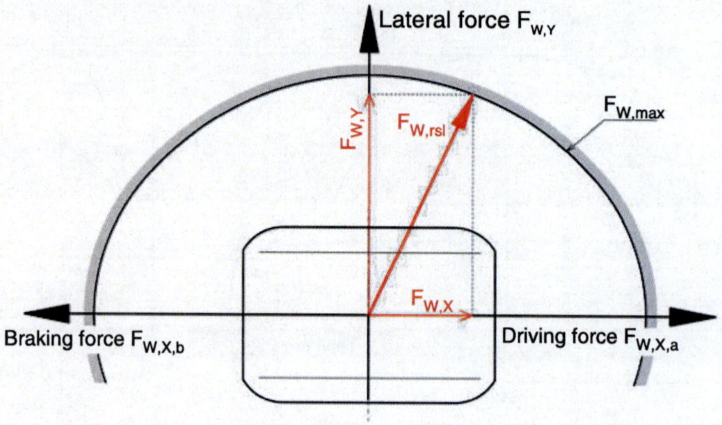

Fig. 1.53 Combination of circumferential and lateral force on a wheel in a corner, schematic. The complete diagram is symmetrical, i.e. the same for positive and negative lateral forces (cornering to the left and right). The resulting total force $F_{W,rsl}$ is composed of a lateral force and a circumferential force component. The vector sum of the two components can never be greater than the limit specified by the tyre ($F_{W,max}$)

Fig. 1.54 Displacement of the tyre contact patch due to the influence of circumferential forces

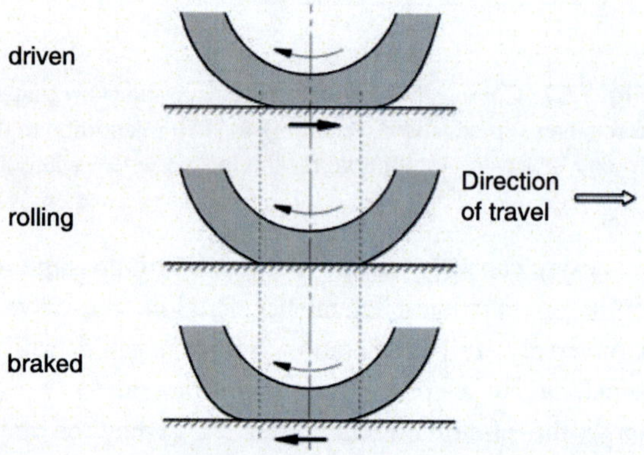

of this asymmetrical distribution of the shear stresses, the slip is greater when braking than when driving.

In a real map the lines of the same slip angle are indeed not symmetrical. In (Fig. 1.56) a clear difference between driving and braking can be seen. Pure braking forces (i.e. $\alpha = 0°$) can be transmitted worse than driving forces. With this tyre, a load change (change from driving to braking) in the curve leads to an increase in the lateral force at a constant slip angle. The greater the slip angle, the greater the increase in lateral force. A vehicle would pull into the corner when braking.

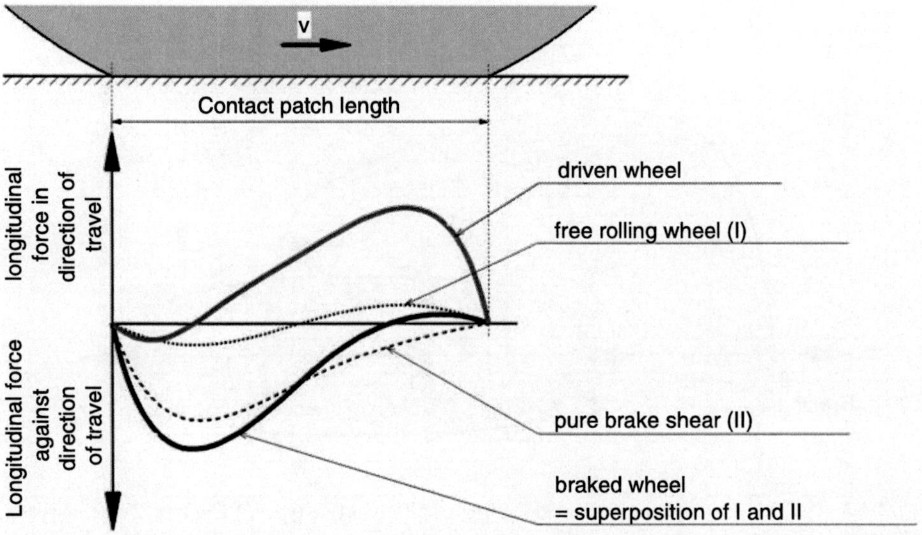

Fig. 1.55 Idealized circumferential shear curve in the contact patch, after [17]. The longitudinal force curve in the tyre contact area at different conditions is plotted over the latitudinal length. The total force acting on the tyre is obtained by integrating the curve for the driven or braked wheel over the slippage length

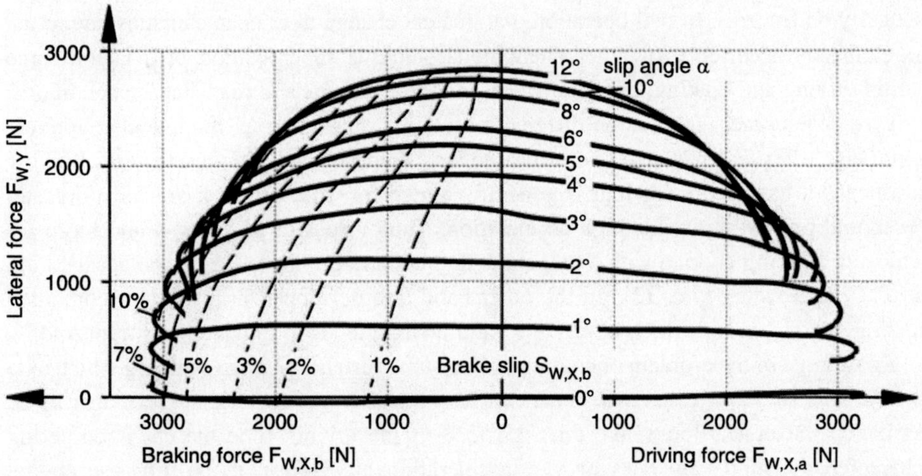

Fig. 1.56 Example of a measured tyre map, after [11]. Tyre tested: radial tyre 185/65 R14 86 S. Inflation pressure: 1.5 bar. Wheel load: 2940 N. The maximum lateral force $F_{W,Y}$ is 2850 N, the maximum braking force $F_{W,X,b}$ is 3130 N

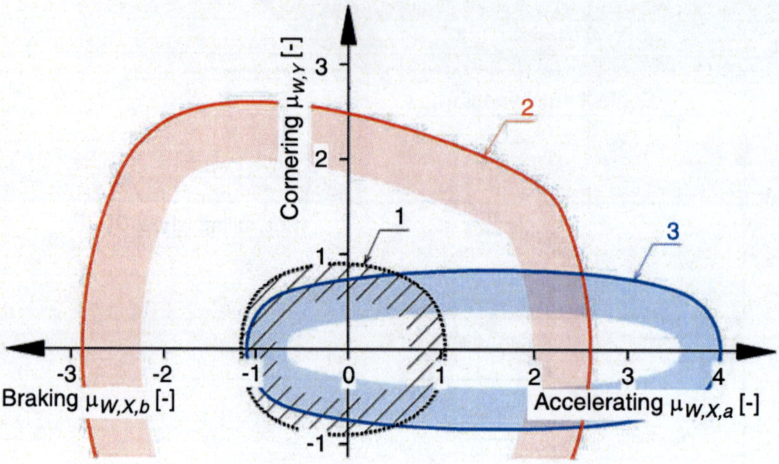

Fig. 1.57 Typical tyre maps for combined loading of different tyre types. 1 Car tyre, 2 Racing tyres, 3 Tyres for dragsters

The difference between the circumferential forces also depends on the type of tyre. Diagonal tyres react more strongly to a change in circumferential force – front-wheel drive thus only became possible with radial tyres.

The behaviour of the tyre when individual parameters change is examined quasi-statically on test rigs. In real operation, parameters change their characteristics due to the simultaneous occurrence of other phenomena, such as superposition of a lateral force during driving and braking, dynamic wheel load fluctuations and road surface conditions.

Tyre development focuses on different objectives depending on the intended application, (Fig. 1.57). For passenger car tyres (1), any compromise is shifted towards braking power and a balanced behaviour is generally aimed for. In contrast, a tire for a dragster (3) should primarily provide high acceleration. Thus values of $\mu_{W,X,a} = 4$ and more are achieved. Braking is done with a parachute and corners do not occur on the acceleration strip. Typical racing tyres (2), on the other hand, are developed primarily for combined braking and cornering, which is why the Kamm circle tends to mutate into a trapezoid.

Knowledge of the problem of combined tyre stress also helps when deciding which axle to drive with single-axle drive. Over the course of a season, the operation of formula cars on permanent racetracks looks like this: 45–55% of the driving time the car is cornering (characteristic: high yaw rates or yaw accelerations), 35–40% of the time the engine delivers full-load torque (acceleration phases) and 10–15% of the time the brakes are in use to decelerate the vehicle [28]. In this process, the front tires are more stressed than the tires on the rear axle during braking and cornering. If the rear axle is the driven one, the tyres on both axles are subjected to similar stresses in total, which, among other things, ensures an approximately constant driving behaviour during a race. This is one of the reasons why most racing cars have rear-wheel drive.

In summary, it can be stated that for the greatest possible transferable circumferential and lateral forces, the contact pressure in the contact patch should be kept low (cf. also Fig. 1.15) and the pressure distribution as homogeneous as possible. For each rubber compound there is a temperature range in which the highest adhesion values are achieved. Outside this range, the coefficients of friction are significantly lower.

1.2.4 Choosing of Tyres

For production vehicles, the axle load or wheel load is used as the basis for determining the minimum tyre size. The load capacity of a tyre increases with its load-bearing air volume (torus) [29]. Further criteria are maximum speed and availability of the tyre dimension.

The tyre size of racing vehicles is usually specified by the regulations within certain limits. Further considerations follow from driving dynamic influences. Small diameter tyres are lighter (unsprung mass!) and have low mass moment of inertia. Larger diameters reduce rolling resistance, but also inevitably lift the centre of the wheel and thus at least parts of the vehicle, which has a negative effect on the centre of gravity height. (One advantage is an increase in contact patch area with increasing diameter for the same tire width, allowing for greater circumferential forces). The larger the tire diameter, the softer the tread compound can be, because the specific load on the rubber decreases and the temperature drops, all other things being equal.

A larger wheel can also accommodate a larger brake disc. With the same outer diameter, reducing the height/width ratio of the tyre achieves the same effect.

A small height/width ratio also leads to more laterally stiff tyres, which the driver perceives as a more precise response to steering movements and which may therefore require a larger steering ratio in extreme cases (see Chap. 5).

Wider tyres increase the lateral force potential up to a certain value, but at the same time increase the air resistance. In single-seaters with open wheels (as an extreme example), the tyres can account for up to 40% of the vehicle's total aerodynamic drag [7].

The tyre width is also determined by the engine power. A dimensioning basis is provided by (Fig. 1.58).

Different tyre diameters due to production tolerances lead to directional driving behaviour in bends. Some drivers want the rear wheels to be staggered. In this case, tires with different diameters are deliberately mounted on the left and right, which favors cornering in one direction due to a cone roll effect. However, this measure is limited to bias-ply tyres and oval stages. However, in general, wheels are selected in sets, i.e. with the same diameter, in order not to change the driving behavior [8]. Tyres with a small height/width ratio are easier to manufacture to the nominal diameter; just as radial tyres are manufactured to the same rolling circumference.

If specific products are selected, the characteristic curves of these tyres can be used. Important evaluation parameters are [29]:

Regarding the circumferential force:

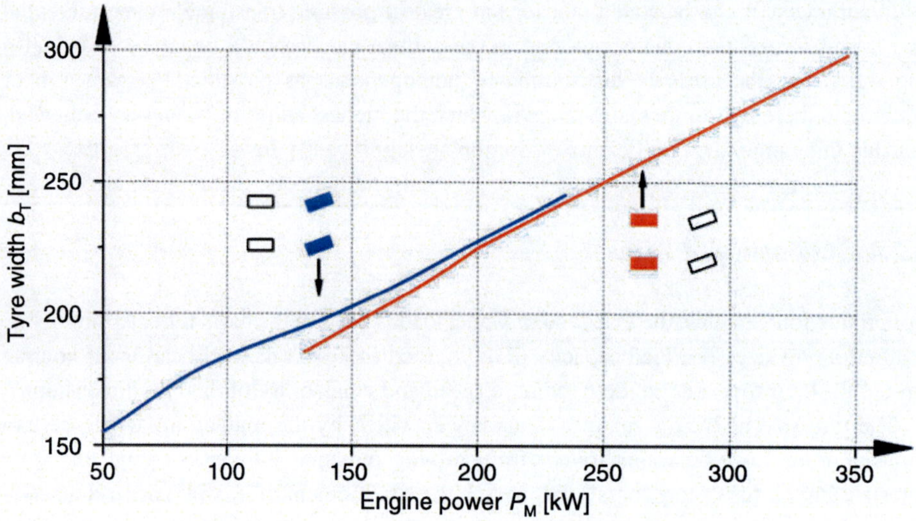

Fig. 1.58 Tyre width as a function of engine power, expanded from [29]. The values differ slightly for front-wheel drive (blue) and rear-wheel drive (red). For permanently all-wheel drive vehicles, the widths can be selected analogously to the distribution of the engine power between the axles

- Increase of the characteristic curve at small slip angles
- Maximum of the characteristic curve for driving and braking (provides $\mu_{W,max}$)
- Maximum end value of the longitudinal force (provides $\mu_{W,lo}$)

Regarding lateral force:

- Zero side force (force at slip angle $\alpha = 0$ °)
- Cornering stiffness (= increase of the characteristic curve at small slip angles)
- Maximum value of lateral force and associated slip angle
- Maximum end value of the lateral force

Regarding the self-aligning torque:

- Zero self-aligning torque (torque at slip angle $\alpha = 0°$)
- Increase of the characteristic curve at small slip angles (self-aligning stiffness)
- Maximum self-aligning torque and associated slip angle
- Slope of the characteristic curve at zero self-aligning torque
- Slip angle at zero self-aligning torque
- Final value of the self-aligning torque
- Tyre caster (= self-aligning moment stiffness/cornering stiffness)

Regarding the selection of tyres via their characteristic curves, it should be said that only certain combinations of wheel load, camber and side slip occur in operation [29]. A typical condition space for passenger car tyres shows (Fig. 1.59). Slip angles of $\pm 12°$ and camber

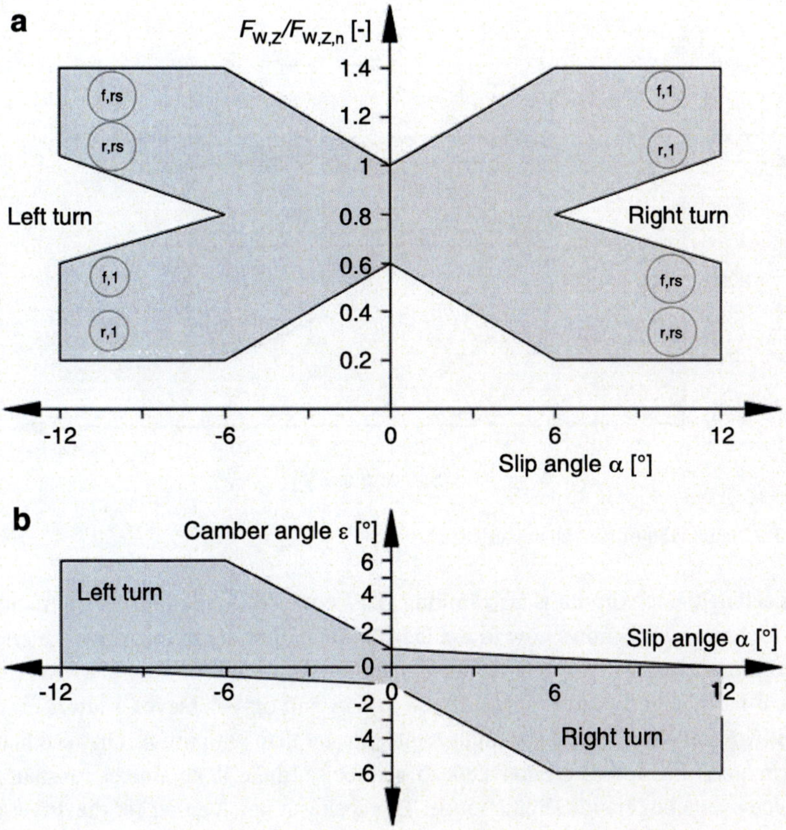

Fig. 1.59 Tyre condition space for passenger cars. The relevant area is shown shaded grey. (**a**) Wheel load side slip plane, (**b**) Camber side slip plane. $F_{W,Z}$ Wheel load, $F_{W,Z,n}$ Tyre load capacity. In addition, the extreme ranges of individual wheels are shown in Fig. a. For example, in a left-hand bend at $-12°$ side slip, the front right tyre (*f,rs*) with maximum wheel load and the rear left tyre (r,l) with minimum wheel load

values up to ±6° are quite achievable. The wheel load is 1.4 times the nominal tire load capacity.

Further considerations are provided by a comparison of characteristics of different tyres, Fig. 1.60. Shown is the lateral force versus slip angle at a given wheel load.

Tyre A shows the steepest increase $dF_{W,\ Y}/d\alpha$ in the linear range away from zero (*cornering stiffness*). It also reaches the maximum lateral force at the smallest slip angle. This means the lowest energy input due to side sliping and thus the lowest temperature increase. The area under the curve (highlighted with hatching) represents the energy lost at a given slip angle (see section on *rolling resistance*). This type of tyre allows the highest cornering speed because the maximum lateral force is well above the maximum values of the other tyres. This makes it the fastest tyre, at least in theory. However, the steep drop-off

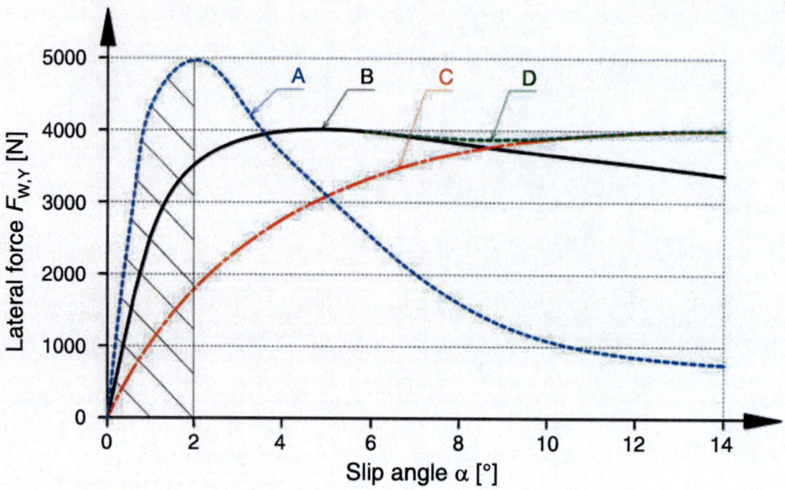

Fig. 1.60 Comparison of tyre characteristics

after exceeding the maximum is also striking. This tyre is difficult to drive and requires an experienced racer who knows how to use the maximum, but not to overdrive the tyre. Tire A is therefore a typical racing tire for the circuit.

Tyre B has a slightly lower slip stiffness than A and only achieves a lower maximum value with greater slip. The energy input is thus greater than with tire A. On the other hand, the maximum is much less pronounced; to be precise, there is an almost constant lateral force value over a larger side sliping range. This makes it much easier for the driver to find and maintain the maximum lateral force.

Tire C provides the driver with more feedback during turn-in, but behaves more sluggishly than type B. The maximum lateral force results from the fact that the tyre slides in the entire slip area, i.e. at extremely large slip angles. This type is the classic passenger car tyre for the normal driver.

Tire D gives the driver virtually no more feedback at large slip angles and also exhibits the greatest energy loss. However, the almost constant lateral force after exceeding the maximum supports the driving style of rally drivers who, in contrast to circuit drivers, always encounter new, unknown corners and therefore take them in a controlled drift (= driving with a large side-slip angle, i.e. large slip angles at the rear axle).

Other considerations when choosing tires include weight, manufacturer restrictions or availability (certain tire dimensions and styles are limited to individual manufacturers or countries), recommended wheel size, and ultimately price.

In addition, tyre selection will take into account whether the maxima of lateral force move to larger or smaller slip angles with increasing wheel load. This characteristic is decisive for the Ackermann component in the steering design (see Chap. 5 *Steering*).

A check whether the tyre parameters (dimension, rubber compound, inflation pressure, wheel load, . . .) have been correctly selected is carried out at the race track by determining

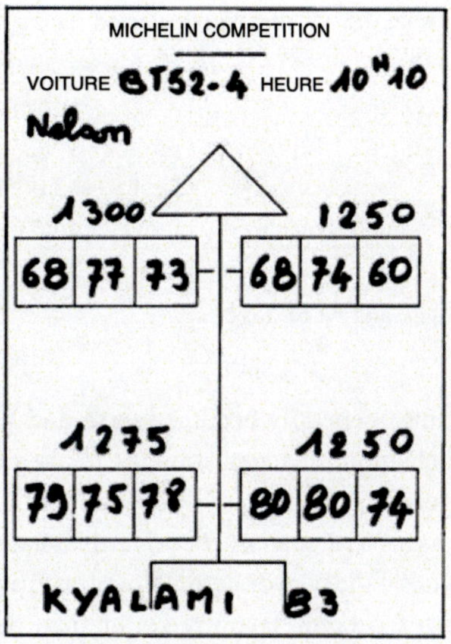

Fig. 1.61 Temperature sheet of a Formula 1 car [8]. The sheet shows the temperatures at three points of all four wheels measured after a practice run. The drive wheels have higher temperatures

the tread temperature over the tyre width, Fig. 1.61.[6] The usual temperature range is approximately between 80 °C and 100 °C [7]. However, rubber compounds that require much lower temperatures are also used. Such tires are used when tire preheaters are prohibited and the track is short. This is the case in hill climbs, for example. The tyre should have the same temperature over the entire width and all tyres on the vehicle should have similar values.

Tires that do not reach the optimal temperature range are oversized and do not transmit the force values that would be possible with optimal tire selection [7].

1.2.5 Tire Data

For comparison, Table 1.5 shows some tire data.

1.2.6 Types of Tyres

Apart from the fact that weather conditions can change, the tyre behaviour in and of itself does not remain constant during a race, but changes due to the influence of the forces to be transmitted. The flexing work increases the temperature (rubber is a poor conductor of heat) and thus the tyre pressure, and the material behaviour also changes. Accordingly, the lap

[6]For more details on the procedure, see Racing Car Technology Manual Volume 4 *Data Analysis, Tuning and Development.*

Table 1.5 Tyre data of selected racing vehicles

Vehicle		Dimension	Inflation pressure bar	Operating temperature, °C	Mass[a] kg
Formula 1 [2]	Front	9.5/25.5–13	1.4–1.65	Approx. 120 (Δp	9–10
	Rear	13.0/26.0–13	1.2–1.3	approx. 0.7 bar)	11–12
Formula Renault [3]	Front	16/53–13	1.45 (warm)		
	Rear	23/57–13	1.6 (warm)		

[a]Tire and wheel together

times generally become worse and worse in the course of a race. In addition, the driving behaviour changes strongly for vehicles with only one driven axle. The driven tyres are generally more heavily loaded and are therefore chosen to be harder than the non-driven ones. With rear wheel drive this leads to understeer in the first part of a race. Towards the end of the race or before the tyres are changed, the car oversteers because the rear axle tyres have become laterally softer than the front axle tyres. In endurance races a continuous control of the tyre pressure has proven to be very useful. Deviations from the target value in the course of the race are indicated to the drivers or the race engineers so that the tires can be changed if damage is imminent [12].

A tire consists of about 150 different components. This leads to a vast number of possible variations. In Formula 1 alone, 150 different tyre solutions are tested annually by a manufacturer. Basically, however, three types of tyres can be distinguished in racing (see Fig. 1.62):

(a) Dry tyres: Consist of a harder rubber compound and are therefore more enduring. Completely treadless tyres (slicks) have been used since 1971 [4]. For wear control, slicks have a few cylindrical depressions with a diameter of approx. 4 mm in the tread, which extend to the end of the usable rubber layer.
(b) Rain tires: Have a softer rubber compound and a water-draining deep tread that can displace up to 90 *l/s* of water at high speeds [2].
(c) Intermediates: are a compromise of the first two types and are used for mixed weather conditions.

Other gradations of a particular type are also offered. For example, there are dry tires in the versions: Hard, medium, soft and extra-soft. These gradations primarily concern the tread compound.

However, the choice of tyre type depends not only on the weather, but also on the circuit. Some circuits are known for their enormous tyre wear, others are known to have even soft compounds that last longer than usual. Other factors influencing the choice of rubber composition are air temperature, track temperature and the length of the race. Extra soft compounds are used when the air and track temperature are low and the track length is short (approx. 50 km).

Fig. 1.62 Types of tyres. (a) Dry tyre (slick), (b) Asymmetric rain tyre for the front axle, (c) Formula 1 dry tyre with grooves, as prescribed years ago to reduce cornering speed

Formula 1 tyres are designed for a mileage of only 200–300 km. The desired service life in many racing classes results from the fact that the wheel change is completed more quickly than refuelling, i.e. it makes no sense to compromise on tyre design so that the tyres last longer than one tank of fuel. For comparison, standard passenger car tyres reach about 25,000–50,000 km, depending on care and driving style. In long-distance races, the tyres are only changed after 500–700 km in favourable conditions (lower temperatures at night) [12].

1.2.7 Valve

Valves are used to inflate the tyres and maintain the desired pressure. On standard vehicles, rubber valves (Fig. 1.63, for dimensions see Table 1.6) are usually found, which can be used up to a pressure of 4.2 bar. At very high wheel speeds, valves experience bending stresses due to inertia, which can lead to leaks. Therefore, on sports cars and racing cars, the valve is supported externally on the appropriately designed rim or bolted metal base valves, (Fig. 1.64), are used. In any case, the valve should not be arranged radially, because in this case the valve body can be opened by its inertia (centrifugal force) at high wheel speeds and air can escape. Adjustable pressure relief valves, on the other hand, provide the desired pressure on some sprint cars by deliberately releasing the internal pressure that has risen due to the tire getting warm.

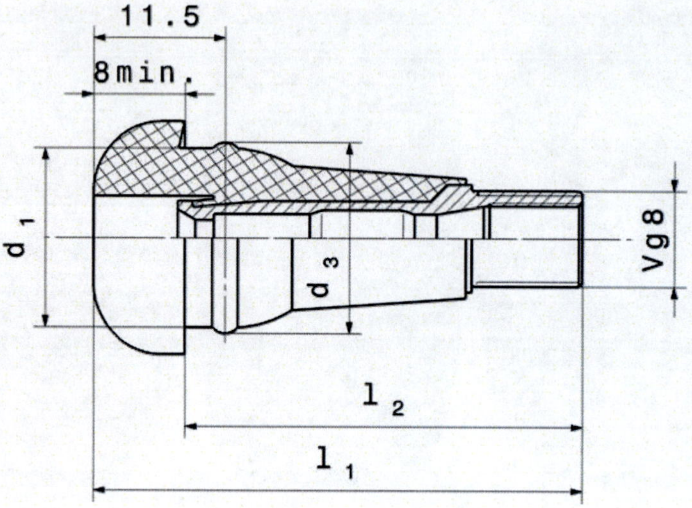

Fig. 1.63 Rubber valve for tubeless tyres to DIN 7780, snap-in valve. The metal valve body is vulcanized into a rubber jacket. The actual valve is screwed into the valve body. The sealing to the rim takes over the outer rubber jacket

Table 1.6 Rubber valves, dimensions in [mm]

DIN	ETRTO[a]	TRA[b]	Rim base diameter +0.4 0	l_1[c]	l_2	d_1 +0.3 0	d_3 +0.6 0
43 GS 11.5	V2-03-1	TR 413	11.3	42.5	34	15.0	16.0
49 GS 11.5	V2-03-2	TR 414	11.3	48.5	40	15.0	16.0
43 GS 16	V2-03-3	TR 415	15.7	42.5	34	19.2	20.2

[a]European tyre and rim technical organisation
[b]The tire and rim association Inc. USA
[c]Dimension designation (see Fig. 1.63)

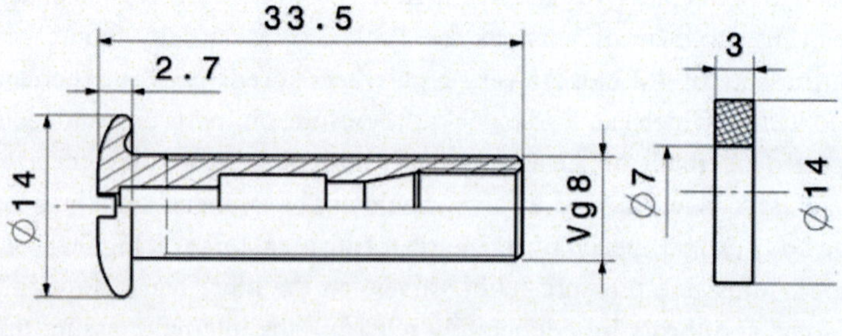

Fig. 1.64 Straight valve with metal base to DIN 7782 for valve hole in rim with 8.3 mm diameter. The actual valve is screwed into the valve body shown. A valve cap is screwed onto the valve body. The seal to the rim is provided by two sealing rings as shown, each of which is in contact with the inside and outside of the rim

1.3 Wheels

1.3.1 Requirements

Wheels belong to the so-called unsprung masses and should therefore be as light as possible. The mass moment of inertia around the axis of rotation should be as low as possible, as it counteracts acceleration and deceleration. The concentricity should be as exact as possible. Wheels should be easy to change. They must transmit the forces introduced by the tyres via the wheel hub to the wheel carrier.

In the case of brakes housed in the wheel, the wheels may also contribute to brake cooling, or at least they must not obstruct brake ventilation.

In addition, especially the wheel bowls or discs are also a styling element.

In the case of tubeless tyres, which are now common in both production and racing vehicles, rims must also be airtight and have a safety contour that prevents the air from escaping abruptly when cornering with a lower inflation pressure.

Even at normal passenger car speeds, wheels reach a high speed that requires balancing. Therefore, space or free space must also be available for balancing masses.

1.3.2 Designation of Wheels

A wheel consists of a rim and a wheel disc. The rim holds the tyre and is responsible for seating and securing it. The internal shape of rims is therefore standardised and tyre/rim combinations are approved for production vehicles. To secure the tyre, rims are designed with a hump (Fig. 1.65). To mount the tyre on the rim, a recess is needed, the drop centre (well).

The possibility of fixing the rim to the hub is the wheel dish (wheel disc). The flange plane does not have to coincide with the wheel centre. On the contrary – due to negative steering rolling radius or space requirement of wheel suspension parts, a distance is required, the offset, (Fig. 1.66). This can be positive or negative. Table 1.7 gives an overview of important connection dimensions.

Designation example of a wheel:

6 ½ J × 16 H2 B ET 45 with:	6 1/2	=	Rim width in inches, corresponds to 165.1 mm
	J	=	Flange version
	X	=	Drop center (well)
	16	=	Rim diameter inch code, See Table 1.8: $d_{nenn} = 405.6$ mm
	H2	=	Double hump, see Table 1.9
	B	=	Asymmetrical well
	ET45	=	Offset is 45 mm

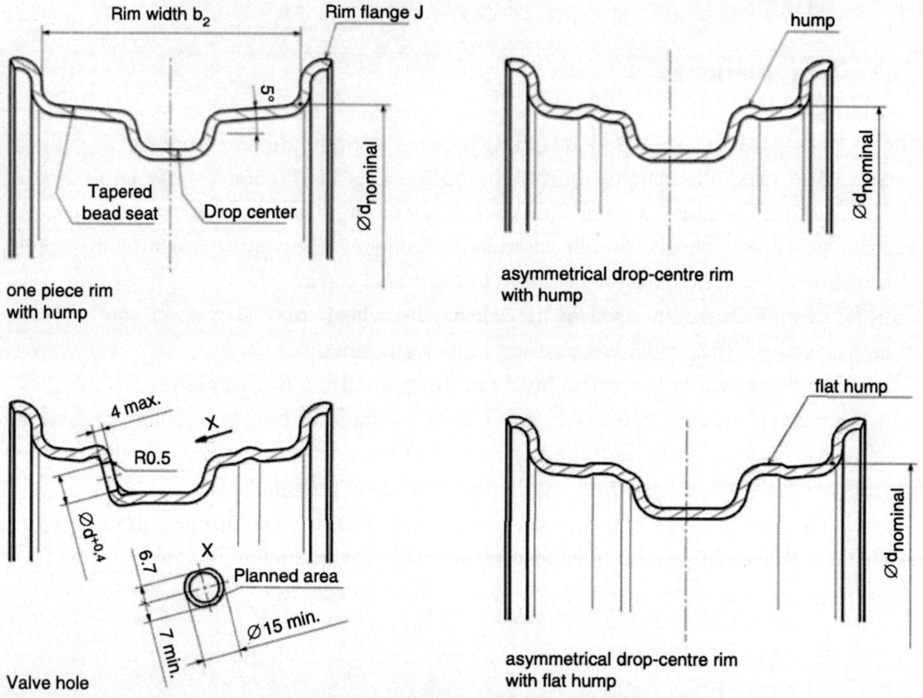

Fig. 1.65 Types of rims and designations. d_{nominal} Rim diameter. Valve hole diameter: 8.8 mm, 11.3 mm or 15.7 mm. For the cross-sections shown, the contour on the tyre side is specified in standards. The seat and securing of the tyre are essential here. The other contours are left to the manufacturers

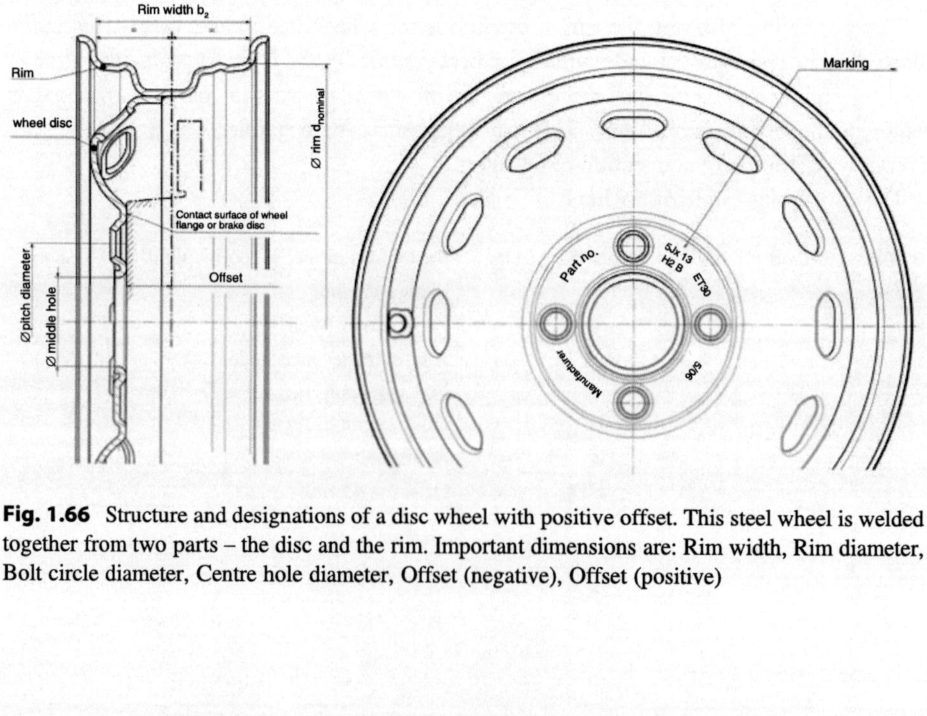

Fig. 1.66 Structure and designations of a disc wheel with positive offset. This steel wheel is welded together from two parts – the disc and the rim. Important dimensions are: Rim width, Rim diameter, Bolt circle diameter, Centre hole diameter, Offset (negative), Offset (positive)

Table 1.7 Important connecting dimensions of standard wheels, according to DIN 74361 T1

Number of screw holes	Ø bolt circle, mm Tolerance: ±0.1	Ø Centre hole mm	Ø Hub collar *d6* mm	Thread of the associated bolt
4	100	57	56.5	M12x1.5 taper collar 60°
		60	59	M12x1.5 taper collar 60° M14x1.5 ball collar
4	130	80	79	M14x1.5 ball collar
5	112	63	62	M12x1.5 taper collar 60°
		66.6	66	M12x1.5 ball collar
				M14x1.5 ball collar
5	120	72.4	72	M12x1.5 taper collar 60°
5	130	90	89	M12x1.5 taper collar 90°
		85	84	M14x1.5 ball collar

Table 1.8 Rim diameter inch codes

	Rim designation (inch code)						
	13	14	15	16	17	18	19
$d_{nominal}$, mm	329.4	354.8	380.2	405.6	436.6	462.0	487.4

Table 1.9 Designations of the humps

	Type of safety shoulder		
Naming	Rim outboard	Rim inboard	Code letter
Unilateral hump	Hump	Normal	H
Double hump	Hump	Hump	H 2
One-sided flat hump	Flat hump	Normal	FH
Double sided flat hump	Flat hump	Flat hump	FH 2
Combination hump	Flat hump	Hump	CH

1.3.3 Types of Wheels

A basic overview of the types of wheels is given in Fig. 1.67. According to this, wheels can be manufactured from one piece or assembled from several parts.

Wheels are made of steel, aluminum and magnesium alloys. In this order, the wheels become lighter but also more expensive. Magnesium wheels are usually preferred in racing. Wheels made of fibre-reinforced plastics (still) enjoy an exotic existence. The main reason for this is that they are not allowed in many racing series. There are some successful developments that have passed the required acceptance tests.

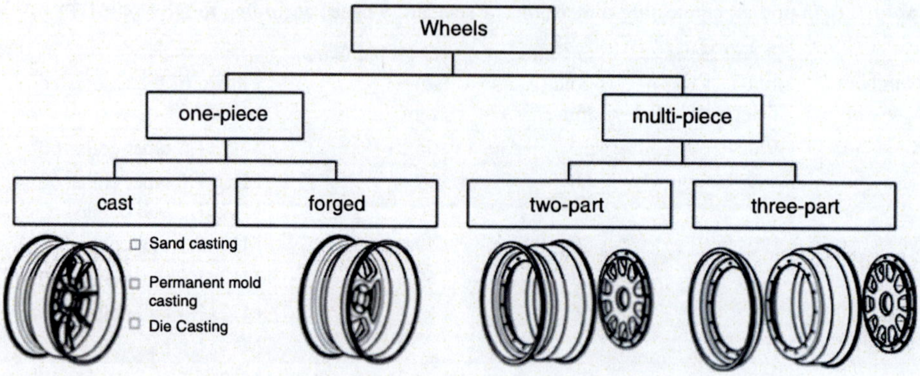

Fig. 1.67 Types of wheels. Two-piece wheels consist of a rim and wheel spider or wheel dish. Three-piece wheels are composed of outer and inner rim, as well as the wheel spider

The thermal conductivity of magnesium is better than that of aluminium. This results in a lower tire temperature for magnesium wheels under otherwise unchanged conditions. Conversely, aluminum wheels are often used in the rain, because it is difficult to get the tire to the desired operating temperature in these weather conditions and the lower heat dissipation is an advantage.

A further differentiation of wheels is offered by the construction method. Wheels can be represented in integral and differential construction. Multi-part wheels have the advantage that only damaged parts need to be replaced in the event of damage. In addition, the rim's jaw width and offset can be changed relatively easily and each section can be made with the most suitable material, e.g. tough aluminium alloy for the rim and light magnesium alloy for the wheel spider, (Fig. 1.70). The disadvantages, apart from the higher mass, are: possible leaks with a three-part construction and increased maintenance effort due to the large number of bolted connections that have to be checked.

One-piece wheels are finished by machining after the blank has been cast or forged. They therefore have a higher concentricity than multi-part wheels. In this category, light alloy wheels run more precisely than wheels made of pressed sheet metal.

A comparison of aluminium wheels according to their type of manufacture is provided in Table 1.10.

A cast wheel for a passenger car is shown in (Fig. 1.68). Figure 1.69 shows a forged wheel of a racing car. The wheel is first forged from a magnesium alloy as a disc wheel. The spokes are then machined out. Shot peening is used to increase the fatigue strength of the material. Before the wheel is delivered, it is inspected for cracks and x-rayed [19] (Fig. 1.70).

Materials

Steel wheels: For rim dual phase steel DP600, for wheel bowl steel HR 60 [5]. Structural steel S235JRG2 (was RSt37).

Table 1.10 Comparison of aluminium wheels, according to [1]

	Production		
Criterion	Cast wheel	Forged wheel	Split wheel
Material	EN AC-AlSi12Mg	AlMgSi1 F31	AlMgSi1 F31
Shaping	Multifarious	Restricted	More limited
Surface	Varnished	Anodized	Anodized
Against chem. attack	Constantly	Vulnerable	Vulnerable
Mass referred to cast wheel	1	0.82	0.75[a]
Production costs	Low	High	High[b]
Picture		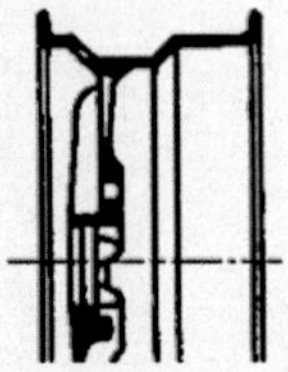	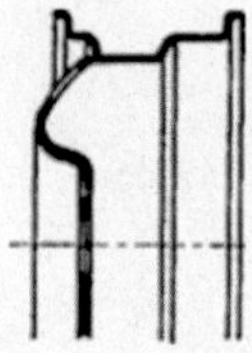

[a]This lowering of the mass of the slotted wheel compared to the forged wheel can be explained by a lowering of the Wöhler line
[b]The split wheel did not find acceptance in the passenger car market due to cost [5]

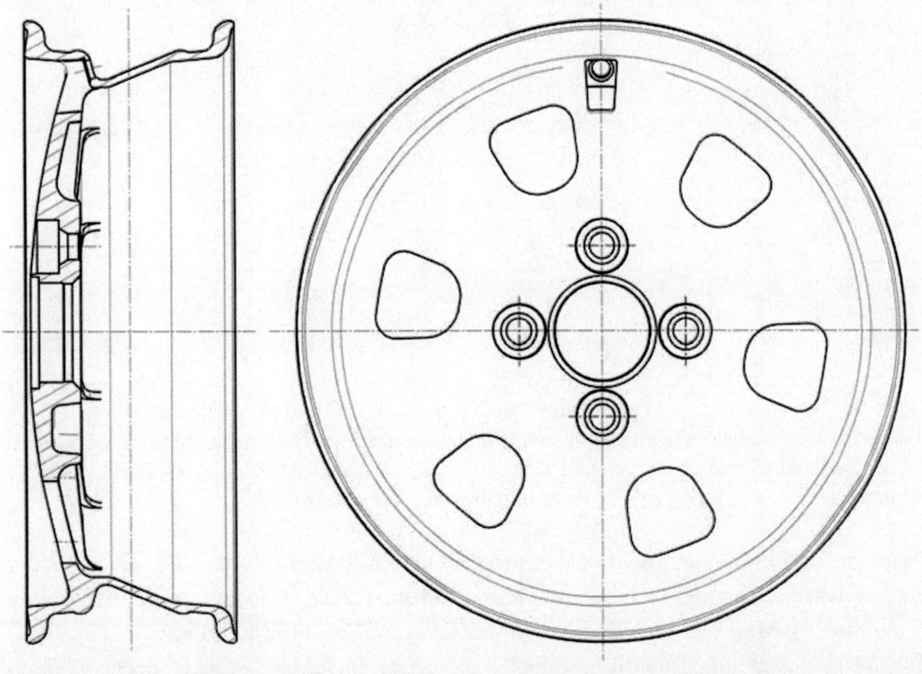

Fig. 1.68 Cast wheel for passenger car with drop-centre cast wheel. The tire is held by a double hump rim. The wheel is centered and fastened with tapered bolts. The wall thicknesses are designed differently due to different strength requirements at different points

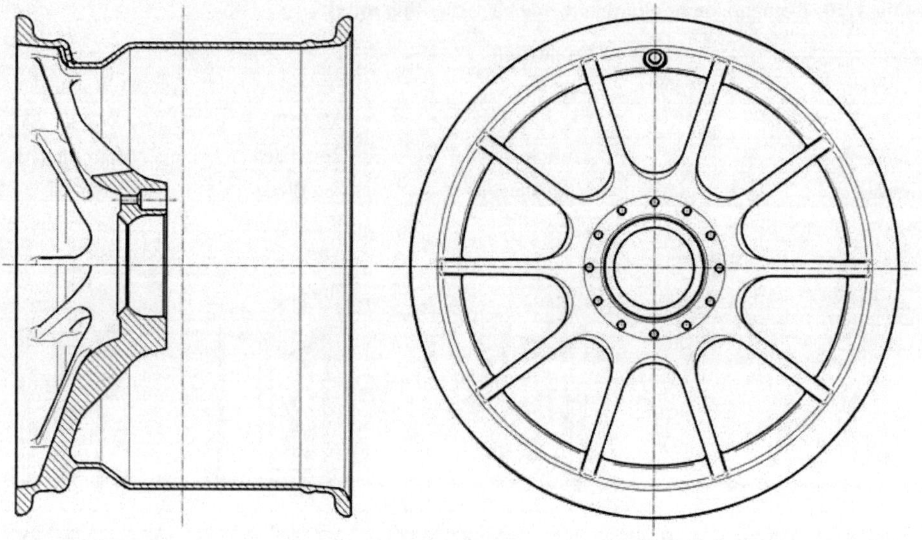

Fig. 1.69 Forged wheel for racing car with drop centre. 13″-F1 wheel with centre lock. The wheel can still get some driving bolts inserted into the holes in the hub area, if they have to be present on the wheel side

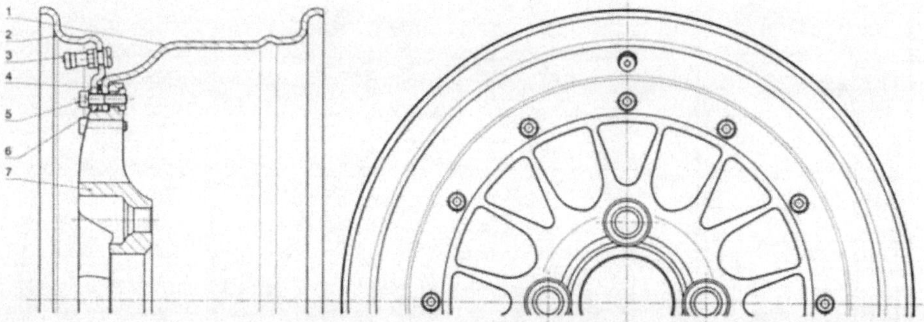

Fig. 1.70 Structure of a three-piece wheel three-piece split rim. The two rim parts (1) and (2) are screwed to the wheel disc (7) with the screws (5). The sealing ring carrier (6) is also clamped between the two parts, which has grooves on both sides for O-rings (4) and with them creates the tightness for the tubeless tyre. The valve (3) is screwed into the outer rim part (2)

Cast wheels: Aluminium gravity die casting EN AC-AlSi12Mg.

Forged wheels: Wrought aluminium alloy AlMgSi1 F31, wrought magnesium alloy AZ80A (= MgAl8Zn to DIN EN 1754), ZK 60 (= MgZn6Zr to DIN).

The fatigue strength properties of light metals are increased by shot peening. With magnesium, the design of the wheel plays an important role. Sharp notches and edges, as well as abrupt cross-sectional transitions must be avoided.

1.3.4 Choice of Wheel Size

On the one hand, the size of the wheel is dictated by the tyre selected; on the other hand, the availability of tyre sizes is a decisive criterion. The tire is the most important single component and its selection should not depend on a limited product range or long delivery times. By far the most common size in racing is 13″. The choice of tyres for these diameters is enormous, as is their stock. So 13″ are the first choice [6]. 10″ wheels, like 12″, have the appeal of lower mass and lower moment of inertia. Disadvantages are just a small selection of different tire widths as well as rubber compounds and the 12″ wheel is a distinct exotic. In addition, a small rim diameter also restricts the installation space for brakes and wishbone connection, see also Fig. 1.3.

For a given wheel size, the mass can still be influenced by the material. However, the shape must be adapted to the material properties. First and foremost, the E-modulus must be mentioned here so that the wheel has the same stiffness. In addition, manufacturing properties dictate the shape (transition radii, draft angles, flow cross-sections, shrinkage cavities, residual stresses, ...). In any case, a lighter wheel has advantages in terms of driving dynamics: Rotational and translational (and unsprung) masses are smaller. The influence on the maintenance of the important road contact is clearly illustrated in Fig. 1.71.

Another important point is the type of centering and mounting. The number of bolts and bolt circle diameter as well as centered or bolt-centered dictate the design of the wheel flange or hub. Individual circumstances can dictate the choice here. If, for example, an existing vehicle is being converted, it is much cheaper to select the wheels according to the existing hubs than to change the hubs and everything that goes with them in order to be able to fit the desired wheels.

The rim width is determined by the tyre width [7]:

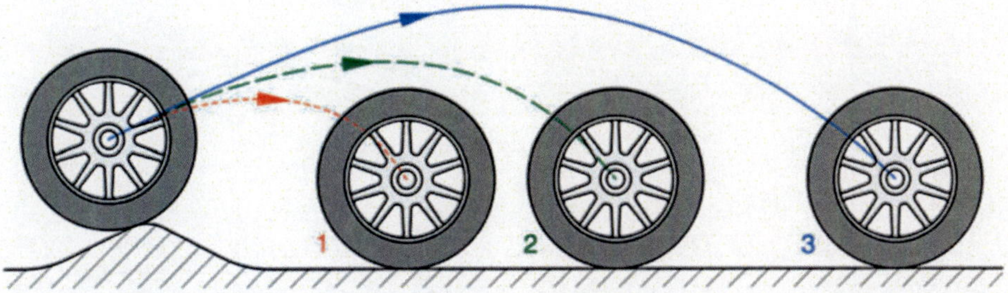

Fig. 1.71 Influence of wheel mass on road contact. When driving over a bump quickly, the wheel only regains contact with the road after a certain distance. The movement is shown for wheels made of three different materials. The compression damping as well as the tyre dimensions have been left the same for reasons of comparison. *1* Magnesium wheel, *2* Aluminum wheel, *3* Steel wheel. The steel wheel has the greatest inertia: It rises the highest and lands last. From the point of view of driving performance, safety and stability, the magnesium wheel is the best

In the case of road vehicles, roughly:	Rim width ≈ tyre width – 0.5″ to 1.5″ (approx. 13–38 mm)
Roughly applies to racing vehicles:	Rim width ≈ Tyre width + 1″ to 2″ (approx. 25–50 mm)

The further apart the rim flanges are, the wider the tyre stands on the rim with the result of improved lateral force transmission. In addition, the internal volume increases and thus theoretically also the load capacity. The combined effect of both is that a widening of the rim width by ½″ is equivalent to an increase in inflation pressure of 0.1 bar. The use of a wider rim therefore results in more favourable cornering and slalom behaviour (generally without any loss of ride comfort) [9]. The section width b_T of the tyre (cf. Figure 1.1) increases by approx. 5 mm per ½″ larger rim width than the measuring rim. In case of doubt, one will go for a wider rim if the regulations allow it and if the tyre fits into the wheel arch when in full motion (steering, bouncing). In any case, the loss of tyre grip due to a too large rim width is less than that due to a too small rim width (Fig. 1.72).

Load assumptions for wheel calculations are based on telemetry data and on measurements with special measuring cells fixed in the centre of the wheel. The same loads are naturally used for the design of the chassis.

For a Formula 1 wheel, typical load cases and values for wheel design are [19]:

(a) Pothole: Vertical load $F_{W,Z}$ of 16.3 kN with an additional force of 68 kN due to the pre-tensioning of the central nut.
(b) Cornering: lateral force $F_{W,Y}$ of 15 kN and vertical force $F_{W,Z}$ of 10.1 kN. Tilting moment $M_{W,X}$ of 4.53 N m. Additional force of 74 kN due to the pre-tensioning of the central nut.

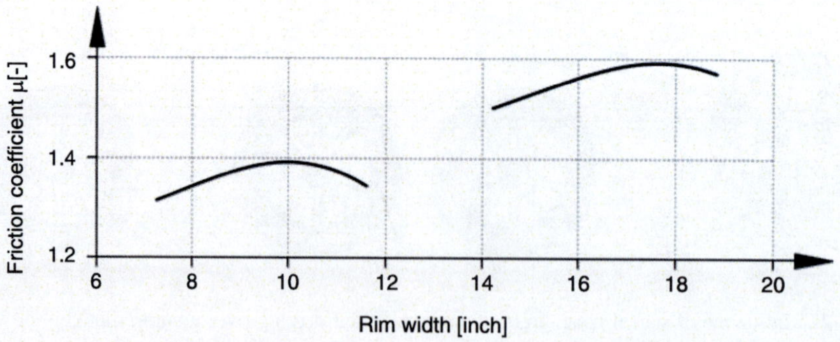

Fig. 1.72 Relationship between rim width and tyre grip. The relationship is shown for two different tyre sizes. It can be seen that wider tyres provide more grip. Furthermore, there is an optimum rim width at which the greatest tyre grip is achieved. The decrease towards larger rim widths is in any case smaller than towards narrower rims

The wheel must withstand these loads at an operating temperature of 120 °C. The service life of these wheels is set at a maximum of 2000 km, 1000 km of which are completed in racing use.

The following areas turn out to be critical: the transition from the spokes to the rim bed, where the greatest tensile stress occurs during vertical impact, and the inside of the spokes, where the greatest (bending) compressive stress is found during cornering.

1.4 Wheel Mounting

Depending on whether a wheel is bolted to the wheel hub or wheel flange with one or more bolts, it is referred to as a central bolt connection or multi-bolt connection.

Multi-Bolt Connection

In series production vehicles, mainly multi-bolt connections are used, where the wheel is centered by the nut or bolt (*bolt centering*), (Fig. 1.73). The contact surfaces of the fasteners are designed as spherical or conical surfaces for this function, (Fig. 1.74). In order to avoid fatigue fractures of steel disc wheels, the contact diameter of the adjacent component (i.e. brake disc or wheel flange) should be larger than that of the wheel disc.

Another possibility for centering is via the tightly toleranced center hole of the wheel (*hub centering*). The wheel hub must of course also have a tightly toleranced locating collar for this.

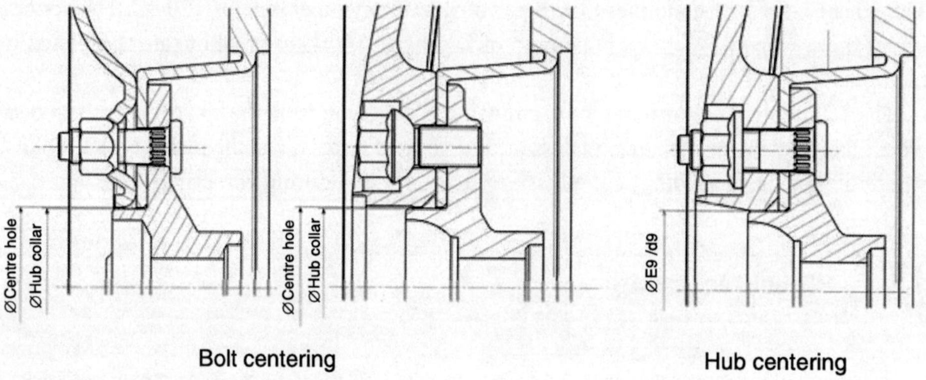

Fig. 1.73 Types of wheel centering for series wheels. With bolt centering, the bolts with taper or ball collar take over the alignment of the wheel. Consequently, a radial clearance is required between the centre hole and the hub collar (see also Table 1.7). In the case of centring, centring is carried out by a collar on the wheel flange (wheel hub), which accommodates the tightly toleranced centre hole. The wheel can be fastened with flat collar bolts or nuts

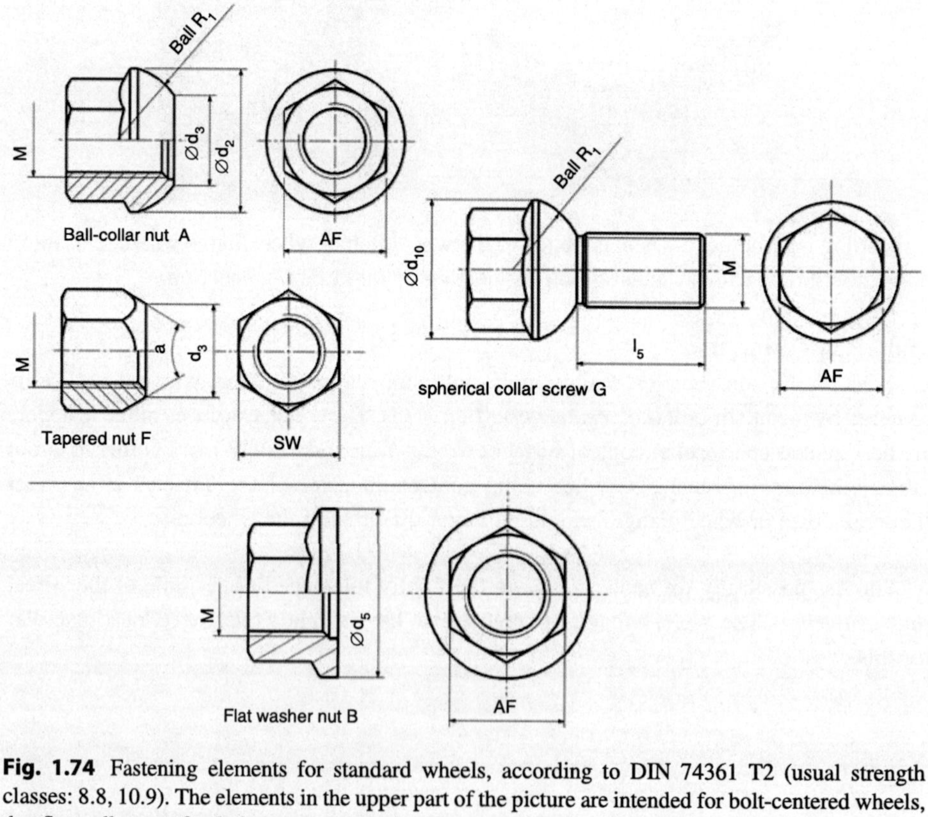

Fig. 1.74 Fastening elements for standard wheels, according to DIN 74361 T2 (usual strength classes: 8.8, 10.9). The elements in the upper part of the picture are intended for bolt-centered wheels, the flat collar nut for hub-centered wheels. The most important measurements are given in the diagrams, which appear in the table

		Version						
		A	A, F	A, G	F	G		B
Thread M	Across hex flats	d_2	d_3	R_1	a	d_{10}	l_5	d_5
M12x1.5	17	23	14.5	12	–	22.5	21	24
M12x1.5	19	–	15	–	60 °	–	–	–
	19	–	15	–	90 °	–	–	–
M14x1.5	19	26	17	14	–	26	24	27
					–	24	18	
M18x1.5	24	28	21	16	–	29	25	29

All dimensions, except angle a, in mm

Central Lock

If the wheels are to be replaced quickly, a central bolted connection is the best solution. The torque to be transmitted between the wheel and the wheel flange or hub (drive and braking torque) is largely transmitted via the axial contact surfaces. Depending on the design and tolerances, the drive pegs (drive pins) also bear part of the load, but the decisive factors for the magnitude of the torque are the friction and the preload force of the central nut. Particular care should therefore be taken when new coatings are introduced to the wheel and/or wheel hub. These can significantly reduce the friction conditions and, if the design is otherwise unchanged, lead to shearing of the drive pins. Conversely, coatings can of course also be helpful in this context.

The driving pins can be fixed in the wheel (Fig. 1.75) or in the hub. Bolts located in the hub are suitable for designs which, starting from series solutions, fasten wheels with central nuts and whose wheels also have holes for the series wheel bolts. The driving bolts engage in these to secure the torque transmission (e.g. Fig. 1.78). The torque is primarily transmitted via friction, caused by the pre-tensioning force of the nut.

A wheel with central locking may be centred by the wheel hub or centred by several driving pins. Figure 1.77 shows a wheel of a typical Formula 1 car with central locking as used on the non-driven front axle. During assembly, the wheel is guided by the extension of

Fig. 1.75 Front wheel of a Formula 1 car. The wheel is secured with a central nut. The torque connection is secured by five bolts (arrow), which are received by corresponding holes in the wheel flange (Fig. 1.76)

Fig. 1.76 Front left wheel flange for the wheel in the above picture (Red Bull Sauber Petronas C17, 1998). The central nut is placed on the wheel bolt in the picture. The holes (arrow) serve the bolts of the wheel (Fig. 1.75) for torque connection. Even if the car is no longer in the race, this type of connection can still be found today

the brake disc cup until the driving pins slide into the locating holes of the hub. The locating holes are of a high oval design so that they only guide the pins laterally. The central nut presses the wheel against the brake disk pot and the hub shoulder via a tapered collar. The

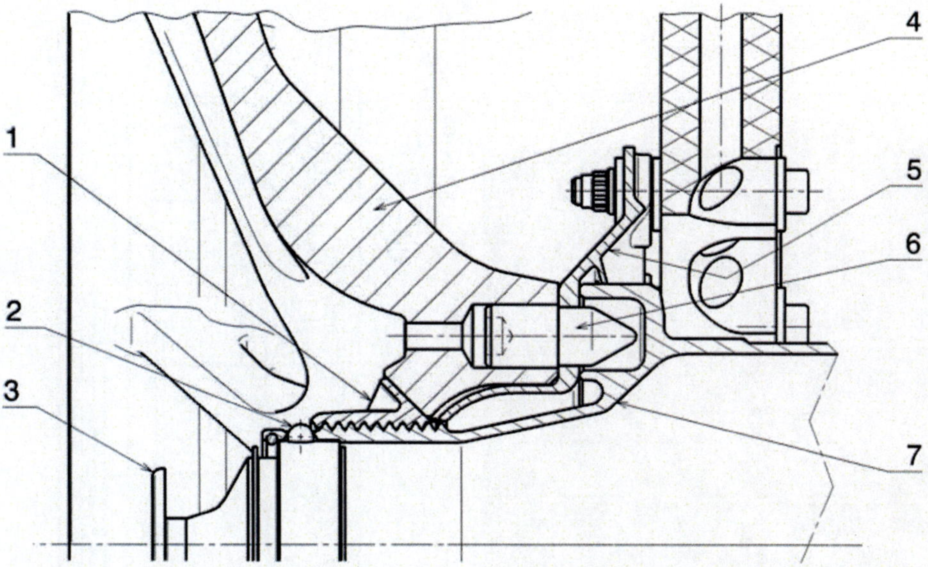

Fig. 1.77 Central bolting of a wheel on a racing car (Formula 1). **1** Centre wheelnut, **2** Locking ball, **3** Slide control, **4** Wheel, **5** Disc bell, **6** Drive peg, **7** Hub

nut itself is prevented from falling off by balls which are pressed outwards by a locking slide.

If a bolt is guided through the wheel hub, a central bolting can also be realised for "conventional" wheel hubs. Figure 1.78 shows an example of this. The wheel nut presses the wheel against the wheel flange via a large washer. The nut and washer are connected to each other in such a way that they cannot be lost. The torque from the brake disc to the wheel is partly transmitted via driving pins which are screwed into the wheel flange. The hub is responsible for centring the wheel. A locking clip prevents the wheel nut from losing its position.

Central Nut

(central lock nut). Wheel nuts with central lock are made of steel, aluminium or titanium alloys (Figs. 1.79 and 1.80). Steel offers the advantage of saving weight on this component and with the pneumatic wrenches these nuts can be held with magnetic inserts. This makes them easier to put on and thus reduces the time needed to change the wheel during a pit stop. For this purpose, aluminium and titanium nuts are provided with a circumferential groove in the hexagon, into which a snap ring in the screwdriver insert engages and thus holds the part during the wheel change. Other designs have recesses in the hexagonal surfaces in the form of a ball cap for this purpose. The thread axis coincides with the rotation axis of the wheel. To prevent the nuts from loosening due to their inertia, the threads on the left-hand side of the vehicle are right-hand threads and on the right-hand side

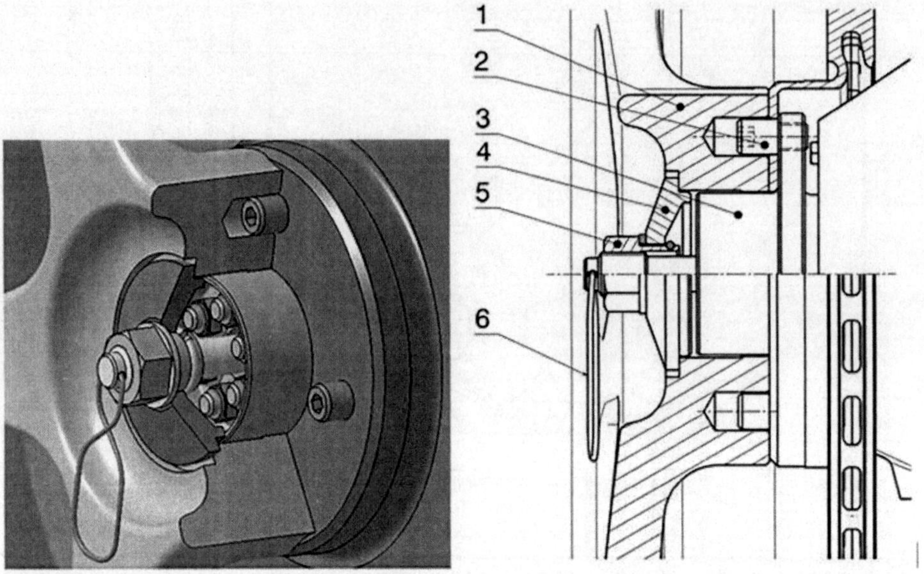

Fig. 1.78 Central bolting of a wheel (Renault 2000 formula). **1** Wheel, **2** Driving bolt, screwed into the flange of the wheel hub, **3** Wheel hub, **4** Washer, **5** Central nut, **6** Locking clip for nut

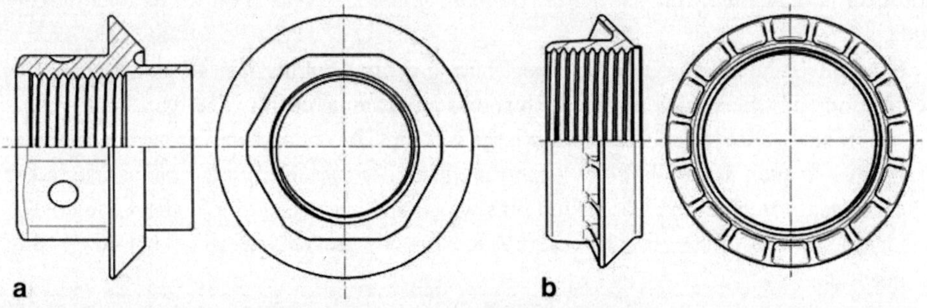

Fig. 1.79 Wheel nuts for central bolting. (**a**) Aluminium, (**b**) Steel

they are left-hand threads. This means that all threads run in the opposite direction to the direction of wheel rotation when starting off. The nuts must therefore be marked to prevent interchanging, e.g. with colour codes: Blue on the right and red on the left. In any case, the nuts must have a high preload, otherwise they could also come loose due to their inertia during heavy braking. There are also many (successful) vehicles that have right-hand threads on all wheels. This simplifies handling when changing wheels and reduces the amount of parts needed. The thread pitch is chosen relatively large: About 2.5–5 mm for nominal diameters of 45–60 mm. For reasons of strength of the wheel hub, trapezoidal threads are sometimes used for the central screw connection instead of the pointed threads

Fig. 1.80 Wheel nut of a central locking system (Formula 1 BMW Williams 1999)

normally used for fastening boltss. The end of the wheel hub is not threaded. On this piece, the nut is also placed and guided effortlessly during the (hectic) assembly in racing use until the first thread engages.

Securing the Central Nut

The torque is also transmitted directly to the brake disc via bolts. A locking device is therefore limited to the nut or bolt alone. The locks work positively and are intended to prevent the nut from being lost. They must be designed in such a way that they cannot be forgotten. If a quick wheel change is required in a race, the best choice is a locking element that is unlocked by the bolting tool itself when it is put on and is re-secured when it is taken off (e.g. Fig. 1.82a, b). If there is a little more time to change the wheel, additional cotter pins or similar prescribed by many regulations, which are connected to the hub, are sufficient. However, mechanics must get into the habit of temporarily storing this type of securing device on the steering wheel or on the gear lever after disassembly, so that the driver immediately sees that they have been forgotten. Some possibilities from the multitude of conceivable solutions for fuses are shown in (Figs. 1.78, 1.81, and 1.82). Figure 1.83 also shows a sectional view in which the operation of an automatic securing device can be seen. The figure shows a driven rear wheel hub, but the nut lock on the front wheel hub looks the same. In principle, the locking mechanism is the same as that shown in (Fig. 1.82d). The wheel nut (4) clamps the wheel (3) against the shoulder of the wheel hub (2), at the same time holding the brake disc cup. The nut itself is prevented from moving axially by the two spring-loaded thrust pieces (5). The thrust pieces slide in a cylindrical receptacle, which in turn is guided in the hollow hub. The thrust pieces protrude through two rectangular openings through the hub shell and thus simultaneously hold their seat in the axial direction. When loosening the nut, the screwing tool first presses the two

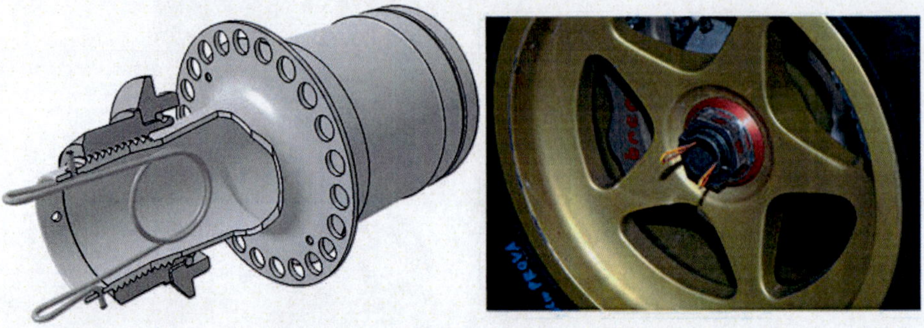

Fig. 1.81 Securing a central screw connection. The spring clip, which must be red or fluorescent orange, is inserted into the hub after the nut has been fitted

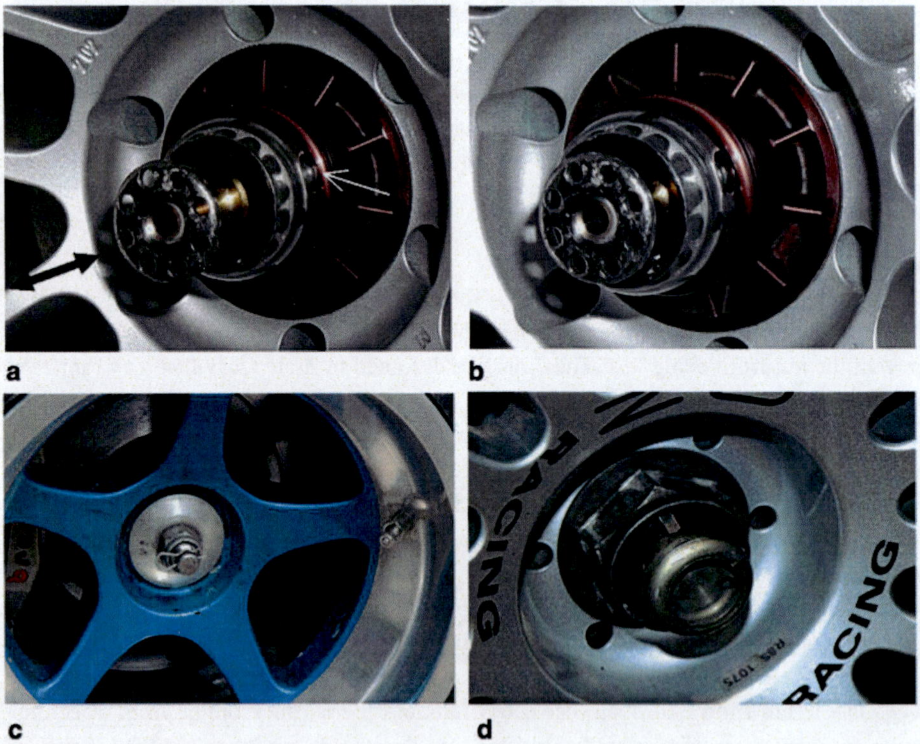

Fig. 1.82 Some securing types of central bolted joints. (**a, b**) Automatic fuse. (**a**) Wheel nut secured (slider outside): Retaining balls protrude over hub (white arrow), (**b**) Wheel nut released (slider inside). (**c**) Cotter pin. (**d**) Automatic securing with wedges

thrust pieces together during the mounting process and only then reaches the nut to turn it. In the opposite case, i.e. when putting on the nut, it presses the two locking elements together before reaching the first thread.

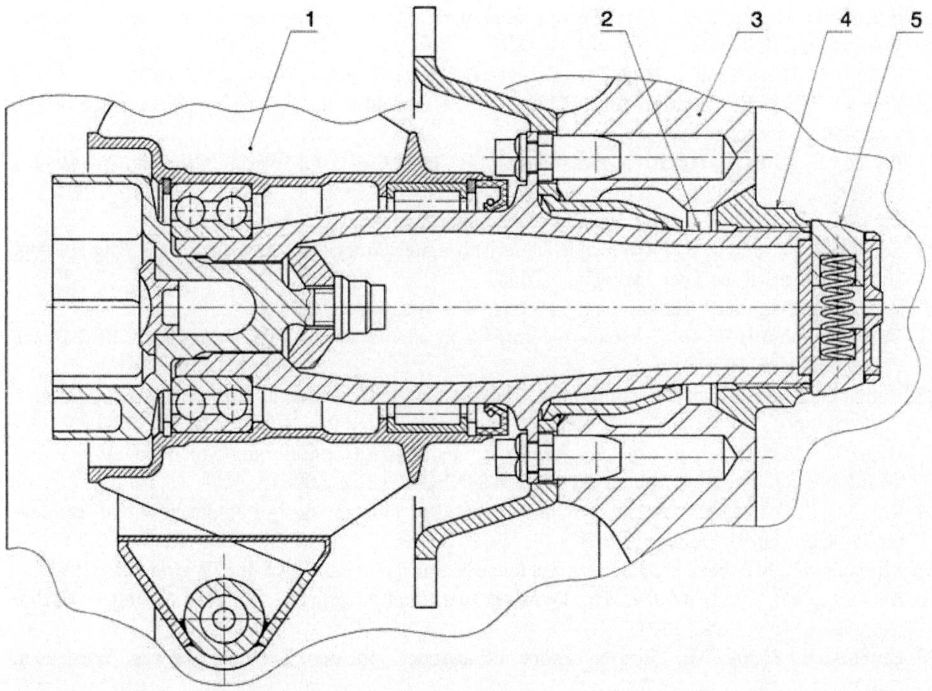

Fig. 1.83 Automatic nut lock for central bolting, Dallara IR8 (Indy Car). 1 Upright, 2 Hub, 3 Wheel, 4 Centre nut, 5 Locking device

References

1. Henker, E.: Fahrwerktechnik. Vieweg, Wiesbaden (1993)
2. Tremayne, D.: Formel 1, Technik unter der Lupe. Motorbuch, Stuttgart (2001)
3. N.N.: Formula Renault 2000 Manual. Renault Sport Promotion Sportive (2001)
4. Andorka, C.-P., Kräling, F.: Formel 1, das Milliardenspiel. Copress, München (2002)
5. Kermelk, W., Schwendemann, H.: Entwicklung des Lemmerz Stahl-Leichtbaurades für Pkw. In: System Partners 97, eine Sonderausgabe von ATZ/MTZ. Springer Vieweg, Wiesbaden (1997)
6. Staniforth, A.: Race and Rallycar Source Book, 4th edn. Haynes Publishing, Sparkford (2001)
7. McBeath, S.: Competition Car Preparation, 1st edn. Haynes, Sparkford (1999)
8. Incandela, S.: The Anatomy & Development of the Formula One Racing Car from 1975, 2nd edn. Haynes, Sparkford (1984)
9. Reimpell, J.: Fahrwerktechnik, 5th edn. Vogel, Würzburg (1982)
10. Wick, J.: Reiner Stickstoff statt Luft im Reifen: Abzockerei oder Aufwertung? In: Automobilrevue, vol. 49, p. 1. Büchler Grafino AG, Bern (2001)
11. Reimpell, J., Betzler, J.: Fahrwerktechnik: Grundlagen, 4th edn. Vogel, Würzburg (2000)
12. Paefgen, F.-J., Gush, B.: Der Bentley Speed 8 für das 24-Stunden Rennen in Le Mans 2003. ATZ. **4**, 280ff (2004)
13. Rowley, W.J.: Race Car Engineering, Book One, 2nd edn. Rowley Race Dynamics, Calgary (2004)

14. Braess, H.-H., Seiffert, U.: Vieweg Handbuch Kraftfahrzeugtechnik, 4th edn. Vieweg, Wiesbaden (2005)
15. Haney, P.: The Racing & High-Performance Tire, 1st edn. SAE, Warrendale (2003)
16. Weber, R.: Kraftfahrzeugreifen. Bildteil zur Vorlesung an der techn. Universität Wien, Universität Hannover (1984)
17. Breuer, B., Bill, K.-H.: Bremsenhandbuch, 1st edn. GWV Fachverlage/Vieweg, Wiesbaden (2003)
18. Internetsite von Continental.: www.continental.de. Accessed 21 Mar 2006 (2006)
19. Lucaora, L., Marconi, P.: F1 Wheel in Material Magnesium Forged, Front and Rear. Engineering. Vortrag auf der Race.Tech, München (2004)
20. Wright, P.: Formula 1 Technology, 1st edn. SAE, Warrendale (2001)
21. Van Valkenburg, P.: Race car engineering & mechanics. Paul Van Valkenburg, Seal Beach (1986)
22. Genta, G., Morello, L.: The Automotive Chassis, Vol. 1: Components Design, 1st edn. Springer, Berlin (2009)
23. Held, D.: Methodik zur streckenspezifischen Fahrbarkeitsoptimierung von Rennmotoren im Teillastbereich. Diplomarbeit an der FH JOANNEUM, Graz (2008)
24. Rouelle, C.: Advanced vehicle dynamics applied to race car design & development, seminar binder. OptimumG, Denver (2014)
25. Milliken, W., Milliken, D.: Race car vehicle dynamics, 1st edn. SAE Inc, Warrendale (1995)
26. Mitschke, M., Wallentowitz, H.: Dynamik der Kraftfahrzeuge, 4th edn. Springer, Berlin/Heidelberg/New York (2004)
27. Heißing, B., Ersoy, M., Gies, S.: Fahrwerkhandbuch, 4th edn. Springer Vieweg, Wiesbaden (2013)
28. Toso, A.: Race cars. In: Mastinu, G., Ploechl, M. (eds.) Road and off-Road Vehicle System Dynamics Handbook, pp. 497–515. CRC Press, Boca Raton (2014)
29. Leister, G.: Fahrzeugreifen und Fahrwerkentwicklung. Strategie, Methoden, Tools, 1st edn. Vieweg+Teubner, Wiesbaden (2009)
30. Rompe, K., Heißing, B.: Objektive Testverfahren für die Fahreigenschaften von Kraftfahrzeugen. Quer- und Längsdynamik. TÜV Rheinland, Köln (1984)
31. Zomotor, A.: Fahrwerktechnik: Fahrverhalten, 2nd edn. Vogel Buchverlag, Würzburg (1991)
32. Reimpell, J., Sponagel, P.: Fahrwerktechnik: Reifen und Räder. Vogel, Würzburg (1986)
33. Olsson, H.: Tire Testing, Tire Data Tools & Reality, Beitrag bei Tire Mathematical Modeling Course, Tire Technology Conference, Hannover., Feb. 13–14 (2017)
34. N.N.: continental formula student Tire. Competition Tire 2018 (C18) – documentation. Continental Reifen Deutschland GmbH, 18. Apr. (2018)

Suspension

2

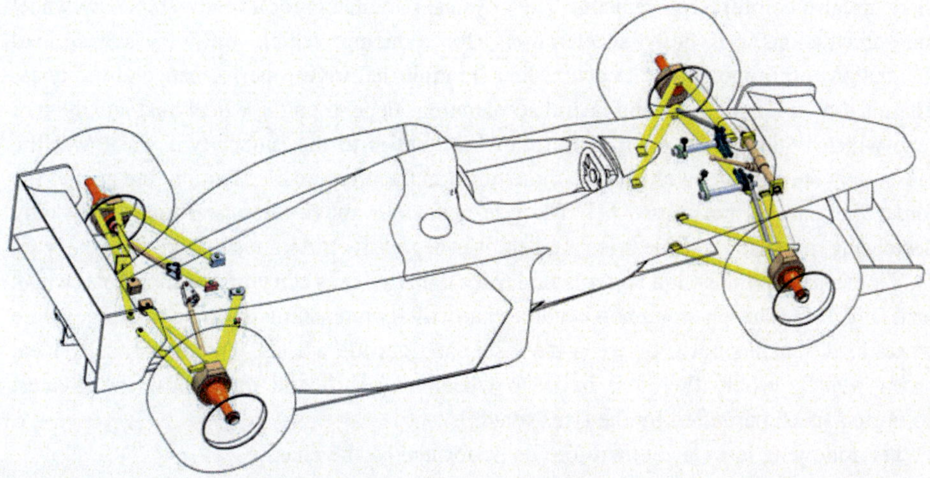

The suspension is the major part of the chassis. It determines the position of the tire and thus becomes the most important assembly of a vehicle, and especially of a competition vehicle, next to the tires. Its performance essentially determines the characteristics of the entire vehicle.

M. Trzesniowski, *Suspension System*,
https://doi.org/10.1007/978-3-658-39847-7_2

2.1 Function

The main function of the suspension is to connect the wheel to the vehicle with a degree of freedom (see Appendix). The degree of freedom is an essentially vertically directed possibility of movement of the wheel. The wheel, or more precisely the tyre, should always make the maximum possible contact with the road surface. All forces coming from the wheel must be directed to the car body and vice versa. Because only these forces (apart from air resistance during braking) make it possible to ensure the desired high accelerations.

The chassis also solves the problem of four wheels distributing the vehicle weight evenly. Just as a table with four legs wobbles on a flat floor as soon as one leg is longer (a plane is completely determined by three vectors, four legs result in static indeterminacy), so would a vehicle,[1] if the suspension did not provide a (length) compensation.

An additional requirement can be that the wheel position changes in a certain way during compression in order to stabilise the driving behaviour (active safety). In addition, there are also comfort requirements. The physical parameters for a road vehicle are wheel load fluctuations and body accelerations; for a racing vehicle only the wheel load fluctuations are important. The actual link limiting the driving performance is the tyres. The maximum longitudinal and lateral accelerations depend on the wheel load and the tyre map which is dependent on it. The following applies to the rubber tyre: the lower the pressure in the contact patch and the more uniform the pressure distribution, the greater the forces that can be transmitted [1]. The tyre behaviour above the wheel load is not only decreasing, but also non-linear (as a result, it is degressive). As a result, two tires carrying the same load can transmit a larger lateral force than two tires carrying the same total wheel load (Fig. 2.1). This phenomenon becomes particularly interesting during cornering, where wheel load transfers occur between the inside and outside wheels. The theoretically ideal racing vehicle would therefore have no wheel load shift and thus allow the greatest cornering speed permitted by the tyres fitted.

The following tasks must therefore be performed by the chassis:

- Maintaining a (preferably) large contact area between tyre and road surface under all driving conditions
- Ensuring a favourable wheel load distribution
- Achieving the greatest possible tyre grip by reducing wheel load fluctuations
- Maintaining driving stability
- Ensuring the desired self-steering behavior
- Allow relative movement between the car body and the track.
- Reduce the effect of impact forces from the road on the body of the wagon.

[1] Just like the table, a vehicle with different spring preloads also wobbles – namely across the diagonal.

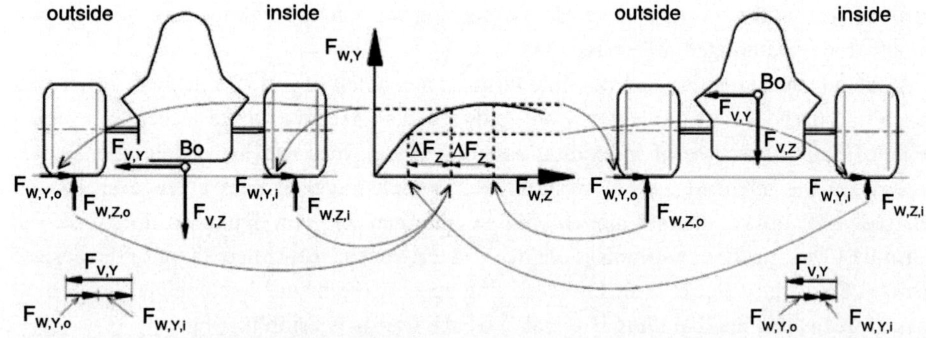

Fig. 2.1 Influence of wheel load displacement on the transmissible lateral force of an axle (schematic). Bo body center of gravity. On the left half of the picture, the vehicle has no wheel load displacement. Both wheel contact forces $F_{W,Z}$ are equal. Because of the downward slope of the lateral force over the wheel load, the total lateral force $F_{V,Y}$ that can be transmitted by the tyres is greater than that on the right half of the picture. Here, a wheel load shift ΔF_Z occurs. Although the tyre on the outside of the bend generates more lateral force $F_{W,Y,o}$ as a result, the drop in force on the inside of the bend is so great that the total $F_{V,Y}$ is smaller

For a competition vehicle the following requirements are added:

- simple and, above all, fine adjustment of wheel position variables, such as camber, toe-in, caster, outside a workshop
- easy adjustment of spring, damper and stabilizer behaviour
- System reliability or rapid replacement of wear parts such as joints, bearings and tyres
- in the case of vehicles with pronounced aerodynamic support: as little as possible disturbing influence of components on the flow around the vehicle and maintenance of the desired position of the car in relation to the road, i.e. as little pitching, rolling and heaving as possible
- low weight
- Use with dry and rain tyres possible
- Controllable behaviour in case of overload
- Predetermined breakage in the event of an accident without damaging the frame or chassis.

At the end of the 1980s, there were even variants of active chassis, i.e. those that loaded or unloaded the wheels depending on the driving condition (Lotus Honda 99 T). At that time, the advantages did not come to bear so strongly in Formula 1. The reasons for this were the low chassis movements on the usual flat circuits, the voluminous tyre itself already being a system capable of oscillating (i.e. its own chassis), and unit tyres. The developers of active wheel suspensions would have liked to have a matching tyre development. But this did not even happen, on the contrary, the active suspension was banned with the later often quoted reason "the unauthorized driver assistance".

Generally speaking, it is not possible to make a generally valid statement as to whether the chassis or the aerodyanmic downforce aids are more decisive for the success of a racing vehicle. The importance of individual assemblies and thus also the development goals depend on the intended use. In formula cars, which travel at high average speeds on relatively flat tracks, chassis movements are minimal and aerodynamics dominate. On Formula 1 cars, major compromises are made at the front suspension in favor of the desired airflow. On cars or tracks with lower driving speeds, the matter is completely different: Downforce is low and handling is dictated by the tire-suspension pairing.

Requirements

The general requirements for a chassis can be summarized in individual categories, within which a wheel suspension can be considered according to tangible design characteristics [2]:

- Driving behaviour, driving safety
- Ride comfort
- Installation space requirement
- Cost.

For racing vehicles, only the driving behaviour and – as far as aerodynamics are concerned – the installation space requirements are of primary importance.

The driving behaviour is characterised by the following characteristic values or their changes during driving operation:

- Characteristic values of the wheel position: wheelbase, track width, camber, toe-in, steering roll radius, kingpin inclination, caster
- Kinematic changes in wheel position: roll centre displacement, wheel displacement curves
- Elastokinematic changes of the wheel alignment: stiffnesses and damping behaviour of the rubber joints (bushes)
- Unsprung masses
- Behaviour in case of overload.

If the ride comfort of a wheel suspension is considered, the following criteria are in the foreground:

- Vibration behaviour: Characterized by spring and damper tuning
- Diagonal and lateral springing
- Acoustic behaviour: influenced, among other things, by the introduction of force into the bodywork
- Unsprung masses

- Accelerating and braking support: influenced by the position of the pitch instant centers and the vehicle centre of gravity
- Steering effort
- Turning circle diameter.

The installation space of a wheel suspension is influenced by those subassemblies of a car with which it has to share space. These are the engine-transmission assembly, the exhaust system and lines. In the case of passenger cars, the tank and boot are also involved, and in the case of racing cars, aerodynamic aids (wings, diffuser, etc.).

Dimensioning

Thus, when designing wheel suspensions, the following must be taken into account:

- Frame stiffness
- Aerodynamic downforces
- Tyre maps
- Vehicle weight and weight distribution
- Wheelbase and track width
- Springs and dampers

These influencing factors are also partly interrelated, but do not necessarily have to harmonize. A chassis development will therefore be a compromise, in which the tuning to the race track and the weather situation is decisive.

The centre of gravity of the entire vehicle should be as low as possible. This means that the wheel contact forces change only slightly during driving manoeuvres (braking, accelerating, cornering) and it is easier to design the best possible chassis. Likewise, the masses should be concentrated around the centre of gravity so that the car's moments of inertia remain as small as possible. The reaction forces required to change the position of the vehicle then remain just as small, or the change in position can be carried out more quickly.

A distinction is made between sprung masses (wagon body, superstructure) and unsprung masses (wheel plus parts suspended from it). Physically, low unsprung masses mean small wheel load fluctuations and should therefore be aimed for. Due to the non-linear behaviour of the tyre rubber when building up and transmitting forces, a large wheel load fluctuation causes losses of lateral forces and longitudinal forces. On undulating road surfaces, the drivable lateral acceleration is thus reduced and the braking distance is increased, whereby the effect is clearly more pronounced in the lateral direction [3]. Conversely, however, it must also be noted that this effect does not come into play on relatively even road surfaces because the wheel load changes caused by the road surface remain small. Numerous studies in the past [4] and present [5, 6] confirm the overestimated influence of the unsprung masses on a level road surface. Additional masses were added to the front wheels and the drivers could not notice any difference in the grip level. The reason for this, apart from the low road excitation, is that tyres themselves also represent a spring (together with a damper). With off-road vehicles, which are exposed to large bumps, it is very important to keep the so-called unsprung masses small.

Wheelbase and track width are concept parameters and are therefore determined in the concept phase.[2]

The rear axle is the more important one for driving stability. In addition, the driver can actively influence the wheel position on the front axle. The following behaviour is aimed for:

- A slightly negative camber during cornering with otherwise little change in camber. Ideally, the camber does not change at all during compression/rebound, i.e. during braking and acceleration.
- As little track change (scrub) as possible, ideally none at all, for good driving safety
- Support against deflection during braking
- Freedom from adverse steering effects due to elastic deformation under driving, braking and lateral forces.

The general design sequence in the construction of racing cars is derived from the regulations and considerations of desired performance.[3] The chassis alone concerns the following points:

1. Tyres
2. Wheels
3. Chassis geometry: For front and rear axle respectively: roll center, instantaneous pole single wheel, instant center distance single wheel,. . .
4. Hubs
5. Brakes
6. Wheel carrier (knuckle, upright)
7. Body supporting springs
8. Damper
9. Stabilizers (anti-roll bars)
10. Steering

Even though the possibilities of simulation by means of computers are far advanced nowadays, the road test proves to be more meaningful, especially in the later development phase. Simulation is (still) too fuzzy to contribute significantly to problem solving. Computational studies are therefore needed at the beginning of the design process so that far-reaching concept decisions can be made in good time. In the further course of

[2] See Racing Car Technology Vol. 2 *Complete Vehicle*, Chap. 2.

[3] For more details see Racing Car Technology Manual Vol. 2 *Complete Vehicle*, Chap. 2.

development, the aim will be to have an experienced driver drive the new axle as soon as possible so that his findings can be incorporated into the design.

2.2 Terms and Kinematic Quantities

See also DIN ISO 8855 (was DIN 70000): Vehicle dynamics and handling, terms.

The position of the wheel in relation to the vehicle or the road is described by various geometric and kinematic terms, Fig. 2.2. The coordinate system used is described in the appendix. These quantities all have their influence on the driving behaviour of a vehicle.

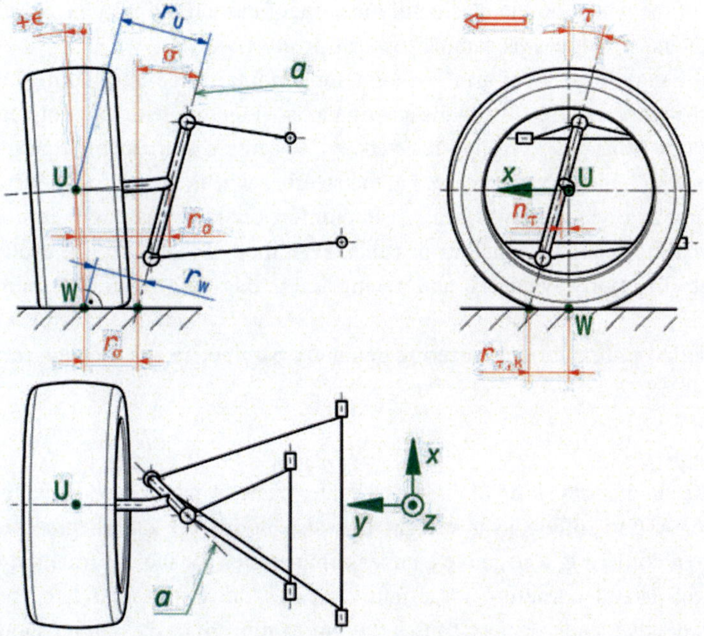

Fig. 2.2 Front wheel position wheel alignment. ε Camber, σ king pin inclination, r_σ scrub radius, steering offset, king pin offset at ground, a Axis of rotation of the knuckle, kingpin axis (also called kingpin inclination axis), its distance on the road surface from the wheel contact point W is in *y-direction* the scrub radius r_σ and in *x-direction* the caster $r_{\tau,k,}$ U wheel centre, W wheel contact point, r_a wheel centre offset, r_U longitudinal force moment-arm, r_W braking moment-arm, τ castor angle, $r_{\tau,k}$ caster offset at ground, trail, n_τ Caster offset at hub

2.2.1 Terms

King Pin Inclination σ

The kingpin inclination is the angle between the steering axis (kingpin axis) and a plane perpendicular to the road surface. With the kingpin inclination, for a given tyre width and a given minimum distance of the steering axle from the wheel centre (brake calliper and brake disc as well as rim are in the way), its penetration point through the road surface can be changed and thus the steering rolling radius r_σ. This distance between the centre of the wheel and the steering axle point on the road has an influence on the feedback of the tyre forces at the steering wheel, although the name is misleading because the wheel does not roll along a curve with this radius when steering, but describes a path curve with increasing radius when kingpin inclination and caster are not equal to zero [7]. Kingpin inclining leads to lowering of the wheel on the outside of the corner (and thus slightly increases its wheel load) and adversely reduces its camber towards positive values.

The disturbance force lever arm[4] r_U (also longitudinal force lever arm) – the normal distance of the wheel centre U from the steering axis – together with circumferential forces on the front tyre (changes in rolling resistance, . . .) cause disturbance torques around the steering axis which the driver perceives at the steering wheel. Brake force fluctuations act on the steering axle via the lever arm r_W. In comfort-oriented passenger cars, this disturbance information is kept as small as possible, even though there is some useful information content. In sporty vehicles and racing cars, this information should be more pronounced.

More detailed information on steering geometry parameters and steering return can be found in Sect. 5.2.2.

Camber Angle ε

Camber is the angular deviation of the wheel center plane from a perpendicular plane to the road surface. A tyre rolling in a straight line also builds up lateral force via negative camber. When cornering, a negative camber compensates for the deformation of the tyre caused by the lateral cornering force and increases the trapezoidal tyre contact area. However, excessive static camber causes the tire to run primarily on one shoulder when going straight. This is to be avoided for several reasons. The tire tread cools on long straights, the temperature of the loaded tire shoulder increases, and the coefficient of friction decreases. The tread itself does not have full contact across the width of the tyre with the road surface, but cants up and traction is reduced for this reason too. Furthermore, this disadvantage comes into play during braking as well as during propulsion. The effects of the edging effect increase with tire width. In addition, too much camber reduces tyre life.

[4]This term is also used in the literature for the kingpin offset at hub. Ultimately, it is the geometric quantity and not the name that is important. Both quantities can be transferred into each other via angular relationships.

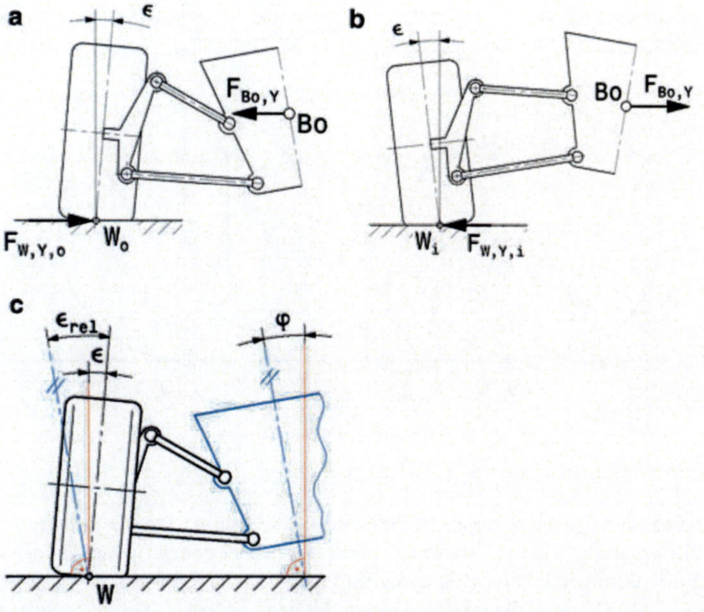

Fig. 2.3 Favourable orientation of the camber angle ε when cornering. $F_{Bo,y}$ Inertia force acting on body at centre of gravity Bo, pointing towards outside of corner. W_o, W_i Wheel contact point outside or inside corner. $F_{W,Y,o}$, $F_{W,Y,i}$ Tyre side force outside or inside corner. (**a**) Laden wheel. (**b**) Unladen wheel. (**c**) Relative camber angle ε_{rel} related to the body with the roll angle φ

In the case of double wishbone axles, the kingpin inclination angle is directly related to the camber angle via the wheel carrier (knuckle, upright).

There are different definitions of the sign of a camber angle. The decisive factor for the lateral force build-up of a tyre when cornering is the inclination of the tyre to the lateral force. In Fig. 2.3, therefore, favourable directions of the camber angle (without sign) are entered for a loaded and a relieved wheel.

Furthermore, a relative camber ε_{rel} (Fig. 2.3c) is often measured in investigations on CAD models, because this is easier to determine when the wheel is compressed. In this context, one also speaks of a *camber gain*; however, the absolute camber angle ε (inclination angle) to the road is ultimately decisive.

Caster

The rolling resistance forces have a direction-stabilising effect on a wheel if they act behind the steering axle in the direction of travel. A caster effect thus generates a caster angle τ or a caster distance $r_{\tau,k}$ (Fig. 2.4). For a caster distance, the steering axis does not have to be inclined, but can also be vertical and lie in front of the wheel centre by the amount of the caster distance.

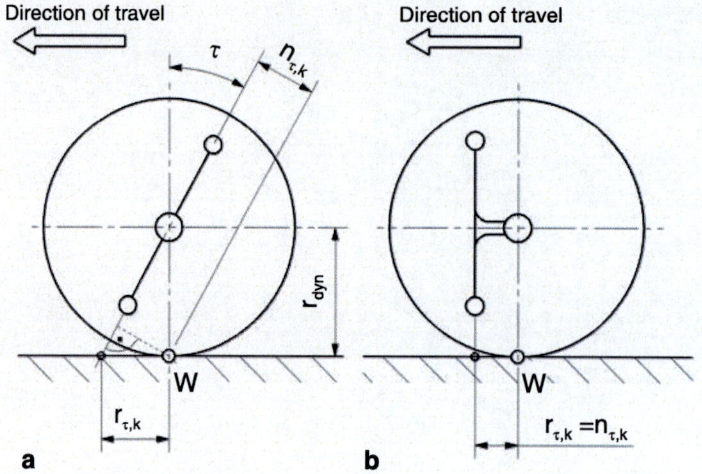

Fig. 2.4 Generation of a positive caster. (**a**) Skewing of the steering axle by the castor angle τ. The kinematic trail distance $r_{\tau,k}$ is the distance between the wheel contact point W and the point of penetration of the steering axle through the road surface. A lateral force acts on the steering axle via the lever arm $n._{\tau,k.}$ (**b**) Offset of the steering axis by the *caster offset* n_τ. side force lever arm $n_{\tau,k}$ and kinematic caster distance $r_{\tau,k}$ are equal to

In some front-wheel-drive passenger cars, you could even find forward motion. In that case, the driving forces pushed the front wheels into the straight-ahead position.

However, the inclined position of the steering axle also causes the inside wheel to be lowered, thus increasing the wheel load on this side. In tight corners, the grip of the front axle is thus noticeably increased by counteracting the dynamic wheel load shift.

A clever combination of positive caster angle τ and negative caster offset n_τ reduces the caster distance $r_{\tau,k}$ and is characterized by the following advantages, (Fig. 2.5):

- the influence of ground unevenness on the steering torque is reduced because the kinematic trail is smaller
- the camber change increases advantageously with the steering angle.

Slip Angle α

Due to the fact that the tread of the tyre gradually deforms as it enters the contact surface with the road when a lateral force is applied, the tyre centre plane forms an angle with the actual direction of movement of the tyre, the slip angle (Fig. 2.6, cf. also Fig. 1.36). The lateral force is also first built up in the contact patch starting at the leading edge, reaches a maximum value and then drops as soon as the tread lifts off the road again. This asymmetrical lateral force distribution has its resultant behind the wheel contact point. This results in a change in caster and the lateral force generates a moment around the wheel contact point, the so-called self-aligning moment.

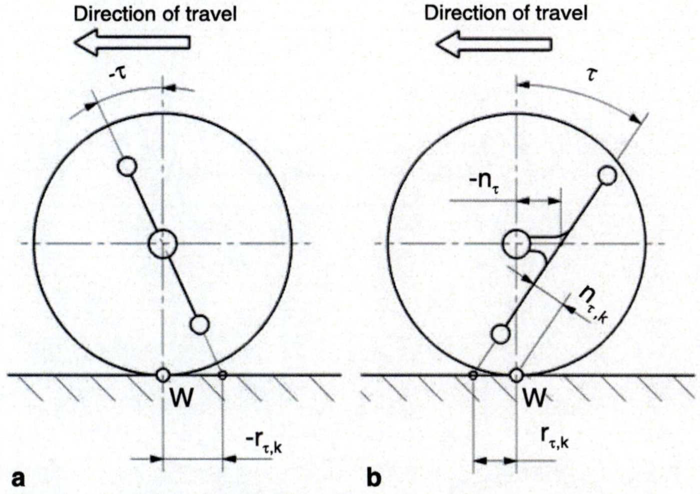

Fig. 2.5 Variants of the caster formation. (**a**) Pre-trail due to negative caster angle τ, (**b**) Negative caster offset n_τ in combination with positive caster angle τ

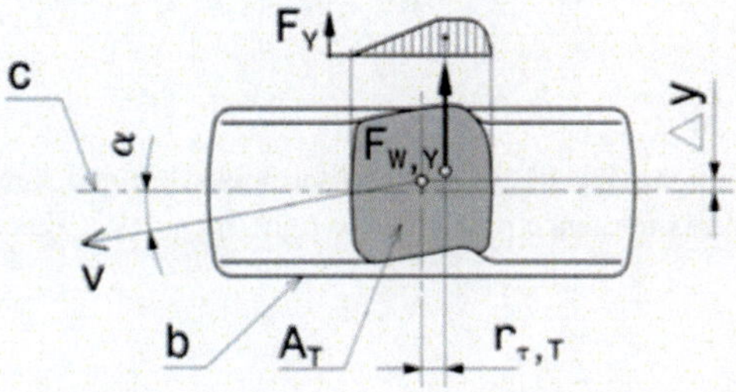

Fig. 2.6 Caster $r_{\tau,T}$, due to tyre (view from below). v Direction of movement (effective track of the tyre on the road, vehicle heading). c geometric trace of the wheel plane on the road (tyre heading). b undisturbed tyre side contour. A_T Tyre contact patch (*footprint*), F_Y Lateral force distribution over the contact patch length, $F_{W,Y}$ resulting lateral force, $r_{\tau,T}$ pneumatic trail: Tyre-related caster dependent on the lateral force distribution in the wheel contact area, approx. 10–40 mm for passenger cars. Δy Displacement of the wheel contact point due to lateral force, α *slip* angle

Wheelbase l

The wheelbase is the distance from the centre of the front axle to the centre of the rear axle, (Fig. 2.7), and has a significant influence on the driving behaviour.[5] Strictly speaking, the wheelbase l is the horizontal(!) distance between the wheel centres, i.e. the distance between the wheel contact forces. However, this fineness is only of importance with different wheel diameters.

[5]See also Racing Car Technology Manual Vol. 2 *Complete Vehicle*, Chap. 2 *Concept.*

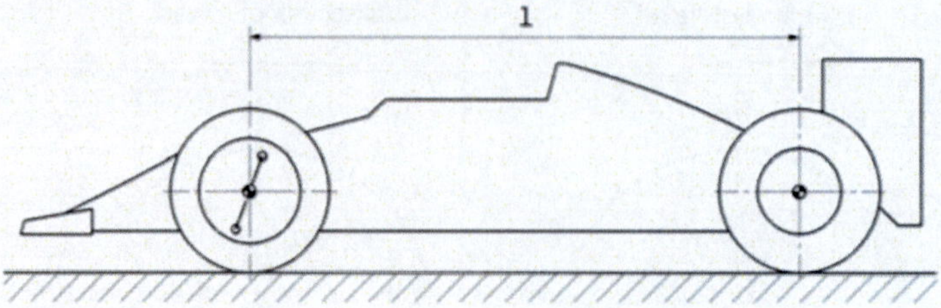

Fig. 2.7 Wheelbase l

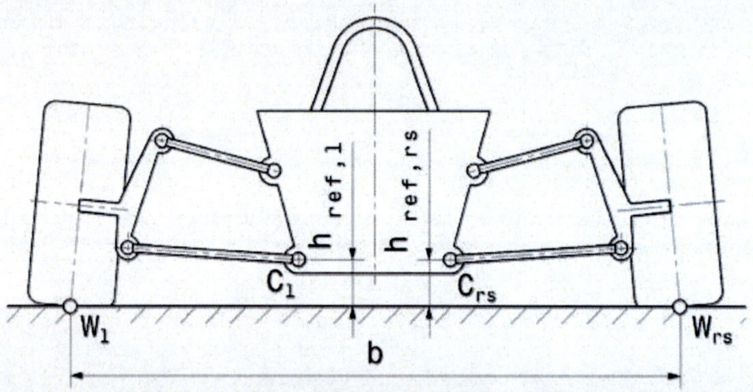

Fig. 2.8 Track and ride height. W_l, W_{rs} Wheel contact points left and right respectively. b Track width. C_l, C_{rs} any frame-fixed reference points left or right, $h_{ref,l}$, $h_{ref,rs}$ Floor distance left and right respectively

Track *b*

Another important parameter in terms of driving dynamics is the track width (Fig. 2.8). The track width b is the distance between the imaginary wheel contact points of an axle measured in the transverse plane of the vehicle. Like the wheelbase, it is determined early in the development of a vehicle.[6]

Ride Height h_{ref}

The ride height is used as an easily measurable reference value when it comes to the vehicle level above the road. This is important for the setup of the chassis or aerodynamics, for example. The ride height is generally not the ground clearance.[7] Any reference points that are easily accessible on the frame are used for measurement.

[6] See footnote 1.

[7] See Racing Car Technology Manual Vol. 2 *Complete Vehicle*, Chap. 7, Fig. 7.1.

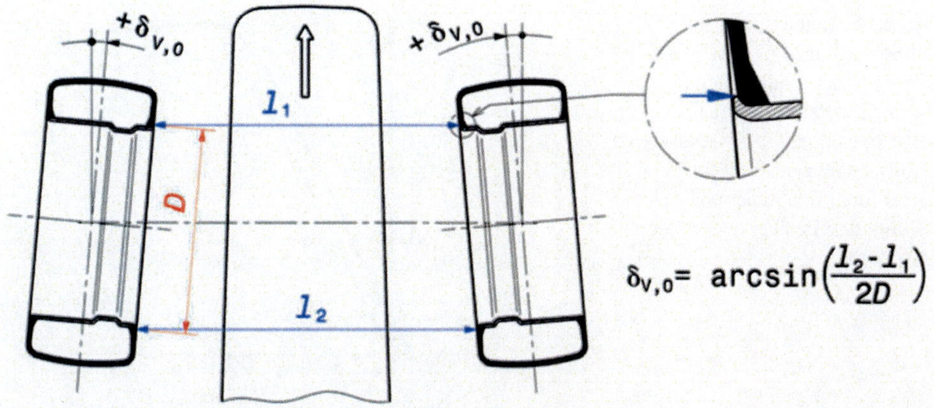

Fig. 2.9 Definition of toe-in angle. $\delta_{V,0}$ Toe-in angle of a wheel, °. Toe-in $r_{\Delta,t} = l_2 - l_1$ *[mm]*: Values measured at the rim flanges (distance *D*) at the height of the wheel centre (DIN 70020), positive values = *toe-in toe-in,* negative values = *toe-out*

Toe-in

The static toe-in angle is the angle, when the vehicle is stationary, between the median longitudinal plane of the vehicle and the intersection of the median plane of a wheel with the plane of the road surface, see Fig. 2.9.

Due to the elasticities in the wheel suspension and in the steering actuation (tie rods, joints, toothed rack,. . .) the non-driven wheels are pushed backwards by the rolling resistance. If they are to be parallel while driving and thus have the lowest rolling resistance, they must be turned in relation to each other when stationary to compensate for the elasticities, i.e. they must be set with positive toe-in. In the case of driven wheels, the driving forces are decisive. These try to push the wheels forward. Such wheels are advantageously given a toe-out so that they are parallel when driven.

Generally applies to the front axle:

- Toe-in for rear-wheel drive vehicles.
- Toe-out on front-wheel drive vehicles with positive steering roll radius.
- Toe-in 0 with front-wheel drive and negative steering roll radius.

By a toe-in position of the wheel, the tyre plane is twisted out of the straight (purely rolling) direction of movement, (Fig. 2.9). The tyre is thus forced to run at an oblique angle even when travelling in a straight line. This results in a lateral force, which is, however, compensated by the symmetrically positioned other wheel of the same axle. This effect of a toe-in position improves the stability of straight running. However, the rolling resistance of this axle – caused by the proportion of lateral forces against the direction of travel – becomes greater.

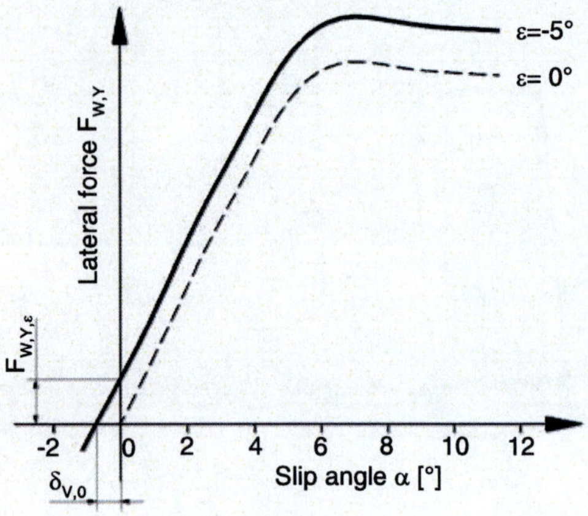

Fig. 2.10 Reduction of the rolling resistance of a inclined wheel by toe-in adjustment. $F_{W,\ Y,\ \varepsilon}$ Lateral force due to negative camber. $\delta_{V,\ 0}$ Necessary toe-out angle per wheel for lateral force-free directional stability despite negative camber

Figure 2.10 shows how toe-out can help to reduce the rolling resistance of a inclined wheel. A tyre with negative camber (example in the diagram $\varepsilon = -5^{\circ}$) generates a lateral force $F_{W,\ Y,\ \varepsilon}$ even when driving straight ahead ($\alpha = 0^{\circ}$). This lateral force is put into equilibrium by the other wheel of this axle, but the rolling resistance increases. If the curve is shifted to the right by the amount $\delta_{V,\ 0}$, no lateral force remains when driving straight ahead. The rolling resistance is reduced accordingly. To reach this state, the wheel must be given the toe-out angle $\delta_{V,\ 0}$.

On racing vehicles, the toe-in on the front axle is selected to be slightly larger than on standard vehicles. This increases the tyre temperature [8] and improves turn-in behaviour. However, an increased rolling resistance must be accepted for this. The values are around 1° [9]. The situation is similar at the rear axle.

At the rear axle, it is important that the toe-in angles on the left and right are the same because of the handling.

To sum up:

- Toe-in at the front: slower steering response, higher straight-line stability, higher tire temperature and wear on the outer shoulders.
- Front toe-out: Quicker steering response, less straight-line stability, higher tire temperature and wear on inner shoulders, higher top speed (front wheel drive).
- Rear toe-in: Higher straight-line stability, better grip on corner exit.
- Rear toe-in: Improved turn-in behaviour, greater top speed (rear wheel drive).

Roll Center (Instantaneous Pole,)

The roll centre *Ro* is the instantaneous pivot point (see appendix) around which the vehicle body rotates relative to the wheel contact points without any elasticities due to kinematic laws [10]. It is therefore the imaginary connecting joint between the body and the wheels. The lateral forces between the tires and the body are also transmitted via this “joint”. The

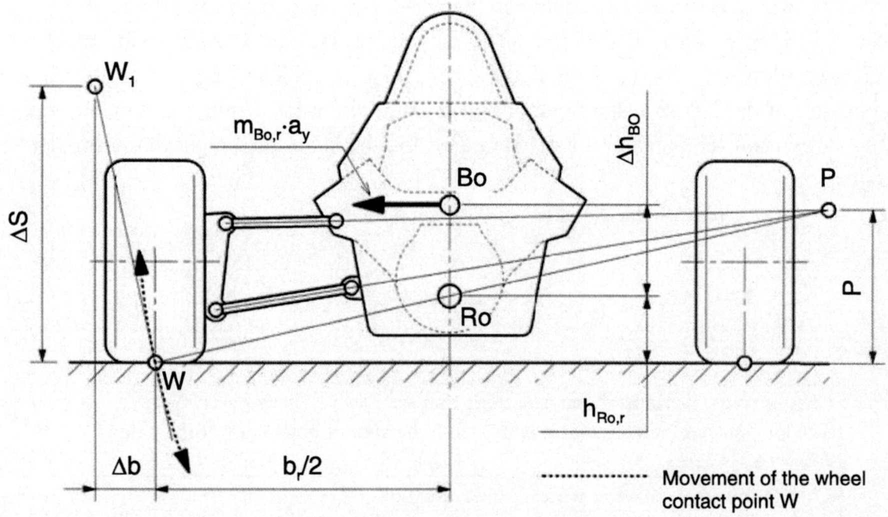

Fig. 2.11 Construction of the roll centre Ro of a double wishbone axle with wishbone axles parallel to the roadway. The rear axle is considered. Bo Body centre of gravity of the mass fraction $m_{Bo,r}$, which rests on the rear axle. The wheel contact point *W* is considered to be another articulation point in a linkage formed by the body, the links and the wheels. *P* Lateral instantaneous center

definitions in DIN ISO 8855 and SAE J670e provide further access: The roll center is the point of the sprung mass in a vertical transverse plane (i.e. parallel to the *y-z plane*) through the wheel centers of an axle at which a lateral force F_y does not cause any roll. This definition enables a static determination of the roll centre via the equilibrium of forces.

For front and rear axle there is one roll centre each, which is located in the design position (see appendix) in the transverse axle plane on the vehicle centre line, (Fig. 2.11).

For the kinematic construction of the roll centre, the wheels' instant centers of rotation in relation to the road (= wheel contact points *W*) and the wheels' instant centers of rotation in relation to the vehicle (= lateral instant centers *P*) are used. The point of intersection of the straight lines through *W* and *P* for the left and right halves of the wagon provides the vehicle's (more precisely: body's) centre of rotation in relation to the carriageway, the roll centre *Ro*.

The distance Δh_{Bo} between the roll center and the body center of gravity determines the moment with which the inertial force causes the body portion of this axle to roll during cornering. The smaller the distance, the smaller the roll moment. A possible disadvantage of body roll is the inevitable change in wheel position. The wheels are coupled to the body via the control arms and thus change the camber angle and/or track width when the body rolls, which affects the possible lateral tire force and driving stability.

Also of interest is the distance between the wheel contact point W and the lateral instant center P. This represents an imaginary (virtual) transverse link (virtual swing arm) with which a wheel moves relative to the body during suspension. The length of this "link" and the position of the lateral instant center P determine the track width and camber change during suspension. The greater the distance $\overline{PW}$ and the lower the height p, the smaller the changes.

Looking at one half of the car, the following applies:

$$\Delta b = \Delta s \cdot h_{Ro,r} \cdot \frac{2}{b_r} \tag{2.1}$$

Δb	Change of track width of one axle side, mm The total change in track width is therefore the sum of both sides of the axle: $\Delta b_t = \Delta b_l + \Delta b_{rs}$
Δs	Change in wheel stroke at which Δb occurs, mm
b_r	Track width rear, mm
$h_{Ro,r}$	Height of the roll Centre of the rear axle, mm

At the front axle the equation applies analogously with the index f.

For high driving stability and good directional control, a low roll centre ($h_{Ro,r}$ <) and a large track width (b_r >) are therefore advantageous.

The distance of the lateral instant center P from the wheel contact point W (*virtual swing arm length*) is approximately between 800 and 4600 mm [11]. Short distances between the lateral instant centers (500–1000 mm) lead to usable static roll centres with a favourable position of the wheel on the outside of the corner during roll, but to unfavourable camber changes during braking or acceleration, i.e. during equilateral suspension.

Large lateral instant center distances (1800–4600 mm) result in low roll centre heights with small track width changes. However, the lateral displacements of the roll centres increase during rolling, which results in rather unfavourable positions of the outer wheel on the turn. In the case of equilateral suspension, a small change in camber is advantageous.

If the distance between the lateral instant centers is selected to be even greater, the transverse links on one side of the axle will approach the parallel position, in which case the wheels will tilt with the body of the vehicle in an extremely unfavourable manner when rolling.

The height of a roll centre is also important for another reason. A high roll centre leads to lifting of the car body by the tyre side force, so-called jacking effect Fig. 2.12. A high roll centre means at the same time high lateral instant centers. If the roll centre is high or if there is a small change in the height of the roll centre during suspension, the following happens. The vehicle deflects less on the outside of the turn than it does on the inside, i.e. the centre of gravity is raised and thus the rolling moment is unfavourably increased. A change in the height of the roll center helps to reduce or completely eliminate this asymmetry of the

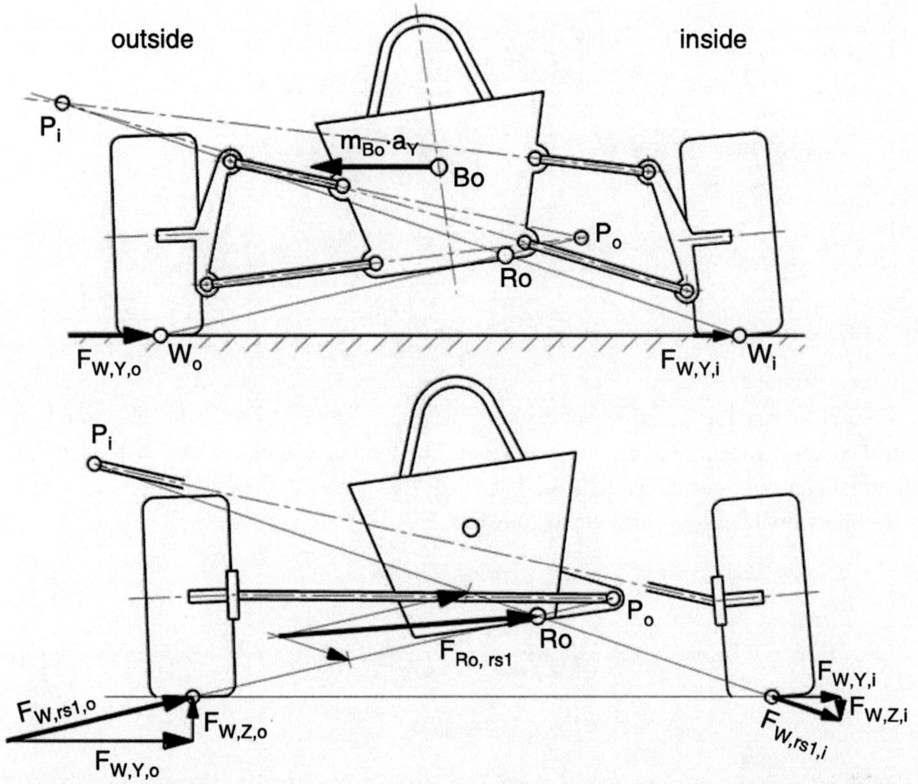

Fig. 2.12 Jacking effect of a high roll center. View from behind, right turn. Top: Due to the higher wheel load, the outer wheel generates the greater lateral cornering force $F_{W,Y,o}$. Both lateral tire forces $F_{W,Y}$ cause the reaction force $m_{Bo} \cdot a_y$ of the body mass as action forces. This reaction force is supported by the wheel links and body supporting springs, and the body rotates about the roll centre. Below: Substitute image for the situation shown above. Only the effect of the lateral forces is considered. The forces are transmitted to the superstructure (body) by the imaginary oscillations resulting from the lateral instant centers P_i and P_o. There they cause the resulting reaction force $_{\mathrm{FRo,rsl}}$ at the roll center *Ro*. The reaction force points upwards, thus lifting the frame. The higher the roll center, the stronger this support effect is

spring force change [1]. In the case of equilateral compression, the roll centre should therefore move downwards.

The inclination of the resultant $F_{\mathrm{Ro,rsl}}$ depends not only on the height of the roll centre but also on the magnitude of the lateral tyre forces. The greater the wheel load displacement, the more the outside wheel dominates the turn and the steeper the resultant points upwards. The figure also shows that when the lateral instant center $\mathrm{P_i}$ is high, the wheel on the inside of the turn makes a contribution $F_{\mathrm{W,Z,i}}$ to the lowering of the car body.

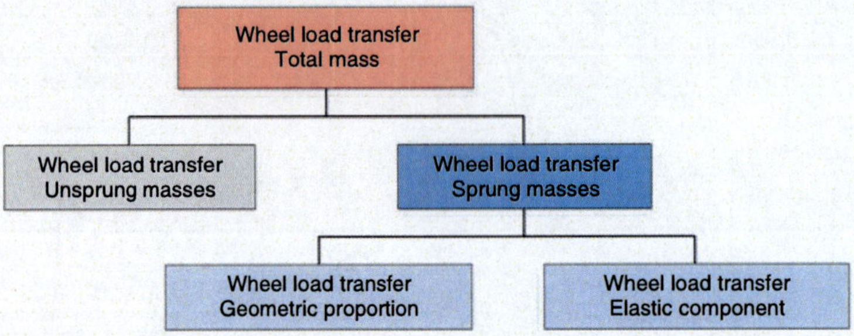

Fig. 2.13 Composition of the wheel load displacement

A closer look at the lateral wheel load displacement opens up further insights into how the roll centre influences the driving behaviour. The total displacement can be divided into individual contributions, (Fig. 2.13).

The individual contributions are as follows:

$$\Delta F_{W,Z,U} = m_U a_y \frac{h_U}{b}$$
$$\Delta F_{W,Z,geom} = m_{Bo} a_y \frac{h_{Ro}}{b} \quad (2.2)$$
$$\Delta F_{W,Z,elast} = m_{Bo} a_y \frac{\Delta h_{Bo}}{b}$$

$\Delta F_{W,Z,U}$	Wheel load displacement of the unsprung masses, N
m_U	Unsprung masses (wheels, links, rods, ...), kg
$\Delta F_{W,Z,geom}$	Geometric part of the wheel load displacement of the sprung masses, N
m_{Bo}	Body mass (sprung mass), kg
$\Delta F_{W,Z,elast}$	Elastic part of the wheel load displacement of the sprung masses, N
h_U, h_{Ro}, Δh_{Bo}, b	Dimensions, see Fig. 2.14, m

The total wheel load displacement $\Delta F_{W,Z}$ results to:

$$\Delta F_{W,Z} = \Delta F_{W,Z,U} + \Delta F_{W,Z,geom} + \Delta F_{W,Z,elast} \quad (2.3)$$

$\Delta F_{W,Z}$	Wheel load displacement of an axle, N

The geometric portion of the wheel load shift is determined by the height position h_{Ro} of the roll centre. This component acts directly during steering. In contrast, the elastic component, which is transferred between the body and the axle via spring/damper elements and stabilizers, only takes effect after a transition period. In the middle of the turn – in the quasi-stationary state – the total wheel load displacement is basically the same as for an

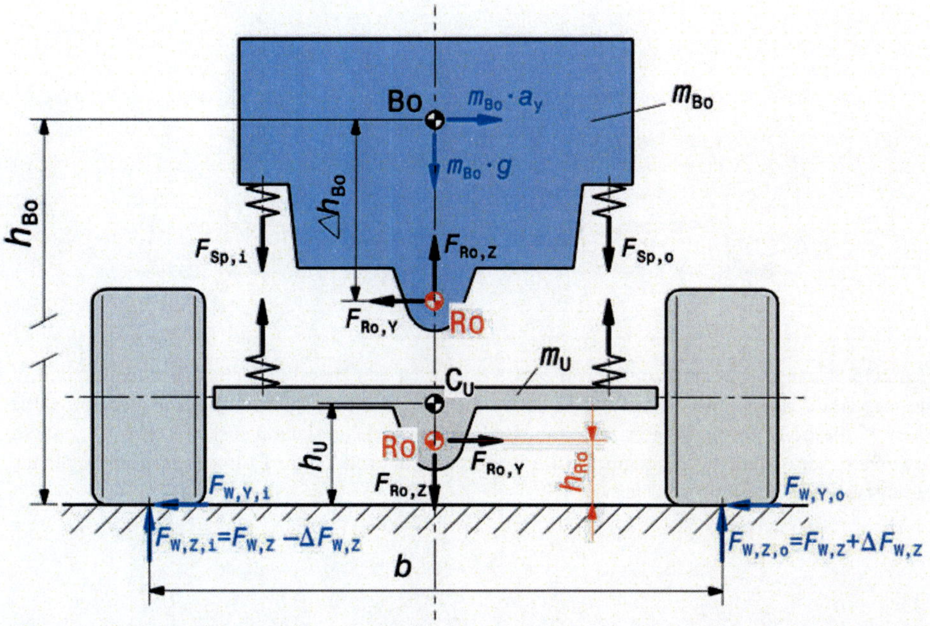

Fig. 2.14 Influence of the roll centre height on the wheel load when cornering (substitute image view from the rear, left turn). The car body m_{Bo} and the unsprung masses m_U are shown separately ("made free"). The connection between the two masses represents the roll centre *Ro*. Further forces F_{Sp} are transmitted via the body supporting springs (and stabilisers). The inertia force $m_{Bo} \cdot a_y$ is ultimately balanced by the lateral forces $F_{W,Y}$, *Bo* centre of gravity (centre of gravity of the sprung masses), C_U Centre of gravity of the unsprung masses. Indices: i inside of turn, o outside of turn, Y lateral, Z vertical, W wheel-related

unsprung vehicle, (2.4). A slight deviation results from the body roll and the consequent lowering (or raising!) of the body centre of gravity.

If the roll centre is at the same height as the body centre of gravity ($\Delta h_{Bo} = 0 \rightarrow \Delta F_{W,Z,\ elast} = 0$), the body supporting springs do not have to transmit any additional forces (caused by inertia) during cornering and the body does not sway, cf. Figure 2.14. In the same sense, however, this also means that dampers and stabilisers cannot have any effect. And further, the smaller the height difference Δh_{Bo}, the smaller the influence of stabilizers. A high roll center also leads to large track width changes during compression and thus has an unfavorable effect on lateral control. Even on passenger cars, roll centers are therefore usually found below 150 mm in height, and below 40 mm on racing cars. The range for the latter is roughly between −25 (i.e. below the road surface) and 50 mm height, see also Table 2.3.

The position of the roll centre influences the driving behaviour in many ways. It should not be forgotten that, depending on the wheel suspension, the position of the instant center changes (sometimes significantly) during compression and roll, (Figs. 2.15 and 2.16).

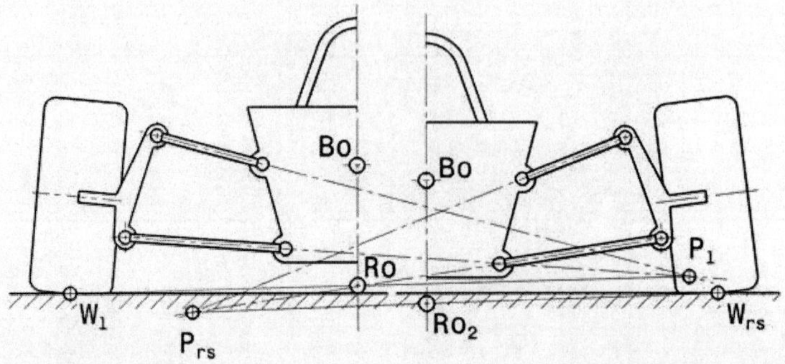

Fig. 2.15 Change in position of the roll centre *Ro* during equilateral deflection. For the left half of the wagon, the lateral instant center P_l is above the track and thus the roll centre is also above the track. If the body of the wagon is compressed (right half), the lateral instant center P_{rs} moves downwards and closer to the centre of the wagon and the roll centre Ro_2 comes to lie under the carriageway. *Bo* body center of gravity

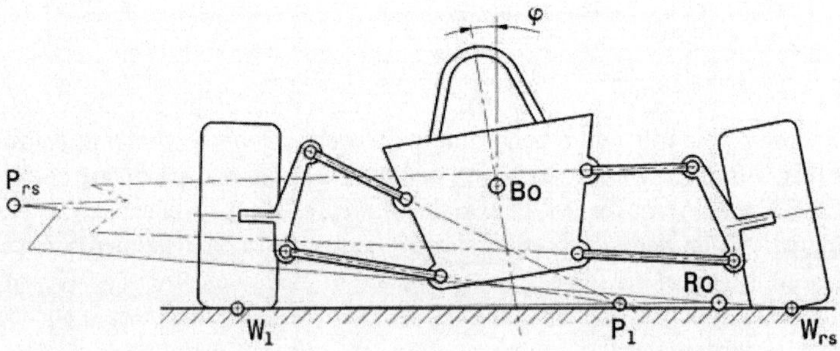

Fig. 2.16 Change in position of the roll centre *Ro in* the case of pure rolling of the wagon body. The car body rolls through the angle φ. In the process, the left wheel bumps and the right wheel rebounds. The lateral instant centers of the left and right wheel suspension, P_l and P_{rs}, occupy very different positions and as a result the roll centre *Ro* lies outside the central planes of the wagon

If the body of the wagon sways strongly, this results in influential camber changes of the wheels. Usually the wheels tilt with the body, i.e. to the outside of the turn. The outer wheel therefore reduces its negative camber or even changes to positive camber and the inner wheel changes to negative camber. This has an unfavourable effect on the lateral force of both tyres. For high lateral force potential, at least the more heavily loaded outer wheel must remain in the negative camber range. This can be ensured by appropriate design of the suspension kinematics. The tilt of the car body should also be kept small for the same reason. This can be achieved by a high roll centre (more precisely: by a small distance between the roll centre and the body centre of gravity). Some reasons mentioned above speak against this. Other more useful options are stiffer body supporting springs, stabilizers and a low centre of gravity. When using stabilizers, however, it must be remembered that they reduce advantages of independent suspension with increasing stiffness.

Table 2.1 Basic effects of roll-center locations

	Location roll center	
Criterion	Low (long distance to *Bo*)	High (small distance to *Bo*)
Mass inertia of sprung masses around roll axis	Large	Small
Vehicle reaction on turn-in	Sluggish	Speedily
Wall angle	Larger	Small
Influence of spring/damper/stabilizer	Pronounced	Lower; none at all for $Ro = Bo$
Tyre temperature	Tends to be lower	Tends to be higher
Support effect	Small (with *Ro* under roadway even sink effect)	Larger
Driving height (ground clearance)	Decreases	Increases

In summary, the following basic statements can be made.

Low roll centres lead to less wheel load transfer to the outside wheel on this axle, smaller to no support effects (constant ground clearance), but to large roll angles of the car body (larger elastic portion of the wheel load transfer). These roll angles must be kept small by stabilizers, but this also means that their influence is large. If the roll centre is below the road surface, the directly acting geometric part of the wheel load difference leads to a wheel load increase(!) on the inside wheel and the steering behaviour is more sluggish because the mass moment of inertia of the sprung masses around the roll axis (depending on $\Delta h_{Bo}{}^2$) is large.

With high roll centers, it's the other way around.

Finally, (Table 2.1) clearly compares the basic effects of the roll-center locations.

A possibility to avoid a too narrow compromise in the choice of the roll center and at the same time to provide a useful tuning tool, is offered by changeable pivot points on the car body or wheel carrier. For example, 3 height positions of the lower wishbone on the wheel carrier are sufficient at the front and rear. This means that 3 roll centre heights and thus 9 roll axis positions can be realised on both axles.

The designs of the rolling centres for some axles are shown below.

The construction of the instant centers for a general double wishbone axle is described in (Fig. 2.17).

A wheel-guiding suspension strut axle (McPherson axle) is the standard for passenger cars on the front axle and for rally vehicles. The design of its roll centre is shown in (Fig. 2.18).

Rear Axles

Among the designs significant for the rear axle, the rigid axle is still interesting despite its age. It is used in some racing series, especially overseas. In the (Fig. 2.19b) it is important to note that the pivot point of the swing arm of the Watt linkage is frame fixed. This design

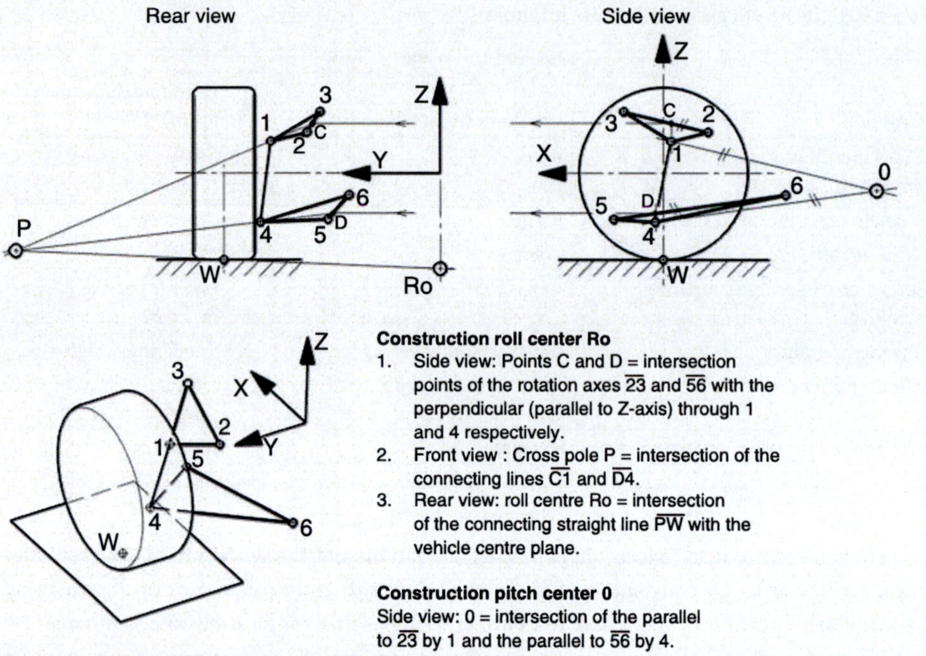

Fig. 2.17 Construction of the roll centre *Ro* and the pitch instant center *O* of a double wishbone axle with double wishbone axles lying obliquely in space. *W* Wheel contact point

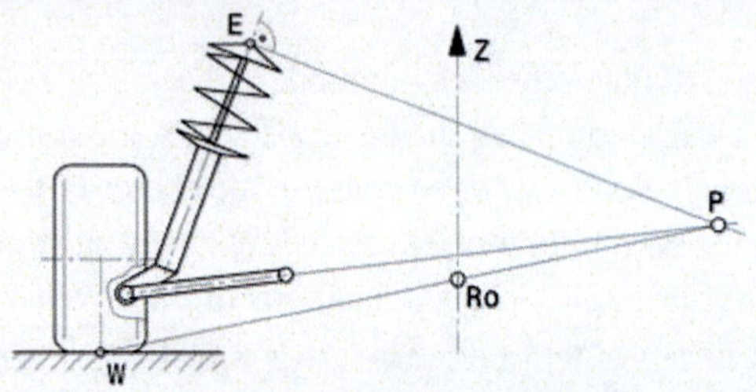

Fig. 2.18 Construction of the roll center *Ro* for wheel leading strut axle (McPherson axle). *E* is the mounting point of the shock strut on the car side. The connecting line is normal to the direction of movement of the damper piston rod.

is much more favourable for racing cars than that commonly used on production cars, where the swing arm is fixed to the axle.

The determination of the roll center for oblique link axles demonstrated (Fig. 2.20).

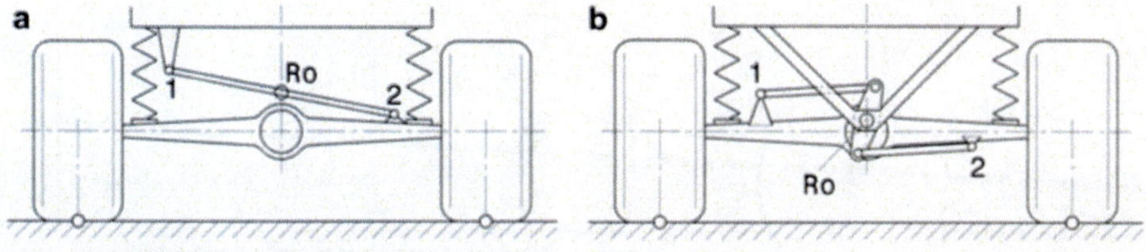

Fig. 2.19 Design of the roll center *Ro* for rigid axles. (**a**) Rigid axle with Panhard rod. Ro is the intersection point of the rod with the vehicle centre plane, (**b**) Rigid axle with Watt linkage. Ro is the pivot point of the connecting lever

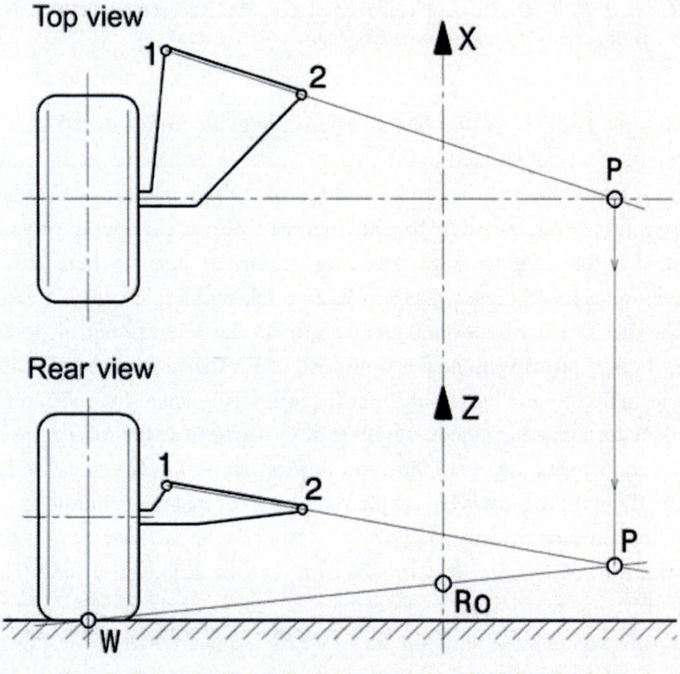

Fig. 2.20 Construction of the roll center Ro for inclined link axles. The construction starts in the top view. The straight line $\overline{12}$ supplies the lateral instant center P. In the rear view follows from the intersection of $\overline{PW}$ with the median plane of the vehicle the Roll center Ro

Roll Axis[8]

The imaginary connection of the roll centres of the front and rear suspension results in the roll axis, (Fig. 2.21).

[8] In some literature, the term rolling axis is avoided because rolling is the natural movement of a wheel and the body of the car, on the contrary, sways. In this book, the terms rolling and swaying axis as well as rolling and swaying centre are used equally, because they are also all heard in everyday language.

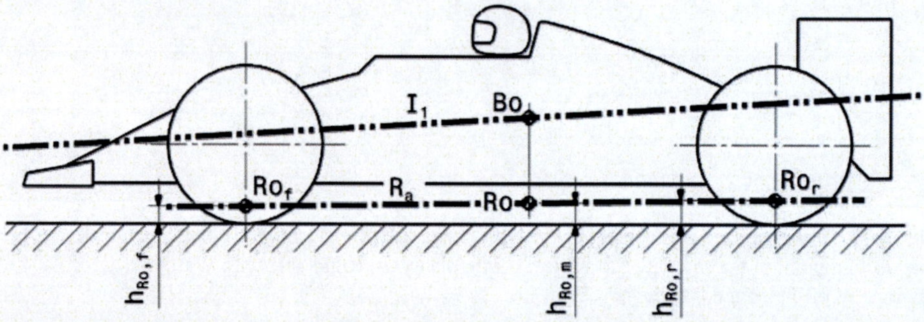

Fig. 2.21 Roll axis. R_a Roll axis, Ro_f Roll center front axle, Ro_r Roll center rear axle, I_1 Main inertia axis in longitudinal direction, Bo body centre of gravity

Theoretically, the distance of the body center of gravity from the roll axis is the lever arm over which the body is pivoted in the event of lateral forces due to the mass inertia. If the centre of gravity of the body is on the rolling axis, there is therefore no lateral tilt. In practical terms, this is contradicted by the fact that the roll axis changes during compression and extension, that the height of the center of gravity is not constant, and that wheel suspensions with a high roll center result in large track width and camber changes during suspension. Camber is a variable that greatly affects the lateral control potential of the wheel and has higher priority in suspension design. Furthermore, the roll stiffness of an axle is also determined by the height of the roll center. The wheel load difference between inside and outside turn wheels increases with rolling stiffness. If the roll centre is above the road surface, it contributes (see geometric proportion above) to the increase in the wheel load difference. If the roll centre is below the road surface, this contribution is negative and the wheels are loaded more evenly [10].

By tilting the roll axis, different roll behaviour can be achieved on the front and rear axles and thus different wheel load displacement on the axles. If the roll centre on the rear axle is higher, the wheel load shift on this axle is greater when cornering and the car oversteers. The total, maximum wheel load shift of the vehicle depends only on the centre of gravity height h_{Bo} (see Fig. 2.14) and track widths b:

$$\begin{aligned} \Delta F_Z &= a_y \frac{m_{Bo} h_{Bo}}{b} \\ F_{Z,o,dyn} &= F_{Z,o,stat} + \Delta F_Z \\ F_{Z,i,dyn} &= F_{Z,i,stat} - \Delta F_Z \end{aligned} \tag{2.4}$$

Compared to the static wheel loads, the wheel loads outside (o) and inside (i) change by the amount ΔF_Z during cornering.

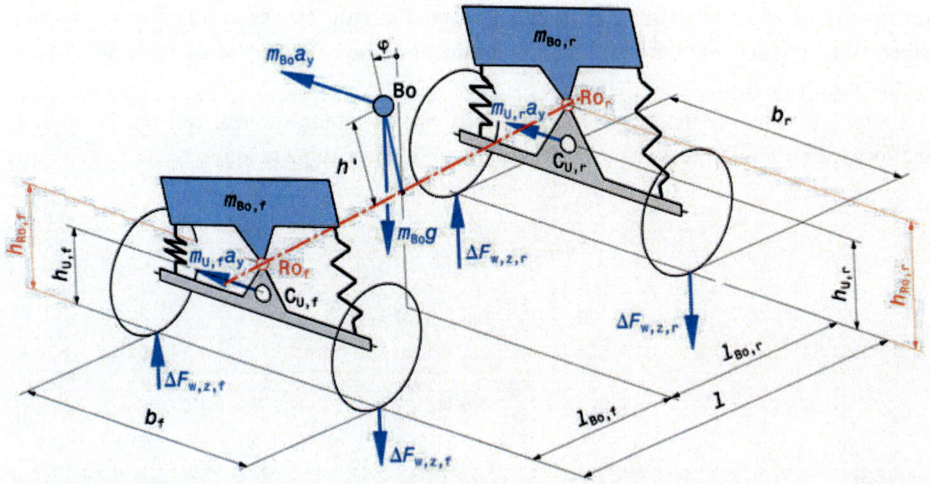

Fig. 2.22 Sketch for calculating the wheel load shift on a complete vehicle. h' Distance of the body centre of gravity Bo from the roll axis. $C_{U,f}$, $C_{U,r}$ Centre of gravity of the unsprung masses at the front and rear respectively. φ roll angle

The position of the rolling axis therefore only influences the distribution of the wheel load displacements between the axles. The factors that play a role in this are shown below. The wheel load differences compared to the static condition on a complete vehicle result from [12], see also Fig. 2.22:

$$\begin{aligned}\Delta F_{W,Z,f} &= m_{U,f} a_y \frac{h_{U,f}}{b_f} + m_{Bo} a_y \left(\frac{l_{Bo,r}}{l} \frac{h_{Ro,f}}{b_f} + \frac{c_{Ro,Sp,f} + c_{Ro,S,f}}{c_{Ro,Sp,f} + c_{Ro,Sp,r} + c_{Ro,S,f} + c_{Ro,S,r} - m_{Bo} g h'} \frac{h'}{b_f} \right) \\ \Delta F_{W,Z,r} &= m_{U,r} a_y \frac{h_{U,r}}{b_r} + m_{Bo} a_y \left(\frac{l_{Bo,f}}{l} \frac{h_{Ro,r}}{b_r} + \frac{c_{Ro,Sp,r} + c_{Ro,S,r}}{c_{Ro,Sp,f} + c_{Ro,Sp,r} + c_{Ro,S,f} + c_{Ro,S,r} - m_{Bo} g h'} \frac{h'}{b_r} \right)\end{aligned} \tag{2.5}$$

$\Delta F_{W,Z,f}$, $\Delta F_{W,Z,r}$	Wheel load transfer to front or rear axle, N
$m_{U,f}$, $m_{U,r}$	Unsprung masses front or rear, kg
$c_{Ro,Sp,f}$, $c_{Ro,Sp,r}$	Roll stiffness due to the body supporting springs referred to the front or rear axle, Nm/wheel. See also (3.31)
$c_{Ro,S,f}$, $c_{Ro,S,r}$	Roll stiffness due to anti-roll bars referred to the front or rear axle, Nm/wheel. See also (3.41)
$h_{U,f}$, $h_{U,r}$, $h_{Ro,f}$, $h_{Ro,r}$, h', b_f, b_r, l	Dimensions, m. see Fig. 2.22
$l_{Bo,f}$, $l_{Bo,r}$	Position of the Centre of gravity of the body in relation to the axles, m

The proportionate front and rear body mass for the geometric part of the wheel load difference follows directly from the position of the centre of gravity of the body mass to $m_{Bo,f} = m_{Bo}\ l_{Bo,r}/\ l$ or $m_{Bo,r} = m_{Bo}\ l_{Bo,f}/l$ respectively. The elastic component, which is determined by the roll stiffnesses $c_{Ro,Sp}$, is extended here by the weight term $m_{Bo}\, gh'$, which takes into account the additional roll moment – caused by the lateral deflection of the centre of gravity *Bo*.

If one now changes the height $h_{Ro,f}$ of the front roll center, the rear roll center remains unaffected, but not the distance h'. The following relationships result:

$$h' = h_{Bo} - \frac{h_{Ro,f} l_{Bo,r} + h_{Ro,r} l_{Bo,f}}{l} \tag{2.6}$$

$$h_{Ro,f,neu} = h_{Ro,f} + \delta h_{Ro,f} \tag{2.7}$$

$$\delta h' = -\delta h_{Ro,f} \frac{l_{Bo,r}}{l} \tag{2.8}$$

$$\delta \Delta F_{W,Z,f} = \delta h_{Ro,f} m_{Bo} a_y \frac{l_{Bo,r}}{l b_f} \left(1 - \frac{c_{Ro,Sp,f}}{c_{Ro,Sp,f} + c_{Ro,Sp,r}} \right) \tag{2.9}$$

$$\delta \Delta F_{W,Z,r} = -\delta h_{Ro,f} m_{Bo} a_y \frac{l_{Bo,r}}{l b_r} \frac{c_{Ro,Sp,r}}{c_{Ro,Sp,f} + c_{Ro,Sp,r}} \tag{2.10}$$

h_{Bo}	Height of the centre of gravity *Bo* above the carriageway (see Fig. 2.14), m
$\delta h_{Ro,f}$	Change in roll centre height $h_{Ro,f}$, m
$\delta h'$	Resulting change in the distance of the centre of gravity from the roll axis, m
$\delta \Delta F_{W,Z,f}$, $\delta \Delta F_{W,Z,r}$	Change in wheel load differences on front or rear axle, N

It can be seen from (2.9) that by raising the front roll centre by the amount $\delta h_{Ro,f}$ the wheel load difference increases, while at the rear axle (assuming the same track width *b*) the difference decreases by exactly this amount. The same applies to the rear roll centre. Raising the roll center therefore increases the roll stiffness at this axle and equalizes the wheel load differences at the other axle. The relative position of the roll centres to each other thus proves to be an adjustment tool for the self-steering behaviour of the car. Figure 2.23 illustrates this possibility for a specific vehicle.

The roll angle φ, which is set for a rigid frame[9] by the roll resistances of the two axes, is

$$\varphi = \frac{m_{Bo} a_y h'}{c_{Ro,Sp,f} + c_{Ro,Sp,r} + c_{Ro,S,f} + c_{Ro,S,r} - m_{Bo} g h'} \tag{2.11}$$

[9] A consideration of the frame influence can be found in the Racing Car Technology Manual Vol. 2 *Complete Vehicle*, Sect. 6.3.

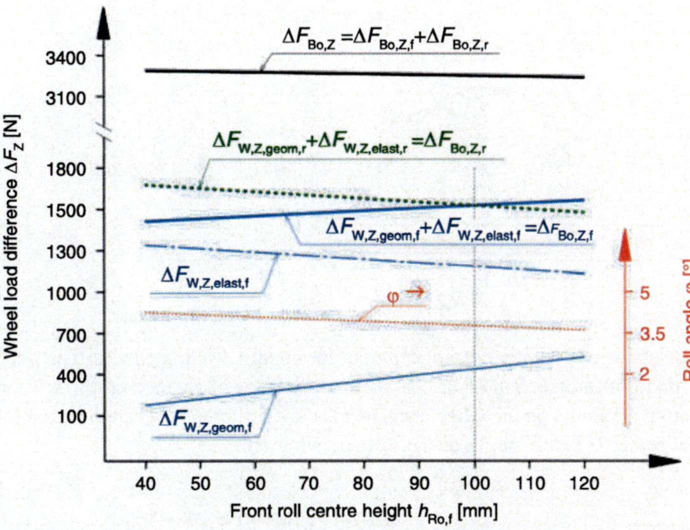

Fig. 2.23 Influence of the roll centre position on the wheel load displacement in the transverse direction. Only the contributions of the body mass m_{Bo} are considered here, because the contributions of the unsprung masses are independent of the roll centres. The front roll centre height $h_{\mathrm{Ro,f}}$ is increased from 40 to 120 mm. The geometric contribution increases and the elastic contribution decreases accordingly. The sum of the two contributions, i.e. the total contribution $\Delta F_{\mathrm{Bo,Z,f}}$ of the sprung masses at the front $m_{\mathrm{Bo,f}}$ increases. The contribution $\Delta F_{\mathrm{Bo,Z,r}}$ of the rear body mass $m_{\mathrm{Bo,r}}$ behaves exactly in the opposite direction. The roll angle changes with the lever arm h' with which the inertial force of the body acts about the roll axis. The fact that the total wheel load displacement $\Delta F_{\mathrm{Bo,Z}}$ – caused by the body – is not constant is related to the weight term $m_{\mathrm{Bo}}\, gh'$. With the lever arm h' the roll moment and thus the reaction moment of the wheel loads also changes. Vehicle data: $l = 2500$ mm, $l_{Bo,f} = l_{Bo,r} = 1250$ mm, $h_{Bo} = 350$ mm, $h_{Ro,r} = 100$ mm, $b_f = b_r = 1200$ mm, $m_{Bo} = 1100$ kg, $c_{Ro,\mathrm{Sp,f}} = c_{\mathrm{Ro,Sp,r}} = 21{,}600$ Nm/rad, $a_y = 1\ g$

Although the lower roll centre is often found at the lighter and/or lower end of the vehicle, this can by no means be taken as a universal recommendation. Rear-wheel drive cars are often designed with a forward rising roll center (not least to compensate for the conceptual tendency to oversteer), while on front-wheel drive cars it drops forward. In rear-wheel drive racing cars, the body supporting springs are generally softer at the rear axle and the mass proportion is higher. Therefore, higher roll centres than at the front are also suitable for compensation here [30].

Pitch Center

Pitching is the rotation of the body around a transverse axis. During the braking process, there is a compression at the front and a rebound at the rear; the nose of the body dips and the rear rises.

Just as independent suspensions have instant centers in the rear view, axles guided by one or two links (or a damper leg) may have pitch instant centers O on both sides in the side view (Figs. 2.24, 2.25, and 2.26). The axles are supported longitudinally on the body at

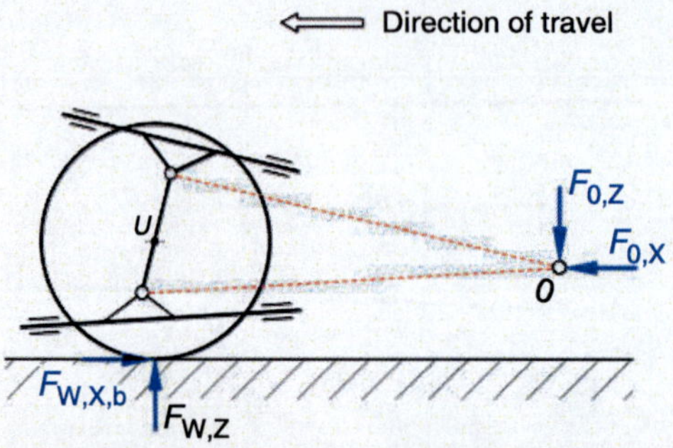

Fig. 2.24 Construction of the pitch instant center *O* for double wishbone axles. The pitching instant center *O* is the point of intersection of the parallels to the axes of rotation of the wishbones on the wagon side through the joints on the wheel carrier side. The distance between the pitch instant center *O* and the wheel center *U* is the length of the virtual swing arm

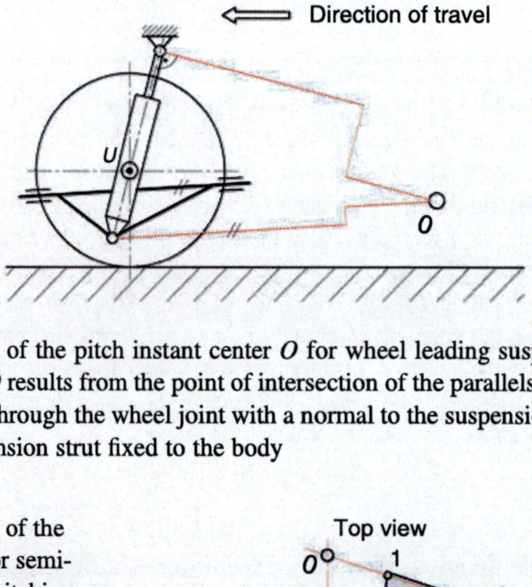

Fig. 2.25 Construction of the pitch instant center *O* for wheel leading suspension strut axles. The pitching instant center *O* results from the point of intersection of the parallels to the wishbone axis of rotation on the car side through the wheel joint with a normal to the suspension strut axis through the connection of the suspension strut fixed to the body

Fig. 2.26 Construction of the pitch instant center *O* for semi-trailing arm axles. The pitching instant center *O* results from the point of intersection of the axis of rotation (connection of joints 1 and 2) of the inclined link through the wheel centre plane

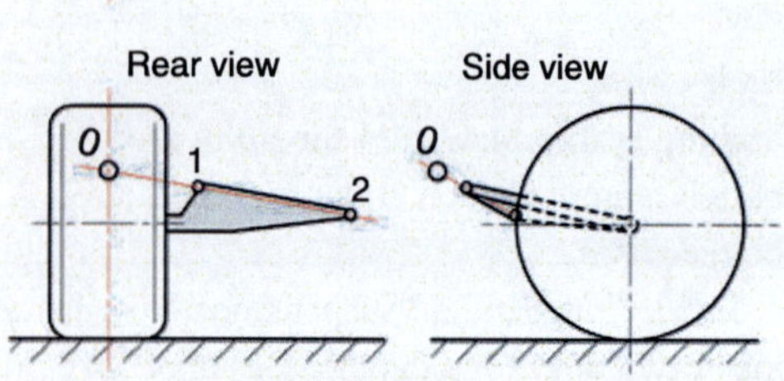

these. The pitch instant center is therefore the (instantaneous) joint with which an imaginary swing arm is connected to the body of the car (virtual swing arm, virtual side view swing arm).

The length of the virtual *side view* swing arm (distance pitch center to wheel center, *virtual side view swing arm length SVSA*) influences the caster change during compression and the generation of reaction forces, which counteract the pitching of the car body during braking or acceleration. The higher the pitching instant center, the greater these reaction forces are. However, this also leads to a large change in caster during compression, which is only desirable to a limited extent on the front axle because the steering restoring force can become too great under certain circumstances during extreme cornering. In addition, the front wheel deflects not only upwards but also forwards during compression (it swings around the pitch instant center), i.e. precisely in the direction of a bump. This leads to an apparent hardening of the suspension and at least worsens the comfort. At the rear axle, however, the wheel moves backwards away from the bump. Here, this so-called diagonal springing is quite desirable, especially with large suspension travel. In rally cars, a large diagonal springing on the rear axle reduces the dreaded kickback. (*Kickback*, catapult effect) during a landing, which in "combination" with a heavy front axle can lead to a rollover during a double jump [13]. The slant suspension angle is now only important for stiff joints (i.e. racing cars). Passenger cars achieve the same effect through elastokinematics.

The pitch axis of the car body results from the position of the instant center *Po of* its total movement during pitching, (Fig. 2.30). The pitch axis is the transverse axis containing the instant center *Po*.

In summary, the described effects of the possible pitch instant centers are summarized in (Table 2.2).

Table 2.2 Effects of the position of the pitch instant center or the length of the virtual swing arm

	Location of the pitch instant center	
Criterion	Low	High
Mass inertia of the car body about pitch axis	Large	Small
Reaction of the body during braking/driving	Sluggish	Speedier
Tyre temperatures by pitching	Small	Larger
	Length of the virtual swing arm (distance between pitch instant center and wheel center)	
	Small	Large
Caster change during bouncing movement	Large	Small
Caster change during pitching	Small	Large

2.2.2 Development Goals

Degree of Quality of the Lateral Force Distribution η_G

Is the ratio of the actual acting lateral force $F_{Y,V}$ per axle to the theoretically possible one:

$$\eta_{Gf} = \frac{F_{V,Y,f}}{\mu_{W,Y} \cdot F_{V,Z,f}} \tag{2.12a}$$

$$\eta_{G,r} = \frac{F_{V,Y,r}}{\mu_{W,Y} \cdot F_{V,Z,r}} \tag{2.12b}$$

$\eta_{G,f}$	Degree of quality of the lateral force distribution at the front, –
$\eta_{G,r}$	Quality of the lateral force distribution at the rear, –
$F_{V,Z,f}$	Front axle load, N
$F_{V,Z,r}$	Rear axle load, N
$\mu_{W,Y}$	Coefficient of friction in transverse direction, –
$F_{V,Y,f}$	Lateral force acting on front axle, N
$F_{V,Y,r}$	Lateral force acting on rear axle, N

An ideal lateral force distribution ($\eta_G = 1$) is obtained with lateral forces distributed proportionally to the wheel loads. Then the same coefficient of friction $\mu_{W,Y}$ is used on the inside and outside wheel. An independent wheel suspension is advantageous for this.

An extremely low centre of gravity height makes it easier to achieve high quality grades, because the dynamic wheel loads then differ only slightly from the static ones.

Toe-in

If you want to improve the quality of the lateral force distribution by means of the wheel suspension kinematics, a steering effect must be created, in which, for example, the outer wheel on the rear axle is turned in the direction of understeering and the inner wheel on the front axle is turned in the direction of understeering when rolling (Fig. 2.27).

If the understeer or at least the neutral behaviour is already sufficiently achieved by other means, then the opposite rotation of the opposite wheels or both together also have a quality-improving effect [10].

The wheel must not be turned to any toe-in position when maximum deflection or rolling occurs. This would otherwise lead to an undesirable steering effect of the outer wheel towards the outside of the turn. This promotes understeer on the front axle and oversteer on the rear axle.

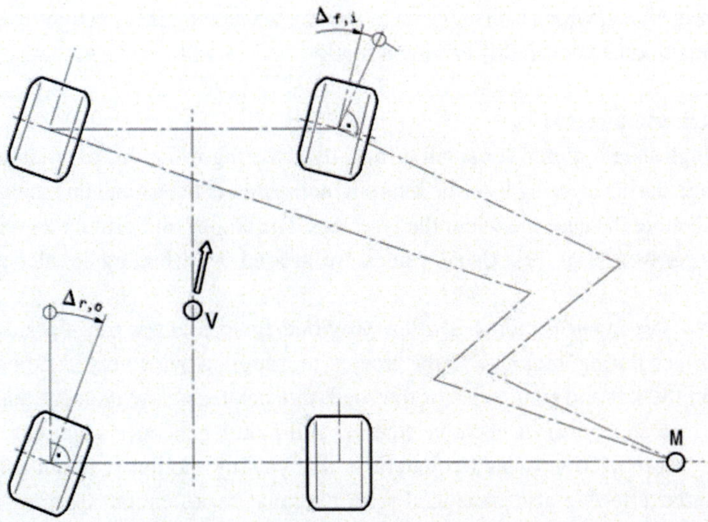

Fig. 2.27 Toe-in angles $\Delta_{f,i}$ and $\Delta_{r,o}$ to improve the quality of the lateral force distribution while promoting the tendency to understeer, after [10]. If the front wheel on the inside of the bend and the rear wheel on the outside of the bend are given toe-in, the quality of the lateral force distribution is increased. Especially on the rear axle, the handling can be adjusted in the direction of neutral or even understeer

Caster

A caster of the front wheels ensures stable straight running of the vehicle but also increases the steering torque. A large positive caster angle leads to a camber change towards negative values on the outer wheel when cornering. This effect is intensified by large kingpin inclination angles. When turning in, this rapid camber change is advantageous in principle, but can also lead to non-linear understeer further down the line. Therefore, a compromise is sought between the camber change caused by caster during steering and that caused by body roll. Aerodynamic downforce increases the wheel load and thus the steering torque. For this reason, the caster angle is chosen very small for racing cars with strong aerodynamic aids. The upper limit for skirted Formula 1 cars in the 1980s was less than 2° [9]. Common values are found in the range 2–6°.

In karts, a large caster angle in conjunction with a large steering roll radius ensures that the inside front wheel is lowered (intentionally) and the outside wheel is raised at the same time. This makes it easier for the driver to lift the inside rear wheel. Karts have a rigid axle at the rear and no differential. This constellation led in narrow turns to strong understeer, remained both wheels on the road. The caster angle for karts is therefore in the range of 12–17°.

On vehicles with movable wheel suspension, the rear wheel does not lift off due to a large caster and the wheel load increase advantageously supports the grip build-up of the inner cornering wheel, which is disadvantaged by the wheel load shift, in tight corners. This helps to combat understeer in this driving situation.

Kingpin Inclination (KPI)

Large KPI angles are required for small or negative steering roll radii. In order to keep the camber change during steering low, the kingpin inclination is kept small on racing vehicles. This is all the more the case the wider the tyres are. The angles on Formula 1 ground effect cars with skirts were 1.5° [9]. Usual values are around 7°, whereby smaller angles are better [11].

A negative steering roll radius is often provided for passenger cars because it has a stabilising effect during braking. If the brakes are applied with very different friction conditions on the left and right sides of the road, this creates a yaw moment that can lead to accidents (see appendix). A negative steering roll radius generates a steering torque on the side with more friction, which counteracts this yawing motion (without driver influence). A similar effect is also beneficial with diagonal brake circuit splitting when one brake circuit fails. In a racing vehicle, the steering roll radius should deviate as little as possible from zero in order to avoid disturbing influences on the steering. A negative steering roll radius has the disadvantage, especially for a racing driver, that when the brakes are applied on one side, the steering pulls to the side with the lower friction and inevitably causes a misjudgement by the driver. The driver will react to this and countersteer in exactly the wrong direction, thus increasing the incipient yawing movement [1].

The kingpin inclination angle is also an important parameter for the steering return, see Sect. 5.2.2 *Steering*.

Braking and Accelerating Pitch Compensation (*Anti Dive and Anti Squat*)

The inertial force of the vehicle causes the body to pitch during braking and acceleration. This movement is influenced by the deceleration or acceleration, the height of the centre of gravity, the stiffness of the body supporting springs, the wheelbase and the chassis geometry. In the case of the chassis, the position of the pitching instant centers is of decisive importance. A strong pitching motion is especially disturbing for vehicles with wings and ground effect, because it inevitably changes the angle of attack of the flow as well as the ground distance and thus the downforce.

The support of the braking forces of an axle is done by the imaginary link, which turns around the pitch instant center. The supporting force therefore passes through the wheel contact point and through the pitch instant center. The angle that the force vector makes with the road is called the brake support angle ε or the starting support angle or *diagonal springing angle* χ.

The axles are therefore supported longitudinally on the body at the pitching instant centers and also the moments caused by the braking force F_B. The prerequisite for this is that the moments arise in the wheel suspension itself, i.e. the brake is located on the outside of the wheel.

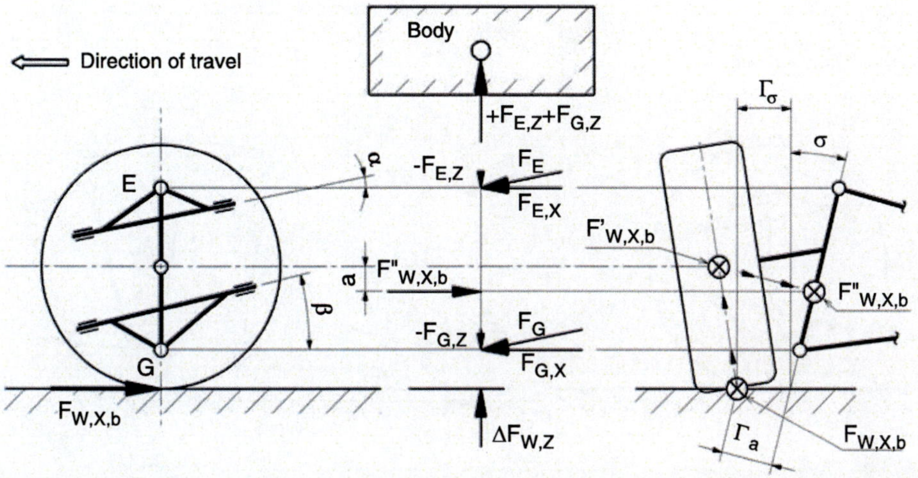

Fig. 2.28 Pitching instant center with internal brake, according to [14]. If the front brake is located on the inside of the differential gear, brake dive can be counteracted by tilting the links in the same direction (left). The braking force should be considered to be below the wheel centre by the amount a (right). Since the wheel bearing does not transmit any moments in the wheel plane, i.e. the wheel and the wheel carrier are not a rigid unit, the braking force $F_{W,X,b}$ must be considered to be in the centre of the wheel. The result is the moment $M_b = F'_{W,X,b} \cdot r_a$. The determination of the forces $F_{E,x}$ and $F_{G,x}$ in the wheel joints E and G requires a further displacement perpendicular to the kingpin inclination axis; $F_{W,X,b}$ thus comes to lie as $F''_{W,X,b}$ by the amount $a = r_a \cdot \sin(\sigma)$ below the wheel centre (r_{dyn}). The brake support angle for this arrangement is $\chi = (\alpha + \beta)/2$

Left and right suspension are generally the same, so that the instant centers determined by the momentary position of the control arms have the same position on both sides and one can speak of a pitching axle O_f at the front and a rear O_r. If O_f is located behind the front axle, the diving bow is pushed up during braking. An O_r located in front of the rear axle ensures that the deflecting rear end is pulled down. If the pitching axle moves to infinity (i.e. it is practically non-existent), the longitudinal force support takes place in the wheel centres, which also applies to the case of the brake arranged on the inside (at the differential gear). In this case, brake compression can be countered by tilting the two double wishbones in the same direction (Fig. 2.28). As can be seen from the figure, the braking force displaced as $F''_{W,X,b}$ from the centre of the wheel perpendicular to the axis of kingpin inclination causes the reaction forces $F_{E,x}$ and $F_{G,x}$ in the control arms, which (due to the skew) give rise to the perpendicular components $-F_{E,z} = F_{E,x} \cdot \tan(\alpha)$ and $-F_{G,z} = F_{G,x} \tan(\beta)$. Forces in one direction of action must be zero, i.e., $+ F_{E,z}$ and $+ F_{G,z}$ oppose front end deflection. Two control arms angled in this way probably have the advantage of no change in caster, but the disadvantage of deflecting forward (i.e. towards the obstacle) when the suspension compresses. Such a link arrangement allows almost 100% brake pitch compensation.

With the brake on the outside, it is also necessary to tilt the control arms to obtain a pitch centre and thus reaction forces in the upward direction; however, both control arms must be

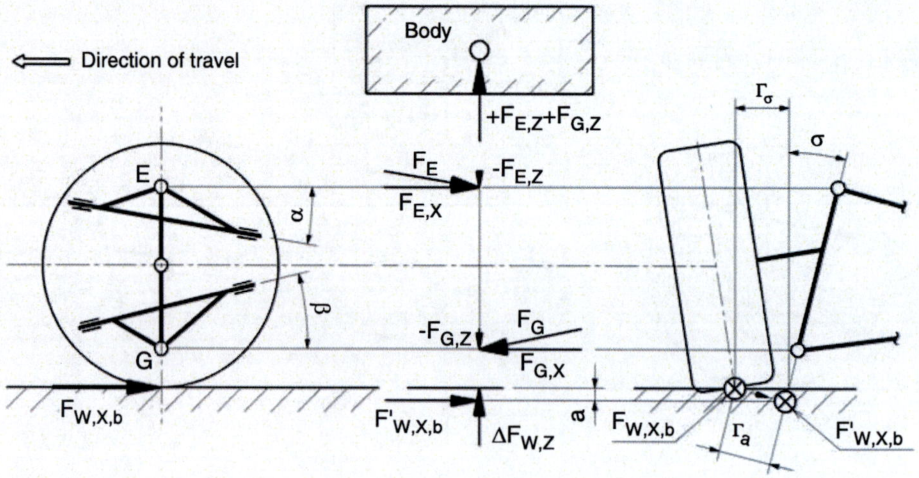

Fig. 2.29 Pitching instant center with position of the brake in the wheel, according to [14]. To reduce brake dive, with the brake on the outside, the control arms must be interlocked with each other (left). Starting from the brake force $F'_{W,X,b}$ which is below the ground by *a,* the forces $F_{E,x}$ and $F_{G,x}$ are to be determined. The components counteracting the diving of the front end are then $+F_{E,z}$ and $+ F_{G,z}$. Due to the brake caliper, the wheel and the wheel carrier form a rigid unit during braking (right). In order to be able to determine the reaction forces in the wheel joints *E* and *G,* the braking force $F_{W,X,b}$ must be displaced as $F'_{W,X,b}$ perpendicular to the kingpin inclination axis; this comes to lie below the ground by $a = r_a \cdot \sin(\sigma)$ or above it in the case of negative steering roll radius

interlocked with each other (Fig. 2.29). The left part of the figure shows the statics with the significantly larger component $F_{G,z}$ (compared to the arrangement with the brake on the inside) caused by the higher force $F_{G,x} = F'_{W,X,b} + F_{E,x}$ with the brake on the outside (with the brake on the inside, $F_{G,x} = F'_{W,X,b} - F_{E,x}$).

The requirement to reduce brake dive demands a pitching axle that is close to the wheel and as high as possible; both, however, result in a strong change in caster. For the front axle, therefore, a compromise must be found between a favourable position and an acceptable change; on rear axles, on the other hand, the situation is different. Here, the pitch instant center O_r can be placed close to the axle. The change in wheelbase associated with the pitch centre should have no effect on driving behaviour; the Renault passenger car models (R4, 5 and 6), which had different wheelbases on the left and right, can serve as proof [15].

If the forces acting on the entire vehicle during braking or acceleration are taken into account, it is possible to determine the braking or acceleration compensation k_ε or $k. \chi$.

The total braking force resulting from the inertia of the car is:

$$F_B = m_{V,t} \cdot a_x \tag{2.13}$$

F_B	Total braking force, N
$m_{V,t}$	Total mass of the vehicle, kg
a_x	Deceleration, m/s^2

This causes the braking force $F_{W,X,B,f}$ on the front axle tyres and $F_{W,X,B,r}$ on the rear axle tyres (Figs. 2.30 and 2.31).

The numerical value for the brake pitch compensation referred to the axle from (Fig. 2.31) also corresponds to the acceleration compensation k_χ, if the drive torque is not supported via the wheel suspension, which is generally the case with frame-mounted axle drives.

Figure 2.32 shows a purely graphical procedure for determining the degree of compensation.

The wheelbase is divided according to the distribution of the braking forces and a vertical line is entered at this point. The distance h_V from the road surface to the vehicle centre of gravity corresponds to 100% brake pitch compensation. The intersection of the straight line through the wheel contact point and the pitching instant center with the vertical line provides the percentage of the pitch compensation for this axle. In the example shown, the front axle has 60% and the rear axle 80% brake pitch compensation.

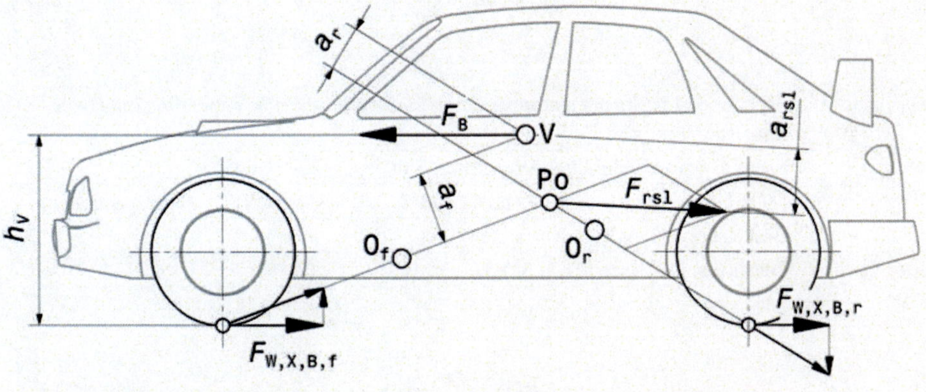

Fig. 2.30 Calculation of brake pitch compensation k_ε in percent with outboard brake

$$k_{\varepsilon,ges} = \frac{F_{rsl}(h_V - a_{rsl})}{F_B \cdot h_V} \cdot 100\%$$

$$k_{\varepsilon,f} = \frac{h_V - a_f}{h_V} \cdot 100\%$$

$$k_{\varepsilon,r} = \frac{h_V - a_r}{h_V} \cdot 100\%$$

V Vehicle centre of gravity

O_f, O_r Pitch instant center front or rear axle

Po pitch instant center of the body

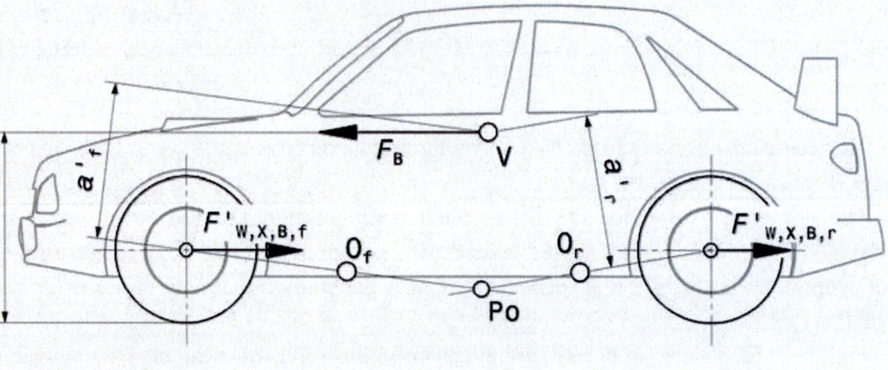

Fig. 2.31 Calculation of the brake pitch compensation k'_ε in percent for inboard brake

$$k'_{\varepsilon,f} = \frac{h_V - a'_f}{h_V} \cdot 100\%$$

$$k'_{\varepsilon,r} = \frac{h_V - a'_r}{h_V} \cdot 100\%$$

O_f, O_r Pitch instant center front or rear axle
Po Pitch instant center of the body

$$k_{\varepsilon,f} = \Phi_f \cdot l \frac{1}{h_V} \cdot \frac{e}{c} \tag{2.14}$$

$k_{\varepsilon,\mathrm{f\ resp.\ r}}$	Front or rear brake pitch compensation (anti dive), – or %, depending on whether Φ dimensionless or in %
$\Phi_{\mathrm{f\ resp.\ r}}$	Front or rear axle braking force ratio, – or %.
	$\Phi_f = F_{W,X,B,f} / F_B$ (times 100 in %) and $\Phi_r = 1 - \Phi_f$ and $100 - \Phi$, respectively.$_f$
ε	Brake support angle, °. $\varepsilon = \arctan(e/c)$
e, c, h_V, l	Distances, mm. (see Fig. 2.32)

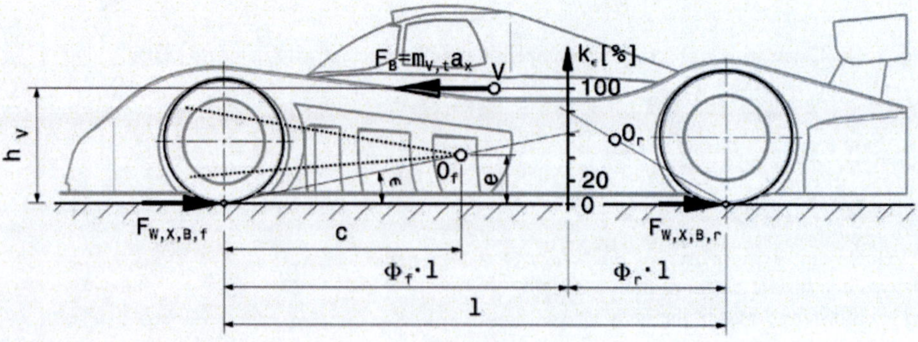

Fig. 2.32 Graphical determination of the brake pitch compensation, according to [16]

A brake pitch compensation of 100% for one axle means that the front axle does not compress during braking and the rear axle does not decompress. If the value for the rear axle is above 100%, the inertial movement is overcompensated and the car lowers at the rear when braking.

On most race cars, the driver can adjust the brake force distribution between the front and rear axles. When the brake force distribution is adjusted, the brake pitch compensation changes accordingly. The reaction forces of the pitch compensation grow comparably with the braking forces. In vehicles with high downforce, enormous braking forces can be generated in the initial phase of braking, which, depending on the pitch compensation, introduce corresponding reaction forces into the chassis. This leads to increased joint friction and reduced suspension effect – in extreme cases the chassis does not move at all. Professional racing drivers use pitching to estimate braking deceleration [17]. For this reason, a maximum of 30% brake pitch compensation is provided on racing cars. On formula cars and sports prototypes, practically no compensation is applied at all, and on touring cars with front engines, about 20–30% [18]. Excessive pitch compensation (over 50%) actually impedes the suspension of the body and thus increases wheel load fluctuation, especially in bumpy braking zones, with the result that the front tyres tend to lock. A further phenomenon of the worsened ground contact is understeer during corner entry on the brakes (*trail braking*).

For passenger cars, brake pitch compensation values can be found between 14 and 48% at the front and 40–120% at the rear.

As with braking, acceleration also results in an axle load shift and a pitching moment. Only in this case to the rear and thus to the sinking in of the rear. This results in an undesirable change in the camber of the drive wheels. The position of the pitch instant center O_r of the rear axle influences the balancing effect of the reaction forces. The degree of compensation is therefore basically determined in the same way as for braking, with the only difference that only one axle absorbs forces and that these forces act on the centre of the wheel because the moment of the drive shafts is supported on the body and not on the chassis (i.e. as with an inboard brake). Figure 2.33 explains a graphical method.

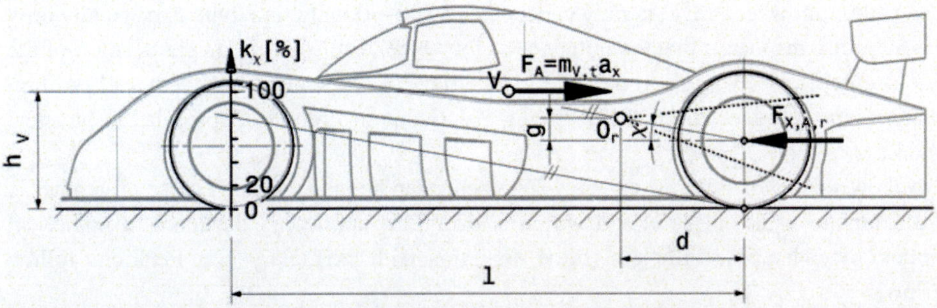

Fig. 2.33 Graphical method for determining the start-up pitch compensation

Analogous to Fig. 2.32, the wheelbase l *is* divided according to the distribution of the drive forces $F_{X,A,r}$. In the case of the single-axle drive here, the vertical straight line is plotted directly at the wheel contact point of the front axle. The height h_V up to the vehicle centre of gravity V represents 100% starting pitch compensation, i.e. the car remains parallel to the road surface when accelerating. A parallel line to the straight line through the pitch instant center O_r and the wheel center of the rear axle intersects the vertical line in the corresponding compensation value. In the example, this is 80% starting pitch compensation.

$$k_\chi = \frac{l}{h_V} \cdot \frac{g}{d} (\text{times } 100 \text{ in\%}) \tag{2.15}$$

k_χ	Start-up pitch compensation (anti squat), – or %
g, d	Distances, mm. (see Fig. 2.33)
χ	Starting support angle, °
	$\chi = \arctan (g/d)$

If the pitching instant center O_r is above the wheel centre, the inertial movement is compensated for by pushing up the rear end when accelerating. If, however, the pitching instant center is below the wheel centre, the rear end is additionally pulled down when starting off. Unlike on the front axle, a high pitching instant center on the rear axle does not interfere with compression. The wheel swings backwards when avoiding a bump, so in the sense of yielding away from the elevation. Nevertheless, on racing cars the pitch compensation is not installed too high, because too high reaction forces cancel the compensating effect of a springy axle and the traction of the tires suffers (load, power oversteer). However, the camber reducing the contact patch must not become too large during compression. The desired degree of compensation depends on the power-to-weight ratio (engine power/vehicle mass) of the car and hardly exceeds 20%. As the power-to-weight ratio decreases, the need for camber compensation also decreases.

Kinematics of Wheel Travel

The maximum wheel travel in rally cars is about 300–500 mm, in circuit cars usually only 25–50 mm. This means that the influence of the wheel displacement on the position of the tyre is comparatively small in the latter. Nevertheless, the tendency of the wheel position change must produce the desired driving behaviour in order to approach the physical optimum.

Likewise, in general only tendency statements can be made, which course of geometric characteristic values is good and which is bad. This applies all the more to numerical values, of course. Nevertheless, some diagrams with exemplary interpretations follow below.

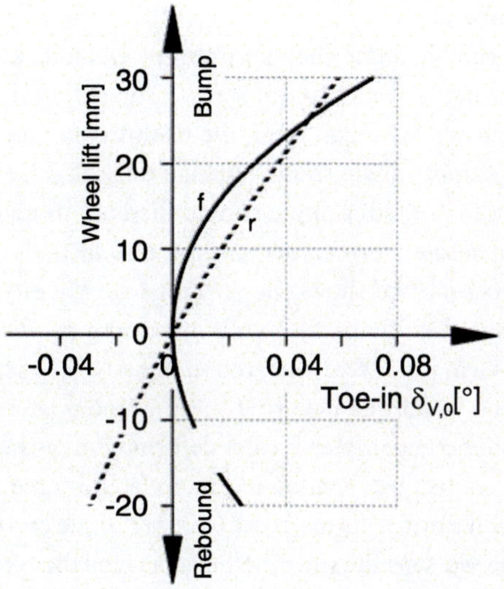

Fig. 2.34 Acceptable toe-in change during spring deflection. The values apply to racing tyres on asphalt and represent the progression for the front wheel (*f*) and the rear wheel (*r*)

Toe-in

For passenger cars, the following applies roughly: no toe-in change should occur on the rear axle during compression (outer wheel during roll) with maximum spring travel. When cornering, however, a toe-in change can be helpful in achieving understeering, i.e. dynamically stable handling: At 3000 N lateral tire force acting about 30 mm in the direction of travel behind the wheel contact point,[10] the toe-in change of the cornering outer wheel should be about 0.3° at the rear axle and 0° at the front axle. The rear axle is the more important in this context. An undesired self-steering of the rear axle is much more difficult for the driver to compensate than that of the front axle, which is directly influenced by the driver through the steering. The latter is especially true for racing cars, which reach high speeds and need a stable rear axle for smooth handling (racing drivers feel toe-in changes from 0.15° on the rear axle). An exception is a vehicle that becomes unstable when braking: In this case, a toe change in the direction of toe-in can provide relief when the suspension is released.

Strong toe-in changes during suspension also increase tyre wear. This is particularly significant in the case of long-distance vehicles, because it noticeably shortens the intervals between tyre changes. This in turn is an aspect that can be decisive for the race. In general, the toe-in angle should change by a maximum of between 0.009 and 0.017° per 10 mm of wheel travel. Figure 2.34 shows typical curves of toe-in changes.

[10] In Chap. 1 Tyres, 2.3 *Influence on driving behaviour* a general estimation under *side slip can be* found.

Camber

The camber angle changes during (unilateral and bilateral) deflection, rolling, steering and deformation under lateral forces. The permissible camber angle depends on the tyre make, the power-to-weight ratio, the use (driven/non-driven) of the tyres and the aerodynamics. The camber angle should change only slightly over the wheel's deflection travel, and if it does, the negative angle should increase in magnitude as the wheel compresses. This also compensates for camber change due to body roll caused by tire spring movement. The camber change should be smaller with wide tyres than with narrower ones. Likewise, the change should decrease with increasing power-to-weight ratio (kW/kg) of the vehicle.

When cornering, the rebounding wheel is the more heavily loaded outer cornering wheel, which can therefore transmit more lateral force through lateral slipping. A negative camber compensates for the deformation of the tyre sidewall and equalises the load on the tyre contact patch, allowing more lateral force to be built up. The deflecting wheel should remain normal to the road (camber angle = 0°). However, an excessively large camber angle worsens the situation because then the tyre contact patch is partially lifted off the road and the possible lateral force cannot be achieved by side slipping.

During body roll, the camber of the rebounding wheel should remain within the permissible limits for the selected tyre with less than 1° change per 1° roll angle of the wagon or approx. 25 mm wheel travel. The rebounding wheel should not change its camber.

For radial tyres, the camber change can tend to be smaller than for cross-ply tyres [9]. However, more (static) camber is set in the design position for radial tyres. Figure 2.35 shows a typical course of the camber change.

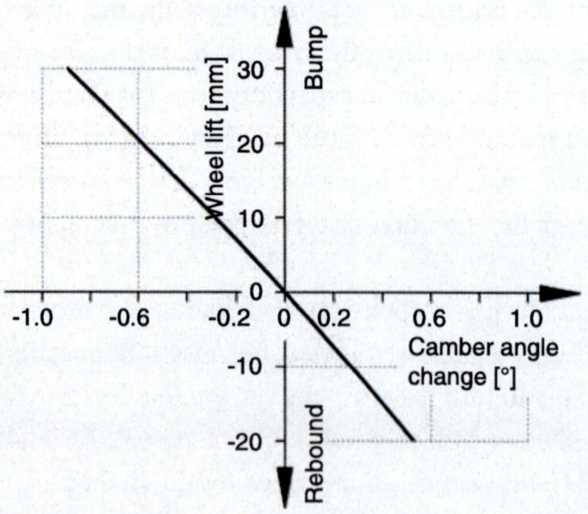

Fig. 2.35 Useful camber change when springing. The deviation from the set camber angle in the design position is plotted. The curve shown is useful for front and rear wheels. During compression, the camber increases towards the negative side, i.e. if it was −3° at wheel travel 0 mm, it becomes approx. –4° at wheel travel 30 mm. When cornering, the wheel that deflects is the inner wheel, which in the (theoretical) ideal case assumes the same inclination as the outer wheel – the camber angle can therefore approach 0° or even become positive

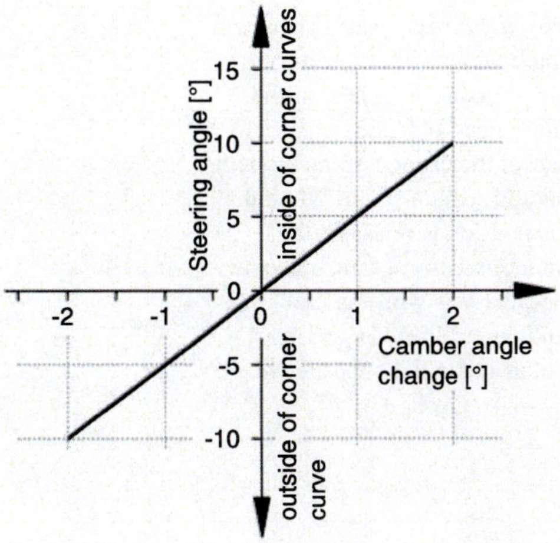

Fig. 2.36 Acceptable camber change during steering. The angle of kingpin inclination inevitably changes the camber during steering. For more details see Chap. 5 *Steering*

For vehicles designed for high lateral acceleration, the camber change during body roll should remain within the permissible camber values. In Fig. 2.36 an acceptable influence of steering on the camber angle can be seen. This influence of steering movement is the reason that the camber change should be greater on the rear axle than on the front axle. When cornering, the front wheels receive an additional camber increase due to the steering angle because of the inclined arrangement of the kingpin axis. However, the aim must be that the absolute camber changes over wheel travel or roll angle are the same at the front and rear so that the car does not change its handling characteristics.

In lap time simulations, it is found that a lateral force gain of the tires has a much more beneficial effect than a longitudinal force gain. For the same magnitude of force gain, the ratio of running time savings on an average circuit is 4 to 1 [19]. On classic circuits, the choice of camber on the front axle is thus easier: a large camber gain during roll, or at least a static negative camber that supports cornering, is more helpful than a camber angle of 0°, which is the ideal tire position during braking.

Track

The change in track width over the wheel stroke is interesting for the chassis design. If the track width changes during suspension, the tyre moves transversely to the direction of travel and thus causes a lateral force via side slip. This worsens the straight-line running, increases the rolling resistance and can influence the steering. In the extreme case of one-sided deflection, this causes a steering effect of this wheel. Ideally, therefore, the track width should remain constant during suspension. An acceptable course of the track width change is shown (Fig. 2.37).

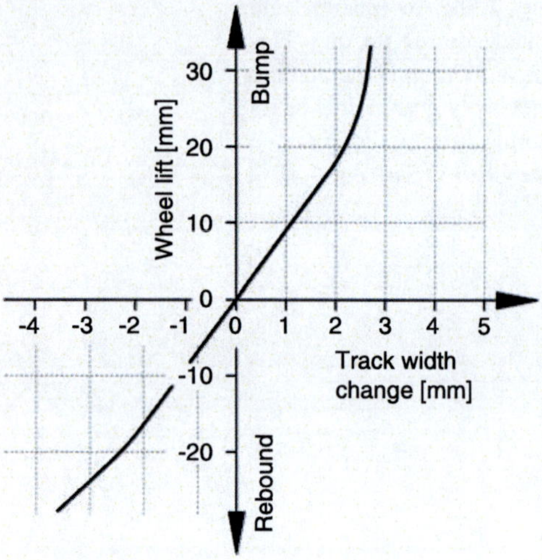

Fig. 2.37 Acceptable change in track width of a wheel during suspension. Especially at the more heavily loaded in-sprung wheel the change of track width should remain small. When the suspension is released, the situation is somewhat alleviated because this wheel is unloaded and contributes less to lateral control than the opposite wheel

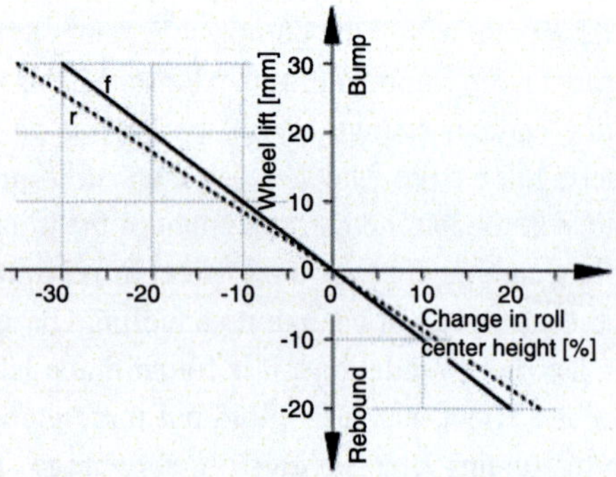

Fig. 2.38 Acceptable change in roll centre height for equilateral springing. Curve for the front axle (f) and the rear axle (r). The roll centers tend to move in height with the vehicle center of gravity with

Roll Centers

The position of the roll centers of both axles in the design position and during suspension is of interest in several ways, as the instant centers influence camber and track width changes as well as wheel load shifts and thus the self-steering behavior of a vehicle.

When designing a chassis, the front roll center – dictated by the track width change – is determined first, and then the rear.

The roll centre should sink with the centre of gravity during equilateral compression. This keeps the roll moment caused by the inertia force constant and the support effects of the wheel links small.

In order to achieve high directional stability (especially with laterally stiff tyres), the lateral instant centers should be far away from the wheel plane and the rolling centres should be close to or on the road surface. This is especially important for vehicles driving on bad roads or unpaved roads. In addition, the roll center should not move through the tire contact patch during suspension. This causes a sudden change in handling that is irritating to the driver. Figure 2.38 represents a useful course of the change in height of roll centres. Typical numerical values are provided (Table 2.3).

Table 2.3 Typical values for static roll centre heights h_{Ro}

Vehicle	Passenger Car[a]	Indy Car[b]	IMSA GTS[b]	Formula Ford[b]	Sports prototype
$h_{Ro,f}$, mm	30 to 100	15	−2.5	−26	15
$h_{Ro,r}$, mm	60 to 130	18	12.5	26.6	40
Indexes:	*f* front, *r* rear		Positive values above the roadway		

[a] Reimpell and Betzler [14]
[b] Crahan [20]

Roll Axis

The roll axis of passenger cars should rise slightly towards the rear. Then parts of the body damping can be used to damp the vehicle yaw motion [14]. This position, which was previously generally recommended to maintain stability (e.g [21]), only applies to vehicles with rigid axles at the rear and those with the engine at the front. In general, the roll axis of vehicles with independent front and rear suspension should be as parallel as possible to the main inertia axis in the longitudinal direction with the aim of achieving equal wheel load changes on the front and rear axles[11] (neutral handling) [15].

With equilateral suspension, the roll axis should sink like the centre of gravity. Pure lifting movements then do not affect the handling and the vehicle remains stable. A simple but solid approach is to start with a suspension design where the roll axis does not change during roll. With independent suspensions, no direct recommendation can be made beyond that. The decisive factor is how the wheel position changes during roll. This will be illustrated by means of a simple model with rigid axles (Fig. 2.39). This vehicle has a horizontal roll axis R_a which passes through the roll centres Ro_f and Ro_r of the front and rear axles. However, it is not the location of the roll axis that is critical to the vehicle's behavior, but rather the behavior of the wheel suspensions during roll. The vehicle has swing axles, which means that the instant centers between the body and the wheel suspension coincide with the centres of roll (= instantaneous poles between the body and the road). However, the rear suspension pivots about a pivot axis p_r, which includes an angle κ to the roll axis R_a. When cornering, the inertial force $m_{Bo}\, a_y$ causes the body to sway about the roll axis R_a by the angle φ. If the relative position of the axis of rotation p_r to the roll axis R_a remains the same – as in the model shown – the axis of rotation p_r must also move out to the side. This also causes the rear axle, which is always normal to the axis of rotation p_r, to rotate by the angle δ_r. In principle, the rear axle behaves like the axle of a skateboard (the front axle of which has an axis of rotation that is inclined backwards, i.e. performs an impact in the opposite direction). In the case shown, the arrangement has an understeering effect (the rear wheels turn in the same direction as the front wheels) and thus a stabilising effect. Transferring this to an independent rear suspension, it can be stated: If the wheel goes into a toe-in position when the suspension compresses, or into a toe-out position when the

[11] For details, see Sect. 2.2 Terms: Roll axis.

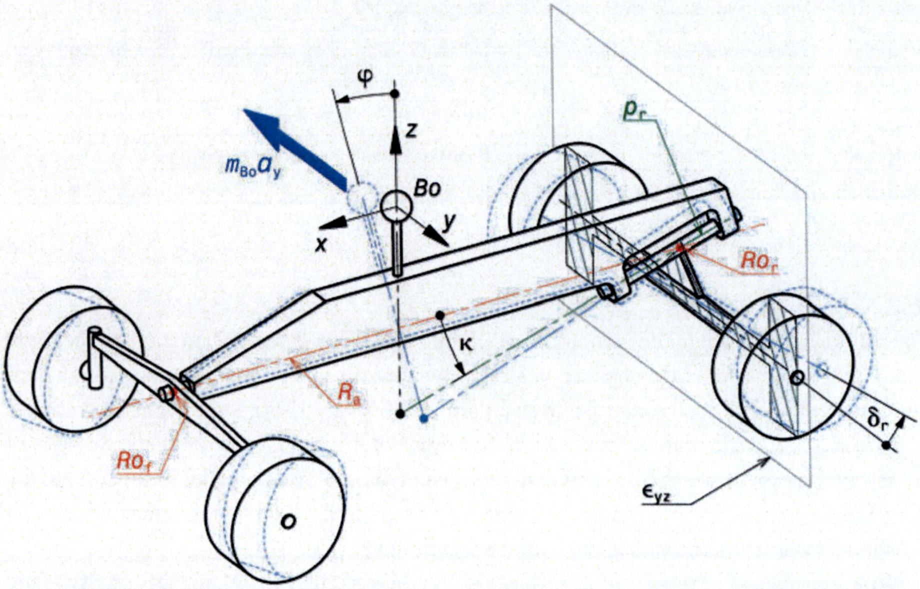

Fig. 2.39 The position of the roll axis. Simple vehicle model driving straight ahead. The situation for a left turn is shown dotted in light blue. *Bo* centre of gravity, Ro_f and Ro_r roll centre front and rear respectively, R_a roll axis, p_r rear instantaneous roll axis, φ roll angle, κ angle between R_a and p_r, δ_r rear steering angle, ε_{yz} transverse plane, m_{Bo} body mass (sprung mass), a_y lateral acceleration. With the orientations of the angles shown, $\delta_r = \kappa\varphi$ (all angles in radians) [7]. On the rear axle, the design of the roll centre Ro_r is also shown. It is the point of intersection of the transverse plane ε_{yz} through the wheel centre with the momentum pivot axis p_r of the rear wheel suspension

suspension decompresses, then this has a stabilizing effect. The opposite is true for the front axle. If the wheel moves into a toe-in position when the suspension is compressed, the steering angle increases and the track radius becomes smaller – the driving behaviour is therefore oversteering.

The higher the average speed of a vehicle and the more decisive the aerodynamic design therefore becomes (Formula 1, Indycar), the stiffer the springs and dampers are set. The change in position of the roll axis thus loses significance because the chassis hardly makes any spring movements anyway.

Anti Dive and Anti Squat

When braking, it is important that the vehicle does not dive too much on the front axle. In the worst case, it could touch down and become unstable. In addition, the pitch angle is aerodynamically important. Acceptable curves of the degrees of compensation are shown in (Fig. 2.40).

Caster

A caster stabilizes the rolling wheel. On the outer tyre of the turn, a caster angle creates negative camber, which has a beneficial effect on the lateral cornering potential. How such a course of the caster looks over the wheel stroke is shown in (Fig. 2.41). The caster should

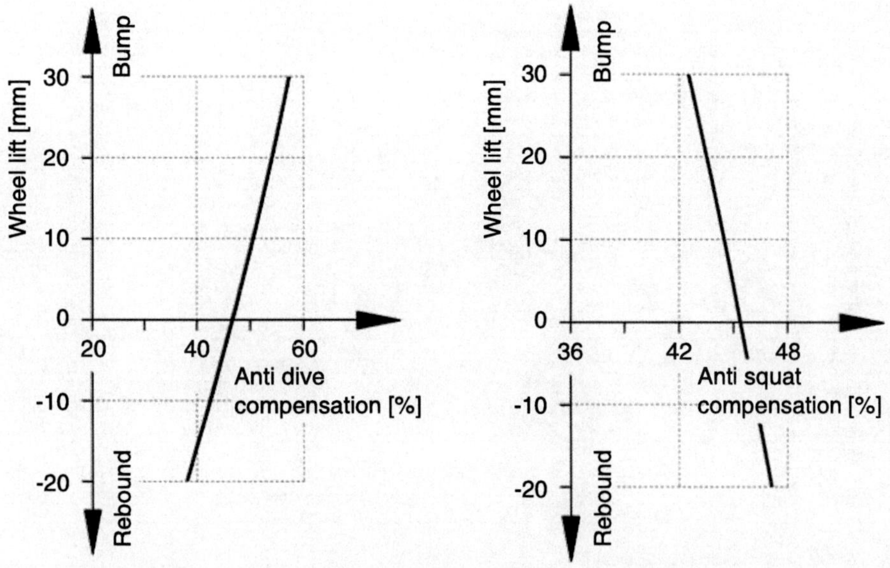

Fig. 2.40 Acceptable curve for braking and starting pitch compensation. As braking deceleration increases, the front axle load increases and the pitch compensation should counteract this and increase on compression. When sinking in due to the drive force, the starting pitch compensation can become smaller

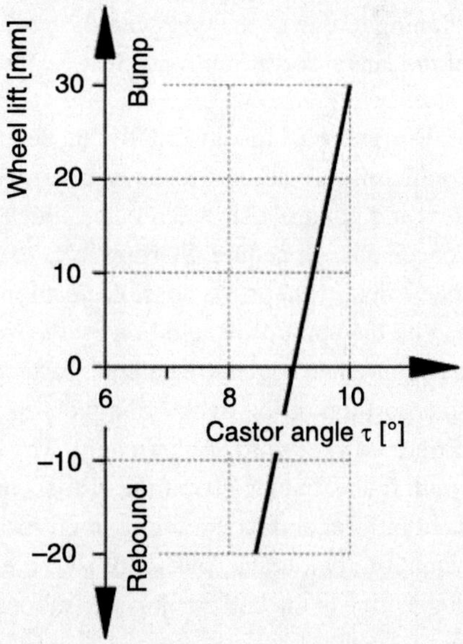

Fig. 2.41 Acceptable course for caster change. During compression, the caster angle increases – a larger negative camber angle is thus generated at the outer wheel when steering

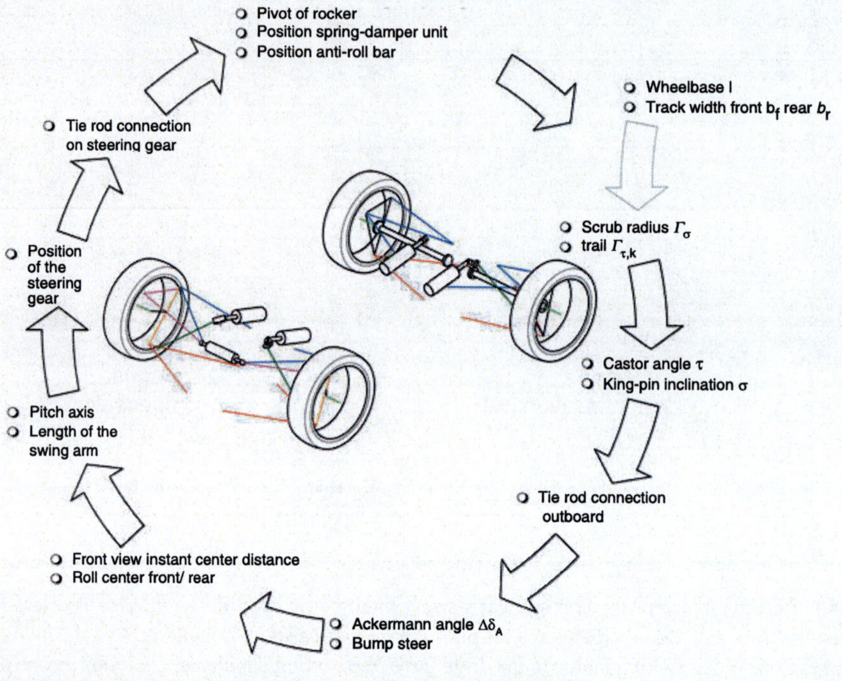

Fig. 2.42 Iterative procedure for defining the chassis kinematics

not become too large, because this would mask the steering feel that comes from the self-aligning torque of the tyres and the professional driver would lose an important indication of the lateral force limit (see Sect. 1.2.3 *Self-aligning torque*).

Designing is fundamentally an iterative process; if only because many – sometimes conflicting – requirements have to be reconciled. It is no different when designing a landing gear and in particular when determining the kinematics. Figure 2.42 clearly illustrates the conceivable procedure. Starting from the values of wheelbase and track widths specified in the vehicle concept,[12] a possible position of the steering axle in space is assumed. The basis can be the point of penetration by the roadway: Steering roll radius and caster dictate this position. Then angles are entered: caster and kingpin angles. This is followed by the wheel-side position of the track rod's pick-up point and an initial estimate of the steering angle difference (keyword: Ackermann). The positions of the roll and pitch instant centers also result in the first lengths of the virtual control arms or swing arms. Once the position of the steering gear and its mounting point for the tierod are fixed, spring/damper units and their actuation (bell crank, push/pull rod) are still missing, and finally the stabilizers and their connection to the bell cranks or wishbones are placed. This now allows investigations to be

[12]See Racing Car Technology Manual Vol. 2 *Complete Vehicle*, Chap. 2 *Concept*.

carried out into the mobility (steering angle, spring travel, wheel travel) and its effects (track difference angle, suspension steering, changes in camber, caster, track widths, wheelbase). There is also a first position of the roll axis. If one is dissatisfied with an effect (which should be the case(!) – an intensive consideration of conceivable alternatives is also the basis of a good design), one moves the corresponding instant center or changes the position of the most effective joint and the described procedure goes into the next round with the goal of getting closer to the ideal solution. Until final positions of all points are found, numerous iteration loops with many compromises will be necessary.

In summary, (Table 2.4) lists the design priorities for some vehicles. The objectives are not the same for all types due to different requirements. Compromises have to be made in the design and if certain important objectives are pursued, other criteria must inevitably be neglected or "sacrificed". With double wishbone axles, for example, it is not possible to optimise the camber behaviour of the wheels when rolling and when springing at the same time. For this reason alone, an analysis before the design is important as to which criteria are decisive and which are subordinate for a particular vehicle.

2.3 Parts of the Suspension

Even though there are sometimes very different requirements for vehicle wheel suspensions, certain components can be found in all designs:

- Axles or wheel-receiving elements (pivot bearings, steering knuckles, axle journals, wheel carriers)
- Connecting links (rods)
- Joints
- Springs and dampers
- Stabilizers.

Springs and dampers as well as stabilizers have their own sections dedicated to them, as well as to the corresponding parts of double wishbone axles, the most important axles for racing cars.

The wheel suspension is essentially intended to ensure a degree of freedom[13] between the wheel and the wagon body. For the basic design, it represents a kinematic transmission chain that connects the web (car body) to the wheel carrier via links. Joints ensure mobility when connecting the links. The resulting degree of freedom of a mechanism depends on the combination of links present. First, let us consider the joints. Of the conceivable types, three are used on racing vehicles, (Fig. 2.43): The ball joint, the pivot joint, and the swivel joint. Depending on their design, joints allow certain rotations and/or translations – the *degrees* of *freedom.*

[13] See attachment.

Table 2.4 Main aspects of the chassis design of some vehicles, partly according to [18, 20]

Vehicle	Label	Targets	Measures
Passenger car	Front engine, front-wheel drive, different payloads, high Centre of gravity	Stable, understeering handling, high traction of the front axle	Very low camber changes of the sprung wheel during roll, high brake pitch compensation
		High lateral cornering forces at the rear axle	Stiff stabilizer at the front axle, masses can be arranged at the rear as well
		High comfort	Large suspension travel, progressive spring rates, moderate damping, large vehicle moments of inertia about longitudinal and transverse axes, large wheels, large pitch compensation (by 30%)
Formula Ford	No aerodynamic downforce aids, low engine power, narrow tyres, limited slip differential prohibited	High braking power	Low camber change of the front wheels during compression (ideal: camber = 0°), avoidance of positive camber angle of the rear wheels during decompression
		High lateral acceleration (cornering)	Increase of the negative camber of the rebounding rear wheel during body roll
		Avoid spinning of the inner drive wheel when cornering.	Large rear suspension travel, low rear wheel load transfer, avoidance of negative camber of the rebounding wheel during rolling (ideal: camber = 0°)
		Slightly oversteering handling, nevertheless stable	Axle load higher at the rear, roll axle moves up and down in parallel during bounce
			Low acceleration penalty
Indy car	High, relatively constant speed (oval course), banked corners, almost flat track, powerful engines	High aerodynamic downforce, therefore low dynamic ground distance changes; almost no pitching motion	Stiff body supporting springs, highly progressive spring rates, low camber change on compression, small change of track width,

(continued)

Table 2.4 (continued)

Vehicle	Label	Targets	Measures
			dynamic roll centers close to the roadway
		Almost no rolling movement	Stiff torsion stabilizers, stiff dampers
		Light understeer in turnd	Stronger wheel load shifting at the front axle due to higher rolling stiffness at the front and large caster angle to compensate this behaviour in slow corners
Formula 1 Can am	Large speed differences, level road with height differences Wide tires with low camber tolerance, powerful engines	High aerodynamic downforce, therefore low dynamic ground distance changes	Highly progressive spring rates, e.g. due to three-spring suspensions.
		High traction on the drive axle	Small camber changes during heave, very wide tires in the rear
		No camber change at the front axle	Long front wishbones
		High braking capacity	
IMSA GTS	Strong engines, wide tyres, no front wing, high dead weight, relatively high Centre of gravity	High aerodynamic downforce at the rear axle, no pitching movements (constant angle of attack rear wing)	Stiffer springs or stronger progressivity on the rear axle so that the downforce from the rear wing, which increases with speed, does not cause pitching. When springing, the roll axis moves like the main inertia axis. Relatively large pitch compensation (25–30%)
		High acceleration capability	Higher axle load at the rear, hardly any camber change at the rear axle during heave guarantees high Tyre adhesion
		Compensation of the axle load shift	High downforce on the rear axle
Rally car, off-road use	Impassable roadway, different nature of the subsoil, large differences in altitude	High ground clearance	Long suspension travel, effective damping
		Maintaining the greatest possible frictional	Large suspension travel, profiled tyres with relatively low inflation

(continued)

Table 2.4 (continued)

Vehicle	Label	Targets	Measures
	Relatively high Centre of gravity	connection, especially of the drive wheels	pressure, small unsprung masses
		Stable flight behaviour when the vehicle is in the air	Balanced mass distribution in the longitudinal direction
		No rebound on landing	Large diagonal springing on the rear axle

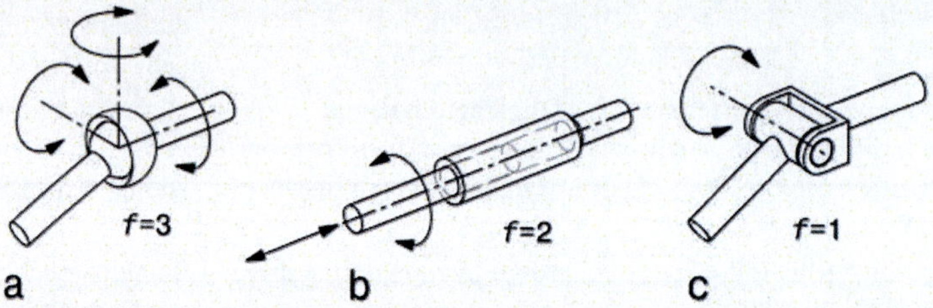

Fig. 2.43 Types of joints. (**a**) Spherical joint, (**b**) Rotary joint, cylindrical joint, (**c**) Revolute joint. In addition, the degrees of freedom *f* of the joints are given

In passenger cars, swivel joints are usually designed in such a way that, in addition to the pure rotary movement about an axis, small rotations are also possible at right angles to it, by ensuring a flexible distance between the bush journal and the bush shell (*conical motion*).

Links provide a connection between the joints. Depending on the combination of individual types of joints at the ends of the links, links with different degrees of freedom are created. Essential combinations are summarised in (Fig. 2.44). The rod link (a) has a total degree of freedom of 6 due to the two ball joints ($f = 3$). However, one degree of freedom remains without external action[14] namely the self-rotation of the rod, so that the rod link effectively has degree of freedom 5. Thus, such a link restricts the possible 6 degrees of freedom of a body connected to it by 1. A triangular link (b), with one pivot and one ball joint, provides degree of freedom 4, or 2 must be subtracted for it when calculating the total degree of freedom of a mechanism. The two axes of rotation of a semi-trailing arm (c) do not have to be parallel to each other. Each wheel-guiding strut of a McPherson front suspension is a swivel link (d).

[14]However, this is not the case with a CAD system. Here, this degree of freedom must also be restricted by a "holder". Problems can also arise in reality if the rod deviates from the stretched shape or a cable is attached to it.

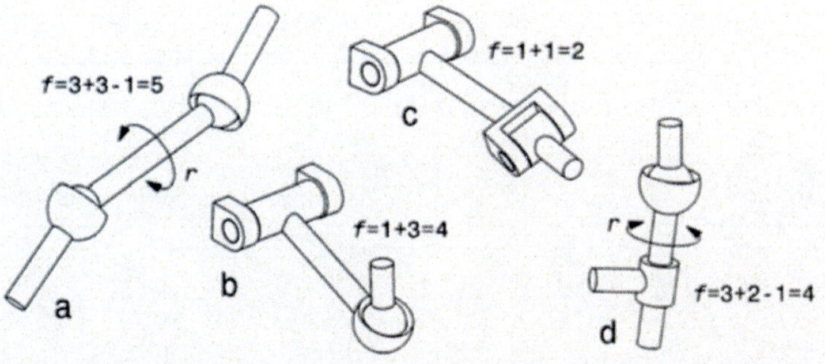

Fig. 2.44 **Link types.** (**a**) Straight link, (**b**) A-arm, (**c**) H-arm (torsion link), (**d**) Rotary slider link (trunnion). *r* Intrinsic rotation

When the above-mentioned parts are assembled to form a wheel suspension, one or more wheel carriers (couplers) are connected to the body of the vehicle (frame) via links. The wheel carrier and link each have 6 degrees of freedom when unattached. Each joint i with the degree of freedom f_i reduces the total degree of freedom of the kinematic chain by $(6-f_i)$. Intrinsic rotations r of links also do not contribute to the mobility of the mechanism and must therefore be subtracted from the total degree of freedom. The total degree of freedom F *of* a kinematic chain is thus calculated as:

$$\begin{aligned} F &= 6(k+l) - \sum\nolimits_1^g (6-f_i) - r \\ &= 6(k+l-g) - r + \sum\nolimits_1^g f_i \end{aligned} \tag{2.16}$$

F	Degree of freedom of a mechanism, –
k, l	Number of wheel carriers or links, –
g	Number of joints, –
f_i	Degree of freedom of the joint i, –
r	Number of self-rotations, –

As an example, the total degree of freedom of the wheel suspensions shown in (Fig. 2.45) shall be determined. In the McPherson suspension (Fig. 2.45a), a wheel carrier k_1 is connected to the frame (frame s) via a suspension strut (swivel link l_1), a triangular link (l_2) and a track rod (rod link l_3). These three control arms are coupled together via a total of 6 joints. Two inherent rotations may not be considered. The balance of degrees of freedom, (2.16), turns out like this:

$$F = 6(1+3-6) - 2 + (3+2+3+3+3+1) = 1$$

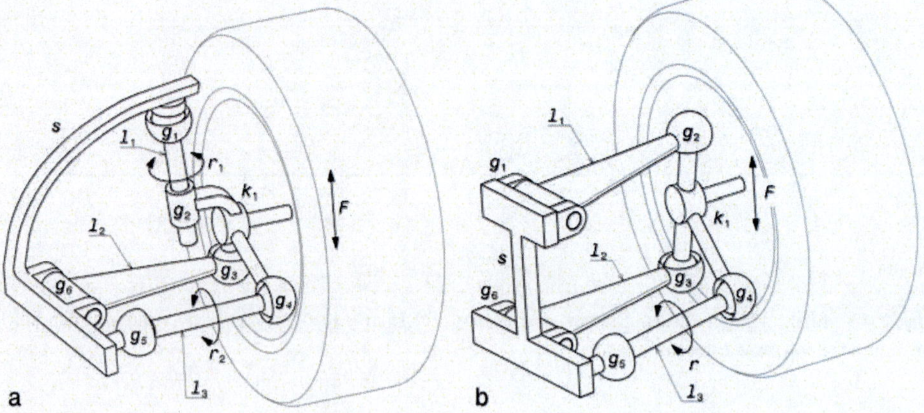

Fig. 2.45 Composition of kinematic chains of links and joints. *s* Frame (fixed link), *k* Wheel carrier (knuckle), *l* link, *g* Joints, *r* Intrinsic rotation (independent rotation), *F* Overall degree of freedom. (**a**) McPherson suspension, (**b**) Double wishbone suspension

Table 2.5 Steering arm combinations used

Combination	Resulting degree of freedom F for wheel carrier
1 trailing arm (on swivel joint)[a]	$F = 6\,(1 + 1\text{–}2) - 0 + (1 + 0) = 1$
2 wishbones + 1 tie rod	$F = 6\,(1 + 3\text{–}2 \cdot 2\text{–}2) - 1 + (2 \cdot 1 + 2 \cdot 3 + 2 \cdot 3) = 1$
1 wishbone + 1 track rod + 1 swivel arm	$F = 6\,(1 + 3\text{–}2 - 2 - 2) - 2 + (1 + 3 + 2 \cdot 3 + 3 + 2) = 1$
1 wishbone + 3 rod links	$F = 6\,(1 + 4\text{–}2 - 3 \cdot 2) - 3 + (1 + 3 + 3 \cdot 2 \cdot 3) = 1$
1 trunnion (rotary push link) + 3 rods	$F = 6\,(1 + 4\text{–}2 - 3 \cdot 2) - 4 + (2 + 3 + 3 \cdot 2 \cdot 3) = 1$
1 semi-trailing arm +1 rod	$F = 6\,(1 + 2\text{–}2 - 2) - 1 + (1 + 1 + 2 \cdot 3) = 1$
5 rods	$F = 6\,(1 + 5\text{–}5 \cdot 2) - 5 + (5 \cdot 2 \cdot 3) = 1$

[a]Wheel carrier connected to trailing arm with fixed joint ($f = 0$)

This leaves exactly the one degree of freedom required for single wheel suspensions. For the double wishbone arrangement (Fig. 2.45b) the calculation, (2.16), looks like this:

$$F = 6(1 + 3 - 6) - 1 + (1 + 3 + 3 + 3 + 3 + 1) = 1$$

The remaining degree of freedom is set in equilibrium by the body spring when the wheel suspensions are designed. It can also be seen that both axles can be designed as steered axles by giving the frame-side joint g_5 a degree of freedom: A translation in the transverse direction of the vehicle leads to a rotation of the wheel carrier.

Finally, it should be noted that rigid axles must have the degree of freedom $F = 2$: An (approximately) vertical translation and a rotation around the longitudinal axis of the vehicle.

(Table 2.5) gives an overview of practically implemented control arm combinations. A non-steered rear axle with the combination of wishbone plus track rod, which is common in racing car construction, offers numerous variation possibilities to attach the wheel carrier to

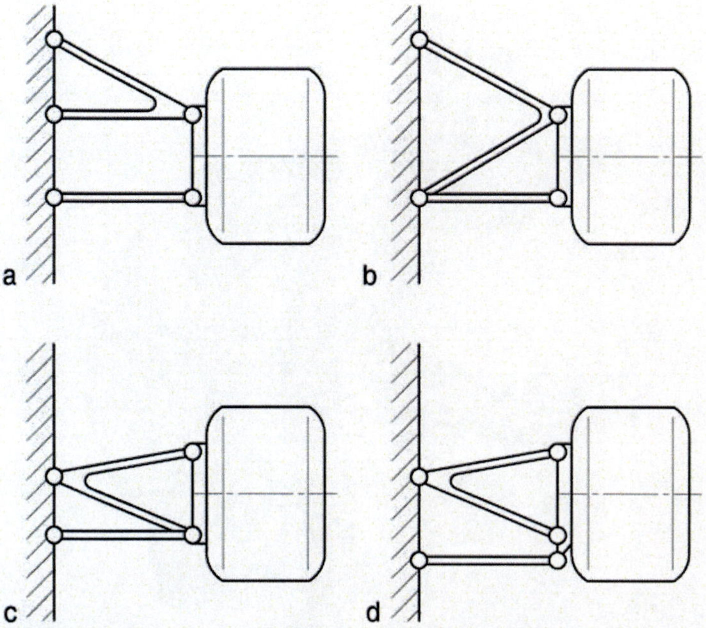

Fig. 2.46 Wishbone plus track rod combinations on rear axles (top view). One of 2 suspension "planes" is shown. In the other plane, a control arm provides the connection between the body and the wheel carrier. (**a**, **b**) Conventional wishbones. (**c**, **d**) Reversed wishbones. It can be seen that wishbones can also be seen as a triangular arrangement of two rod links (A arm)

the car with one remaining degree of freedom. Figure 2.46 shows some possibilities. Variations (a) and (b) can also be used on steering axles. Essentially, the suspensions in the Fig. consist of links in two planes, with only the one with more bindings shown. However, the track rod can also be arranged between these planes.

2.3.1 Wheel-Carrying Elements

Depending on whether the wheel is driven or not, the wheel is connected to the drive shaft via a wheel flange or mounted on an axle journal. In any case, a component must accommodate the stationary parts of the wheel bearings and transmit the forces to the frame via connecting links. Examples of such parts are shown in (Figs. 2.47 and 2.48) a wheel carrier and an axle journal.

2.3.2 Linkages

The wheel-receiving parts must be movably connected to the frame. This task is performed by the connecting links. Two typical representatives of this category of parts are shown in the (Figs. 2.49 and 2.50) with a control arm and a trailing arm.

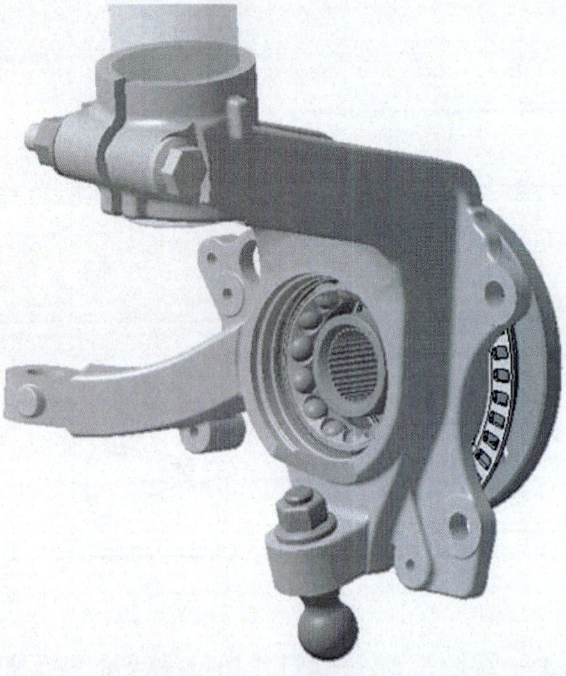

Fig. 2.47 Swivel knuckle (wheel carrier) of a front wheel-guiding suspension strut axle. On wheel-guiding suspension strut axles, the wheel carrier holds the lower end of the suspension strut in a clamp and connects to the transverse control arm via the guide joint. On the front axle, the wheel carrier rotates around the guide joint and the suspension strut mounting point on the body side during steering. For this reason, it is also called a swivel knuckle in this application. The knuckle receives the outer rings of the wheel bearings and contains the steering lever to which the track rod is connected. It also provides two eyes for mounting the brake caliper

Fig. 2.48 Axle stub (cf. Fig. 2.73). An axle journal accommodates the inner rings of the wheel bearings. It can only be used for non-driven axles. This design is bolted to the control arm, e.g. of a twist-beam axle, with four bolts

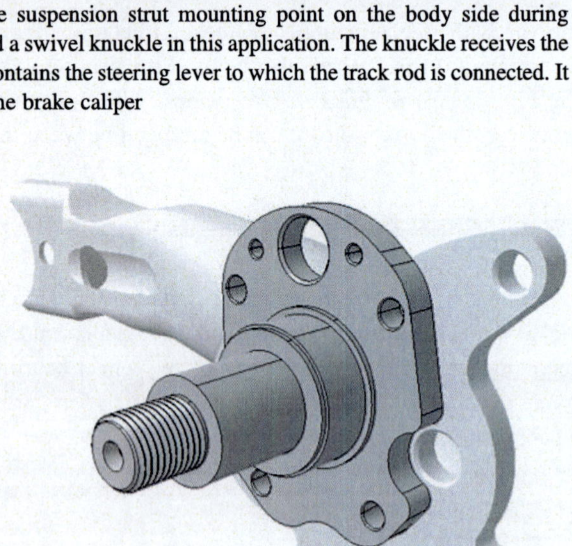

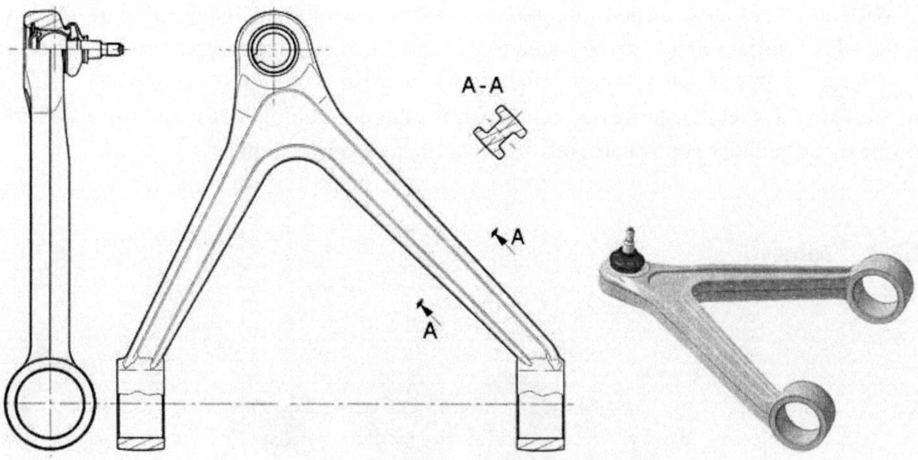

Fig. 2.49 **A-arm** (triangular link) with built-in guide joint. Material: Aluminium alloy EN AC-AlSi7Mg0,3 (was GK AlSi 7 Mg) ($R_m = 260$ N/mm^2, $R_e = 220$ N/mm^2). The triangular control arm is manufactured using gravity die casting. The wheel joint (outboard) is integrated into the control arm. On the car side (inboard), the control arm is held in rubber bushings

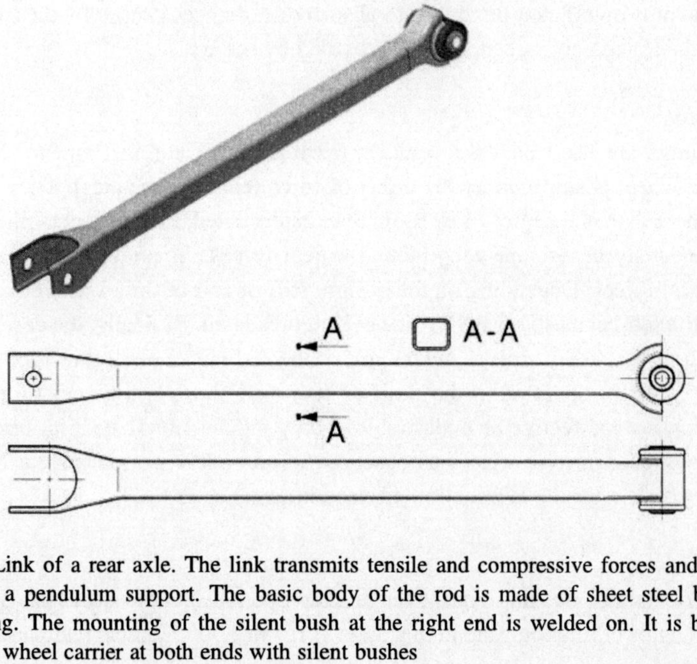

Fig. 2.50 Link of a rear axle. The link transmits tensile and compressive forces and is therefore designed as a pendulum support. The basic body of the rod is made of sheet steel by means of hydroforming. The mounting of the silent bush at the right end is welded on. It is bolted to the subframe or wheel carrier at both ends with silent bushes

With an independent suspension, the control arms must restrict five degrees of freedom of the wheel carrier relative to the frame so that the one desired movement option remains. In extreme cases, this means that five individual sway bars (push/pull forces only) perform this function. Generally, however, control arms take on multiple constraints at once. For example, a triangular control arm restricts two degrees of freedom.

2.3.3 Joints

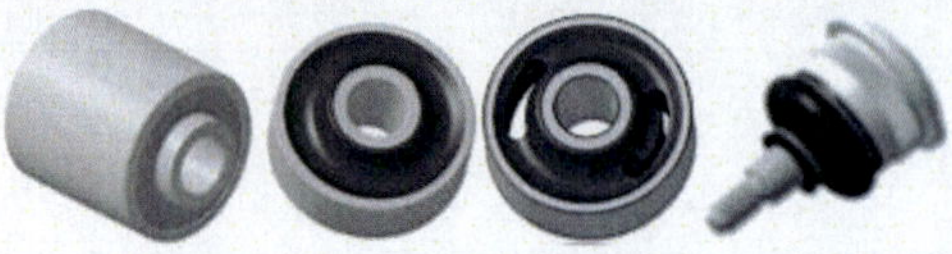

Joints are required between the rigid control arms, wheel carriers and the body of the vehicle so that the wheel can move in the desired manner. In production vehicles, elastic bushings have become established for reasons of comfort and manufacturing (tolerances!). In racing vehicles for paved tracks, where steering precision and exact wheel directional control are of prime importance, the exact opposite is the case. Here, only play-free joints are used. The only significant flexibility to absorb disturbances caused by the road surface and (unavoidable) chassis inaccuracies is provided by the tyres.

Elastic *Mount*

Elastic bushings are used on road vehicles for reasons of comfort and to deliberately influence the wheel position under the effect of force (elastokinematics). They also allow joints with more than one degree of freedom to be represented, such as for stabilizer mounts that simultaneously absorb the longitudinal wheel forces. These bushings can support radial and axial forces. Depending on the design, stiff or soft bushings as well as symmetrical and asymmetrical elasticity behavior can be represented. Basically, the design consists of two metal parts, usually sleeves, which are connected via an elastomer part.

In racing vehicles operated at the limit of tyre side slip, elastokinematic changes in wheel position are ineffective or undesirable if they reduce lateral steering force [17]. In addition, a reduction in tyre wear was observed in an endurance touring car based on a production vehicle after the elimination of elastokinematics [22].

Silentbloc

The silentbloc (silent bushing) has been a common design for decades, (Fig. 2.51). Permissible angles of rotation β are in the range $\pm$ 15° to $\pm$30°. Conical (cardanic) rotations κ are possible from about $\pm$1° to $\pm$6°. Axial displacements are achieved up to $\pm$3 mm. Radially, the bushings yield approximately up to $\pm$0.5 mm. The permissible surface pressure depends on the ratio of the length to the wall thickness of the rubber and increases

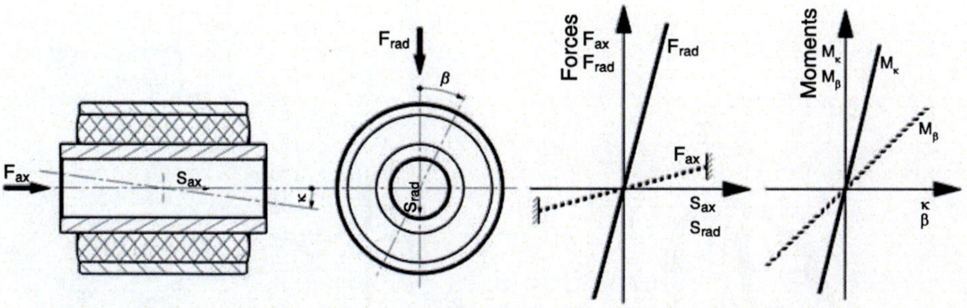

Fig. 2.51 Silent bush or silent bloc. F_{ax} Axial force, F_{rad} Radial force, s_{ax}, s_{rad} Displacement travels. In this design, the elastomer part is pressed in between the metal sleeves under great deformation. The bushing is therefore relatively stiff and the spring characteristics for displacement and torsion are steep

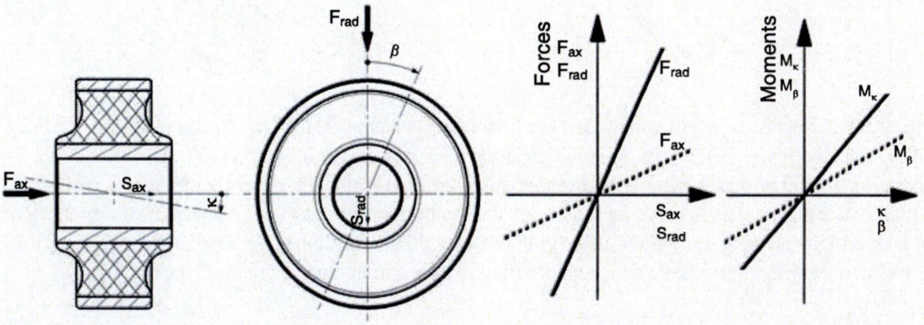

Fig. 2.52 Rubber bushing. Designations as Fig. 2.51. In this design, the elastomer part is vulcanised between the metal sleeves. Due to the lack of preload, the bushing is much more elastic than that of (Fig. 2.51) and only smaller deformations are allowed compared to the latter

somewhat with it. The bushing design can be based on the static forces. If the surface pressure remains below 1–3 N/mm^2, the bush can also withstand load peaks caused by driving, braking and cornering [10].

Rubber Mount

In contrast to the silent bush, the elastomer part of the rubber bushing (ultra bush) is not inserted under great deformation, but vulcanised in (Fig. 2.52). This design therefore has a more elastic effect.

Elastic bushings can also be deliberately designed asymmetrically and thus, for example, cause different bushing reactions during braking or driving, as is required for longitudinal springing (Fig. 2.53). Longitudinal springing is used in production vehicles to reduce the drone caused by steel-belted tyres (harshness).

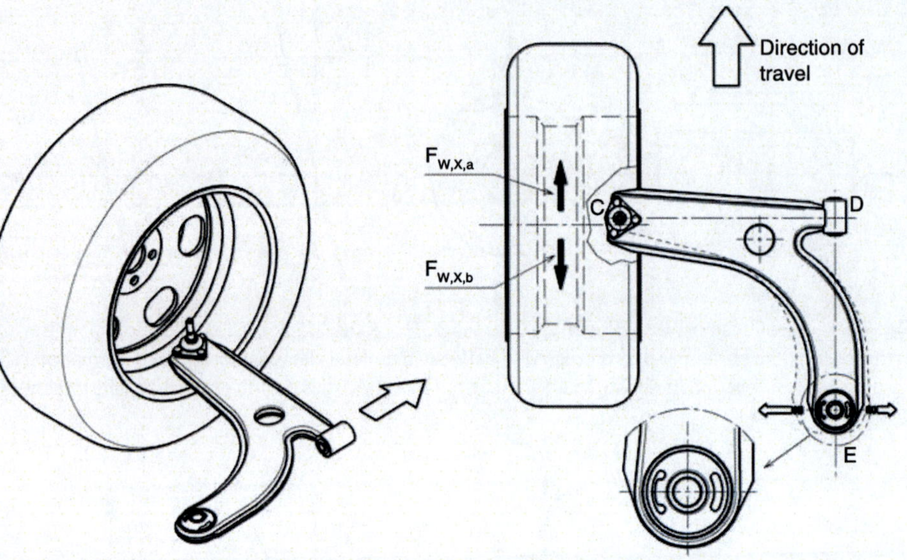

Fig. 2.53 Effect of an asymmetrical rubber bushing (elastokinematics). Shown is a left driven front wheel with the lower sickle arm. This has the following bushing types (top view): A wheel joint *C*, a silent bushing D and an asymmetrical rubber bushing *E*. Bushing *E* ensures longitudinal springing, which reacts more softly to drive forces $F_{W,X,a}$ than to braking forces $F_{W,X,b}$. The control arm thereby rotates around bushing *D*, which absorbs the lateral tire force and must be stiff in this direction. Unwanted steering movements are prevented by an appropriate arrangement of the track rod

Zero-Clearance Joints

Where an exact transmission of movements is required, joints are used which do not allow any play even under load. In the chassis area, where spatial movements result from superimposed movements, ball joints must be used. A classic application is the transmission of the steering movement to the wheels. Wheel joints for this purpose are found on many vehicles (Figs. 2.54 and 2.55). Wheel joints also serve as supporting and guiding joints, an example is shown in Fig. 2.74. In addition, spherical plain joints are often installed on racing vehicles, Fig. 2.56.

Spherical plain jointss are also available in a bolt-on mount. These so-called rod ends are used at the ends of tension and compression struts as well as tie rods and wishbone ends.

More detailed information on spherical plain joints can be found in (Sect. 4.2.2) Spherical *plain joints* on components of double wishbone axles.

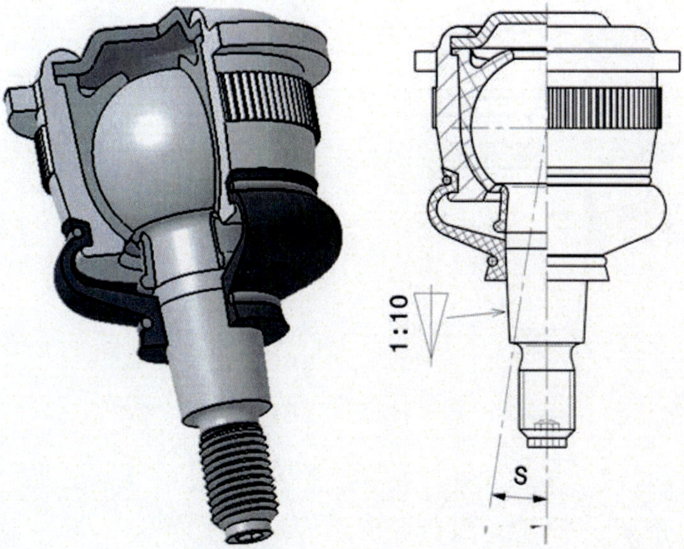

Fig. 2.54 Wheel joint for press-fitting. Connection cone 1:10 (DIN 71831). *S* max. Swing angle. The housing, which holds the joint ball, is closed with a sheet metal cap. This sheet metal cap is rolled in after assembly. The actual joint area is protected from dirt and moisture by an elastomer bellows. The max. Extraction force of the ball stud is 14 kN

2.3.4 Wheel Bearing

Function

The wheel bearings ensure the rotatable connection between the wheel hub and the wheel carrier (steering knuckle, pivot bearing). They must therefore also support all

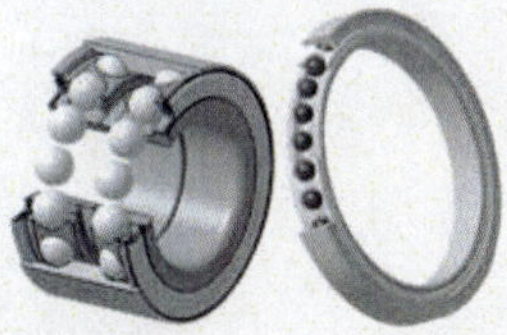

forces transmitted from the tyre to the frame. In (Fig. 2.57) a typical front wheel bearing of a racing vehicle is shown.

Requirements

The following requirements are placed on wheel bearings in the motorsport sector with decreasing importance [23]:

- Low weight: Low unsprung mass, low mass moment of inertia.

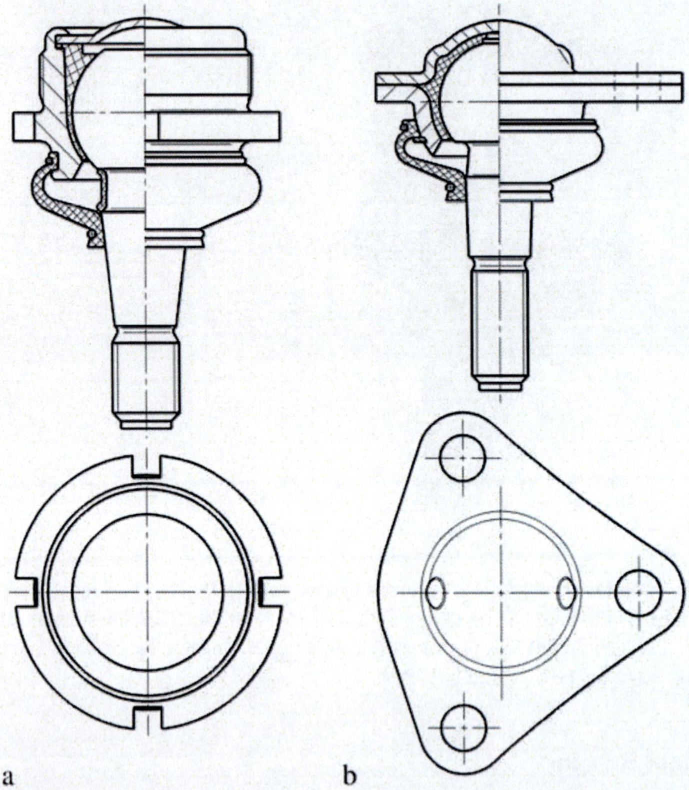

a b

Fig. 2.55 Wheel joints. These joints basically work like the one from (Fig. 2.54) only their construction and attachment are different. (**a**) Joint for screwing in, (**b**) Split sheet metal housing with flange for screwing on

Fig. 2.56 Rod ends (*Rose Joint*). The rod ends have spherical plain joints and can be connected to components via internal or external threads. These joints transmit radial and axial forces without play. Depending on the size and design, the permissible swivel angles are between 6 and 15°

- High stiffness especially camber stiffness, no axial play: "zero play" at the brake, constant brake pedal position.
- Low frictional torque: Low power loss.

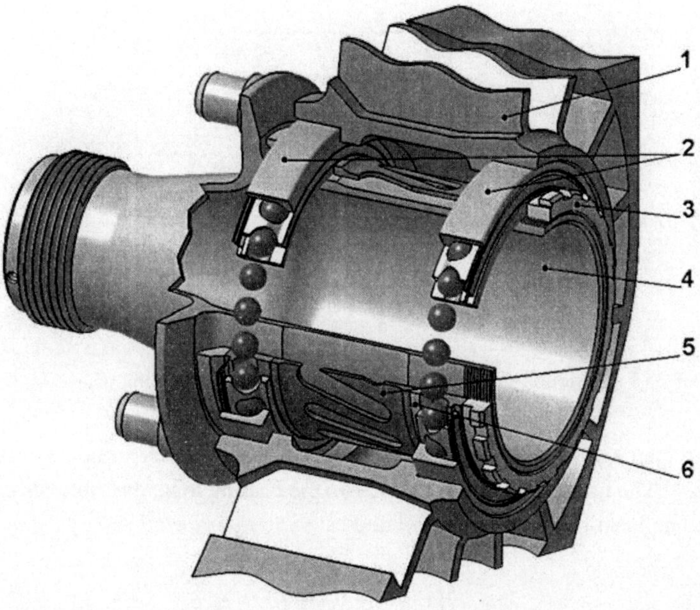

Fig. 2.57 **Wheel** bearing of a non-driven front wheel (partially cut open). **1** Wheel carrier (*upright*), **2** Two single bearings as wheel bearing, **3** Retaining nut, **4** Wheel hub (hollow), **5** Spacer, **6** Spacer ring for fine adjustment of the bearing distance

- Smaller installation space: More free space for brake and especially its ventilation.
- High load capacity: Safety against incalculable loads such as curbs, crashes, etc. Designed for cornering and braking.
- Highest reliability: Dimensioning not according to load rating.
- Adjusted service life: Example Formula 1 approx. 2000 km.

High rigidity in the wheel bearing area makes tyre wear calculable on the one hand and prevents play on the brake disc on the other. The latter leads to a reproducible brake response, which makes the driver's job easier. The wheel bearings are designed exclusively for cornering and braking. In production vehicles, the load plays a significant role. Although racing cars have aerodynamic downforce at high speeds, the loads caused by lateral acceleration and braking are many times higher. The service life target for Formula 1 is around 2000 km, for sports cars in racing 35,000 km (100,000 km in road use). Incidentally, this clearly illustrates the difference in load compared to an everyday vehicle. If a Formula 1 front wheel bearing were used in a mid-size car, it would be able to withstand 200,000 km of everyday stresses.

Friction

Friction in the bearing is caused by rolling resistance between the rolling elements and raceways, by partial sliding of the rolling elements on the raceways, by sliding of the cage

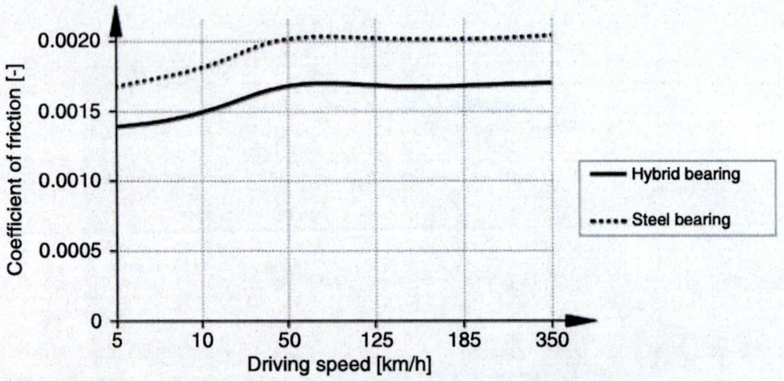

Fig. 2.58 Course of the friction values of a Formula 1 front wheel bearing [23]

on the rolling elements, by the resistance of the lubricant and by grinding of the seal in sealed bearings. The frictional torque depends on the bearing load, the lubrication condition and the speed and can be roughly determined as:

$$M_{fr} = \mu_L \cdot F \cdot d_L/2 \tag{2.17}$$

with:	M_{fr}	Bearing frictional torque, Nm
	μ_{L}	Coefficient of friction bearing, –
	F	Resulting bearing load, N. $F = \sqrt{F_r^2 + F_a^2}$. F_r, F_a (see Fig. 2.65)
	d_{L}	Bore diameter of the inner ring, m

A comparable larger bearing therefore has a larger frictional torque.

Hybrid bearings have lower friction than conventional steel bearings over the entire speed range, (Fig. 2.58).

Designs

Table 2.6 gives an overview of single bearings that can be used in pairs as wheel bearings.

Bearings that can be dismantled have the advantage that the inner ring and outer ring can be fitted separately, thus facilitating mounting.

Operating Temperatures

Steel ball bearings and tapered roller bearings can be used at operating temperatures up to approx. 120 °C (short-term up to 270 °C), cylindrical roller bearings up to approx. 150 °C (short-term up to 300 °C). At higher temperatures, the bearing material must be stabilized, which is associated with a decrease in hardness and thus with a reduction in service life, (Table 2.7). A minimum temperature of −40 °C is required for road vehicles. This value can also be used for rally vehicles.

Table 2.6 Bearing types for wheel bearing arrangements

Design		Advantages	Disadvantages	$\Delta\beta$	μ_L [×10−3]
Tapered roller bearing		+ dismountable + high axial and radial load carrying capacity due to linear contact of the rolling elements	– friction – Axial forces can only be transmitted on one side (installation in pairs required) – Bearing clearance must be adjusted	4′	1.8…2.5
Spindle bearing (shoulder ball bearing)	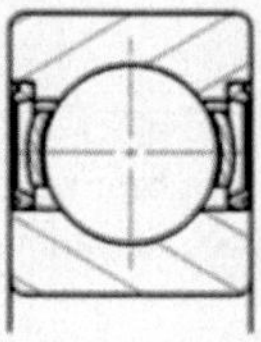	+ dismountable + high guiding accuracy + suitability for high speeds	– relatively low load carrying capacity due to one-sided osculation on the outer ring – axial forces can only be transmitted on one side (installation in pairs required)	10′	1.5
Angular contact ball bearing	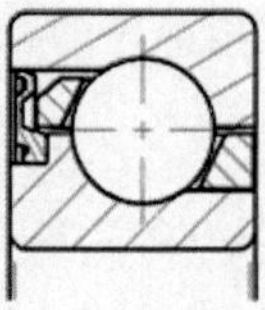	+ absorbs radial forces + can support higher axial forces than deep groove ball bearings + suitability for high speeds	– axial forces can only be transmitted on one side (installation in pairs required) – sensitive to shaft misalignment	10′	2…2.5
Deep groove ball bearing	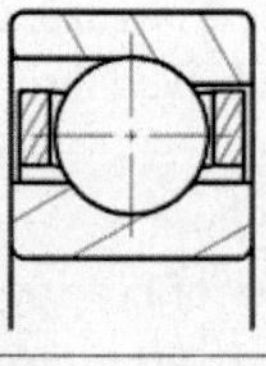	+ can absorb relatively high radial and axial forces in both directions	– cannot compensate for shaft misalignment and therefore require aligned bearing positions	16′	1.5…2
Cylindrical roller bearing	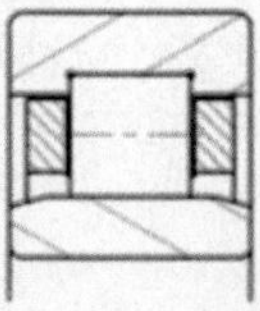	+ dismountable + low friction + high load capacity	– can only absorb radial force – sensitive to shaft misalignment	7′	1.1

$\Delta\beta$ max. setting angle for shaft misalignment
μ_L Coefficient of friction of the bearing, –

Table 2.7 Temperature factor f_t, which reduces the dynamic load rating C

Operating temperature, °C	< 150	200	250	300
Factor f_t, –	1	0.73	0.42	0.22

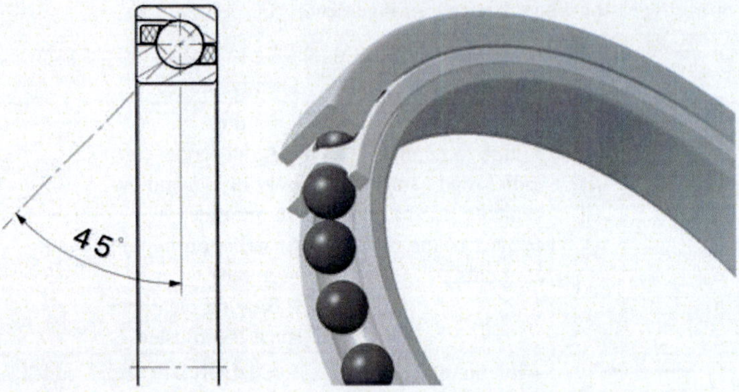

Fig. 2.59 Thin-walled angular contact ball bearing as hybrid bearing. The inner and outer ring are made from rolling bearing steel, the balls are made from silicon nitride. The bearing does not have a seal. An external seal must therefore be provided in the wheel carrier. The pressure angle is 45°

Hybrid Bearing

In Formula 1, so-called hybrid bearings (designation due to different materials) are used. Here, the two rings are made of steel and the rolling elements are made of ceramic (silicon nitride). The bearings are designed as angular contact ball bearings, which have a large number of relatively small balls, (Fig. 2.59).

In addition to the hybrid bearings, pure ceramic bearings would also be available, in which the rings are also made of silicon nitride. However, these are used sparingly in racing. The reason for this is the coefficient of thermal expansion of the ceramic, which is far below that of metals. However, the bearing must remain securely coupled and preloaded in the metal wheel carrier over a wide temperature range. This can only be achieved through extensive testing and the appropriate design of all the parts involved.

The bearing shown in (Fig. 2.59) weighs approx. 160 g as a hybrid bearing with a shaft diameter of 45 mm, and approx. 70 g as a ceramic bearing. A pair of bearings, as required for the complete bearing arrangement, therefore weighs 320 and 140 g respectively. For comparison, front wheel bearings of production cars weigh about 800 to 1100 g.

One way out of the compromise between stiffness and friction is offered by the future development of "intelligent" bearings (Fig. 2.60). Here, the bearing adjusts its raceway forging to the current bearing load within milliseconds via a sensor layer and a piezo element. Thus, when driving straight ahead, the bearing sets a wide forging for low friction. When cornering or braking, the bearing stiffens by placing the raceway against the rolling balls and thus increases the load carrying capacity.

An alternative to two bearings is the *compact bearing*, which can also be found in production vehicles. Figure 2.61 shows e.g. a double-row angular contact ball bearing. With an inner diameter of 45 mm, such a bearing weighs approx. 830 g.

In passenger car construction, an integrated bearing unit is gaining acceptance. The inner ring is combined with the hub to form a functional unit (third generation wheel

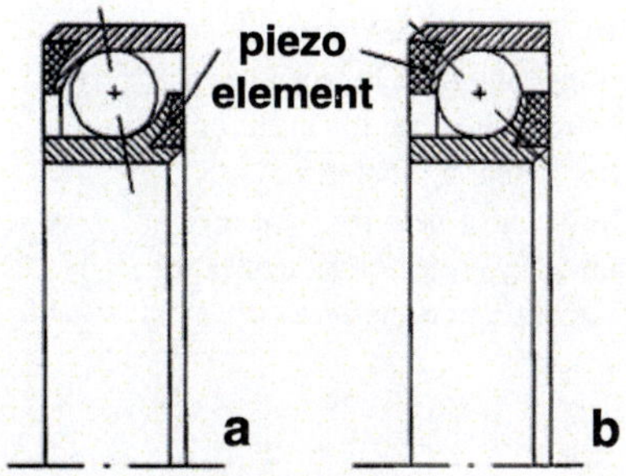

Fig. 2.60 "Intelligent" wheel bearing [23]. Two extreme positions of the piezo element are shown: (**a**) Wide bevel for minimum friction, (**b**) Narrow bevel for maximum stiffness and maximum load capacity

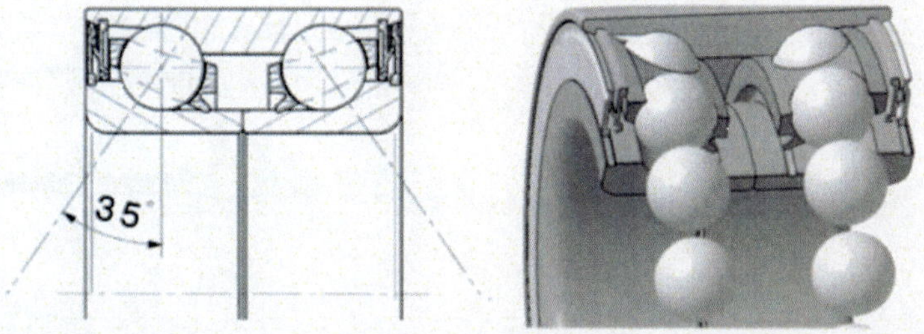

Fig. 2.61 Two-row angular ball bearing. Rolling elements, inner ring and outer ring are made from rolling bearing steel. The bearing has a lifetime grease filling and an internal seal. The pressure angle is 35°

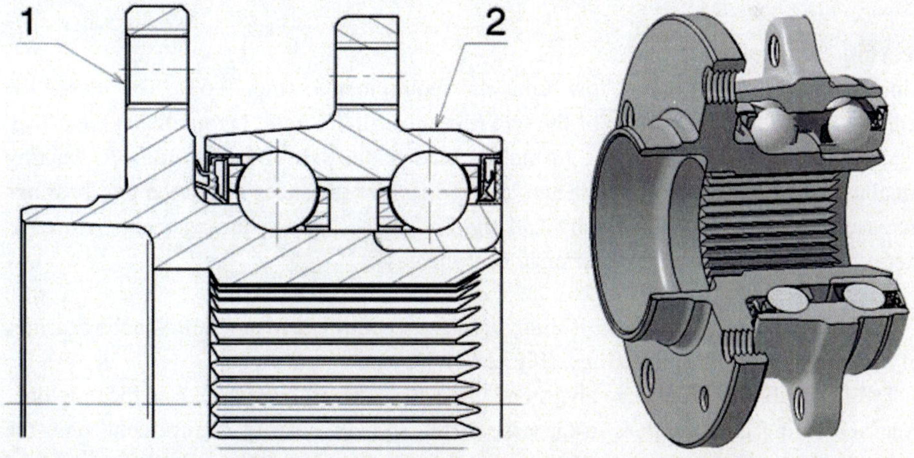

Fig. 2.62 *Wheel bearing* third generation. This bearing unit is used for a driven axle. The inner ring has additional splines for this purpose. **1** Wheel flange, **2** Mounting for swivel bearing

bearing, Fig. 2.62). This unit is bolted directly to the wheel carrier or the swivel bearing. This integral design is used in production vehicles because of the following advantages. The higher rigidity with the same installation space leads to an increase in service life and

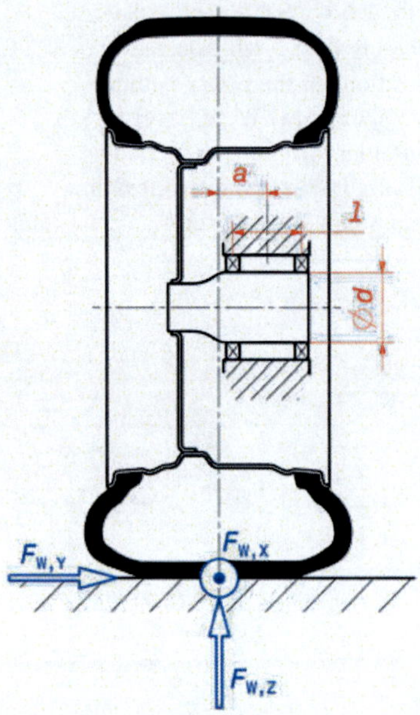

Fig. 2.63 Wheel bearing arrangement. *a* Distance from centre of bearing to centre of tyre. *l* Bearing distance, *Ød* Diameter of bearing (seat on hub), $F_{W,X,Y,Z}$ Forces from the road acting on the tyre

the very small clearance tolerance, which is practically independent of the assembly because the clearance is specified by the supplied bearing unit, simplifies the assembly process.

Bearing Arrangement

Some basic considerations follow from the requirements, (Fig. 2.63). The closer the bearings are to the centre plane of the tyre (distance *a*), the more favourable is their load, because additional moments due to the distance of the wheel forces from the bearing remain small. Larger bearing diameters *Ød* and further distances *l* between two bearings increase the load carrying capacity and rigidity of the bearing arrangement. However, friction and mass of the bearing arrangement increase.

The offset of the wheel or the position of the tyre centre plane in relation to the bearings influences the load on the bearings, (Fig. 2.64).

Two individual bearings can always be fitted as a support bearing or as a locating/non-locating bearing combination. In the case of the support bearing arrangement, only the variant with an axially clearance-free adjusted bearing arrangement is used. The preload makes the bearing arrangement stiffer.

Adjusted Bearing Arrangement

If two angular contact ball bearings or tapered roller bearings are arranged in mirror image, the bearing clearance can be adjusted by moving one bearing ring. Such bearings are advantageously installed in a rigid O-arrangement (back-to-back arrangement) (Fig. 2.65).

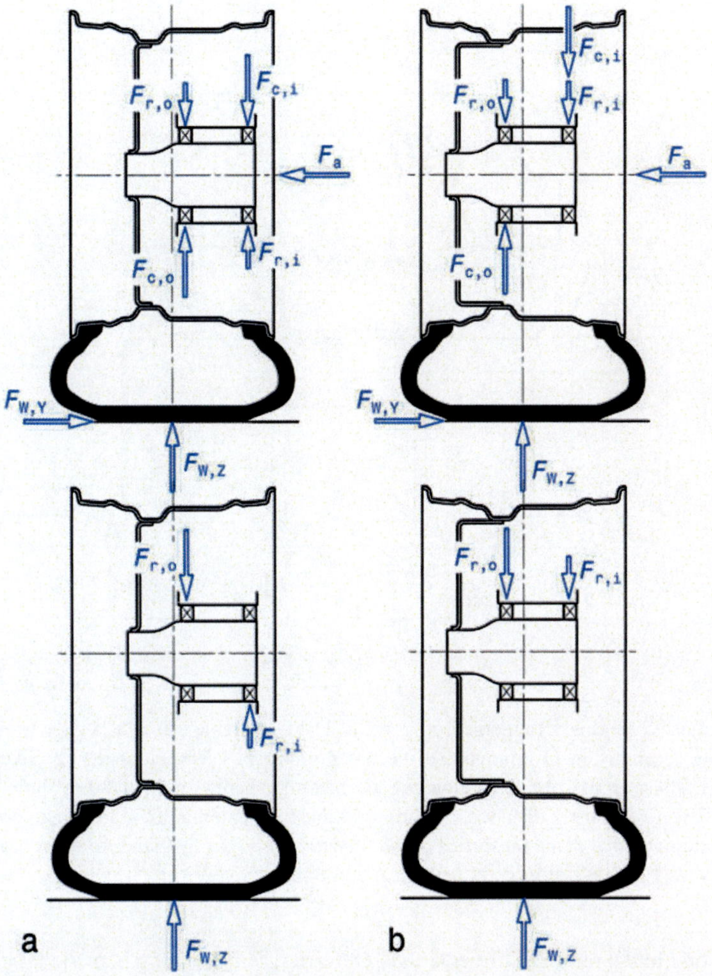

Fig. 2.64 Load on the wheel bearings depending on their arrangement. The bearing forces caused by the wheel contact force $F_{W,Z}$ and by the lateral force $F_{W,Y}$ are shown (top) and the situation when the contact force acts alone (bottom). (**a**) Tyre centre plane outside the bearing planes, (**b**) Tyre centre plane inside the bearing planes. Indices: r radial, a axial, c proportion due to cornering, i internal, o external. In case a, the bearings are unevenly loaded by the weight force alone when travelling straight ahead (bottom). When cornering, the additional lateral tyre force has a balancing effect (top). In the case of bearing arrangement b, the situation is exactly the opposite: when driving straight ahead, the wheel contact force is distributed between the two bearings (below). When cornering, the radial force of the inner bearing increases and that of the outer bearing decreases (top)

The other possibilities, X and tandem arrangement, need not be considered here. The bearing clearance can be adjusted by means of an adjustable clamping screw or nut (e.g. crown nut with split pin). For racing vehicles, adjustment by means of a spacer

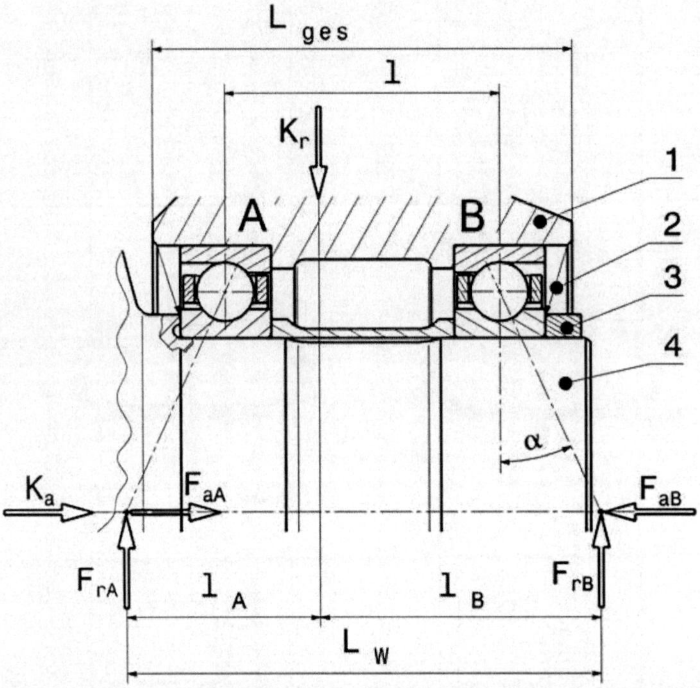

Fig. 2.65 Adjusted bearing arrangement in general. The favourable effect of a large pressure angle on the bearing assembly in O arrangement becomes obvious. **1** Wheel carrier, **2** (External) seal, **3** Element for adjusting the play or preloading the bearings (screw, nut), **4** Wheel hub. α Pressure angle, °, L_W Effective bearing distance, l Bearing distance, K_a external axial force (supports bearing A only), K_r external radial force (absorbed by both bearings), F_{rA}, F_{rB} Radial force in bearing A and B respectively, F_{aA}, F_{aB} Axial force in bearing A and B respectively

bushing ground to the required dimension is preferred. This bushing is made longitudinally elastic and the bearing has no play, but a *preload* (adjusted bearing). When the preload is adjusted, the bushing is compressed. The preload forces should remain below half of the external axial forces, then the bearing life will hardly be affected. An increase in the preload force above this no longer brings any significant advantage in terms of higher rigidity. The resilient bush maintains the preload even in the event of large temperature differences.

Due to the contact angle α, the effective bearing spacing L_w is considerably greater than the structural spacing l *of* the two bearings. As a result, the moment support is greater with increasing contact angle (camber rigidity!), the bearing arrangement is stiffer and the cornering of the wheel is improved. As a rough guideline for the design can be considered:

Table 2.8 Resulting axial force F_a with adjusted bearing (see also Fig. 2.65) [25]

Balance of power	Axial force F_a, to be used for the dynamically equivalent load	
	Bearing *A*	Bearing *B*
$\frac{F_{rA}}{Y_A} \leq \frac{F_{rB}}{Y_B}$	$F_{aA} = K_a + 0.5 \cdot \frac{F_{rB}}{Y_B}$	$F_{aB} = 0^a$
$\frac{F_{rA}}{Y_A} > \frac{F_{rB}}{Y_B}$ and $K_a > 0.5 \cdot \left(\frac{F_{rA}}{Y_A} - \frac{F_{rB}}{Y_B}\right)$	$F_{aA} = K_a + 0.5 \cdot \frac{F_{rB}}{Y_B}$	$F_{aB} = 0^a$
$\frac{F_{rA}}{Y_A} > \frac{F_{rB}}{Y_B}$ and $K_a \leq 0.5 \cdot \left(\frac{F_{rA}}{Y_A} - \frac{F_{rB}}{Y_B}\right)$	$F_{aA} = 0^a$	$F_{aB} = 0.5 \cdot \frac{F_{rA}}{Y_A} - K_a$

[a]Not taken into account in the calculation

$$L_w = 0.2 \times \text{tyre radius } [1] \tag{2.18}$$

However, the distance between the bearings should not be selected too large, since temperature-related changes in the length of the hub are then more pronounced. The rolling contact of the rolling elements has the greatest influence on the camber rigidity with approx. 40% of the tilting angle. This is followed by the wheel hub with approx. 29% and the outer ring (depending on the design) by 10% [24]. The remaining contribution to a change in the camber angle is made by the wheel carrier. The bearing environment (mounting in the wheel carrier) naturally also influences the radial bearing stiffness.

The axial rigidity of the bearings is also improved by a large contact angle. This makes the bearing assembly less sensitive to errors in the axial preload, as can occur due to mounting errors or unforeseeable thermal expansion of the conversion components. The latter is often the result of a pace car phase when the cooling of the brake is reduced due to low driving speed.

If the bearings themselves contain a seal, the external seal (2) can be omitted and the required overall length L_{ges} of the bearing arrangement becomes shorter and therefore lighter with the same effective distance L_w. An integral seal also allows the bearing to be provided with friction-minimising minimum greasing.

In the figure, the two bearing positions are marked *A* and *B*. For Table 2.8 below, the bearing that supports the external axial force K_a is bearing *A*.

The distribution of the external radial force K_r results from the equilibrium conditions to

$$F_{rA} = K_r \frac{l_B}{L_W} \text{ and } F_{rB} = K_r \frac{l_A}{L_W} \tag{2.19}$$

Due to the inclined raceways in tapered roller bearings and angular contact ball bearings, axial reaction forces F_{aA} and F_{aB} arise in the bearings as a result of external radial forces, which must be taken into account for the equivalent bearing load, see Table 2.8.

The dynamically equivalent load *P of* a bearing follows from this to:

Table 2.9 Factors X and Y for some bearing types

	Pressure-	e	Load ratio			
	angle (contact angle)	–	$F_a/F_r \leq e$		$F_a/F_r > e$	
Bearing type	°		X	Y	X	Y
Angular contact ball bearing O-arrangement	40	1.14	1	0.55	0.57	0.93
Spindle bearing	25	0.68	1	0	0.41	0.87
Tapered roller bearing O-arrangement	10–30	0.4	1	1.66	0.67	2.49

$$P = X \cdot F_r + Y \cdot F_a \tag{2.20}$$

The factors *X* and *Y* depend on the bearing type and the load ratio (F_a/F_r). They can be taken from a bearing catalogue. Table 2.9 provides a brief selection.

The rating life of a bearing is calculated from the equivalent bearing load *P* generally to:

$$L_{10} = \left(\frac{C}{P}\right)^p \tag{2.21}$$

with:	L_{10}	Service life in millions of revolutions
	C	Basic dynamic load rating, kN. Dependent on bearing type and size; can be taken from a bearing catalogue
	P	Equivalent dynamic bearing load, kN
	p	Lifetime exponent, – For ball bearings: 3; for roller bearings: 10/3

With an average bearing speed, the service life in hours follows from this:

$$L_h = \frac{L_{10} \cdot 10^6}{n_m \cdot 60} \tag{2.22}$$

with:	L_h	Service life, h
	n_m	Average bearing speed, min^{-1}

In fact, the bearing load will not remain constant but will show a time-varying course. To determine the total service life, data from measurements of comparable vehicles or from initial test vehicles are therefore analysed. The loads are divided into stages and their frequency (the time fraction q_i) is determined. From the load stages of axial and radial forces, the corresponding equivalent bearing loads P_i are calculated with (2.20), (Fig. 2.66). From the individual service life values $L_{h,i}$ according to (2.22) for the load classes P_i the achievable total service life $L_{h,t}$ follows to:

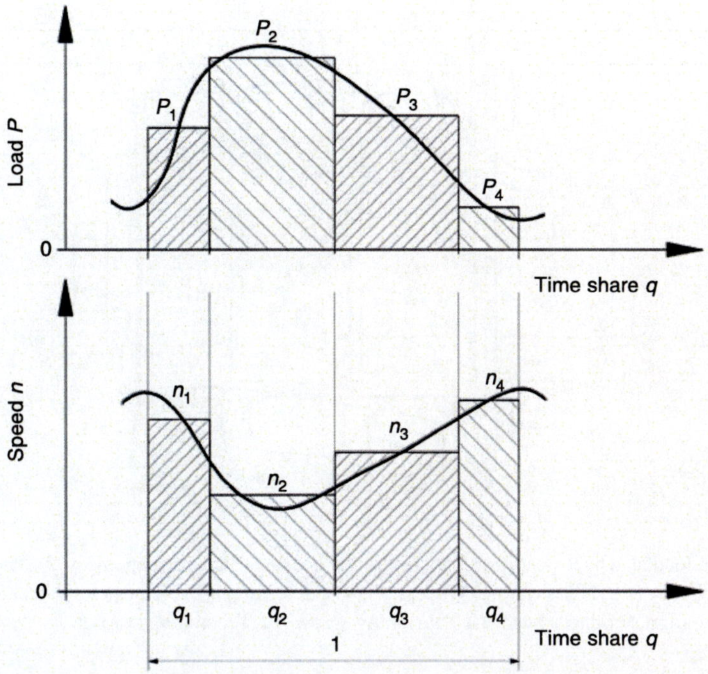

Fig. 2.66 Classification of loads and speeds in steps. P_1, P_2, ... equivalent partial loads from F_{a1}, F_{r1}; F_{a2}, F_{r2}; ..., N, n_1, n_2, ... associated speeds, min^{-1}

$$L_{h,t} = \frac{1}{\frac{q_1}{L_{h,1}} + \frac{q_2}{L_{h,2}} + \frac{q_3}{L_{h,3}} + \dots \frac{q_n}{L_{h,n}}} \tag{2.23}$$

$L_{h,t}$	Total service life, h
$L_{h,i}$	Service life for load class i,-
q_i	Time components of load class i, – Where $\sum_1^n q_i = 1$

While the number of rolling elements (balls, rollers) has a linear influence on the rating life, their diameter enters into the calculation with an exponent. Large rolling elements are therefore used for high load ratings, while small rolling elements are used when high speeds must be achieved.

Figure 2.67 shows taper roller bearings in O-arrangement. Since this bearing type has no seal, a separate seal (2, 7) must be provided. For example, axially resilient stainless steel washers (e.g. Nilos rings) can be used. This is quite sufficient for grease lubrication. The bearings are only moderately lubricated with grease in any case.

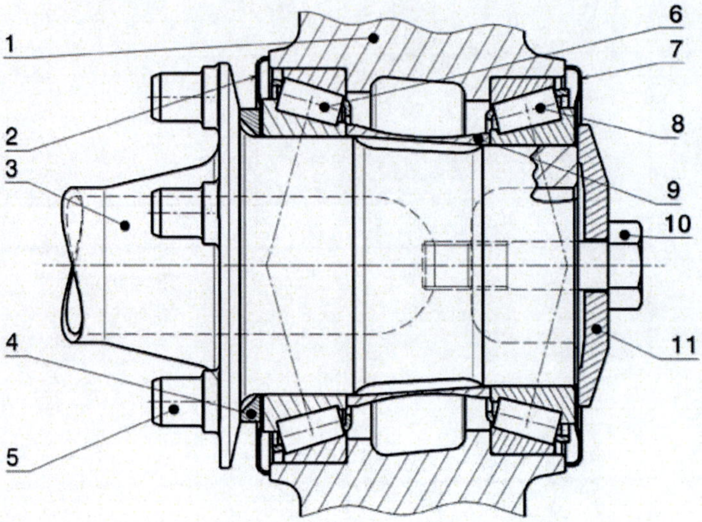

Fig. 2.67 Preloaded wheel bearings: Tapered roller bearings in O arrangement. **1** Wheel carrier, **2, 7** Sealing washer (e.g. Nilos ring), **3** Wheel hub, **4** Spacer ring, **5** Drive pegs for wheel, **6, 8** Outer/inner wheel bearing, **9** Spacer bush, **10** Tensioning screw, **11** Tensioning washer

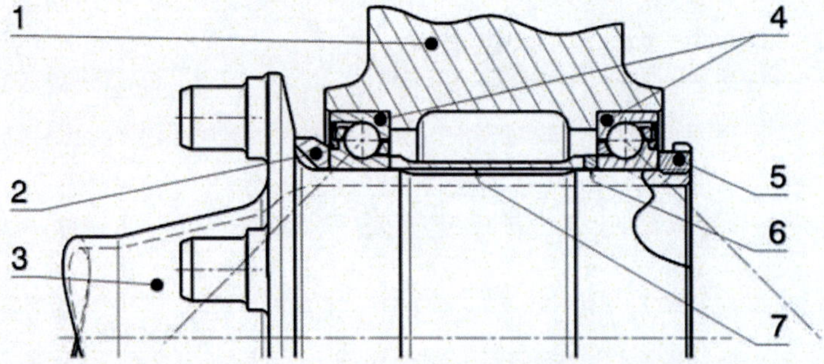

Fig. 2.68 Preloaded wheel bearings: angular contact ball bearings in O arrangement. **1** Wheel carrier, **2** Spacer ring, **3** Wheel hub, **4** Angular contact ball bearing, sealed, **5** Shaft nut, **6** Adjusting ring, **7** Spacer bush

A spacer ring (4) ensures that the inner bearing ring of the outer bearing is not pressed onto the fillet of the wheel hub (3).

A solution as used in Formula 1 [23], among others, can be seen on (Fig. 2.68). A pair of thin-walled angular ball bearings (4) is preloaded via a shaft nut (5). The axial bearing preload is adjusted by grinding off the adjusting ring (6). Since the bearings themselves contain a seal, no external seal is required and the bearing arrangement can be made correspondingly shorter. A large contact angle ensures a large effective bearing spacing.

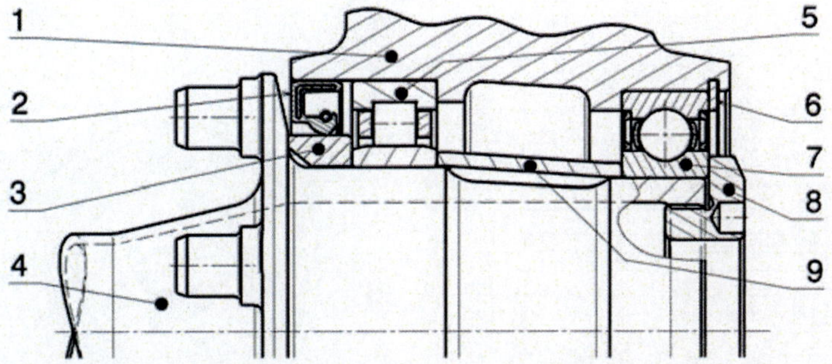

Fig. 2.69 Wheel bearing in locating/non-locating bearing arrangement: **1** Wheel carrier, **2** Oil seal, **3** Spacer ring, **4** Wheel hub, **5** Cylindrical roller bearing as floating bearing, **6** Circlip for bearing, **7** Ball bearing as locating bearing, **8** Tension screw, **9** Spacer bush

Non-Locating Arrangement

This type corresponds to the "classic" statically determined bearing arrangement with two support points. In addition to the radial force, one bearing also supports the axial forces (locating bearing). The second bearing position can only support radial forces (non-locating bearing) and thus allows expansion in the axial direction, such as occurs due to thermal expansion or manufacturing tolerances. Only those designs that can support forces in the axial and transverse directions, such as deep groove ball bearings, can be considered as locating bearings. An example can be seen in (Fig. 2.69). The cylindrical roller bearing (5) can only support radial forces. The ball bearing (7) operates as a locating bearing and thus also supports all axial forces. A tensioning screw (8) ensures that the inner rings are fitted together without clearance. The ball bearing is self-sealed, therefore no external seal is required as is the case with cylindrical roller bearings.

Further Examples

Compact bearings can be easily replaced on split hubs. Figure 2.70 shows an example of a non-driven wheel. The two hub halves (6 and 7) are braced with six bolts (4) and thus hold the inner ring of the bearing (no double fit, i.e. the two hub parts do not touch each other). The bearing is held axially in the wheel carrier (2) by the retaining ring (1).

This concept can also be used for a driven axle, Fig. 2.71.

Design of Wheel Bearings

The inner rings are subjected to a point load, therefore the bearing seat on the hub may be loose (tolerance position approx. k6 to m6). The outer rings experience a circumferential load and must therefore be seated firmly in the wheel carrier to prevent them from moving (tolerance N7 to P7). If the wheel carrier is made of light metal, the tolerance that gives a tighter fit should be selected due to its greater thermal expansion. The bearing can then only

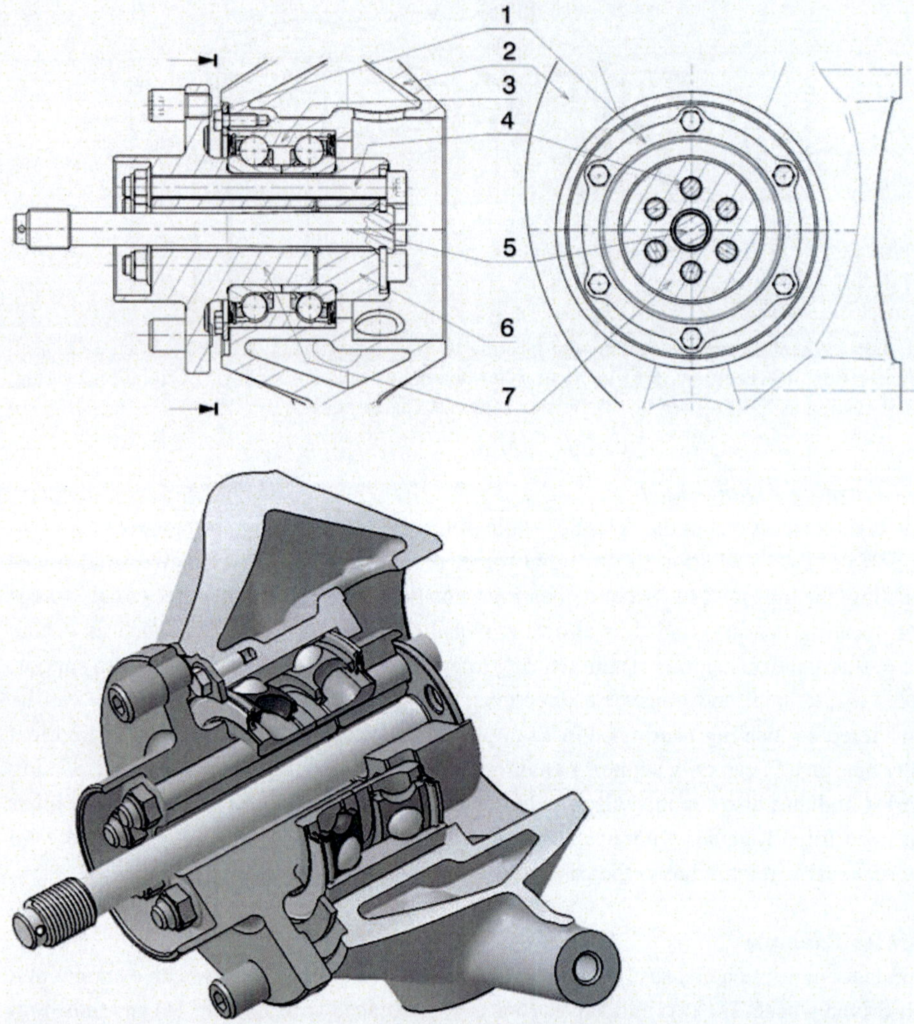

Fig. 2.70 Compact bearing on split hub for non-driven front wheel (Formula Renault 2000). **1** Retaining ring for compact bearing, **2** Wheel carrier, **3** Compact bearing, **4** Hub screw connection, **5** Central screw for wheel screw connection, **6** Inner hub part, **7** Outer hub part with wheel flange

be joined with a temperature difference (around 125 °C), but it does not move when heated by the brake system.

The pros and cons of the basic bearing options are summarized in Table 2.10.

Figure 2.72 shows the bearing arrangement of a non-driven wheel with two separate bearings. For the outer rings, two separate fits with a tight fit are required (circumferential load). For the inner rings, one loose fit is sufficient (point load). This is also necessary for

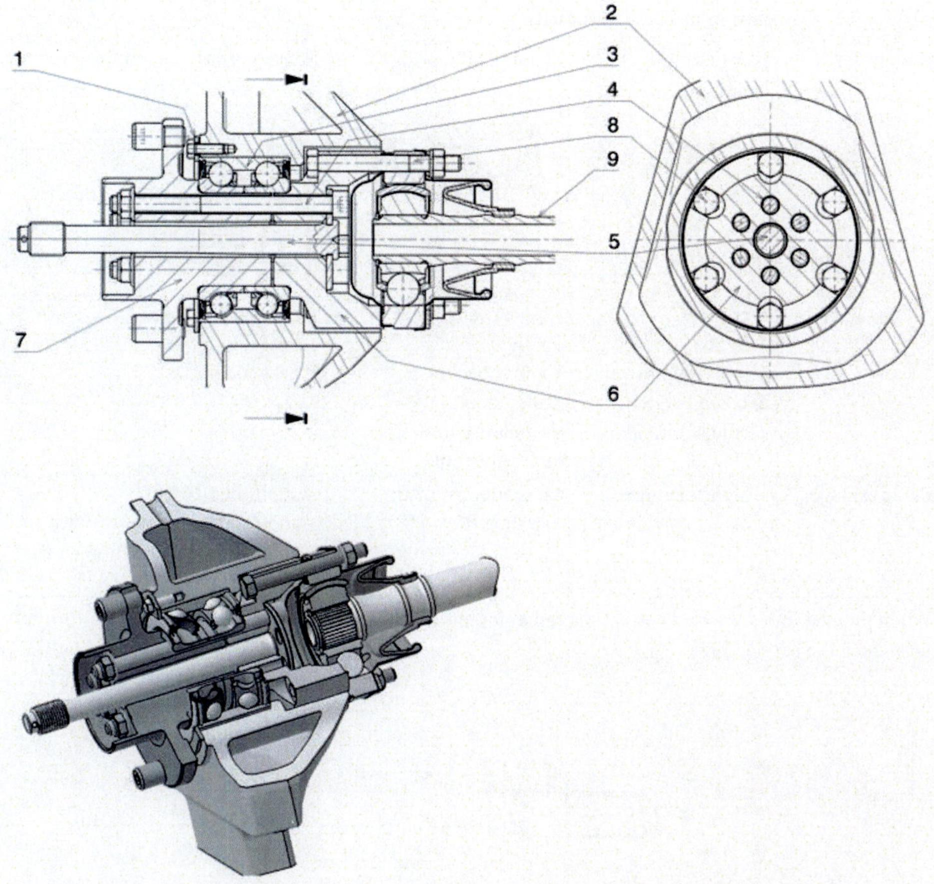

Fig. 2.71 Compact bearing on split hub for driven rear wheel (Formula Renault 2000). **1** Retaining ring for compact bearing, **2** Wheel carrier, **3** Compact bearing, **4** Hub screw connection, **5** Central screw for wheel screw connection, **6** Inner hub part, **7** Outer hub part with wheel flange, **8** Screw connection with constant velocity joint, **9** Drive shaft

the outer bearing, because the axial force of the castle nut is used to adjust the preload or bearing clearance (Fig. 2.73). The effective bearing distance is about 20% of the tire radius. The bearing arrangement of a driven front wheel can be seen in (Fig. 2.74).

Sealing

The seals fulfil two tasks in bearing arrangements:

- Prevent lubricant from escaping
- Protection against external contamination

Table 2.10 Comparison of bearing variants, according to [26]

Variant	Pair of single bearings	Compact bearing	Bearing third generation
	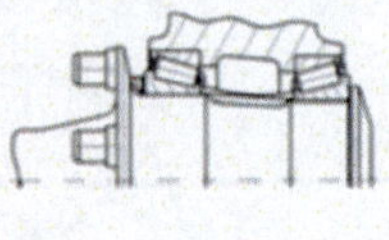	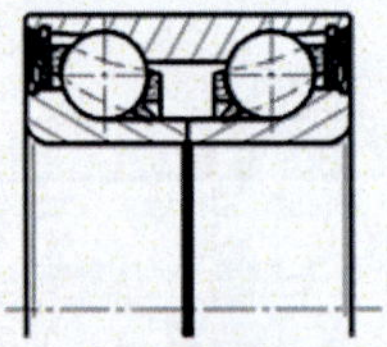	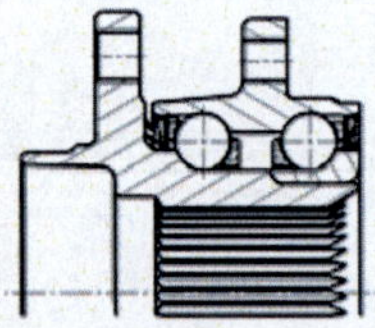
Advantages	+ cheap + light + individually exchangeable + adjustable	+ inexpensive + light + low backlash tolerance + bearing unit exchangeable	+ best play tolerance + best durability + easy assembly
Disadvantages	– high tolerance	– scattering of the service life[a]	– additional cost – additional weight – adaptation construction of the wheel carrier

[a]When pressed into the wheel carrier, the radial forces are not evenly distributed. This non-uniformity can reduce the service life

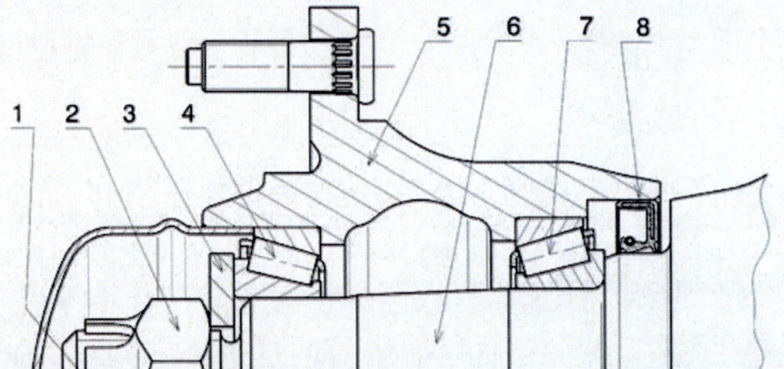

Fig. 2.72 Bearing arrangement of a non-driven wheel. The running clearance of two tapered roller bearings (3, 7) in an O arrangement is adjusted by the castle nut (2). The nut itself is secured by a split pin (1). In addition, the washer (3) has a lug which engages in a groove machined in the thread. A shaft seal (8) seals the bearing arrangement from the outside. An additional dust protection lip provides protection against the ingress of dirt from the outside. **1** Split pin, **2** Castle nut, **3** Washer with inner link, **4** Tapered roller bearing, **5** Wheel flange, **6** Axle journal, **7** Tapered roller bearing, **8** Oil seal with dust protection lip outside

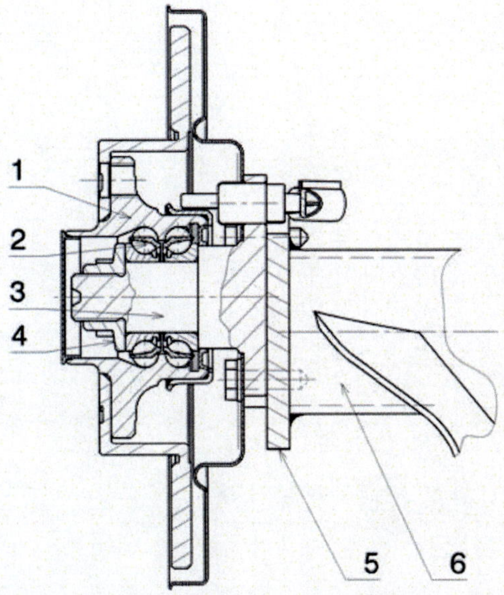

Fig. 2.73 Bearing of a non-driven wheel of a passenger car rear axle (cf. Fig. 2.48). **1** Wheel flange, **2** Wheel bearing, **3** Axle journal, **4** Nut, **5** Support plate, **6** Trailing arm. The wheel flange (1) forms the outer running surface of the wheel bearing. The inner rings of the bearing are braced against a shoulder of the axle journal (3) via the nut (4). The journal itself is bolted to the support plate (5), which is welded to the trailing arm (6). A cap pressed into the wheel flange seals the bearing from the outside

The seals can be arranged in the bearing itself or outside it. The seals can be designed to be contacting or non-contacting. Contact seals have a greater sealing effect, but generate friction. With non-contact seals, on the other hand, lubricant can leak out.

Seals integrated in the bearing not only help to reduce the overall length of the bearing but also to keep the grease requirement in the bearing low, which reduces friction losses. For a typical Formula 1 front wheel bearing, 0.8 g of grease is sufficient over the entire service life [23].

In passenger cars, if separate seals are required, shaft seals are used which, in addition to the actual sealing lip on the inside, also have a dust protection lip on the outside (Fig. 2.72).

A simple type of bearing with a compact bearing is shown (Fig. 2.75).

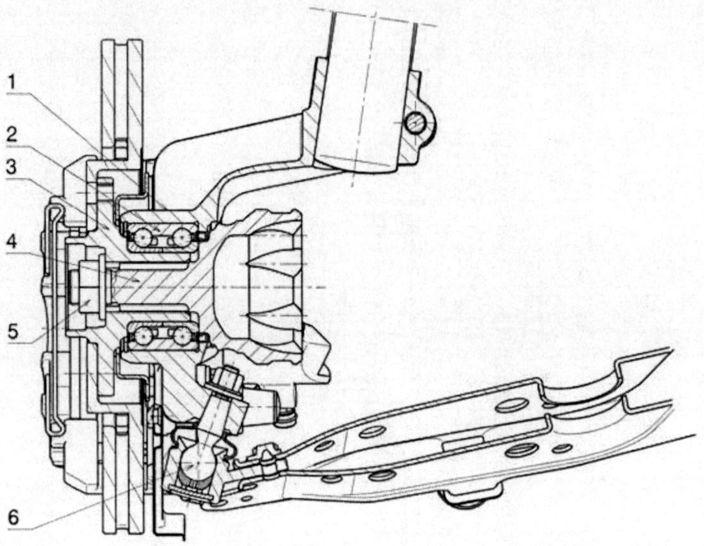

Fig. 2.74 Bearing arrangement of a driven wheel of a front wheel-guiding suspension strut axle. **1** Wheel carrier (swivel bearing), **2** Wheel bearing, **3** Wheel flange, **4** Shaft journal with receiving cup of the fixed joint of the cardan shaft, **5** Nut, **6** Guide joint. The inner ring of the wheel bearing is preloaded with the nut (5) between the wheel flange (3) and the shaft journal (4). The outer ring is supported by the swivel bearing (1). The wheel is driven by the cardan shaft, the illustrated journal (4) of which has a splined connection to the wheel flange (3)

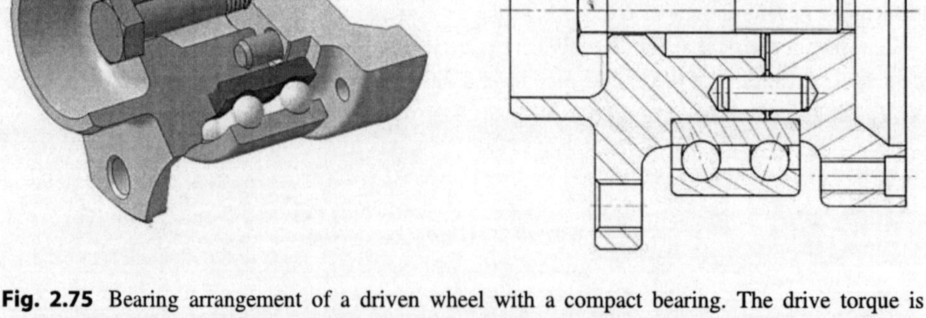

Fig. 2.75 Bearing arrangement of a driven wheel with a compact bearing. The drive torque is transmitted to the wheel flange (1) via six dowel pins (4). The inner ring of the wheel bearing is preloaded between the flanges (1) and (5) by the screw (2). **1** Wheel flange, **2** Screw, **3** Angular contact ball bearing, double row, **4** Dowel pin, **5** Flange for constant velocity joint

2.3.5 Calculation

Load assumptions for the calculation of chassis parts are based on measurement data (telemetry, data acquisition).[15] The same loads are naturally used for the design of wheels and tyres. If no load collectives (see appendix) or data of comparable vehicles are available, one can roughly assume the following extreme load cases, according to [10, 27]. If the vehicle has aerodynamic aids, the vertical air forces are added for individual load cases. In addition to the wheel load $F_{W,Z}$, the downforce per wheel $F_{W,L,Z}$ $(= F_{L,Z,f}/2$ *or* $F_{L,Z,r}/2)$ *is added:*

(a)	Maximum vertical impact (pothole): $F_{W,Z} = k_{dyn}\, F_{W,Z,0} + F_{W,L,Z}$ mit *k*dyn = 4 bis 5 $F_{W,Z,0}$ Static wheel load of the fully fuelled vehicle with driver, N $F_{W,L,Z}$ Aerodynamic downforce per wheel, N $F_{W,L,Z} \approx 1.3 F_{W,Z,0}$ (formula 1) or $0.25 F_{W,Z,0}$ (formula Student)	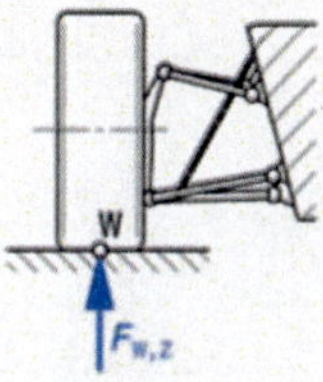
(b)	Extreme cornering, cornering outer wheel: $F_{W,Z} = 2F_{W,Z,0} + F_{W,L,Z}$ $F_{W,Y} = \mu_{W,Y} \cdot F_{W,Z}$ $\mu_{W,Y}$ coefficient of friction in the transverse direction, –	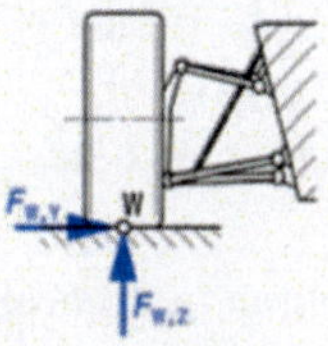
(c)	Extreme cornering, inside the turn wheel: $F_{W,Z} = F_{W,Z,0} + F_{W,L,Z}$ $F_{W,Y} = -\,\mu_{W,Y} \cdot F_{W,Z}$	
(d)	Extreme braking: Front: $F_{W,Z,f} = 2F_{W,Z,f,0} + F_{W,L,Z}$; $F_{W,X,b,f} = \mu_{W,X} \cdot F_{W,Z,f}$ Posterior: $F_{W,Z,r} = 0.8F_{W,Z,r,0} + F_{W,L,Z}$; $F_{W,X,b,r} = \mu_{W,X} \cdot F_{W,Z,r}$ $\mu_{W,X}$ coefficient of friction in longitudinal direction, –. For calculation assume min. 1.2–1.7	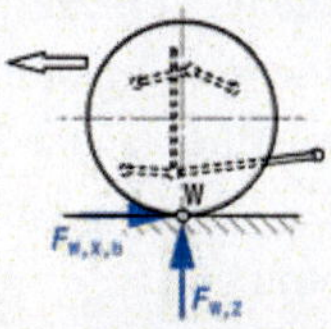
(e)	Extreme approach: Front-wheel drive (front wheels only): $F_{W,Z,f} = 1.5F_{W,Z,f,0}$; $F_{W,X,a} = \mu_{W,X} \cdot F_{W,Z,f,0}$ Rear wheel drive (rear wheels only): $F_{W,Z,r} = 1.8F_{W,Z,r,0}$; $F_{W,X,a} = 1.5\mu_{W,X} \cdot F_{W,Z,r,0}$	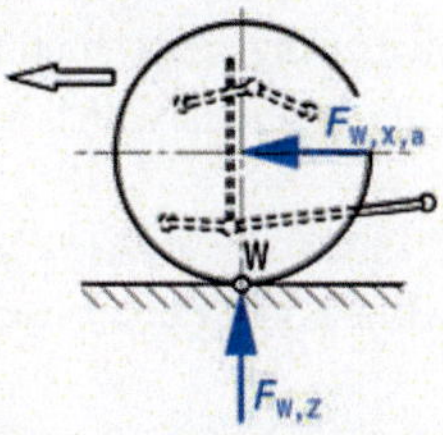

Further load cases for a static design or recalculation can be used:

[15] For more details, see Racing Car Technology Manual Vol. 5 *Data Analysis, Tuning and Development.*

(f) Deflection into the rebound stop: $F_{W,Z} = -2F_{W,Z,0}$

(g) Impact of trackside barriers (kerbs):

$$F_{W,Y} = k_{dyn} F_{W,Z,0}$$

$$F_{W,Z} = 1.5F_{W,Z,0}$$

with $k_{dyn} = 2.5$–4.

(h) For the rear axle: braking with 0.5 g when reversing (may occur, for example, after a spin):

$$F_{W,Z,r} = F_{W,Z,r,0} + \frac{0.5 g m_{V,t} h_V}{2l}$$

$$F_{W,X,b,r} = -\mu_{W,X} \cdot F_{W,Z,r}$$

$m_{V,t}$	Total mass of the vehicle, kg
h_V	Centre of gravity height of the vehicle, mm
l	Wheelbase, mm

(i) For the front axle: braking in potholes:

$$F_{W,Z,f} = k_{dyn} F_{W,Z,f,0} + F_{W,L,Z}$$

$$F_{W,X,b,f} = \mu_{W,X} \cdot F_{W,Z,f}$$

with $k_{dyn} = 3$–5.

For a Formula 1 wheel, typical load cases and values for wheel design are [28]:

(a) Pothole: Vertical load $F_{W,Z}$ of 16.3 kN with an additional force of 68 kN due to the pre-tensioning of the central nut.

(b) Cornering: lateral force $F_{W,Y}$ of 15 kN and vertical force of 10.1 kN. Tilting moment $M_{W,X}$ of 4.53 Nm. Additional force of 74 kN due to the pre-tensioning of the central nut.

Measuring cells mounted in the centre of the wheel have specifications for Formula 1 applications which can be found in Table 2.11.

Table 2.11 Value range of wheel load cells for Formula 1 application [29]

		Axle:	Front	rear	
Static	Wheel load	$F_{W,Z,stat}$	1.5	2.2	kN
Maximum values	Longitudinal force	$F_{W,X,max}$	6	8	kN
	Lateral force	$F_{W,Y,max}$	7.5	10	kN
	Vertical force	$F_{W,Z,max}$	8	10	kN
	Tilting moment	$M_{W,X,max}$	3	4	kN m
	Drive/braking torque	$M_{a,max}$ /$M_{b,max}$	1.5	2.2	kN m
	Self-aligning torque	$M_{W,Z,max}$	3	4	kN m

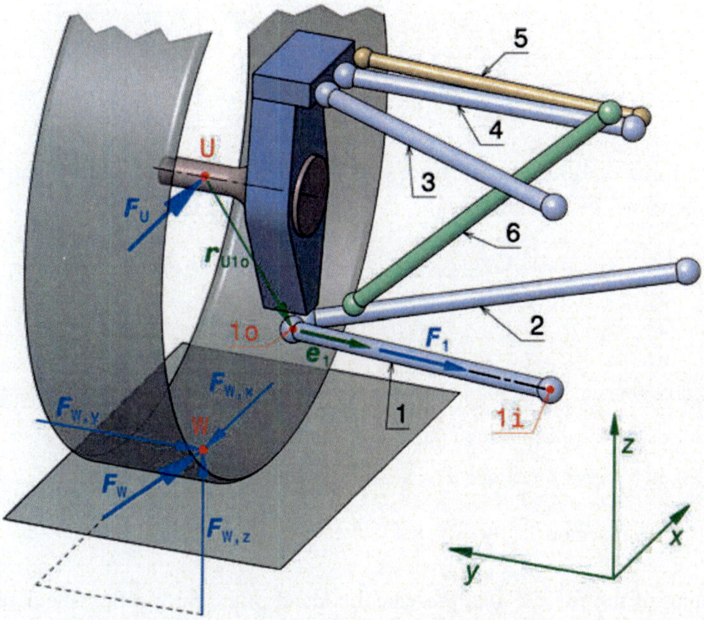

Fig. 2.76 Calculation of the rod forces of a wheel suspension. Forces act at the wheel contact point W, which are combined to form the force $\boldsymbol{F}_W$. A driving force can act at the wheel centre point U, which is transmitted via a side shaft. The forces are absorbed by 6 links or rods (1–6). The designation of the rod end points (1o, 1i, 2o,. . .) is composed of the rod name and an index for the position. Index i for inside and o for outside

The safety factor for calculations is between 1.1 and 1.2 for racing vehicles.

The above load cases describe load peaks that have to be endured (for a short time) but do not represent a permanent load. If permanent loads are to be included in the design calculation, load cases such as average braking and average cornering can be considered. The input variables for calculating the forces on the tyre are mean braking deceleration (= half maximum deceleration) or mean lateral acceleration (= half maximum lateral acceleration), mean aerodynamic downforce (= downforce at half maximum speed) and mean friction values (= half maximum values). The tyre forces follow from (6.8) and (2.4).

The forces concentrated at the wheel contact point are absorbed by all chassis parts and transmitted to the body. For the spatially arranged rod-shaped control arms of a double wishbone suspension, a vectorial approach is suitable when determining the rod forces (Fig. 2.76).

The wheel suspension is loaded by the external forces $\boldsymbol{F}_W$ at the wheel contact point W and, if applicable, $\boldsymbol{F}_U$ at the wheel centre point U. $\boldsymbol{F}_U$ acts when circumferential forces on the tyre are transmitted by a shaft and are not supported on the wheel suspension. This is the case with drive forces and with inboard brakes. In component notation, these force vectors read:

$$\vec{F}_W = \begin{pmatrix} F_{W,x} \\ F_{W,y} \\ F_{W,z} \end{pmatrix}, \vec{F}_U = \begin{pmatrix} F_{U,x} \\ 0 \\ 0 \end{pmatrix} \tag{2.24}$$

For the static force equilibrium, all forces and moments about a point are required. The reaction moments $\boldsymbol{M}_j$ of the individual rod forces $\boldsymbol{F}_j$ around U follow:

$$\vec{M}_j = \vec{r}_{Ujo} \times F_j \cdot \vec{e}_j = F_j \cdot \vec{\xi}_j \tag{2.25}$$

$\vec{M}_j$	Reaction moment of the rod j about the wheel center U caused by the rod force $\vec{F}_j$
$\vec{r}_{Ujo}$	Vector of the force application point of the bar force $\vec{F}_j$ referred to U: $\vec{r}_{Ujo} = \vec{r}_{jo} - \vec{r}_U$
$\vec{e}_j$	Unit vectors in bar direction: $\vec{e}_j = \frac{1}{\lvert\vec{r}_{j,oi}\rvert}\vec{r}_{j,oi}$ with the location vectors $\vec{r}_{j,oi} = \vec{r}_{ji} - \vec{r}_{jo}$
$\vec{\xi}_j$	Unit moment vectors: $\vec{\xi}_j = \vec{r}_{Ujo} \times \vec{e}_j$

The moment of the force $\boldsymbol{F}_W$ at the wheel contact point around the wheel centre U is: $\mathbf{r}_{UW} = \mathbf{r}_W - \mathbf{r}_U$

$$\vec{M}_W = \vec{r}_{UW} \times \vec{F}_W \text{ with the radius vector of the force application } \vec{r}_{UW} = \vec{r}_W - \vec{r}_U \tag{2.26}$$

For the sake of clarity, the external forces are also added up

$$\vec{F}_{rsl} = \vec{F}_W + \vec{F}_U$$

The system of equations of the static equilibrium conditions

$$\sum \vec{F} = \vec{0} \text{ and } \sum \vec{M} = \vec{0} \tag{2.27}$$

can be summarized in matrix notation:

$$\begin{pmatrix} e_{1,x} & e_{2,x} & e_{3,x} & e_{4,x} & e_{5,x} & e_{6,x} \\ e_{1,y} & . & . & . & . & . \\ e_{1,z} & . & . & . & . & . \\ \xi_{1,x} & \xi_{2,x} & \xi_{3,x} & \xi_{4,x} & \xi_{5,x} & \xi_{6,x} \\ \xi_{1,y} & . & . & . & . & . \\ \xi_{1,z} & . & . & . & . & . \end{pmatrix} \cdot \begin{pmatrix} F_1 \\ F_2 \\ F_3 \\ F_4 \\ F_5 \\ F_6 \end{pmatrix} + \begin{pmatrix} F_{rsl,x} \\ F_{rsl,y} \\ F_{rsl,z} \\ M_{W,x} \\ M_{W,y} \\ M_{W,z} \end{pmatrix} = \begin{pmatrix} 0 \\ 0 \\ 0 \\ 0 \\ 0 \\ 0 \end{pmatrix} \tag{2.28}$$

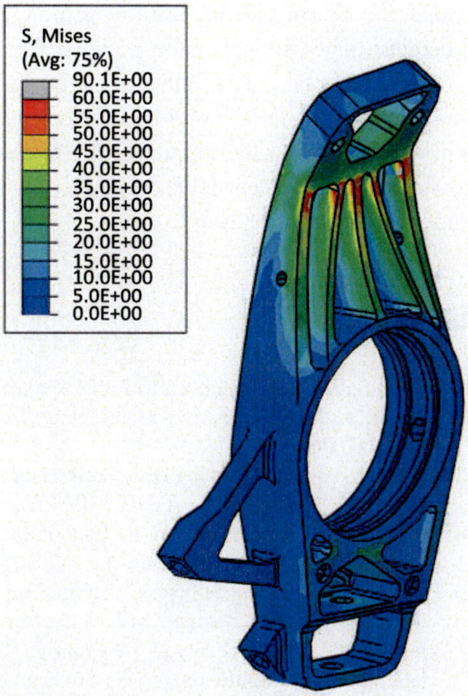

Fig. 2.77 Graphical representation of an FEM calculation result. The left front wheel carrier (in installed position) is shown under the load case "extreme cornering, outer wheel on turn". The von Mises equivalent stresses are shown as a color scale in MPa and the deformation with a scaling factor of 100 (Abaqus software). The wheel carrier is connected to the car via unequal-length wishbones and tie rods. The greatest deformation and equivalent stress occurs in the area under the attachment of the upper wishbone (red areas with approx. 60 MPa)

In short form, this equation is presented in the same way as it can be solved elegantly with the help of a matrix-oriented computer program:[16]

$$\mathbf{E}\vec{F}_{Rd} + \vec{L} = \vec{0} \Rightarrow \vec{F}_{Rd} = -\mathbf{E}^{-1}\vec{L} \tag{2.29}$$

More extensive calculations are made possible by finite element software tools (FEM).[17] With these, the stresses occurring under load (equivalent stresses, principal normal stresses) in the components can be calculated numerically, approximately and, among other things, displayed graphically. In addition, deformations can be visually highlighted by means of a magnification scale. Figure 2.77 shows the graphically prepared result for an upright as an example.

Particularly interesting for racing vehicles is the possibility to perform optimizations based on FEM calculations. For example, topology optimizations can be created automatically by specifying boundary conditions such as loads, limiting geometries and connection points to other components to the computer, which then removes material from the unloaded and low-loaded points until the remaining ones are roughly evenly (and highly)

[16]E.g. MATLAB or Octave.

[17]Finite element method, see appendix.

loaded. It is no coincidence that the resulting shapes resemble organic systems. In nature, structures (bones, trees,. . .) also grow only where they are needed.

In general, a fatigue calculation[18] is carried out for parts that are to have low mass and whose functional safety is essential. Here, not only the temporal course of the load is considered, but also the fatigue of the material. As a result, in addition to the strength verification, a statement is also obtained as to when (operating hours, load cycles) the component must be replaced.

References

1. Braess, S.: Vieweg Handbuch Kraftfahrzeugtechnik, 4th edn. Vieweg, Wiesbaden (2005)
2. Heißing B.: Moderne Fahrwerksauslegung, Vortrag im Rahmen der ÖVK-Vortragsreihe. Graz 12. Mai (2004)
3. Rompe, K., Heißing, B.: Objektive Testverfahren für die Fahreigenschaften von Kraftfahrzeugen. Quer- und Längsdynamik, 1st edn. TÜV Rheinland, Köln (1984)
4. Bennett, N.: Inspired to Design. F1 Cars, Indycars & Racing Tyres: the Autobiography of Nigel Bennett. Veloce Publishing, Poundbury (2013)
5. Bickerstaffe, S.: Going full circle. Autom. Eng. **38**, 37 (2013)
6. van Schalwyk, D.J., Kamper, M.J.: Effect of Hub Motor Mass on Stability and Comfort of Electric Vehicles. University of Stellenbosch, Stellenbosch (2013)
7. Matschinsky, W.: Radführungen der Straßenfahrzeuge, 3rd edn. Springer, Berlin (2007)
8. McBeath, S.: Competition Car Preparation, 1st edn. Haynes, Sparkford (1999)
9. Incandela, S.: The Anatomy & Development of the Formula One Racing Car from 1975, 2nd edn. Haynes, Sparkford (1984)
10. Henker, E.: Fahrwerktechnik. Vieweg, Wiesbaden (1993)
11. Staniforth, A.: Competition Car Suspension, 3rd edn. Haynes, Sparkford (1999)
12. Mitschke, M., Wallentowitz, H.: Dynamik der Kraftfahrzeuge, 4th edn. Springer, Berlin/ Heidelberg/New York (2004)
13. Schweighart, T.: Das optimale Dakar-Rallye-Fahrwerk. Diplomarbeit an der FH JOANNEUM, Graz (2006)
14. Reimpell, J., Betzler, J.: Fahrwerktechnik: Grundlagen, 4th edn. Vogel, Würzburg (2000)
15. Reimpell, J.: Fahrwerktechnik, 5th edn. Vogel, Würzburg (1982)
16. Breuer, B., Bill, K.-H. (eds.): Bremsenhandbuch, 1st edn. GWV Fachverlage/Vieweg, Wiesbaden (2003)
17. Milliken, W., Milliken, D.: Race Car Vehicle Dynamics, 1st edn. SAE Inc, Warrendale (1995)
18. Smith, C.: Tune to Win. Aero Publishers, Fallbrook (1978)
19. Rouelle, C.: Advanced Vehicle Dynamics Applied to Race Car Design and Development. Seminar Binder. Optimum G, Denver (2014)
20. Crahan, T.C.: Modelling steady-state suspension kinematics and vehicle dynamics of road racing cars, part II. SAE paper 942506. In: Smith, C. (ed.) Racing Chassis and Suspension Design. SAE International, Warrendale (2004)
21. Zomotor, A.: Fahrwerktechnik: Fahrverhalten, 2nd edn. Vogel Buchverlag, Würzburg (1991)

[18] See also Racing Car Technology Manual Vol. 2 *Complete Vehicle*, Sect. 2.5.

22. Lützow, J.: Anpassung und Optimierung eines Serienfahrwerks für den Einsatz im Motorsport. Beitrag auf der Tagung chassis.tech, München (2009)
23. Wemhöner, J., Bergrath, E.: Keramik in Getriebe- und Radlagern. Von der Formel 1 zum Serieneinsatz? ATZ. **11**, 1080–1087 (2001)
24. Heißing, B., Ersoy, M., Gies, S. (eds.): Fahrwerkhandbuch, 4th edn. Springer Vieweg, Wiesbaden (2013)
25. FAG Wälzlager, Katalog WL 41 520/3 DB, Schweinfurt 1999. FAG-Katalog WL 42 520/3FAG-Katalog WL 42 520/3 (1999)
26. Seubert, J.: Höhere Leistungsfähigkeit der Radlagereinheit HBU1. System Partners Mai 2003 (Sonderausgabe ATZ/MTZ). Springer Vieweg. Wiesbaden. (2003)
27. Happian-Smith, J. (ed.): An Introduction to Modern Vehicle Design. Butterworth-Heinemann, Oxford (2002)
28. Lucaora, L., Marconi, P.: F1 Wheel in Material Magnesium Forged, Front and Rear. Engineering. Vortrag auf der Race.Tech, München (2004)
29. Barz D., Herrmann M.: RoaDyn Applications, Measuring Wheel Systems for Use on Race-Car. Ingegneria dell'Autoveicolo, Heft 5/6 Mai/Juni 2007, **60**, 34–42 (2007)
30. Haney, P.: The Racing & High-Performance Tire, 1st edn. SAE, Warrendale (2003)

Springs and Dampers

3

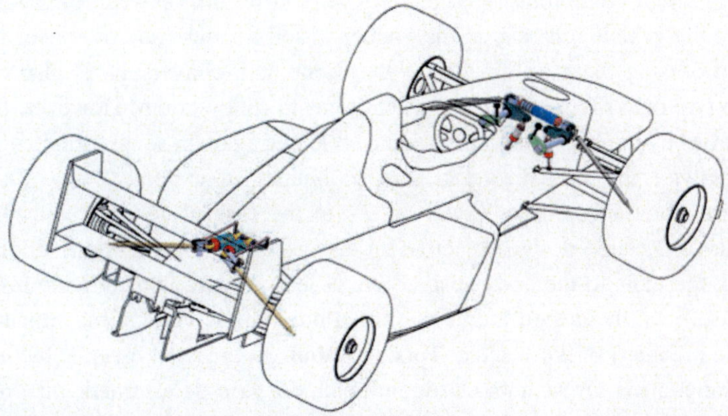

The body spring is the balancing element between the body, which should remain as still as possible, and the wheels, which should follow the road surface. The spring absorbs force peaks without introducing them directly into the car body. Dampers release the energy stored in the spring in a controlled manner and also act as dynamic springs. In this way, both also contribute to lightweight construction, because load-bearing parts do not have to be designed for the load peaks caused by bumps. Only a well-tuned spring-damper system enables (long-lasting) fast driving.

M. Trzesniowski, *Suspension System*,
https://doi.org/10.1007/978-3-658-39847-7_3

3.1 Springs

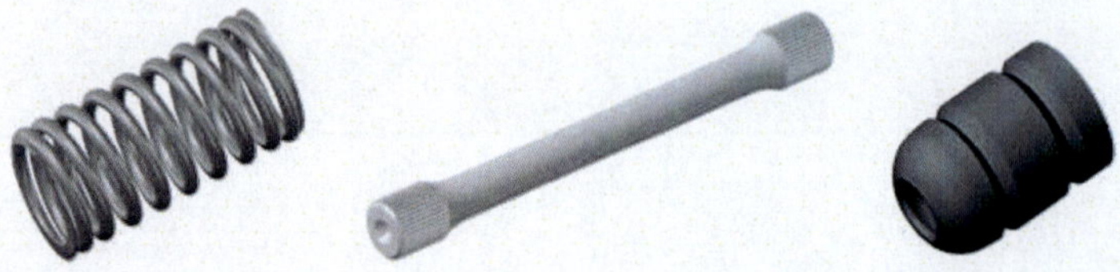

3.1.1 Function and Choice

As with all other chassis components, the main task of the suspension is to maintain the greatest possible contact between the tyres and the road surface under all operating conditions and road conditions. This requires an elastic link between the tyres and the rigid body of the vehicle (chassis, frame), which absorbs shocks and overloads for a short time without passing them on directly to the frame and, conversely, is also capable of pressing the tyre onto the road surface when it has to follow a dip. However, the impact energy absorbed by the elastic member must be released again in some form in a controlled manner, otherwise the vehicle would start to bounce in extreme cases. This task is performed by vibration dampers (Sect. 3.2.2). In the case of series-produced vehicles, comfort is also a declared design objective for the suspension in addition to driving safety. Another task that falls to the body suspension, at least in part, is to provide resistance to body roll. However, the design is initially based only on the vertical movements (lifting/lowering and pitching) of the vehicle. Torsion stabilizers are used for roll stabilization.

Active components for such an elastic link include: gases, elastomers, oils, metals and plastics. These components can be leaf springs, torsion bars, rubber blocks, air bellows or coil springs, among others. In racing, metal springs, which have no wheel-guiding functions, have become established.

Useful elastic restoring forces can be represented by all types of stress, i.e. rebound/compression, shear, bending and torsion.

The following consideration of the work absorption capacity of springs will be helpful in deciding which type of spring can be used to solve the task in the most weight-efficient way.

The work capacity *A of* an elastic material can be expressed by the following equation for tensile, compressive or flexural loading [1]:

$$A = K \cdot V \cdot \frac{\sigma^2}{E}$$

ÖHLINS
ADVANCED SUSPENSION TECHNOLOGY
MATERIAL FÜR SIEGER
AUTOMOTIVE TTX FLOW
TTX 46 TCR
TTX 40
TTX 36 ILX
ÖHLINS STOSSDÄMPFER IM HIGH-END MOTORSPORT
Rennteams wollen stets das beste Material, um siegen zu können. Wir verwenden ausschließlich extrem leichte und zugleich hochfeste Werkstoffe. Wir achten auf exakte Produktionsverfahren mit geringen Toleranzen und gleichbleibender Qualität. Seit mehr als 40 Jahren entwickeln und fertigen wir Fahrwerke, deren Performance neue Maßstäbe setzt. Unsere Leidenschaft für die Rennstrecke und unsere Erfahrung im Motorsport helfen uns dabei, den hohen Ansprüchen von Teams und Fahrern weltweit gerecht zu werden. Erfahren Sie mehr unter www.ohlins.eu.
OHLINS.EU OHLINSRACING
ÖHLINS DTC • Gottlieb-Daimler-Straße 25, 53520 Meuspath, Germany • Tel +49 (0)2691-93 77 80 • info odtc@ohlins.com

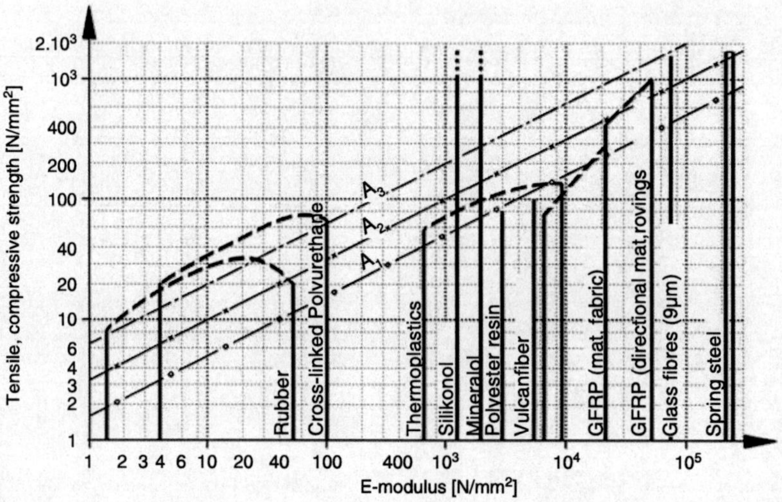

Fig. 3.1 Comparison of different spring materials, according to [1]. The representation is in double logarithmic scale, thus lines with constant work absorption capacity are represented as straight lines. The straight lines behave like. $A_1 : A_2 : A_3 = 1 : 4 : 16$ e.g. in J. From this progressively increasing ratio, it becomes clear both what large effects even seemingly small differences have, and the great influence of the permissible stresses

A	Working capacity, N mm or J·10-3	σ	Stress, N/mm²
K	Constant, –	E	E-modulus, N/mm²
V	Volume, mm³		

Figure 3.1 shows the working range of some materials for comparison.

If one considers in particular springs that are bent or twisted, one obtains:

$$\text{Bending spring} A_{r,b} = \frac{1}{6} \cdot \frac{\sigma_{zul}^2}{E} V_b$$

σ_{zul}	Permissible bending stress, N/mm²
E	Modulus of elasticity, N/mm²

$$\text{Torsion spring} A_{r,ts} = \frac{1}{4} \cdot \frac{\tau_{zul}^2}{G} V_{ts}$$

τ_{zul}	Permissible shear stress, N/mm²
G	Shear modulus, N/mm²
V	Volume involved in the work, mm³

$$\text{for steel springs follows}: \frac{A_{r,ts}}{A_{r,b}} = 2.6 \text{ bis } 2.8$$

i.e. made of steel, torsion springs are more favourable in terms of lightweight construction; torsion bars and coil springs are therefore to be preferred to leaf springs.

For a rough estimation of the required installation space of a helical spring (Fig. 3.2) can be used. The required volume V_{Sp} of the spring results from the force F which it is to absorb during the spring stroke s_{Sp} and its material. If the outer diameter D_o or the overall height l_{Sp} is selected, the other dimension can be determined from the volume.

The choice of the spring type is influenced by the following sizes:

- permissible variations in ground clearance during the journey.
- permissible wheel travel or that required by the regulations during compression and rebound
- desired wheel eigenfrequency.

In general, the spring travel is divided in such a way that, starting from the design position, one third can be deflected (up to the rebound stop) and two thirds can be compressed (up to the bump stop), i.e. approximately in the ratio 35/65% (Fig. 3.3). Table 3.1 shows some typical values of wheel strokes. In off-road and rally vehicles, the maximum deflection angle of the cardan shaft limits the wheel travel.

The rule of thumb for passenger cars is that the spring should not go to its limit at 2.5 times the static wheel load. A progressive increase in the spring rate avoids abrupt deflection, which can lead to overloading of the tire with an equally sudden loss of lateral force when cornering, and also absorbs the aerodynamic downforce that increases with speed. In any case, the downforce is a problem in itself. The vehicle level should remain as constant as possible, so that e.g. during braking the vehicle does not swing because the suspension travel has been used up by the air forces, and so that the downforce forces remain predictable for the driver. A change in vehicle level of only 3 mm resulted in a change in downforce of 450 N in a 1987 Formula 1 car [4]. Apart from this, a variation in vehicle level inevitably leads to a movement of the suspension and thus a change in the position of the tyre relative to the road. The bump stop is usually formed by a supplementary spring that contains the compression limit and determines the remaining spring travel in the compression direction. The *rebound stop* is an elastic spring limit that can be included in the shock absorber and determines the remaining spring travel in the direction of rebound. It must be dimensioned in such a way that the dynamically occurring negative spring forces ($F_{Sp} < 0$) can also be absorbed.

Helper Spring

In certain designs it may happen that the spring force becomes zero when the wheel is fully deflected or that the spring even becomes loose before this. In such cases, a short auxiliary

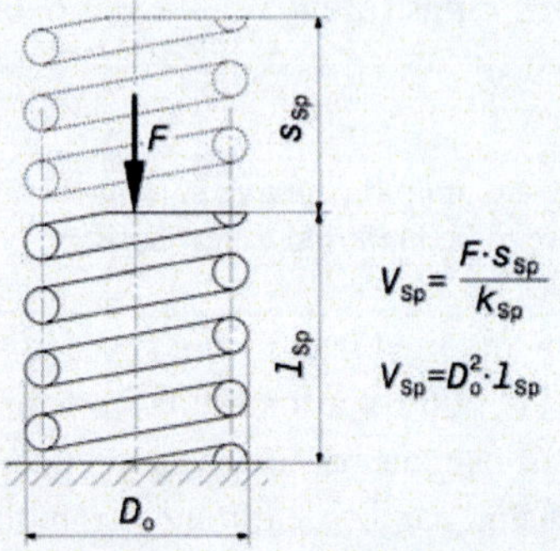

Fig. 3.2 Estimation of the required installation space of a helical spring, according to [2]. *F* Spring force, N, s_{Sp} Spring stroke under force *F*, mm, l_{Sp} Length of the spring at force *F*, mm, D_o Outer diameter, mm, V_{Sp} Construction volume of the spring, mm³. Some values for the spring characteristic value k_{Sp} [N/mm²]: Spring steel $k_{Sp} = 0.4$; AlSi1 Mg $k_{Sp} = 0.04$; Steel S235JRG1 $k_{Sp} = 0.01$

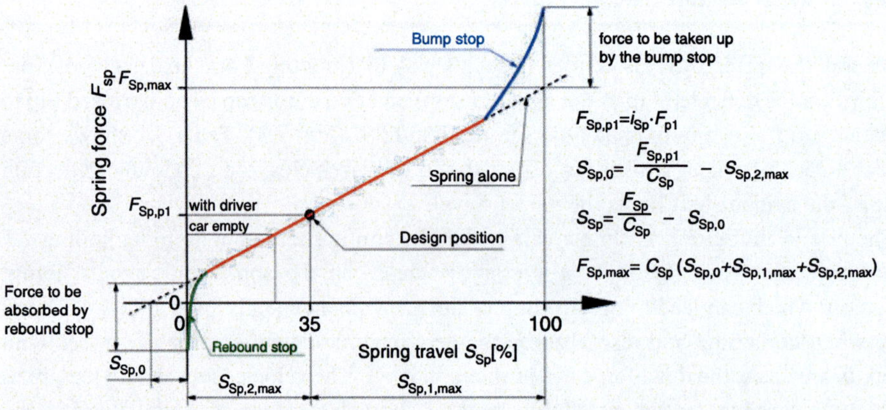

Fig. 3.3 Spring characteristic curve for a vehicle suspension. Starting from the design position, the wheel can deflect about 1/3 to the rebound stop ($s_{Sp,2,max}$) and 2/3 to the bump stop ($s_{Sp,1,max}$). In order to match the desired spring deflections and the force $F_{sp,pl}$ in design position, the spring must be preloaded by $s_{Sp,0}$ during installation. c_{Sp} Spring rate. $c_{Sp} = F_{sp}/s_{Sp}$

Table 3.1 Values for maximum wheel travel *s*

Vehicle	Passenger Car	Indy Car[a]	IMSA GTS[a]	Formula Ford[a]	Rally	Dakar, Baja
Front s_1/s_2, mm	120/80	38/0	63.5/25.5	51/25.5	325/175	650/350
Rear s_1/s_2, mm	130/90	63.5/25.5	76/38	70/25.5	325/175	650/350
Indexes:	1 ... compression, 2 ... extensions					

[a] Crahan [3]

spring is connected in series (Fig. 3.4). The considered spring Sp_b (red characteristic curve) alone would no longer provide any force when the wheel suspension goes into the traction stop. Combining an auxiliary spring Sp_h with spring Sp_b, results in a softer identifier than with Sp_h alone (Sp_h + Sp_b, magenta). The creation of this combined identification ($1/c = 1/c_h + 1/c_b$) is solved graphically: The auxiliary spring goes coil bound after the wheel stroke $i_{Sp}\ l_{h,Bl}$ and its stiffness goes almost to infinity. Both springs absorb the same force F_{Bl}. The total deformation of both springs is now $s* = i_{Sp}\ l_{h,Bl} + s_b$. From this follows directly the combined spring rate ($c = F_{Bl}/s*$). From the wheel stroke $s*$ on, the characteristic curve is determined only by spring Sp_b.

For comparison, the characteristic curve of a standard body spring Sp_a is also entered. In the fully deflected state, the wheel force $F_{W,Z}$ is also zero, but this is because the rebound stop 2 absorbs the pre-tensioning force of the spring.

In (Fig. 3.5) an installed auxiliary spring can be seen in the unloaded state.

In (Fig. 3.6) the most important influences are compiled, which are included in the considerations when determining the body supporting spring rates.

Progressive Spring Rate

Strictly speaking, it is not the body spring but its effect on the wheel that is decisive for the driving behaviour. The wheel-related spring rate (see appendix) should remain constant over the wheel stroke or increase slightly during compression. If the spring rate is not constant, but increases when the spring is loaded, this is referred to as a progressive spring rate. This behaviour prevents the spring from breaking under heavy loads and allows oscillations to decay more quickly. Thus, the realization of a transmission between the wheel stroke and the spring stroke is of great importance. Figure 3.7 shows the difference of basic values. The motion ratio between the wheel and spring strokes results from this to:

$$i_{Sp} = \frac{s_1}{s_{Sp,1}} \tag{3.1}$$

i_{Sp}	Ratio of wheel stroke to spring stroke (spring *ratio*, *motion* ratio), –
s_1, $s_{Sp,1}$	Paths according to (Fig. 3.7), m

Usual values for i_{Sp} are between 1.25 and 1.7. The spring ratio does not have to be constant and – depending on the geometry – can also change over the wheel stroke. The larger the gear ratio selected, the stiffer the springs/dampers and torsion stabilizers must be. Large gear ratios also mean relatively small damper travels. This complicates the tuning of conventional dampers, which build up their force via the piston speed and have a certain breakaway force due to friction. A way out is offered by the separate arrangement and actuation of spring and damper with individual ratios.

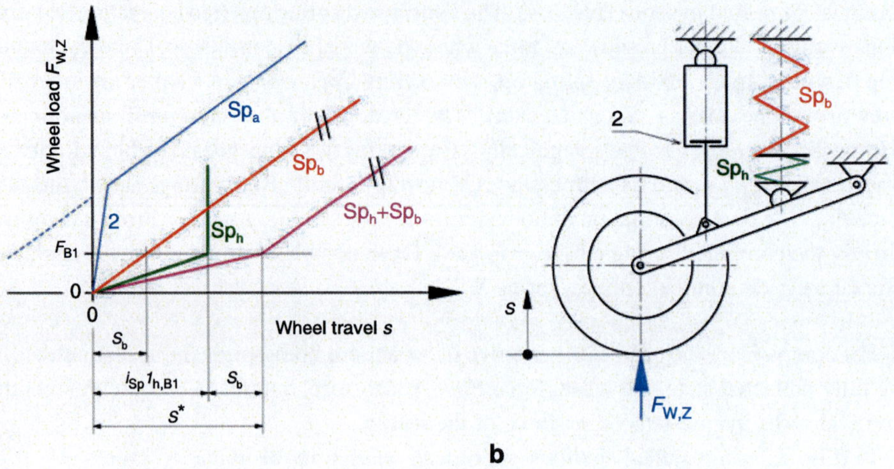

Fig. 3.4 Body spring with auxiliary spring. **a** Spring characteristics, **b** General sketch, fully deflected state is shown (2 rebound stop). Sp_a Characteristic curve of a common body spring, Sp_b Characteristic curve of a spring without preload, Sp_h Characteristic curve of an auxiliary spring, $Sp_h + Sp_b$ Characteristic curve of series connection of Sp_h and Sp_b, i_{Sp} Spring ratio (see (3.1)), $l_{h,Bl}$ Coil-bound length of auxiliary spring

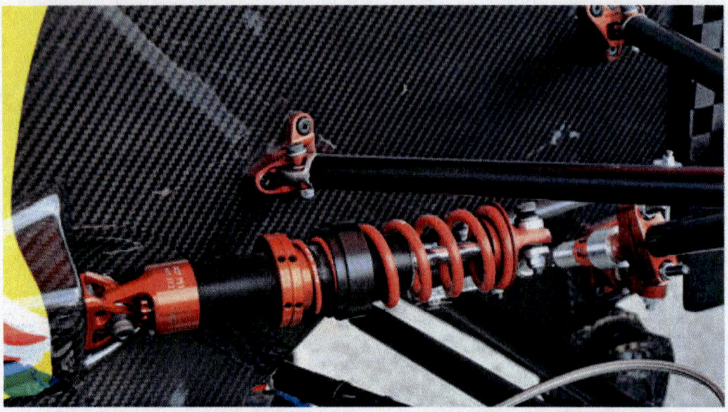

Fig. 3.5 Auxiliary spring on a shock absorber strut. You can see a part of the right front suspension with pull rod actuation of the spring. The car is jacked up, the wheel removed. The weight of the wheel carrier together with the wheel brake is held by the damper which has gone into the rebound stop. The damper strut carries the body spring (1) and the auxiliary spring (2) acts on this via a floating sleeve

This displacement ratio naturally also acts as a force ratio (i.e., spring force = wheel contact force × i_{Sp}), so it enters quadratically into the spring rate relationship:

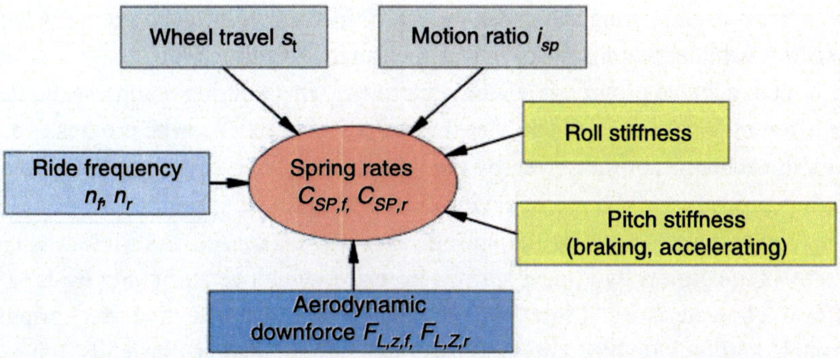

Fig. 3.6 Compilation of the most important influences on the choice of spring rate

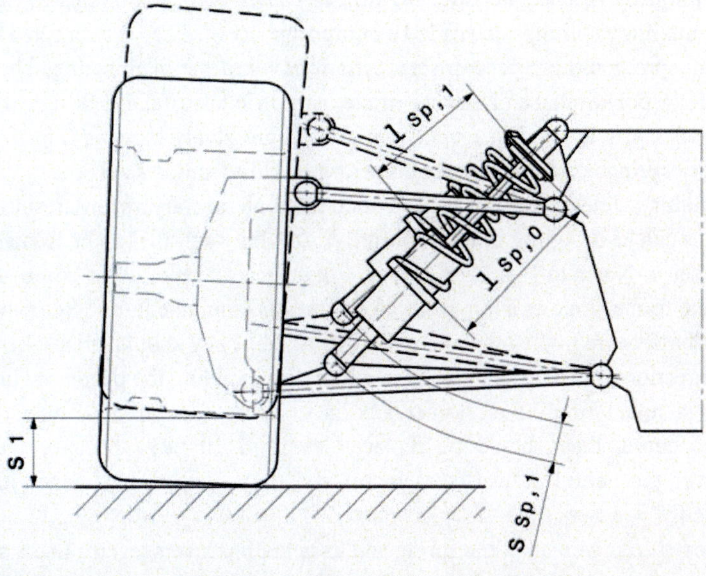

Fig. 3.7 Difference between wheel displacement and spring deflection with independent wheel suspension. If the wheel compresses by the distance s_1, the spring is shortened from its original length $l_{Sp,0}$ to the length $l_{Sp,1}$. The spring is therefore only compressed by the amount $s_{Sp,1} = l_{Sp,0} - l_{Sp,1}$. Accordingly, there is a path transmission between wheel and spring stroke

$$c = \frac{c_{Sp}}{i_{Sp}^2} \tag{3.2}$$

c	Wheel related spring rate, *wheel center rate*, N/m
c_{Sp}	Actual spring rate of the body spring, suspension spring rate, N/m

The wheel travel and spring travel can be determined from a drawing or a CAD model. For an existing vehicle, the distances can be measured directly.

If you want to achieve progressive behaviour of the wheel during compression, there are basically different possibilities. On the one hand, the spring itself can be progressive, on the other hand, the actuation of the spring by the wheel suspension can cause a variable travel ratio and thus produce the desired behavior even with a linear spring.

A progressive characteristic of the spring itself can be caused in different ways. The simplest way is to connect two linear springs in series, which actually only leads to a kink in two linear characteristics. More complex but also more effective are springs with variable pitch and/or variable wire cross-section. When loaded, the coils lying closer together go coil bound and the effective (remaining) spring length becomes continuously shorter, so the spring becomes stiffer. Changing the coil diameter (e.g. conical spring) also results in a progressive characteristic. Figure 3.8 shows two possibilities of progressive springs. The auxiliary spring – also called a bump stop – (Fig. 3.8b) is connected in parallel with the main spring and only acts from a certain travel of the main spring. The spring is made of cellular polyurethane elastomer (trade name: Cellasto) and is designed in such a way that it sets softly at the first touchdown and progressively hardens (Fig. 3.8c). Such supplementary springs can absorb large forces when fully compressed.

By combining a linear steel spring and such a supplementary spring, resulting spring characteristics can be designed almost arbitrarily. Auxiliary springs can be located within a coil spring (see a. Notes to Fig. 3.15), act in a damper as a rebound or bump stop, or be attached to the frame alone as a bump stop for a chassis component. The Shore hardnesses of elastomer buffers are in the range between 40 and 65. They should at least be guided or held in the direction of expansion, e.g. by annularly surrounding the piston rod or by being enclosed by a metal ring. If elastomer auxiliary springs are used to tune the spring characteristic curve, they should be replaced every 6–10 races because their length decreases over the period of use (settlement) and the tuning thus changes its original characteristic [6].

With a clever arrangement of the spring and its actuating levers, even a linear spring can show a progressive wheel-related rate, (Fig. 3.9). However, an undesirable degressive behaviour is also possible by careless arrangement of the spring.

Move	0 to 1	1 to 2	2 to 3	3 to 4
Wheel stroke s, mm	30	30	30	30
Spring stroke s_{Sp}, mm	17.7	18.3	18.9	19.7
i_{Sp}, –	1.69	1.64	1.58	1.52
c, N/mm	10.5	11.2	12.0	13.0

With springs on the outside, an increase in spring rate of about 10–15% can be achieved. Stronger increases are achieved by operating the springs via levers.

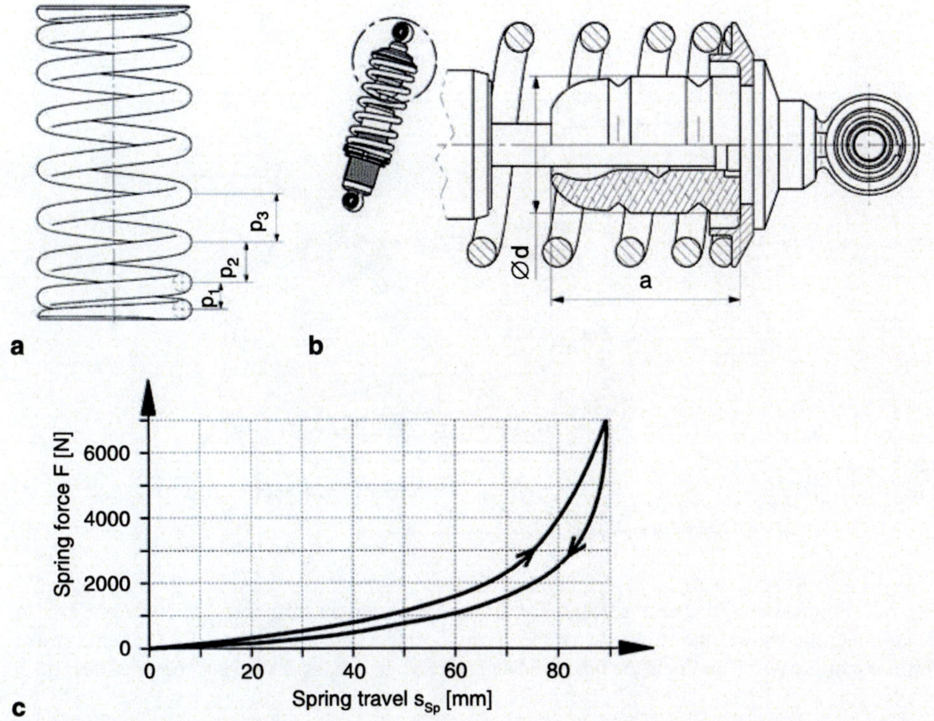

Fig. 3.8 Progressive springs. **(a)** Variable spring pitch. **(b)** Additional spring connected in parallel. **(c)** Characteristic curve of a supplementary spring with dimensions $a = 115$ mm, $\varnothing d = 70$ mm [5]. The total spring rate $c_{\mathrm{Sp,t}}$ with parallel connection follows from the individual rates to: $c_{Sp,t} = c_{Sp1} + c_{Sp2}$

The ideal progressivity of both axles will not be the same due to the different axle loads and achievable acceleration values. For a basic design, an increase in spring rate of no more than 20% is recommended for the front axle and a flatter increase of about 5% for the rear axle [7]. In the case of extremely undulating race tracks, the car needs above all appropriate wheel travel and therefore the spring rates will not be increased, but the bump stops will first be selected more forcefully and the ground clearance will be increased.

In any case, the progressivity should not be chosen too large, otherwise there is a risk that during pitching (braking, accelerating) the difference in the wheel-related spring rate between the axles (one compresses, the other extends) is too large and thus the distribution of the roll rates between front and rear is changed too much. The balance of the car would thus be severely disturbed.

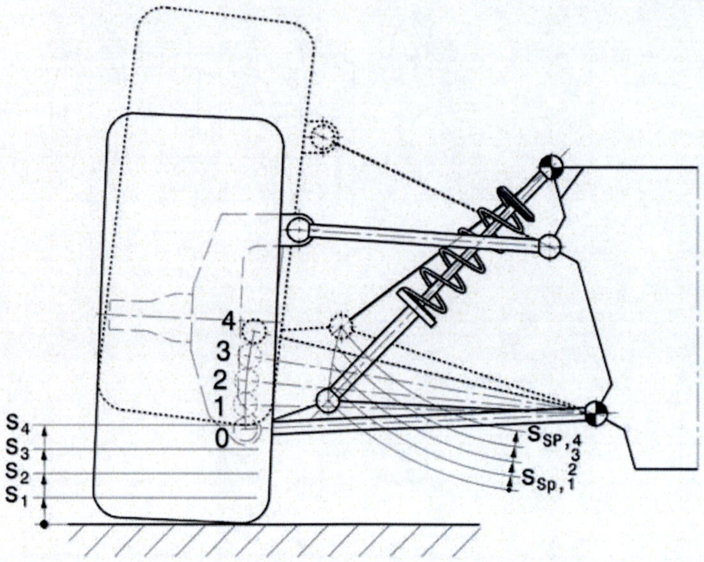

Fig. 3.9 Progressive spring arrangement. The spring has a constant stiffness $c_{Sp} = 30$ N/mm. Due to the arrangement shown, the stroke s_{Sp} of the spring becomes larger and larger for the same stroke sections s of the wheel and thus the wheel-related spring rate c increases during compression

3.1.2 Spring Actuation

The wheel stroke must result in a defined spring stroke. The spring could therefore directly support the wheel as is the case with a front fork of a motorcycle. The spring can also be hinged to a wishbone arm (e.g. Fig. 3.9). However, this creates a bending moment in the link as well as a longitudinal force in the case of an obliquely articulated spring, and the link must be designed accordingly and thus becomes heavier. The closer the pivot point is to the wheel carrier, the smaller the bending stress on the wishbone. Springs can also be actuated directly or indirectly via levers, (Fig. 3.10). The lever versions offer almost innumerable possibilities to realize an increasing spring characteristic and at the same time to place the suspension strut at a convenient place inside the car. In addition, the strut in this case also makes room for the air flowing past between the wheel and the body of the car, which is good for aerodynamics. The reaction forces at the other end of the spring must be introduced into the frame at a suitable location. Suitable locations are those that can transmit forces as directly as possible: Nodes of a lattice tube frame, attachment points of the spring on the other side of the car, articulation points of wishbones, etc. With this type of actuation, the ride height of the vehicle can be easily adjusted by changing the length of the actuating rod. If the upper wishbone is designed as a rocker arm (c), the spring can also

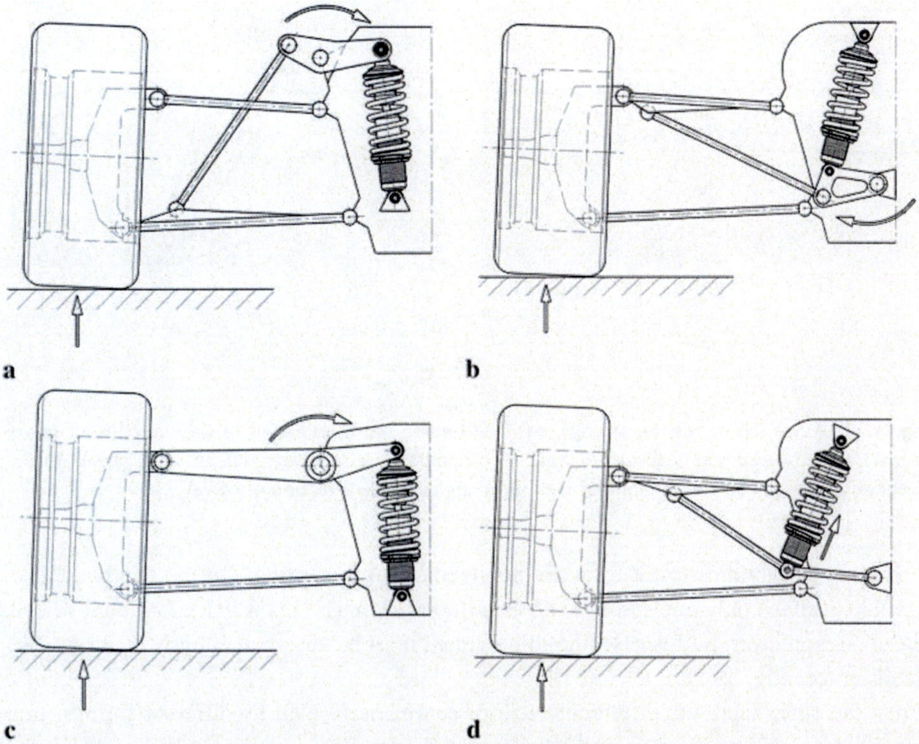

Fig. 3.10 Possibilities of spring actuation. (**a**) Push rod with bell crank *(rocker)*. (**b**) Pull rod with bell crank. (**c**) Wishbone as rocker. (**d**) Knee lever arrangement

be accommodated inside the car. The lower wishbone, which is in principle subjected to higher loads, is then free of bending moments, but the upper wishbone must be designed as a bending beam. This is unfavourable in terms of lightweight construction and makes it heavy. The lever bearing must also be designed to be strong.

In Formula 1 cars with extreme utilization of downforce by sealing the underside of the car with so-called skirts, the spring rates were between 350 and 600 N/mm, which was criticized by the majority of drivers because the behavior was already reminiscent of an unsprung vehicle [8]. In fact, the ground clearance of the cars should hardly change in order to maintain the sealing function of the skirts. The spring diameter was approx. 57 mm on the inside [8].

Torsion-bar springs can be elegantly accommodated to save space and actuated by levers. The dampers are thus freed from unwanted forces (but possible due to the tolerance-related misalignment of the coil spring) and are actuated by the same lever (Fig. 3.11).

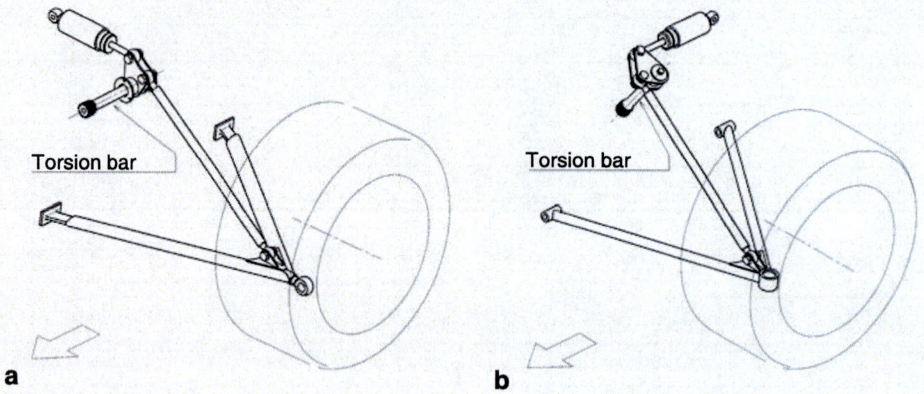

Fig. 3.11 Actuation of torsion springs. (**a**) The torsion bar is mounted in the direction of travel (arrow) and also represents the lever axis. The damper is also horizontal, at right angles to the direction of travel. (**b**) The torsion bar is at right angles to the direction of travel

Further possibilities result from the arrangement of the spring. On the one hand, it is possible to attach only one spring to an axle (mono-spring). This is loaded by both wheels via a common lever. In this case, the roll moment must be absorbed entirely by a (special) stabiliser (see Fig. 3.53).

On the other hand, the different functions can be performed by different springs. In a three-spring system, two springs take over the suspension of one wheel each in the conventional manner and, in addition, a third spring is brought into play via a common coupling in the case of equilateral springs (see Fig. 3.54). Such systems are suitable for vehicles with a high aerodynamic component, which also cover a large speed range. In these, the ground clearance should ideally remain constant over the driving speed, although downforce increases disproportionately as speed increases. If one were to simply provide stiffer springs to solve this problem, for example, one would have an unnecessarily hard suspension at low speeds, which would cause traction problems and place greater stress on the driver and components.

3.1.3 Spring Calculation

The determination of the required spring rate is described in (Sect. 3.2.1).

The following relationships can be used to predict the spring dimensions at a known spring rate:

Coil spring coil spring: $c_{Sp} = \frac{G \cdot d_{Sp}^4}{8D_m^3 \cdot j_{Sp}}$	c_{Sp}	Spring rate, N/mm	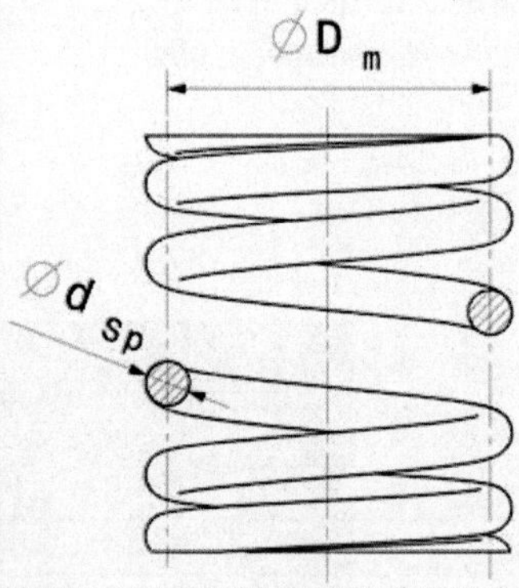
	D_m	Mean average coil diameter, mm	
	d_{Sp}	Wire diameter, mm	
	j_{Sp}	number of active coils, –.	
	$j_{Sp} = j_t - 1.5$		
	j_t	Total number of coils s, –	

Torsion bar spring:		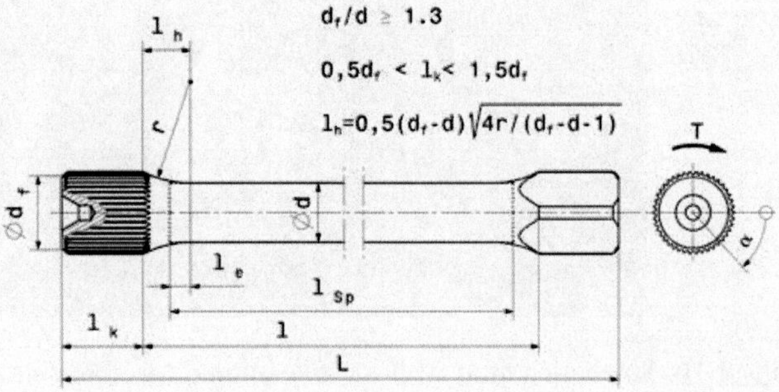
$c_\alpha = \frac{\frac{\pi}{32} G \cdot d^4}{\frac{180^\circ}{\pi} l_{Sp}}$	$T = c_\alpha \cdot \alpha$	
c_α	Torsion spring rate (angular (twist) rate), N mm/°	
G	Shear modulus, N/mm^2	
d	Bar diameter, mm	
l_{Sp}	Active length, mm	
T	Torque, N mm	
α	Twist angle, °	
r	Fillet radius, mm	
d_f	Root diameter of the head spline, mm	

The lengths follow from:

$$l_{Sp} = l - 2(l_h - l_e)$$

l	Free shaft length, mm
l_h	Fillet length, mm
l_e	Replacement length, mm. $l_e = \upsilon - l_h$, υ according to Table 3.2
l_k	Head length, mm L Total length of torsion bar, mm

Table 3.2 Ratio υ (= l_e/l_h, replacement length to fillet length), after [9]

d_f/d		1.3	1.5	1.7	1.9	2
	1	0.731	0.638	0.581	0.534	0.522
r/d	2	0.728	0.632	0.568	0.511	0.502
	50	0.725	0.625	0.555	0.5	0.481

Fig. 3.12 Suspension strut with variable spring rate (Formula König). The lower spring plate receives the spring in a helical groove. By screwing this plate, the number of springing coils and thus the spring rate can be changed. If the number of active coils decreases, the spring becomes stiffer, cf. also equation for predetermination above

The moment T is introduced into the torsion bar springs via serrations, square, hexagon, eccentric end or similar.

Figure 3.12 shows a practical application of how the spring rate can be changed with the number of spring coils.

Torsion springs can be elegantly connected in series by a torsion bar actuating one or more torsion tubes, (Fig. 3.13). The actuating levers can thus also be arranged side by side, keeping the bearing forces concentrated in one place.

The exact design of springs is generally carried out by the spring manufacturer. For this purpose, he requires data as shown in Fig. 3.14.

Common spring diameters are 47 and 57 mm.

In the case of coil springs for racing, the spring rate is usually specified by the manufacturer on the spring. This is in the Anglican unit *pound-force per inch* [1 lbs./in = 0.175 N/mm], i.e. a spring with a characteristic value of 80, for example, has a spring rate of 14 N/mm.

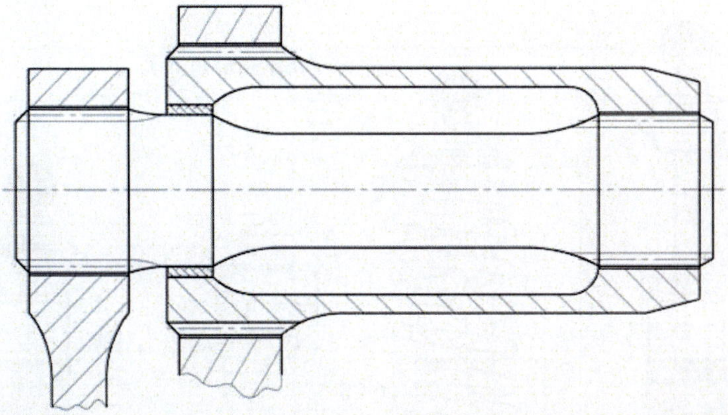

Fig. 3.13 Series connection of torsion springs. A torsion bar is mounted in a torsion tube in such a way that the reaction forces of the actuating lever are also introduced into the tube bearing. The total overall length of this arrangement is short. The total spring rate $c_{\alpha,t}$ follows in series connection from the rates of the individual springs: $\frac{1}{c_{\alpha,t}} = \frac{1}{c_{\alpha 1}} + \frac{1}{c_{\alpha 2}}$

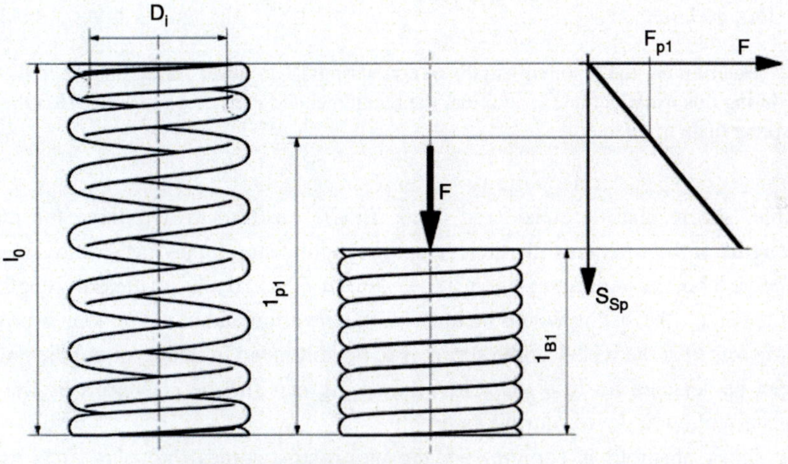

Fig. 3.14 Ordering data for coil springs. l_0 free length, mm, D_i Inside diameter, mm, l_{pl} spring length at design position, mm, F_{pl} spring load at length l_{pl}, N, l_{Bl} spring length with coils in contact, coil-bound length, mm, c_{Sp} Spring rate, N/mm

The required lengths must be coordinated with the wheel suspension and the damper strut, Fig. 3.15. The maximum spring stroke $s_{\mathrm{Sp,t}}$ results from the desired maximum wheel stroke and the travel ratio wheel/spring i_{Sp}.

The total spring stroke $s_{\mathrm{Sp,t}}$ is divided approximately in such a way that, starting from the design position, the wheel can deflect 1/3 and compress 2/3 (cf. Fig. 3.3). The lower

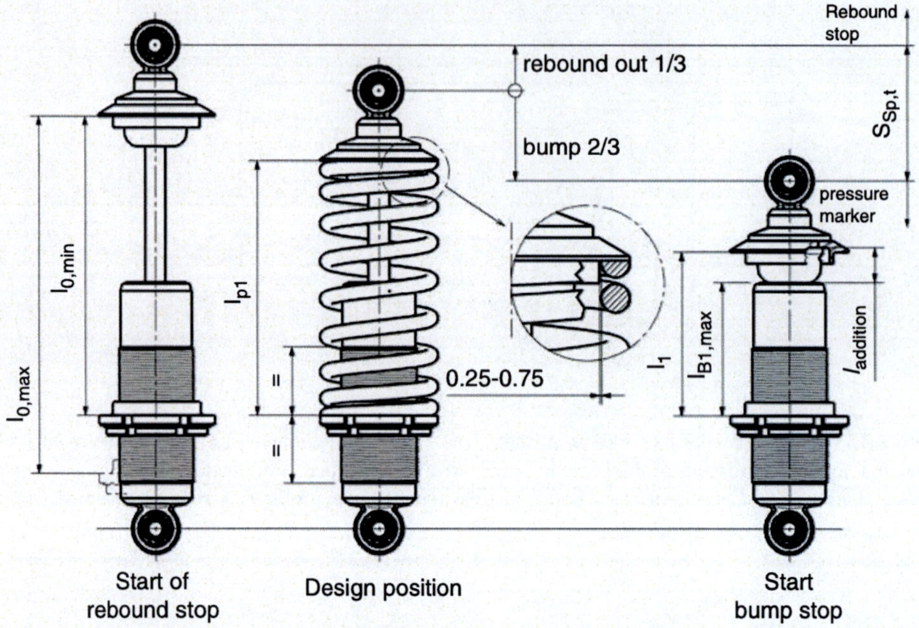

Fig. 3.15 Determining the required lengths of coil springs. The lower spring plate is in its middle position during this examination. In addition, the recommended clearance range between the spring and the spring plate is entered

(adjustable) spring plate is considered in its middle position to determine the required spring lengths. If the wheel is fully deflected, the rebound stop limits further movement and the spring reaches its minimum untensioned length $l_{0,min}$. If the unstressed length were shorter than $l_{0,min}$, the spring would be loose in this position and could no longer provide a supporting force for the wheel. If the spring is to be preloaded in this position, the unloaded length can be selected up to $l_{0,max}$. Then the spring can still be preloaded by the lower spring plate without a device during assembly.

If the shock absorber is compressed, the bump stop comes into play. It is itself an additional spring and extends the possible spring travel by $l_{addition}$. The coil spring with the length l_1 must not have reached its coil-bound length l_{Bl} yet. The coil-bound length must therefore not restrict the travel of the additional spring and must be smaller than $l_{Bl,max}$.

Materials: spring steel (Tables 3.3 and 3.4), titanium, cellular polyurethane elastomer

Table 3.3 Hot rolled steels for heat treatable springs (according to DIN 17221, 17,224)

Steel type	Abbreviation	Material number	Tensile strength R_m, N/mm^2	Yield strength $R_{p0.2}$, N/mm^2	E, G[a] N/mm^2
Quality steels	58Si7	1.0903	1320–1570	1130	
	60SiCr7	1.0961	1320–1570	1130	$E = 2.0 \cdot 10^5$
Stainless steels	50CrV4	1.8159	1370–1670	1180	$G = 80{,}000$
	51CrMoV4	1.7701	1370–1670	1180	
Stainless	X12CrNi17 7	1.4310	1320–1570		$E = 1.9 \cdot 10^5$
					$G = 73{,}000$

[a]E E-modulus, G shear modulus

Table 3.4 Steels for cold formed springs (according to DIN 17223)

Steel	Variety	Tensile strength R_m, N/mm^2	E, G[a] N/mm^2
Spring steel wire	C	1370–1670	$E = 2.06 \cdot 10^5$
	D	1370–1670	$G = 81{,}500$
54SiCr6	FD SiCr		$G = 79{,}500$

[a]E E-modulus, G shear modulus

3.2 Damper

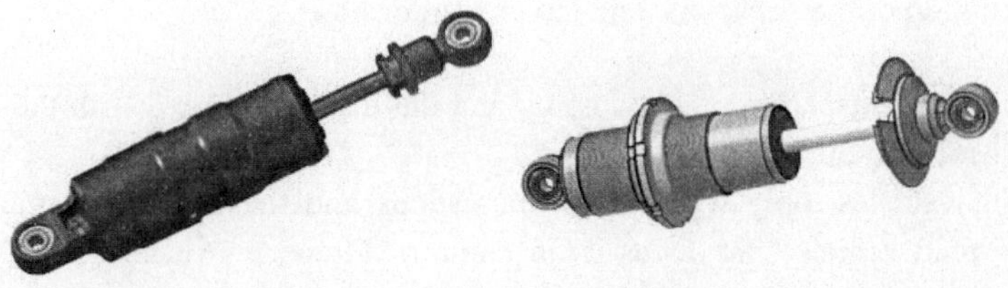

3.2.1 Oscillations

Natural Frequencies, Spring *Calculation*

Vibrations are of general interest in vehicle design for the following reasons:

- Influence on wheel load fluctuations and thus on road contact and driving safety
- Vibration load on the occupants and/or the cargo

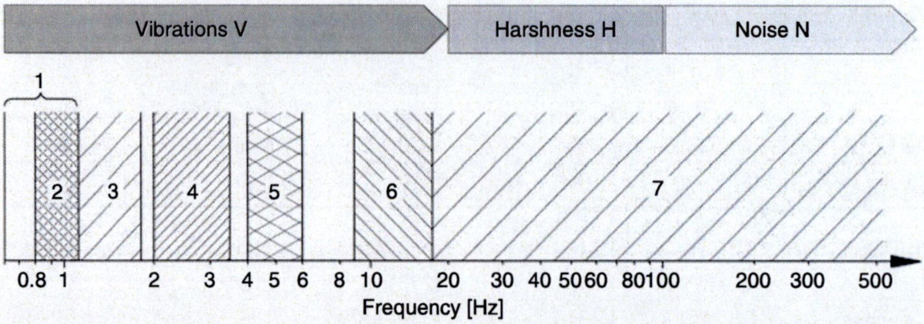

Fig. 3.16 Position of the natural frequencies on a logarithmic scale, after [1]. In vehicle-related vibration technology, this range is also referred to as NVH (*noise, vibration and harshness*), in that a distinction is made between perceptible, audible frequencies and a transition range (roughness). (1) Very low tuned system seat-human. This low tuning is only recommended in special cases. It requires a seat suspension system in which the level is adjustable, as otherwise the differences in the static deflection become too great for people of different weights. (2) System vehicle body suspension springs in the case of suspension with level control. Theoretically, these natural frequencies would also be possible for vehicles with very low permissible payloads without level control. (3) System vehicle body suspension springs without level control. When fully loaded, the vehicles are all closer to the lower limit. For the larger passenger cars and for vehicles with progressive spring characteristics, a natural frequency below 1.4 Hz is generally always achieved. (4) Frequency range proposed for the seat-human system. (5) Human natural frequency to be avoided by the other systems (vertical). (6) Natural frequency of the axle mass or the mass connected to the wheel. For most passenger cars it is close to 10 Hz. (7) Audible frequenciesThey range from 16 to 20,000 Hz (field stopped at 500 Hz)

- Vehicle and road stress
- Ground clearance variations influence the aerodynamic downforce of aero vehicles
- Noise pollution of the occupants and the environment.

Of the chassis, tire suspension and damping and the mass associated with the wheel are critical to wheel load fluctuations.

For racing vehicles, only wheel load fluctuations and thus tire-road contact are of interest. For road vehicles, the focus is on comfort. Figure 3.16 shows comfort-driven target areas of the design of road vehicles.

Natural frequency of the undamped harmonic oscillation:

$$f_n = \frac{1}{2\pi}\sqrt{\frac{c}{m}} \text{ [Hz] or : the natural angular frequency } \omega_0 = \sqrt{\frac{c}{m}} \left[\text{s}^{-1}\right] \tag{3.3}$$

e.g. wheel travel vibration with independent wheel suspension [1]:

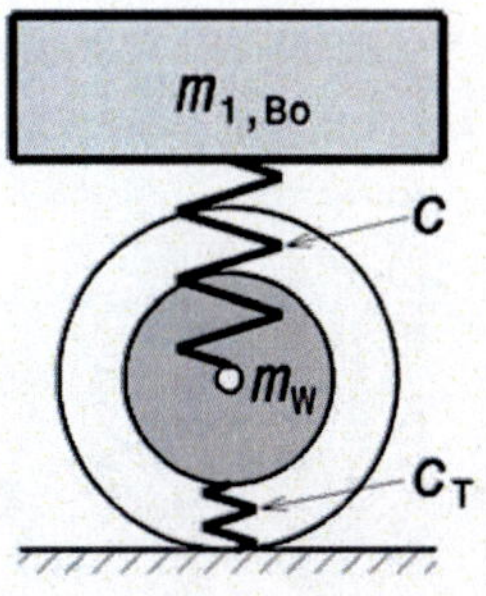

$$f_{n,W} = \frac{1}{2\pi}\sqrt{\frac{c_T + c}{m_W}} \quad (3.4)$$

$$f_{n,Bo} = \frac{1}{2\pi}\sqrt{\frac{c_T \cdot c}{m_{1,Bo}(c_T + c)}} = \frac{1}{2\pi}\sqrt{\frac{c_{Bo}}{m_{1,Bo}}} \quad (3.5)$$

c_T	*Tire rate*, N/m
c	*Wheel center rate*, N/m
m_W	Mass associated with the wheel, kg
$f_{n,W}$	Natural frequency of the unsprung masses m_W, Hz
$m_{1,Bo}$	Body mass per wheel (sprung mass), kg
$f_{n,Bo}$	Natural frequency of the sprung mass $m_{1,Bo}$, *ride frequency*, Hz
c_{Bo}	Spring rate body-related (*ride rate*), N/m
Values for c_T: Passenger cars: 120,000–200,000 N/m; racing cars up to 480,000 N/m	

The general position of the natural frequencies of the essential mass accumulations shall be chosen in such a way that they do not coincide with the natural frequencies of the human body, so that the vehicle occupant does not become a vibration absorber or, to put it more drastically, so that it does not coincide with the "vomit frequency".

Typical values of natural frequencies are summarized in Table 3.5.

In vehicle design, the vibration number n [min^{-1}] is often used instead of the frequency f [Hz]. The following therefore applies: $n = 60f = (30/\pi)\,\omega$.

Numerical values of natural frequencies (body frequencies $n = 60f_{n,Bo}$) [4]:

Road vehicle (comfortable) 60–80 min^{-1} (1–1.4 Hz); road vehicle (sporty) 80–100 min^{-1}.

Racing vehicle (without wing or ground effect) 100–125 min^{-1}, racing vehicle (with ground effect) up to 500 min^{-1}; racing vehicle (average) 200–350 min^{-1}.

Typical ranges of bodywork vibration rates of passenger cars are illustrated in Fig. 3.17.

The recommended starting value for the design of a circuit vehicle is 130 ± 15 min^{-1} [4].

The natural frequency of the vehicle body above the rear axle should be 10–20% higher than that of the body above the front axle in road vehicles. This results in a heave vibration

Table 3.5 Typical values of natural frequencies from (3.4) and (3.5)

Vehicle type or race series		Unsprung masses $f_{n,W}$, Hz	Sprung masses $f_{n,Bo}$, Hz
Passenger car		10–12	1–2
Formula student, FSAE		15–19	2.5–3.5
Rally, WRC	On tarmac	13.5–15.5	2–2.6
	On gravel	13–13.5	1.6–1.8
NASCAR		15–17[a]	1.5–4
GT3		21–25	2.8–4
Formula 3		19–25	3–5
Formula 1, LMP 1		20–37	3.5–6
IndyCar, IRL		23–27	5–7

[a]All 4 wheels have different values!

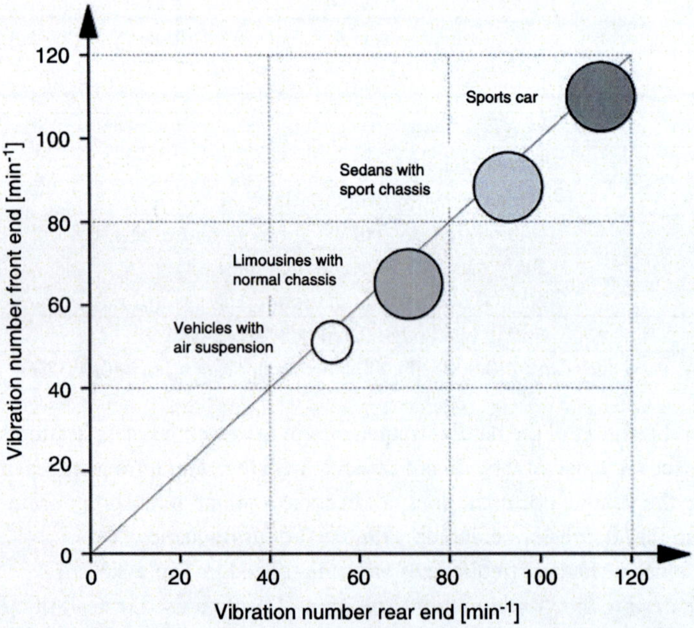

Fig. 3.17 Vibration figures (frequencies) of passenger cars, according to [10]. The frequency of the rear axle tend to be higher than those of the front axle. The higher the comfort level, the lower the frequency figures. Conversely, the values increase with increasing sportiness

of the vehicle instead of a pitching vibration during road excitation, because the movements of the rear axle, which is excited offset by the wheelbase *l*, "catch up" with those of the front axle [5]. For a certain driving speed v_{ref} (e.g. typical, average speed), the required natural frequency at the rear axle $f_{n,Bo,r}$ can be calculated exactly:

$$\Delta t = \frac{1}{f_{n,Bo,f}} - \frac{1}{f_{n,Bo,r}} = \frac{l}{v_{ref}}$$
$$\Rightarrow f_{n,Bo,r} = \frac{f_{n,Bo,f}}{1 - \Delta t \cdot f_{n,Bo,f}} \quad \text{or}\; n_r = \frac{n_f}{1 - \frac{1}{60}\Delta t \cdot n_f} \tag{3.6}$$

Δt	Time difference of the excitation between front and rear axle, s
v_{ref}	Reference or design speed at which no pitching should occur, m/s
l	Wheelbase, m

For racing vehicles, $n_r = 1.05$–$1.1\ n_f$ is generally recommended. For circuit racing vehicles with high natural frequencies, a different design may be advantageous for flat tracks [4]. Experience with rear-wheel-drive racing vehicles shows a higher natural frequency of the front axle caused by the greater roll stiffness at the front (which makes the contact forces at the rear comparable) [11]. The aforementioned comfort-related vibration behaviour is not an issue here, and racing cars are in any case much more heavily damped than passenger cars, so that no pitching vibrations develop.

However, the vibration frequency and its amplitude cannot be seen completely independently of each other. Human perception dictates an acceleration-dependent limit above which the comfort of the occupants suffers. Figure 3.18 shows this limit for excitations in vertical direction. At low frequencies the vibration excitation can be greater and at higher frequencies only a very small amplitude is tolerable for humans, although for the racing driver the limit can be shifted somewhat further into the "uncomfortable" range.

Spring Rates

The wheel-related spring rate c_f for the front axle follows from Eq. (3.8) for the *ride frequency* n_f (see below) without or with consideration of the tyre spring rate:

$$c_f = 0.011 \cdot n_f{}^2 \cdot m_{1,Bo,f}$$
$$c_f = \frac{1}{\frac{91.19}{n_f^2 \cdot m_{1,Bo,f}} - \frac{1}{c_{T,f}}} \tag{3.7}$$

c_f	Spring rate *wheel center rate for* the front axle, N/m

The wheel-related spring rate c_r for the rear axle follows from the equation above with the corresponding values for the rear axle (index r instead of f), therefore in the following only the equations for the front axle are written.

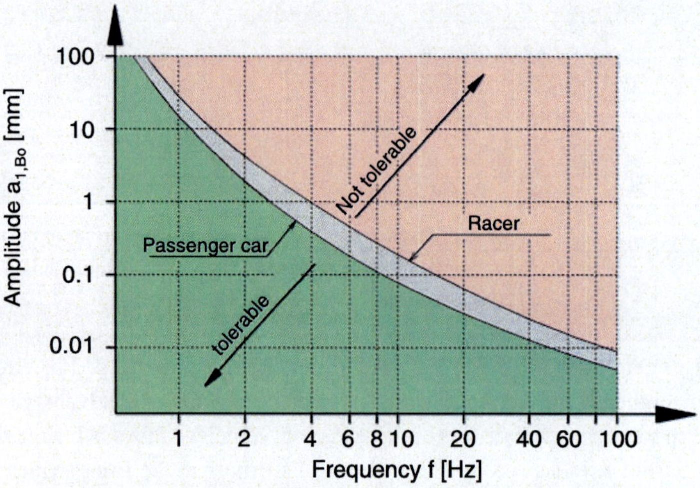

Fig. 3.18 Limit range for occupants during vertical vibrations, according to [12]. For vertical vibration excursions, frequency dictates the amplitude acceptable to humans. For racing drivers, the load can be designed rougher than for the passenger car driver who expects comfort

Experience values for initial values of wheel-related spring rates for the front axle c_f can also be taken directly from Table 3.6:

Multiplying this ratio by the body mass per wheel $m_{1,Bo,f}$ in [kg] gives the initial value of the development of the wheel-related spring rate c_f in [N/m].

$$n_f = 9.55 \cdot \sqrt{\frac{c_f}{m_{1,Bo f}}} \quad \text{or} \qquad (3.8)$$
$$n_f = 9.55 \cdot \sqrt{\frac{c_f c_{Tf}}{m_{1,Bo f}(c_f + c_{Tf})}}$$

n_f or n_r	Front or rear body vibration coefficient (ride frequency), min-[1]
c_f or c_r	Rate of the front or rear body spring in relation to the wheel contact point, N/m
m_{Bo}	Body mass, kg. $m_{Bo} = 2(m_{1,Bo,f} + m_{1,Bo,r})$
$m_{1,Bo,f}$ or $m_{1,Bo,r}$	Part of the front or rear body mass acting on a wheel, kg
$m_{1,Bo,f} = 0.5 \cdot (m_{V,f} - m_{U,f})$	
$m_{V,f}$ or $m_{V,r}$	Axle load front or rear, kg
$m_{U,f}$ or $m_{U,r}$	Unsprung mass front or rear, kg

The masses involved are illustrated (Fig. 3.19).

Table 3.6 Initial values of front spring rates (wheel-related), partly according to [13]

Vehicle type or race series	Ratio $c_f/m_{1,Bo,f}$ 1/s^2
Formula student, FSAE	247–484
Formula ford	386–425
Trans-am	463–502
Formula 2000	618
Sports car	657
ALMS LMP (Le Mans prototype USA)	772–850
CART, IRL	888–927

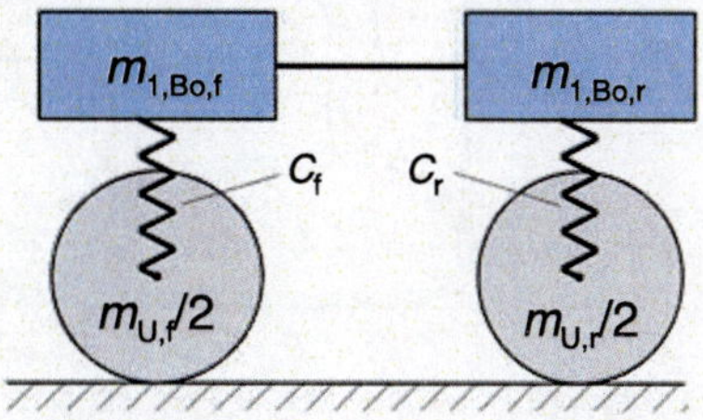

Fig. 3.19 About the equations: Illustration of a simplified oscillation system of a sprung body. Only one lateral half of the vehicle is considered, thus two quarter vehicles

The unsprung masses can be determined by measurements on the real vehicle[1] or approximately by calculation from design data, (Fig. 3.20).

Where here it has been assumed that the centre of gravity of the lower link is at a distance r_1 from the joint. The sprung mass $m_{Bo,i,t}$ is the complement to the total mass: $m_{Bo,i,t} = m_{Bo,i} + 2(0.5m_{l1} + 0.5m_a + \frac{r_2 - r_1}{r_2} m_{l2} + \frac{2r_2 - r_1}{2r_2} m_D + \frac{2r_2 - r_1}{2r_2} m_{Sp})$

$$s_{0,f} = \frac{m_{1,Bo,f} \cdot g}{c_f}$$

$s_{0,f}$ or $s_{0,r}$ static front or rear wheel suspension, m

$$c_{Sp,f} = c_f \cdot i_{Sp,f}^2$$

$c_{Sp,f}$ or $c_{Sp,r}$ rate of the body spring (suspension spring) at the front or rear, respectively, N/m
$i_{Sp,f}$ or $i_{Sp,r}$ lever ratio wheel to spring front or rear (spring ratio), –; see (Fig. 3.21)

This ratio makes the essential difference between body spring rate and wheel related spring rate. $i_{Sp} = \frac{b}{a}$

[1] See Racing Car Technology Manual, Vol. 5, *Data Analysis, Tuning and Development*, Chap. 4

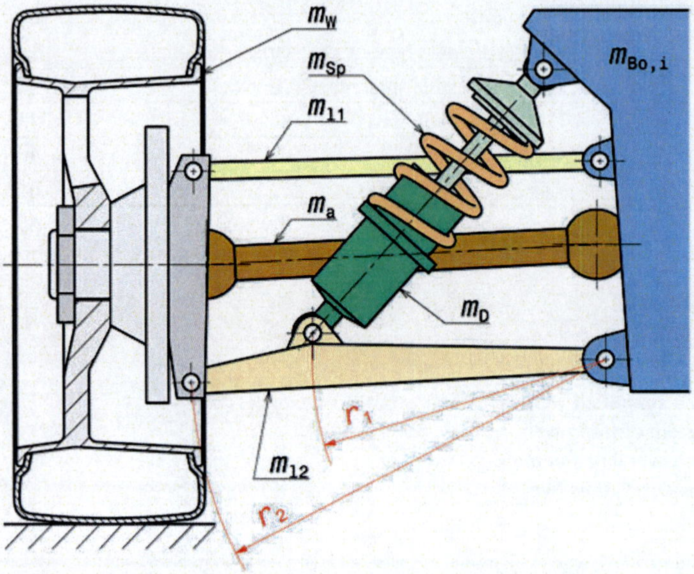

Fig. 3.20 Calculated division of the axle masses. In the case of the wheel suspension shown, the unsprung masses include the wheel (m_W) together with the tyre and all parts that move in the same direction: wheel hub with wheel nut, wheel brake and wheel carrier. Symmetrical connecting parts, the ends of which are connected on the one hand to the wheel and on the other hand to the body of the vehicle (sprung mass $m_{Bo,i}$), are each added to half of the corresponding masses. In the example, these are the upper control arm (m_{l1}) and the drive shaft (m_a). Asymmetrical parts are divided into two masses, which are assigned to the two joints. The division is made using their static moment (mass times centre of gravity distance). In the example, this concerns the lower link (m_{l2}). Parts that move indirectly with the wheel are added proportionally to the unsprung masses. The damper (m_D) only moves in the ratio r_1/r_2. In addition, the other end of the damper is attached to the body of the vehicle. Therefore, only half of its mass is taken into account in the calculation: Its unsprung portion is thus r_1/r_2–$0.5m_D$. If the damper is extremely asymmetrical, the mass of the heavier part (dark green here) is taken instead of half. However, the total balance, i.e. the total mass must correspond to the vehicle mass. The remaining mass $(1 - r_1/r_2\ 0.5)m_D$ must therefore be assigned to the sprung mass in any case. Half of the body spring (m_{Sp}) is also taken into account in the unsprung and half in the sprung masses. The situation is different for a typical leaf spring. Approximately 90% of it is included in the unsprung masses. The unsprung masses of this suspension (left and right side) are summarised as follows: $m_{U,i} = 2(m_W + 0.5m_{l1} + 0.5m_a + \frac{r_1}{r_2}m_{l2} + 0.5\frac{r_1}{r_2}m_D + 0.5\frac{r_1}{r_2}m_{Sp})$

$$s_{Sp,f} = \frac{s_{0,f}}{i_{Sp,f}} = \frac{m_{1,Bo,f} \cdot g \cdot i_{Sp,f}}{c_{Sp,f}}$$

$s_{Sp,f}$ or $s_{Sp,r}$ static compression of the spring at the front or rear, m

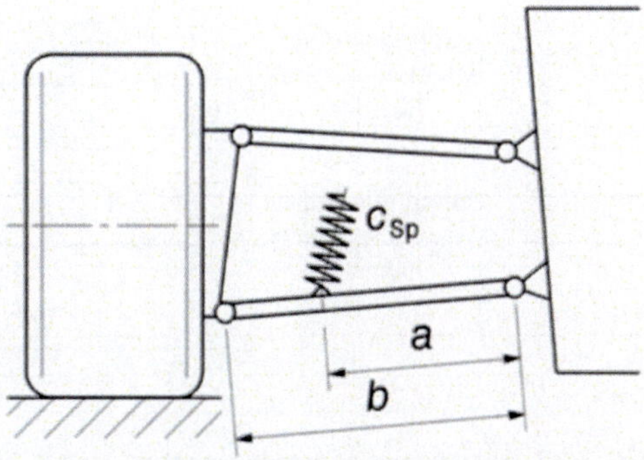

Fig. 3.21 Example of lever ratio wheel to body spring (spring ratio), cf. also (Fig. 3.7) or (3.1)

Spring Travel

The total suspension travel $s_{Sp,t}$ (i.e. from fully extended to fully compressed) is dictated by the course of the race track (the more undulating, the more suspension travel), but cannot be determined independently of the commercially available damper struts. The common total travels are approximately between 80 mm (e.g. for single-seaters, small sports cars) and 150 mm (e.g. for sports cars, touring cars) [4].

For the distribution of the available spring travel (see Fig. 3.3) and Table 3.1.

As an extreme example, the following numerical values resulted for a Formula 1 car of the "wing era" (before 1983) at low driving speed, i.e. the influence of the downforce is still very low, according to [4]:

Vehicle mass: 580 kg, i.e. total mass including driver (74 kg) and half tank (73 kg) is 727 kg.

With a front/rear mass distribution of 45:55, the axle loads are $m_{V,f} = 327$ kg and $m_{V,r} = 400$ kg respectively.

	Compound Oscillation Figure	Body mass per wheel	Leverage ratio	Wheel related rate of the spring	Static deflection	Rate of the spring	Static deflection of the spring
	n	$m_{1,Bo}$	i_{Sp}	c	s_0	c_{Sp}	s_{Sp}
	min-1	Kg	–	N/mm	mm	N/mm	mm
Front	410	143	2.0	264	5.3	1056	2.65
Rear	504	175	1.2	489	3.6	704	3.0

The body supporting springs were therefore designed to be extremely stiff, because these vehicles had to maintain ground clearance even at high driving speeds with an enormous downforce, so that the then customary skirts for undertray sealing remained effective. The spring travel was also kept correspondingly small at approx. 38 mm.

The spring rates should not be chosen too high in the basic design. At the rear axle one will start with such soft springs that the rear does not touch the ground at the selected

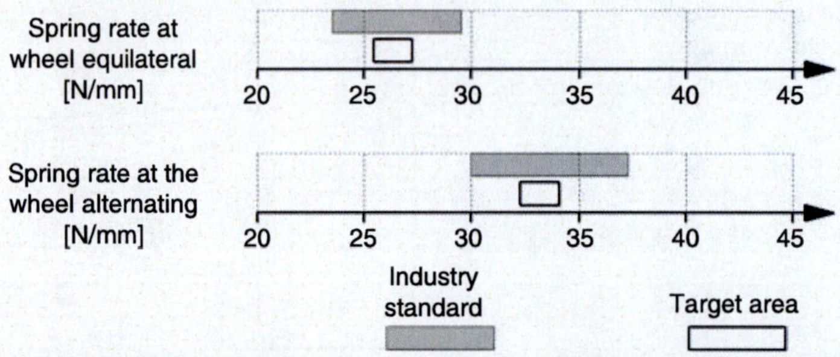

Fig. 3.22 Spring rates of passenger car rear axles, after [10]. With reciprocal springing (roll motion), the stiffening effect of the torsion stabilizer is added and the rates increase compared to heave motion (equilateral springing)

ground clearance. With the suspension of the front axle the basic driving behaviour can then be adjusted.

Rough guide values for spring rates for comparison are shown (Fig. 3.22) for passenger car rear axles. Heavy touring cars, such as those operated on oval tracks in the USA, have rates around 175 N/mm.

Wheel load fluctuations follow from the excitation frequency f *of* the road surface, which depend, among other things, on the wheel-related spring hardness c and the damper constant k_D, (Fig. 3.23).

3.2.2 Vibration Dampers (Shock Absorbers)

Basics

Due to the course of the more or less uneven road surface, a vehicle is stimulated to vibrate via the wheels when driving. Dampers are needed to ensure that these quickly subside and that the wheels do not lose contact with the ground due to bouncing. In vehicle design, these vibration dampers are also called shock absorbers. In passenger cars, this inevitably results in a conflict of objectives. Driving safety demands the greatest possible road contact of the tires, i.e. a tight damper, while at the same time comfort is maintained by low body accelerations, i.e. small damper forces and large suspension travel. On racing vehicles, on the other hand, the dampers can be specifically designed for the lowest wheel load fluctuation.

Common hydraulic telescopic dampers always act when they change their length. In driving operation, this has several effects:

- Dampers absorb part of the spring energy during rebound and thus control vibrations (especially of the sprung mass).

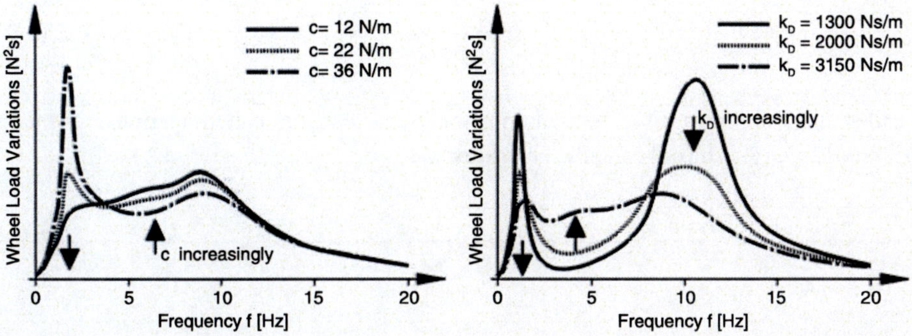

Fig. 3.23 Influence of the spring rate on the wheel load fluctuations [14]. *c* wheel-related spring rate, k_D damper constant

- During spring compression, they harden the spring, so to speak, because they briefly provide a force that is equal to the spring force. This means that softer body springs can be used. They also delay the lift-off of the wheels. So they also affect wheel load variation and thus driving safety. The effects of a variable wheel load on the tyre are dealt with in Sect. 1.2.3, subsection Wheel load.[2]
- In transitional phases, they reduce the body movement, i.e. the pitching during braking and acceleration as well as the swaying and diagonal tilting of the car body at the entrance and exit of corners.
- They slow down body movements, which is particularly advantageous for vehicles with aerodynamic downforce aids, because the air forces also change more slowly as a result.

Body movements take place in the frequency range below 5 Hz. The unsprung masses oscillate at about 5–20 Hz and the deformation of the tyre contact patch or tread occurs at over 20 Hz.

An initial estimate of the damping force can be made using the differential equation of the single mass system in the *z*-direction. The accelerated mass *m is* acted upon by a velocity-proportional damping force and a displacement-proportional spring force:

$m\ddot{z} + k\dot{z} + cz = 0$	m	Mass, kg
	z	Position of mass, m
Trägheits-kraft (→ $m\ddot{z}$), Dämpf-kraft (→ $k\dot{z}$), Feder-kraft (→ cz)	k	Damping rate (*damping coefficient*), ns/m
	c	Spring rate, N/m

The value

[2] In the Racing Car Technology Manual, vol. 5 *Data Analysis, Tuning and Development,* Chap. 6 shows the effects of different damper effectiveness in road tests.

$$D = \frac{k}{2\sqrt{cm}} \tag{3.9}$$

is called the *damping ratio*. The denominator $2\sqrt{cm} = k_{ap}$ is called aperiodic *damping* (*critical damping*). Thus D can also be written as:

$$D = \frac{k}{k_{ap}} \tag{3.10}$$

D	Damping, *damping ratio*, –

D is therefore the ratio of the current damping rate to the aperiodic one.

Borderline cases	$D = 1$	Aperiodic movement (*non-oscillatory movement*), i.e. after deflection, the mass moves directly back to its rest position without oscillating
	$D = 0$	Undamped oscillation, i.e. after deflection the mass oscillates eternally around the rest position with the natural frequency (*resonant frequency*) $\omega_0 = \sqrt{c/m}$

The following values occur with vehicle vertical movements:

Passenger cars $D =$ approx. 0.3–0.4, sports cars $D = 0.5$, formula cars $D = 0.7$ and for some racing cars with extremely high downforce $D > 1$. These values apply to the rebound of a damper, i.e. to the damping of the sprung mass (see below).

Note: If there is a transmission ratio between the wheel travel and the spring or damper travel, this ratio is squared in k and c, cf. (3.2) (see also Fig. 3.7)

The fundamental influence of damping on oscillations can be clearly seen (Fig. 3.24). The undamped periodic motion ($D = 0$) would theoretically never decay in a frictionless state, because the potential energy of the spring and the kinetic energy of the mass would remain constant in total and increase or decrease in opposite directions. If damping comes into play, part of the energy in the damper is converted into heat and is lost to the oscillation – the amplitude decays over time. Interesting is on the one hand the aperiodic limit case ($D = 1$). The mass creeps back to the rest position ($z = 0$) after deflection without overshooting. On the other hand, the behaviour at an even greater damping is interesting ($D > 1$, overdamped). The mass moves even slower back to the equilibrium position. For a wheel suspension, the obvious case of aperiodic damping actually proves not to be an ideal solution. Although the wheel does not overshoot, it takes longer to reach the initial position $z = 0$. A good compromise between speed (reaction of the system to disturbances) and damping is $D = 0.7$. The wheel moves faster through the initial position and is thus "ready" earlier for the next bump. Race cars with overdamped settings use the high damper forces primarily to minimize roll and pitch without hard springs or anti-roll bars, which can help with extreme aerodynamic designs at some tracks. In any case, a damper design that is too

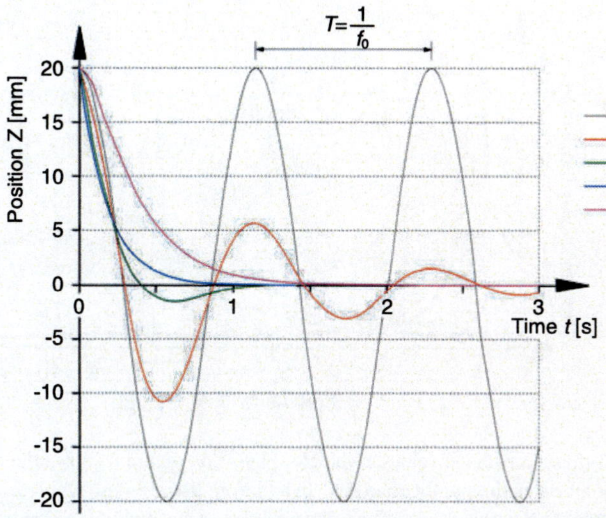

Fig. 3.24 Influence of damping on vibrations. A spring-loaded mass is deflected by 20 mm and released. The following movement is plotted against time. The damping D varies between 0 (undamped), 1 (aperiodic limit case) and greater than 1 (overdamped)

stiff is bad because it increases wheel load fluctuations, which reduces the grip level of the tires. Generally, this occurs with damping from $D = 1.25$ to 1.75.

The handling at extreme damping settings can be characterized like this: In the overdamped setting, the wheels rattle over undulating road surfaces and the car feels unsteady as well as nervous. In the underdamped case, the body floats and the vehicle reacts only sluggishly to steering inputs. The tyres easily lose contact with the ground in waves.

In fact, the tyre itself acts like a spring. A more detailed analysis of the vibration behaviour of a vehicle is thus obtained by the model of a two-mass oscillator, (Fig. 3.25). A wheel and the proportional body mass are considered. The wheel represents the unsprung masses, which account for about 8–10% of the vehicle mass in a passenger car [15]. In a racing vehicle, the proportion is higher because the wheels have similar masses but the vehicles are lighter. Thus, the proportion is from 12 to over 15%. The wheel mass m_W is sprung by the tire with the spring rate c_T. A part of the body mass $m_{1,Bo}$ is supported on the wheel by the body spring with the spring rate c. The damping is in proportion to the speed. The damping is proportional to the speed with the rate k. The tire damping is very small and is neglected here. Excitation is from the road (footpoint excitation) according to $h(t) = \hat{h} \cdot \cos \omega t$ ($\hat{h}$ is the amplitude of the road wave, ω is the excitation angular frequency). The excitation frequency results from the "wavelength" L *of* the roadway and the travel speed v_V:

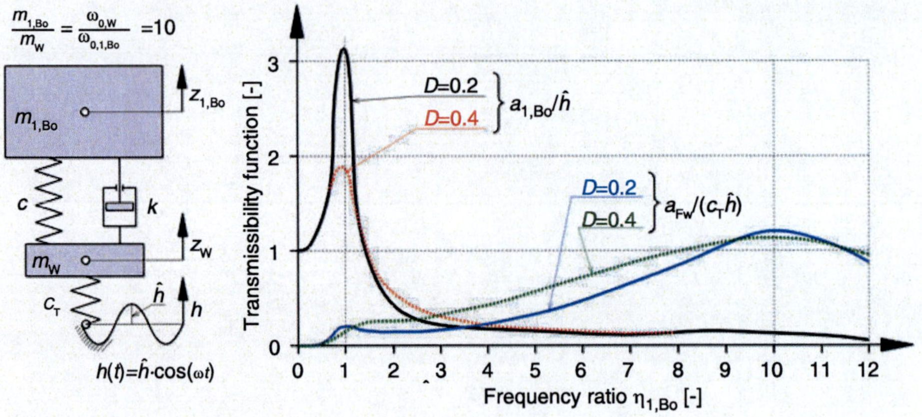

Fig. 3.25 Damped dual-mass transducer (quarter vehicle) [15]. The diagram shows the response of the body mass to a footpoint excitation by the roadway. In addition, the resulting wheel load variation is entered. $m_{1,\mathrm{Bo}}$ Proportionate body mass, m_{W} Wheel mass, $a_{1,\mathrm{Bo}}$ Amplitude of the body movement $z_{1,\mathrm{Bo}}$, a_{Fw} Amplitude of the wheel load variation, $\eta_{1,\mathrm{Bo}} = \omega/\omega 0_{,1,\mathrm{Bo}}$, ω Excitation frequency, s-^{1}, $\omega_{0,1,\mathrm{Bo}}$ Natural frequency of $m_{1,\mathrm{Bo}}$, s-^{1}, $c/c_{\mathrm{T}} = 1/10$

$\omega = 2\pi \frac{v_V}{L} = 2\pi \cdot f$	ω	Excitation circuit frequency, s-^{1}	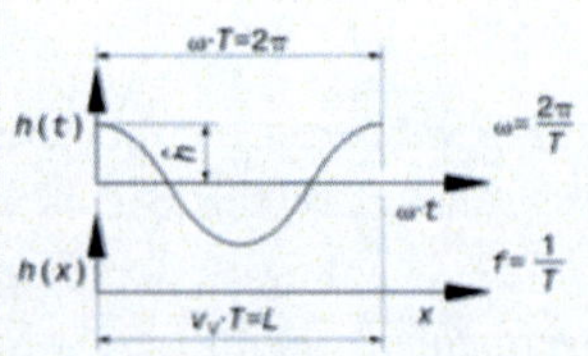
	f	Excitation frequency, Hz	
	v_V	Driving speed, m/s	
	L	Distance between roadway waves, m	
	T	Period duration, s	

The magnification function $a_{1,Bo}/\widehat{h}$ has two maxima: Once at the body resonance ($\eta_{1,\mathrm{Bo}} = 1$) and once at the natural frequency of the wheel mass ($\eta_{1,\mathrm{Bo}} = 10$). The ratio of the two masses is 10 and the corresponding spring stiffnesses behave in exactly the opposite way. From this follows with (3.4) and (3.5) the given relation of the resonance frequencies. As the damping ratio increases, the amplitude of the body motion decreases significantly. The diagram also shows the course of the wheel load variation as a magnification function $a_{Fw}/\left(c_T\widehat{h}\right)$ (blue and green curves). This also shows the two maxima mentioned, although much less pronounced. A stronger damping hardly reduces the amplitudes of the wheel load fluctuation in the resonance case, but widens the affected area (green curve). The small reduction of the wheel load amplitude in the vicinity of the vehicle resonance ($\eta_{1,\mathrm{Bo}} = 1$) is to be evaluated higher for a real vehicle, because for long road waves (i.e. small excitation frequency) the excitation amplitude $\widehat{h}$ is considerably larger than for short waves.

The damper forces and the accelerations caused by them have a considerable influence on the driving behaviour:

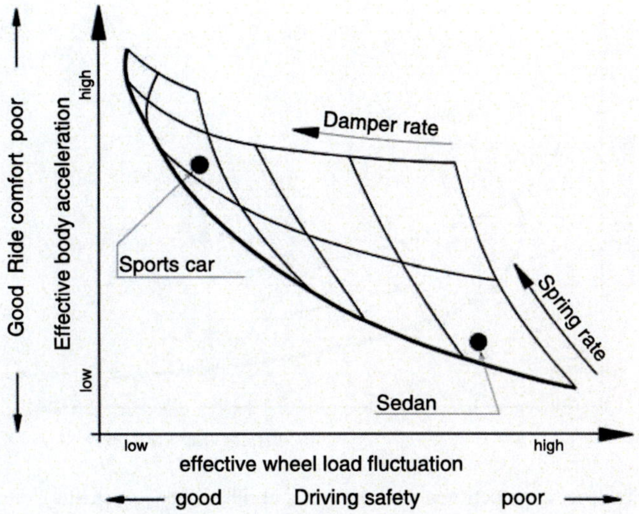

Fig. 3.26 Influence of spring stiffness and damper rate on driving behaviour, after [16]. A stiff body spring and a hard damper increase driving safety enormously, but reduce driving comfort due to the high body accelerations caused. For racing vehicles, the choice of spring and damper rate is easier in this respect. For them, only driving safety counts

Damping force adjustment:	Low damping force	→ High driving comfort, low body acceleration
	High damping force	→ low ride comfort, but lower wheel load fluctuations – Higher safety with sporty driving style

This results in a conflict of objectives for series-production vehicles. On the one hand, ride comfort should be high and on the other hand, driving safety should not suffer too much. Variable dampers provide a solution to this conflict of objectives (Fig. 3.26).

The fact that a high damper force does not increase driving safety without restrictions is demonstrated by (Fig. 3.27). If the damping force becomes too high, the wheel load fluctuation exceeds a minimum and then increases again (!), while comfort deteriorates drastically. If the damping is too tight, the spring no longer manages to track the wheel to the bumps, and the wheel bounces from one bump to the next [15]. A harder spring tuning (dashed line) basically does not change this. The road grip improves slightly, but if the damping is too strong, the wheel load fluctuation also increases again.

It is more advantageous to use softer body springs. With a desired damping D, this results in a smaller damping rate k (D is directly proportional to the damping rate k and indirectly to the square root of the spring rate c). This in turn is reflected in lower damping forces and thus lower stresses on the vehicle and its occupants.

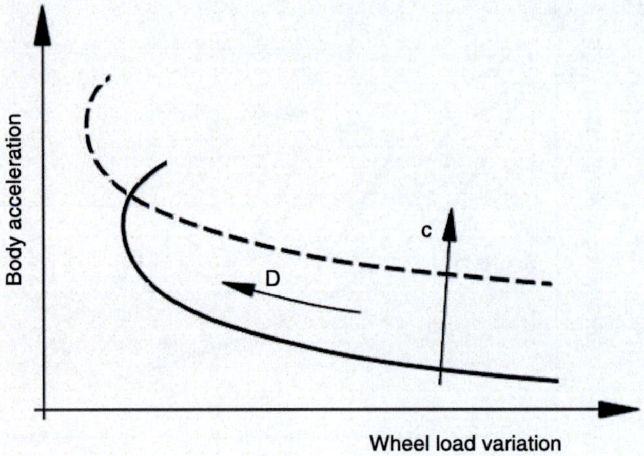

Fig. 3.27 Coordination scope between ride comfort and ride safety (schematic), after [15]. For the meaning of the axes (see also Fig. 3.26). Damper rate. *c* Rate of the body spring

Damper *Designs*

Of the various basic designs, the telescopic damper has prevailed. With this in turn there are two different versions. The older twin-tube damper and the single-tube or gas pressure damper developed from it.

Twin Tube Damper

Figure 3.28 shows the principle of operation of a twin-tube damper. When the piston rod (1) is retracted, the oil flows through valve A of the damper piston. However, this valve does not have a large throttle because the oil in the working chamber (5) is practically only under atmospheric pressure and vapour bubble formation (cavitation) could be the result in the event of large pressure drops. In the bottom valve (10), the main throttling occurs in the pressure stage, as the retraction of the piston rod is called. This means that the throttle D of the bottom valve must produce a greater resistance to the oil flow than the bore A. The oil flows into the bottom valve. The oil then flows into the equalizing chamber (8). The equalizing chamber must be approximately half full. This prevents air from being sucked through the bottom valve into the working chamber during extreme driving conditions. The oil level in the compensation chamber drops even further when the damper is installed at an angle at the top relative to the bottom valve. Therefore, there are limits to a slanted installation.

In the opposite movement, the rebound stage, the oil must pass through the throttle bore B. Above the piston, the oil volume is smaller than below, because the piston rod occupies a space. The oil that is therefore missing below the piston is sucked out of the expansion chamber through the hole C in the bottom valve. The oil is compressed above the piston

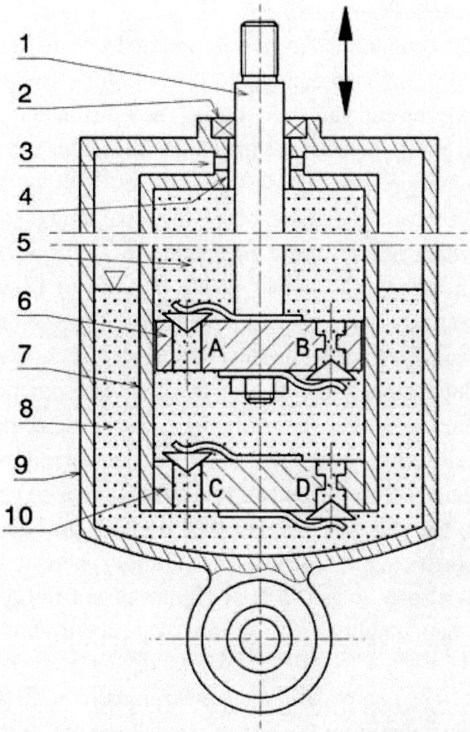

Fig. 3.28 Mode of operation of a twin-tube damper. **1** Piston rod. **2** Piston rod seal. **3** Return hole. **4** Piston rod guide. **5** Working chamber. **6** Piston with valves A and B. **7** *Inner* tube. **8** *Reservoir*. **9** Outer tube. **10** Foot valve with valves C and D

and inevitably a small amount of oil is also forced through the sealing gap of the piston rod guide (4). This oil reaches the equalizing chamber via the return bore (3).

The twin-tube damper also works with low oil loss, but has the disadvantage that its installation position is restricted by a 45° range around the vertical when compressed.

Pros:

- simple design
- cheap
- low friction due to piston rod seal with relatively low contact forces
- small overall length.

Disadvantages:

- effective oil cooling only in the annular compensation chamber
- smaller piston diameter with given damper outer diameter leads to higher working pressures and risk of fading.
- air and oil can mix: foam formation leads to reduced damping
- installation position strongly restricted
- sinking of the oil column in the working chamber when the vehicle is stationary for a longer period of time.

Mono *Tube Damper*

The common design for production and racing vehicles is the gas pressure damper, (Fig. 3.29). In contrast to the twin-tube damper, in this design the damper piston (3) has two throttle valves A and B. The oil is under pressure to prevent vapour bubble formation in bump. In order to prevent the formation of vapour bubbles in the pressure stage, the oil is under pressure. For this purpose, the damper has a gas volume (6) which is separated from the working chamber (4) by the separating piston (5). The gas used is usually nitrogen (N_2), which has an upstream pressure of between 10 and 30 bar. Without a compressible gas chamber, the damper piston would not be able to retract at all. This gas spring also compensates for the volume released by the piston rod (1) in the rebound stage. However, the high gas pressure must be sealed accordingly. The seal (2) is therefore more strongly dimensioned than that of the twin-tube damper. As a result, the breakaway forces of the monotube damper are higher. On the other hand, it can be installed in any position and because of the preload of the oil, this damper responds more quickly. The different surface ratios on the top and bottom of the piston do not have a noticeable effect on the throttle due to the gas pressure, but they do have a noticeable effect on the outside. The differential pressure of the gas towards the atmosphere acts on the piston rod cross-section. This results in a force (e.g. 196 N at 10 mm piston rod diameter and 25 bar) that can lift the vehicle when simply changing from twin-tube to monotube dampers.

The compensation chamber together with the separating piston can also be structurally separated from the rest of the damper and is then connected to the working chamber by a pressure line. In this way, the disadvantage of the longer design compared to the two-pipe design can be eliminated.

The single-pipe version has a number of advantages over the two-pipe version:

- good cooling, because the cylinder tube can give off the heat directly to the environment
- a larger piston diameter is possible with the same outer diameter
- mounting position any
- no foaming of the oil (because of the pressure pre-tension)
- multi-part design possible (damper element and compensation chamber structurally separate). This facilitates the accommodation.

The disadvantages are:

- greater breakaway forces (response of the damper)
- temperature-dependent piston rod extension force
- higher costs.

Types *of* Installation

Dampers – like the springs – can either be arranged directly between the frame or chassis and the wheel carrier or a link, or they can be actuated indirectly via bell cranks. Normally the dampers are installed in such a way that bump is effective when the suspension is

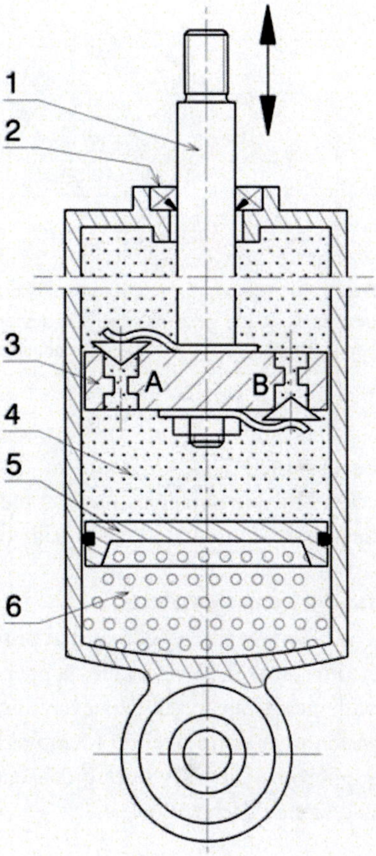

Fig. 3.29 Mode of operation of a single-tube damper (gas pressure damper). **1** Piston rod. **2** Piston rod seal. **3** Piston with valves A and B. **4** Working chamber. **5** Separating piston (*floating piston*). **6** *Reservoir*

compressed and rebound when the suspension is released. If the mounting position of a damper is optional, which is the case with a gas pressure damper, one will prefer the arrangement where the unsprung masses are smaller. This means that the heavier cylinder tube with the compensation chamber (reservoir) should be fixed to the frame, while the lighter piston rod moves with the other unsprung masses.

A typical bracket for mounting a damper strut eye is shown in (Fig. 3.30).

The choice of the appropriate damper ratio can become a problem for dampers with conventional valves. The occurring velocity range of the damper piston becomes too large. The tendency towards low spring travel combined with stiff body springs leads to low damper piston velocities. If the valves in the damper are now tuned for these speeds, which is suitable for the majority of vehicle speeds on the circuit, they are overtaxed in the event of shocks. Such shocks occur on the circuit during tight line selection with crossing curbs or on road courses through banking and potholes. Conversely, the valve setting for high piston speeds becomes useless for the rest of the conditions. The solution is offered by dampers with additional compression valves for high piston speeds.

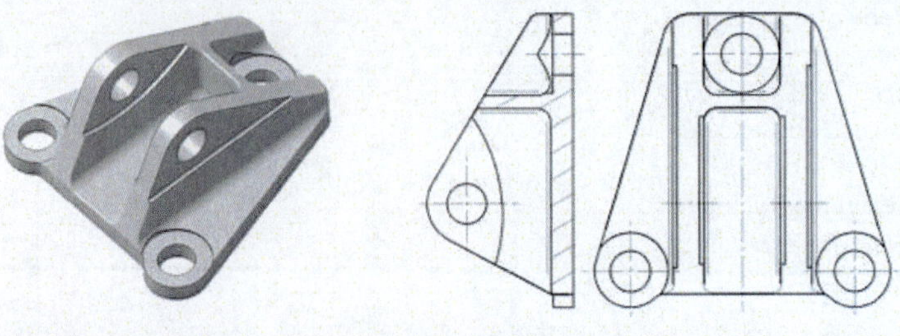

Fig. 3.30 Bracket for one damper. The console is installed on the car side, i.e. bolteded to the frame or chassis, to the gearbox, etc. It is designed for the preferred pull/pressure direction of the damper – approximately parallel to the contact surface

The dampers absorb work due to their function. They convert this into heat, which must be dissipated to the environment. If dampers are located inside the car, they must be force-cooled in some vehicles (e.g. on the rear axle of production sports cars), e.g. through air ducts that are routed to the outside of the working chambers.

Damper Characteristics

The damping force, such as that provided by hydraulic dampers, is proportional to speed. At low stroke speeds of the damper, the reaction forces are small and the forces increase with increasing speed. However, this also means that if the piston rod or more precisely the piston does not move, no (damping) force is provided. If such forces are needed, e.g. to support the car body during constant cornering, they must be provided by body springs and/or stabilisers.

The damper force results from a pressure difference between the top and bottom of the damper piston and the effective piston area. In the static case, however, the pressures are equal and the following applies (see also Fig. 3.31):

$$\begin{aligned} & p_1 = p_2 = p_N \\ & F_D = F_0 = F_1 - F_2 = p_1 D^2 \frac{\pi}{4} - p_2 \left(D^2 - d^2\right) \frac{\pi}{4} \\ & \rightarrow F_0 = p_N d^2 \frac{\pi}{4} \end{aligned} \tag{3.11}$$

p_1, p_2	Pressure in oil chamber, N/mm²
p_N	Pressure in the compensation space, N/mm²
F_D	Damper force, N
F_1, F_2	Auxiliary forces, N
F_0	Static damper force, N
D	Damper piston diameter, mm
d	Piston rod diameter, mm

Fig. 3.31 Calculation sketch for damper force F_D of a gas pressure damper

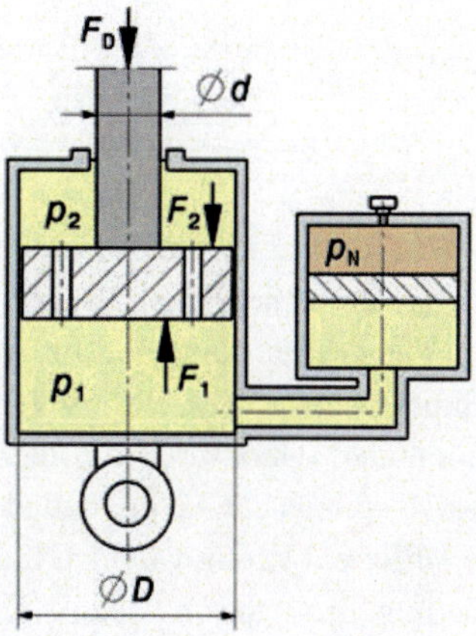

If the damper is at rest, the gas pressure acts on the piston rod and the damper moves to the rebound stop.

If the damper changes its length due to the external force F_D, oil is pressed through the throttle holes, causing a pressure difference in the working chambers. The damper force is now represented as follows:

$$\begin{aligned} p_1 &= p_N p_1 \neq p_2 \\ \Delta p &= p_1 - p_2 \\ F_D &= F_1 - F_2 = F_0 + \Delta p \left(D^2 - d^2\right) \frac{\pi}{4} \end{aligned} \tag{3.12}$$

Δp	Pressure difference, N/mm^2

Note that the force entered in the figure is the damping force of bump. At rebound, $p_2 > p_1$ and the pressure difference becomes negative. Therefore, the damper force also becomes negative. It therefore points in the opposite direction.

The pressure drop Δp together with the thereby shifted volume ΔV of the damper oil represents the work W_D, which the damper absorbs and converts into heat Q_D. From this equality, the temperature increase ΔT (neglecting heat dissipation to the environment) can be determined for one damper stroke:

$$W_D = \Delta p \Delta V \equiv Q_D = m_D c_p \Delta T$$

$$\Rightarrow \Delta T = \Delta p \frac{\Delta V}{m_D} \frac{1}{c_p} = \Delta p \frac{1}{\rho_D} \frac{1}{c_p}$$

m_D	Mass of the damper oil, kg
ρ_D	Density of damper oil, kg/m^3. $\rho_D = 850–1200$ kg/m^3 (at 40 °C)
c_p	Specific heat capacity of the damper oil, J/(kgK). $c_p \approx 2000$ J/(kgK) (at 0–100 °C)

For $\Delta p = 1$ N/mm$^2 = 1$ MPa, this results in a temperature increase of approx. 0.6 °C per stroke in the heat-sealed state.

The desired course of the pressure drop Δp and thus of the damper forces is set via the throttle valves. There are several ways of influencing this in principle, which can also be combined. Uncovered throttle bores and those with spring-loaded valves are used. The valve springs can be preloaded or several throttle bores can be released one after the other by different strong valve springs.

In general, the characteristic curve of a complete damper for one direction of movement (compression or rebound) will look like in (Fig. 3.32). There are three distinct areas that are noticeable: Nose, knee, and slope (rise). The first region (nose) describes the damping of body movements (roll, pitch, heave) and the last (slope) describes that of wheel movements. The transition area in between (knee) can be more or less pronounced. The slope of the curve corresponds to the damping rate k ($k = dF_D/dv_D$). It can be seen that the damping in the low-speed range (body movements) is considerably greater than in the high-speed range (wheel travel movements).

Recent developments enrich the adjustment possibilities by a decisive parameter, namely the viscosity of the damper oil itself. So-called electro-rheological fluids made of silicone oil and polymers as well as magneto-rheological fluids, a suspension with iron-containing particles, change their properties under the influence of an electric current. Thus,

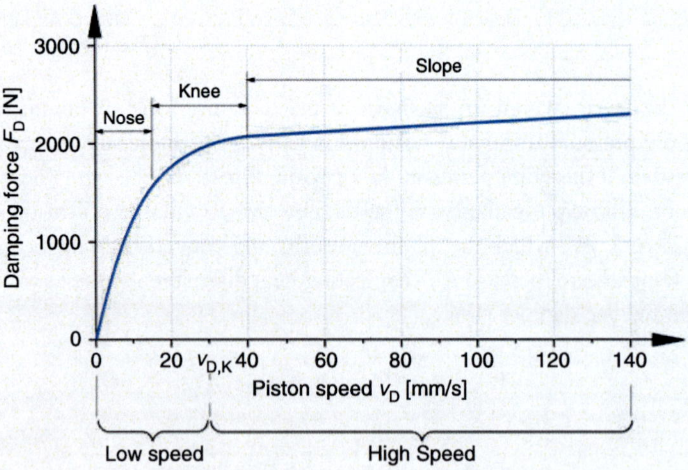

Fig. 3.32 General damper characteristic curve. The course of the characteristic curve can typically be divided into three areas: Nose, *knee*, *slope*. $v_{D,K}$ knee speed. The knee speed represents the boundary between the low and high speed ranges

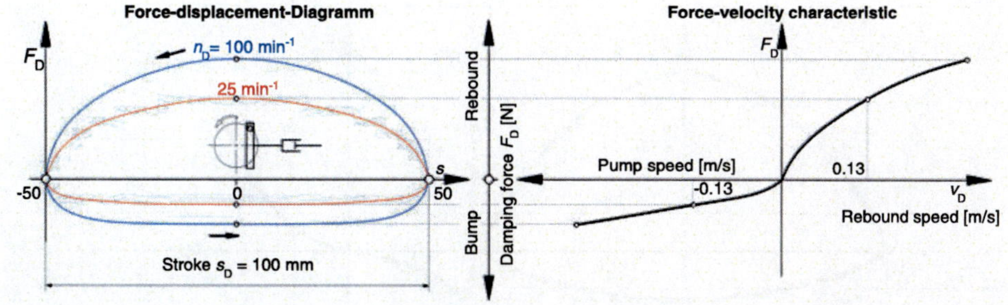

Fig. 3.33 Determination of damper characteristics, according to [5]. The values from the testing machine (left) are plotted as a force-velocity curve

an electrical signal in the millisecond range can be used to change the viscosity and thus the reaction force of the damper. Dampers of this type are already in series use in luxury-class cars.

To determine damper characteristics (force-velocity characteristic), the damper is operated on a testing machine[3] with constant stroke and the speed n_D of the machine is varied in steps and thus the damper speeds. This gives the force-displacement diagram, (Fig. 3.33) on the left. The largest forces measured in the direction of rebound and bump are entered in the diagram on the right at the speed associated with the rebound. The largest values follow the relationship:

$$v_{D,\max} = \frac{\pi \cdot s_D \cdot n_D}{60}$$

$v_{D,max}$	Max. piston speed, m/s
s_D	Damper stroke, m
n_D	Speed of the testing machine, min-[1]

Testing machines are also used in which the stroke is changed in steps at a fixed speed.

Piston speeds for road and circuit vehicles can be roughly divided into these ranges [4]:

Low	0–0.05 m/s
Medium	0.05–0.13 m/s
High	0.13–0.20 m/s
Particularly high	0.23–0.30 m/s
Extremely high	Above 0.33 m/s

[3]See Racing Car Technology Manual Vol. 5 *Data Analysis, Tuning and Development*, Chap. 6 *Development.*

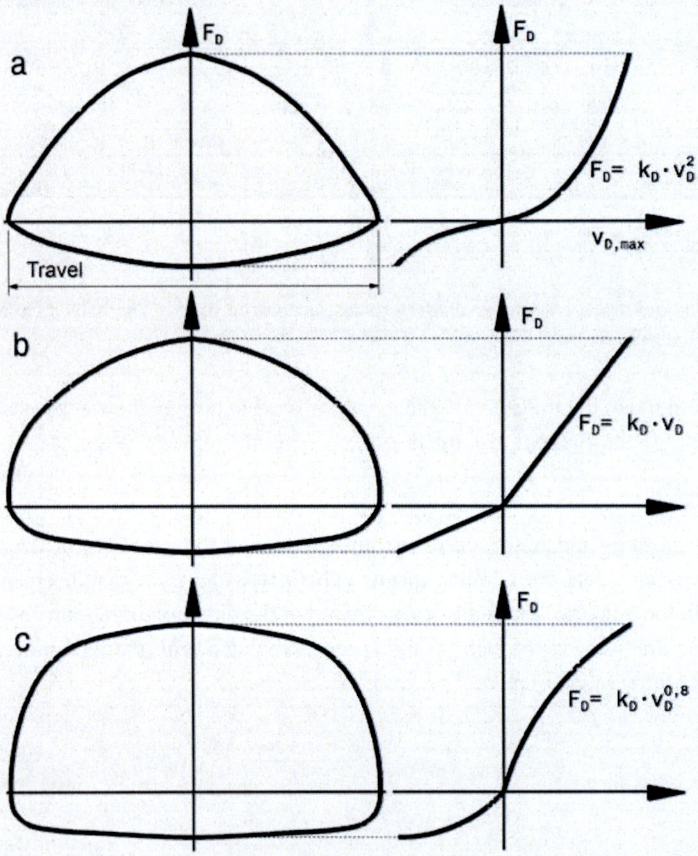

Fig. 3.34 Damper diagrams [5]. The damping characteristic F_D (v_d) can be progressive (**a**), linear (**b**) or degressive (**c**). Curve shape and diagram shape F_D (stroke) are directly related. The diagram belonging to a progressive curve has the smallest area and thus the lowest damping, and the diagram belonging to degressive damping has the largest area

Rally and off-road vehicles reach up to 3 m/s with correspondingly larger damper strokes. In extreme cases – during hard landings – 7–8 m/s have also been measured [17].

The basic curve of a resistance force through a throttle bore increases with the square of the oil velocity (i.e. progressively). However, linear and decreasing (degressive) characteristics can also be achieved by appropriate design of bores and valves, Fig. 3.34.[4]

[4] See also Racing Car Technology Manual Vol. 5 *Data Analysis, Tuning and Development*, Chap. 6 *Development*.

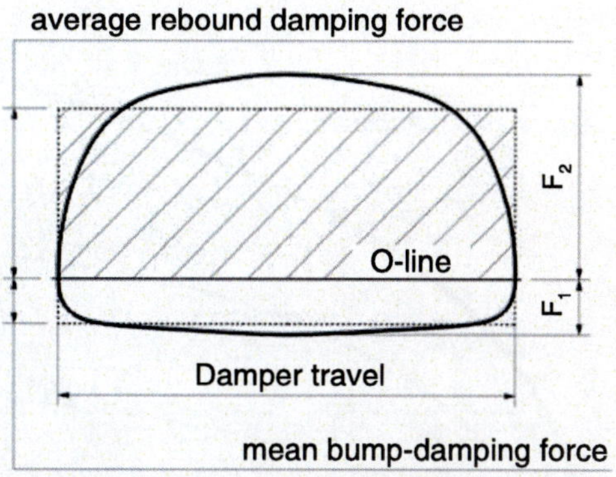

Fig. 3.35 Mean damping force, according to [5]. The mean damping force in the direction of rebound is the height of the rectangle with length = stroke, which has the same area as the section under the damper curve above the 0-line. The mean damping force in the compression direction is obtained in the same way below the 0-line

$$F_D = k_D v_D \tag{3.13}$$

The course of the damping characteristic can be expressed by the exponent n in an equation: $F_D = k_D \cdot v_D^n . n > 1 \rightarrow$ progressive, $n = 1 \rightarrow$ linear, $n < 1 \rightarrow$ degressive.

F_D	Damper force, N
k_D	Damper constant, ns/m
v_D	Piston speed of the damper, m/s

To determine the wheel and body damping, only the maximum piston speed $v_{D,max}$ and the maximum damper forces in rebound F_2 and in compression F_1 are taken into account. Both are easy to measure. This simplification does not take into account the shape of the diagram. The mean forces could be determined from the height of the equal-area rectangle, (Fig. 3.35).

Adjustable dampers can be used to solve the conflict between comfort and safety in series production vehicles. In (Fig. 3.36) three different characteristic curves of a damper are shown, which are achieved by adjusting two control valves. It can be seen that considerably lower forces occur in the comfort setting than in the sport setting and that higher forces are provided in the rebound stage, i.e. when springing out, than in bump.

This basic insight that rebound damping is more important than compression damping is found in all vehicles regardless of their intended use. The actual distribution of damper forces varies, of course, depending on the application. The ratio of rebound force to compression force is about 3:1 for road vehicle dampers, (Fig. 3.37). For racing vehicles the ratios are 2:1 to 1.5:1 in extreme cases 1:1 [4]. In rally cars in particular, however, a ratio of 1:2 is often set so that the compression stage supports the body springs on landing as a “dynamic spring” after a jump.

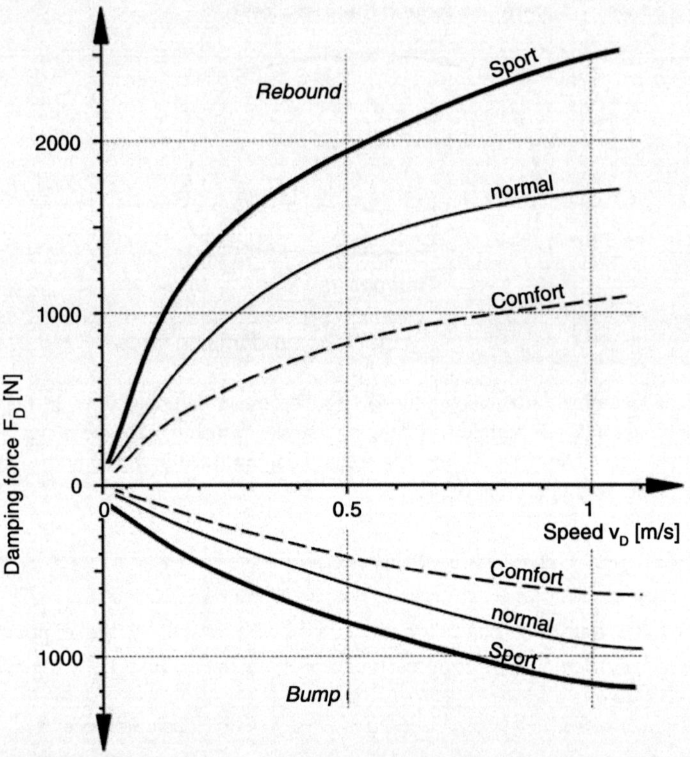

Fig. 3.36 Variable damping of a two-pipe damper with two control valves. The "Sport" setting is the basic characteristic. The two other characteristic curves result from different opening cross-sections of the two control valves. For the basic characteristic curve, the following results for the rebound stage: $n = 0.5$ and $k_D = 2773.5\ \mathrm{N(s/m)^{0.5}}$ and for the pressure stage: $n = 0.64$ and $k_D = 1214\ \mathrm{N\ (s/m)^{0.64}}$

$$\varepsilon = \frac{F_{D,rebound}}{F_{D,bump}} \tag{3.14}$$

ε	Ratio of rebound to compressive forces of a damper, –.

Very roughly, the rebound controls the movement of the sprung mass and the compression controls that of the unsprung mass: When driving over a bump, the height of the sprung body remains practically the same due to its inertia and the spring absorbs the height difference for a short time. The uncontrolled recoil of the spring and thus the lifting of the body is slowed down by the rebound damping. The relatively light, unsprung wheel could lose contact with the road when rapidly passing over a bump. This tendency is limited by the compression damping. In this sense, the ratio of rebound to compression forces also

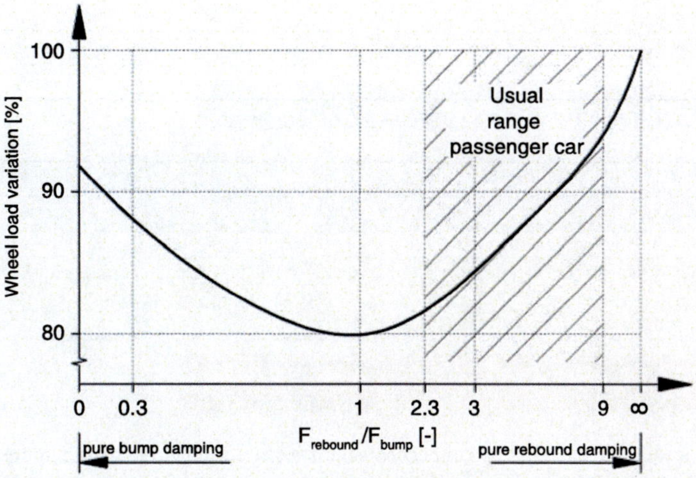

Fig. 3.37 Influence of the damper design on the wheel load fluctuations. With a rebound to compression ratio of 1, wheel load fluctuations reach a minimum. The road grip is therefore best with this design. However, the suspension comfort is not optimal, which is why the hatched area is preferred for passenger cars

reflects the ratio of these masses (more precisely: the root value of the mass ratio, see definition of damping D (3.9)).

An asymmetric distribution of forces between inward and outward deflection inevitably leads to vehicle sinking under uninterrupted vibration excitation. This dynamic lowering results approximately to [15]:

$$\Delta h_{ref} \approx \frac{4 s_D \eta D}{\pi} \cdot \frac{k_{D,2} - k_{D,1}}{k_{D,2} + k_{D,1}}$$

Δh_{ref}	Change of ride height, mm
s_{D}	Damper stroke, mm. For this equation it is assumed that the excitation amplitude is equal
η	Frequency ratio, –. $\eta = \omega/\omega_0$. ω_0 is the natural frequency. The above assumption about the damper stroke is true above $\eta = 5$
D	Lehr's damping factor, –

With small excitation amplitudes or approximately symmetrical damper design, this lowering remains negligibly small. However, this phenomenon can very well be observed in practical operation if the damper is designed too tightly for the body spring.

Too much rebound damping is bad in any case, because then the damped wheel receives too little contact force when rolling in the corner entry and thus builds up less lateral force than would be possible. This leads to understeer if the front rebound damping is too strong and to oversteer if the rear rebound damping is too tight.

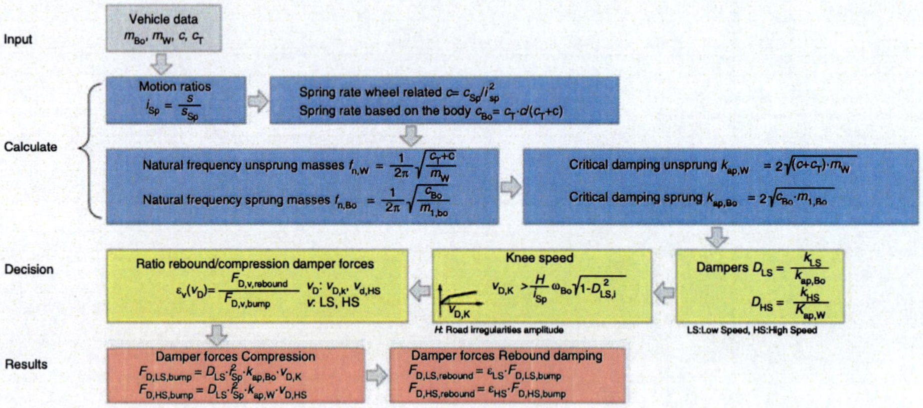

Fig. 3.38 Procedure for calculating damper characteristics, according to [18]. The figure summarises the (Sects. 3.2.1 and 3.2.2)

An overview of the entire procedure for calculating the damper characteristics is shown in Fig. 3.38. Basically, the illustration (Sects. 3.2.1 and 3.2.2) summarizes and extends the procedure. Based on data from the vehicle, first the required spring rates are calculated, (3.1), (3.2). This allows the natural frequencies of the unsprung and sprung masses to be determined, (3.4), (3.5). Next, the aperiodic damping rates of these masses are calculated. Now some decisions have to be made. First, the damping rates for high and low speed ranges are determined, (3.10). In this decision the body related spring rate, the pitch frequency and the roll frequency are included as well as driver wishes. These damping rates are the result of the conflicting demands placed on the damper by limiting the movements of unsprung and sprung masses induced by body displacements and bumps. Typical starting values for racing vehicles are $D = 0.5–0.7$ for low speed (body movements) and $D = 0.3$ to 0.5 high speed (wheel movements). The piston speed at which the transition from low to high speed takes place, the so-called knee speed $v_{D,K}$ (see Fig. 3.32), is then determined.[5] A speed above the resonance speed of the sprung masses is usually selected: Around 30 mm/s for asphalt, around 500 mm/s for rally cars on gravel. An approximate formula for this is given in the box in the figure. The ratio of tensile to compressive forces, (3.14), is typically set for the knee speed and the resonance speed of the unsprung masses. Thus, the basic values for the damper design are fixed and the damper forces for rebound and compression at knee speed and at speed at natural frequency of the unsprung masses (high speed) can be calculated to create a damper diagram with (3.13). The whole procedure is carried out for front and rear axle.

[5] See also Racing Car Technology Manual Vol. 5 *Data Analysis, Tuning and Development*, Chap. 6 *Development*.

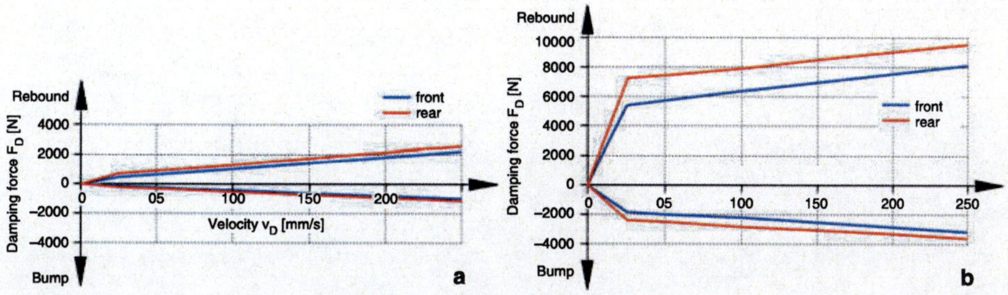

Fig. 3.39 Comparison of two damper designs, according to [18]. (**a**) Vehicle without aerodynamic downforce aids, (**b**) Vehicle with downforce aids (Indy Car). Vehicle mass 726 kg, axle load distribution v/h 40/60%, unsprung masses front 54.4 kg, unsprung masses rear 63.6 kg, spring rates vehicle **a**: front 280, rear 364 N/mm, tyres 525 N/mm; vehicle **b**: front 525, rear 612 N/mm, tyres 787 N/mm. The knee speed for both versions is 25 mm/s

A note on the classification of the calculation results: Differences in force of up to ±50 N can occur with identical dampers at the same piston speed.

For comparison of damper designs, two different setups are compared in (Fig. 3.39). Both vehicles have the same unsprung and sprung masses as well as static axle loads for comparison purposes. The difference lies solely in the aerodynamic downforce aids. However, the car with aerodynamic design (b) has stiffer body springs as well as stiffer tires. Its design aims at a parallel position of the car body towards the road. The low-speed damping is therefore chosen to be strongly overdamped ($D = 6$). The car without downforce design (a) runs in the low-speed range with $D = 0.71$. In the high-speed range, both have underdamped setups. The Indy Car, however, has $D = 0.9$ at the front and 0.7 at the rear, the comparison vehicle 0.4 at both axles.

Dampers not only act on the lifting movements of a wheel, but also on any change in the position of the vehicle body in relation to the road surface. A distinction is therefore made between heave, roll and pitch damping for the vehicle as a whole. For sports cars, target values for the decay of these movements are $D = 0.25$ for heave, $D = 0.55$ for roll and $D = 0.7$ for pitch vibrations [19]. In these considerations, the mass of the car body and the total spring rate or damper rate, formed from all four wheel-related individual variables, are decisive.

The analytical consideration of these movements is basically analogous to the consideration of the single-mass oscillator with the natural frequency according to (3.3) and the damping according to (3.9). In general, these relationships can be formulated as follows:

$$\omega_0 = \sqrt{\frac{c_{ref}}{m_{ref}}}, D = \frac{k_{ref}}{2\sqrt{c_{ref} m_{ref}}} \tag{3.15}$$

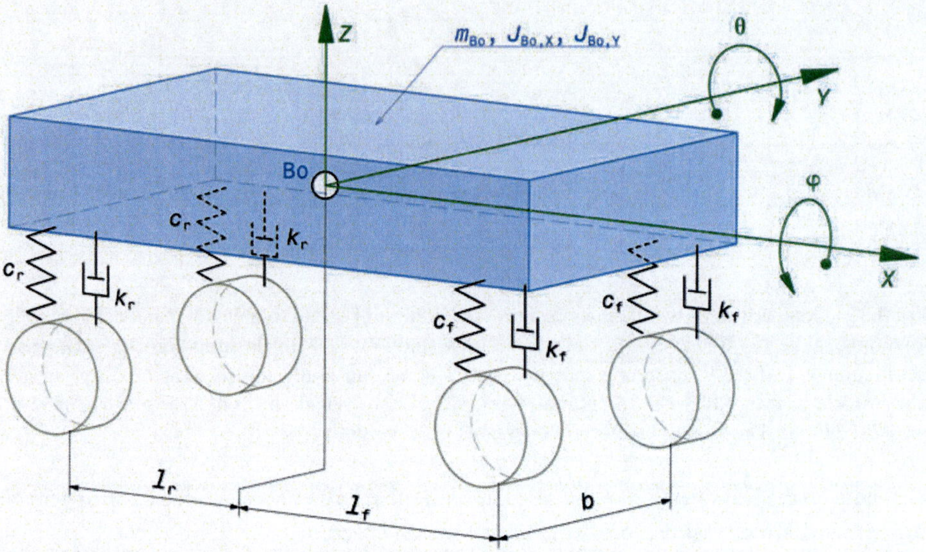

Fig. 3.40 Sketch of vehicle vibrations. The car body, mass m_{Bo} with centre of gravity Bo, is mounted on the wheels via springs and dampers. The spring and damper effect of the tyres is neglected here. Depending on the mode of vibration considered, different combinations of springs and dampers act on the mass. The equations of motion follow from center of gravity and twist theorems. The terms required for (3.15) for mass m_{ref} as well as for spring rate c_{ref} and damping coefficient k_{ref} can be taken directly from these equations
Heave: $m_{Bo}\ddot{z} + 2(k_f + k_r)\dot{z} + 2(c_f + c_r)z = 0$
Roll: $J_{Bo,X}\ddot{\varphi} + 2(k_f + k_r)\left(\frac{b}{2}\right)^2\dot{\varphi} + 2(c_f + c_r)\left(\frac{b}{2}\right)^2 = 0$
Pitch: $J_{Bo,Y}\ddot{\theta} + 2\left(k_f l_f^2 + k_r l_r^2\right)\dot{\theta} + 2\left(c_f l_f^2 + c_r l_r^2\right)\theta = 0$

The index ref. indicates that the reference quantities decisive for the considered movement are used, see also Fig. 3.40. Applied to the changes in position of the car body, these relationships are concretely represented as follows.

Bouncing oscillations (*heave mode*, periodic movement in *Z-direction*):

$$\omega_{heave} = \sqrt{\frac{2(c_f + c_r)}{m_{Bo}}}, D_{heave} = \frac{2(k_f + k_r)}{2\sqrt{2(c_f + c_r)m_{Bo}}} \tag{3.16}$$

Rolling oscillations (*roll mode*, periodic movement around *X-axis*):

$$\omega_{roll} = \sqrt{\frac{2(c_f + c_r)\left(\frac{b}{2}\right)^2}{J_{Bo,X}}},$$
$$D_{roll} = \frac{2(k_f + k_r)\left(\frac{b}{2}\right)^2}{2\sqrt{2(c_f + c_r)\left(\frac{b}{2}\right)^2 J_{Bo,X}}} \tag{3.17}$$

Pitching oscillations (*pitch mode*, periodic movement around *Y-axis*):

$$\omega_{pitch} = \sqrt{\frac{2(c_f l_f^2 + c_r l_r^2)}{J_{Bo,Y}}}, D_{pitch} = \frac{2(k_f l_f^2 + k_r l_r^2)}{2\sqrt{2(c_f l_f^2 + c_r l_r^2) J_{Bo,Y}}} \tag{3.18}$$

For racing vehicles with extreme design for aerodynamic downforce, the goal is to keep the car body as parallel to the road as possible when driving. The damping should therefore have such a strong effect that stationary end positions (e.g. max. Roll angle, max. Pitch angle) are not reached during typical maneuvers. In addition, heave and roll movements are additionally damped independently of the body dampers. In the case of the heave motion, this is done by a three-spring suspension (e.g. Fig. 3.54), in which not only a third spring but also a third damper is provided at this point.

For the additional damping of the rolling motion, these systems essentially function in such a way that the heave and the roll motion are decoupled, (Fig. 3.41). The core element is a central rocker which, in contrast to conventional designs, has two degrees of freedom. The pivoting movement about axis *b* corresponds to the usual deflection of the wheel contact force in the case of equilateral springing. The lever operates the heave damper together with the heave spring (8). In the version shown, however, the lever is connected via the lower(!) spring plate, i.e. the piston rod is pulled during compression. When the springs are alternately compressed, the bell crank rotates around axis *a and* is supported on the roll damper (6) via the hollow shaft with extension arm (4). This damper thus operates exclusively during rolling motions. The restoring forces are generated by a torsion bar (3) which is connected on the one hand to the hollow shaft (4) and on the other hand to the mounting (5) on the transmission housing. The hollow shaft is rotatably mounted in the bearing bridge (1) at the top and in the gearbox mount (5) at the bottom.

Designs

Dampers are offered in a wide variety of designs. For racing cars, adjustable dampers with a separate spring (*coil over damper*) are standard (Fig. 3.42). This makes it easy to change the spring, which is particularly advantageous during the development phase of a new car. A disadvantage of this otherwise extremely compact design is the bending moment that the spring introduces into the piston rod. This increases friction and (one-sided) wear of the rod seal. To avoid this disadvantage, springs and dampers must be arranged separately. On the one hand this increases the design effort, but on the other hand it is worthwhile because

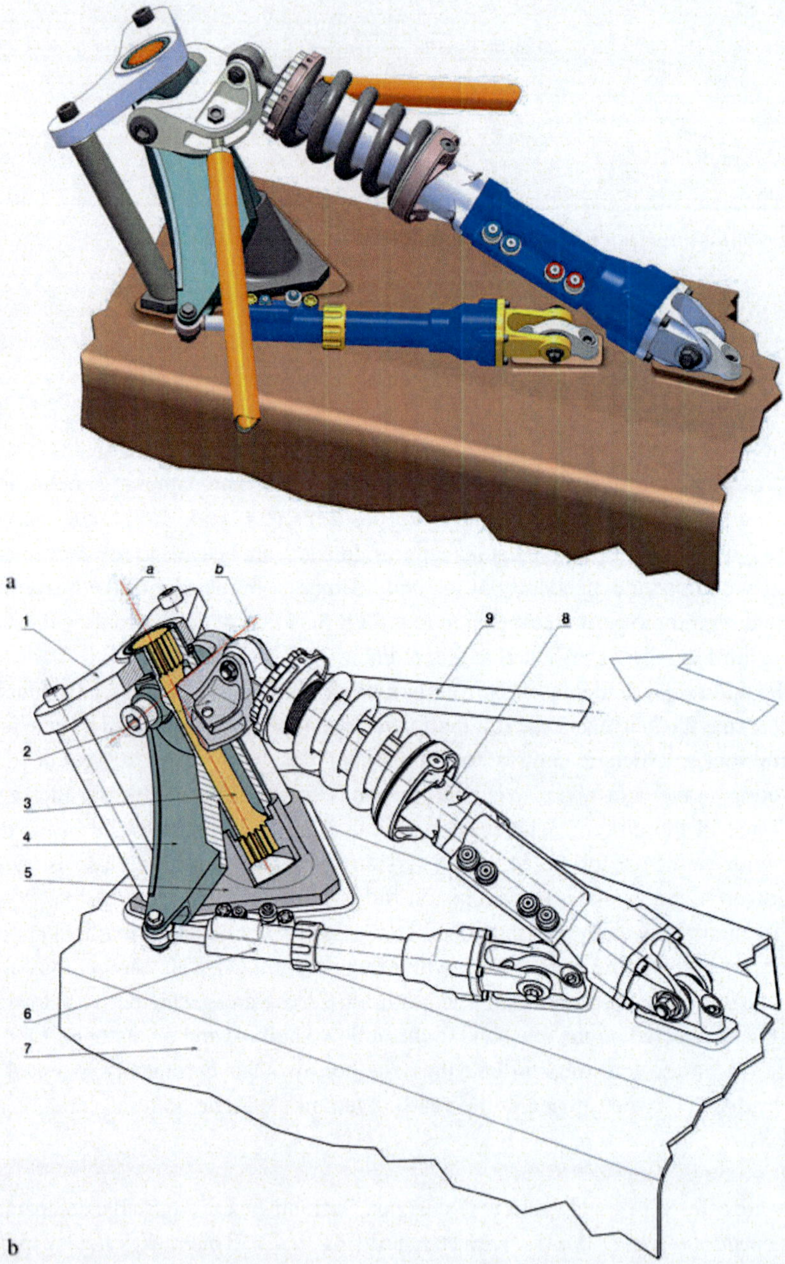

Fig. 3.41 Rear suspension with separate roll damper (LMP1 Porsche 919 Hybrid). (**a**) axonometric representation, (**b**) sectional representation, left compression bar removed. The central suspension strut (8) is shown, which is arranged above the transmission housing (7) in the direction of travel (arrow). The central bell crank (2) takes over the wheel contact forces of the two rear wheels via the push rods (9). When rolling, the roll damper (6) comes into play. *a* Axis for roll, *b* Axis for heave. 1 Bearing bridge, 2 Rocker, 3 Torsion bar, 4 Rocker bearing with cantilever, 5 Mounting on gearbox housing, 6 Roll damper, 7 Gearbox housing, 8 Heave damper, 9 Pressure bars

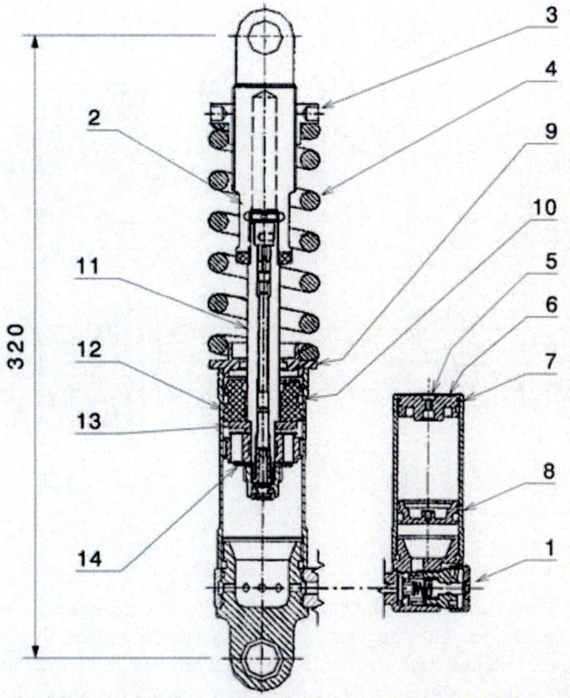

Fig. 3.42 Damper strut of a racing vehicle, after [20]. The element is operated as a monoshock on the front axle via push rod. Rebound and compression damping are separately adjustable. **1** Bump (Pressure stage) adjustment via rotary knob. Turning the screw in = harder setting. **2** Rebound adjustment by turning the piston rod. **3** Spring plate adjustable (spring preload). **4** Spring, stiffness between 120 and 210 N/mm. **5** Screw plug in plug 6. It is used to release pressure, to measure pressure and to fill the tank with nitrogen. Inflation pressure 8 to 9 bar. **6** Closing plug nitrogen container. **7** Snap ring, holds stopper 6. **8** Separating piston between working chamber and nitrogen container. **9** Spring seat. **10** Snap ring, secures guide 12. **11** Piston rod. **12** Piston rod guide. **13** Valve plates for compression stage (3 pieces). **14** Valve plates for rebound stage (6 pieces)

different installation ratios (*motion ratio*) are possible. Due to its function, the damper will be assigned the larger ratio (3.1). In this way, it responds effectively even with small wheel strokes.

Separate adjustment of rebound and compression damping is made possible by adjusting wheels with detents. This allows fine adjustment without disassembly work, provided the damper is not too "hidden" in the vehicle. For coarse presetting, entire valve assemblies can be replaced on some types, but this requires the damper to be completely drained and disassembled.

Figure 3.43 shows a damper/strut with the most important connection dimensions. The maximum stroke is predetermined by the design and must cover the entire spring travel

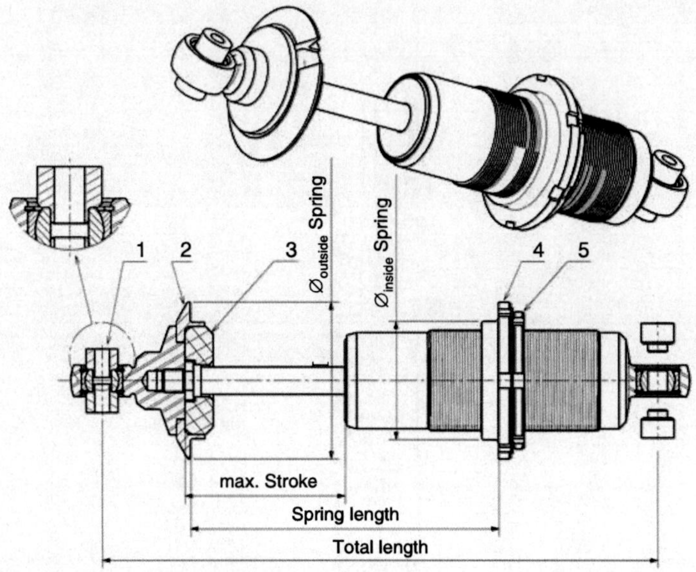

Fig. 3.43 Mounting dimensions of a damper. The coil spring is obtained from another supplier and is not yet present on the damper. The upper spring plate has a recess and can thus be removed without dismantling the damper. The lower spring plate is designed as a nut and allows the spring preload to be changed. The coil spring is slid over the damper, then the upper spring plate is mounted. The maximum stroke is measured without pressure stop. **1** Spherical bearing with spacer bushes. **2** Top spring perch. **3** Bump stop. **4** Lower spring platform. **5** Lock nut

(compression and rebound). However, the bump stop can be adjusted by replacing the elastomer piece (3). The total length is determined by the maximum stroke and the design. For the same design, an increase in stroke will result in an increase in overall length of twice the extension. For example, a typical monotube damper with integrated compensating volume with 50 mm stroke has an overall length of 250 mm. If a damper with 130 mm stroke is installed from the same series, the mounting points must be 410 mm ($= 250 + 2 \times 80$) apart.

Damper pistons have a diameter in the range of 30–46 mm with a piston rod diameter of 11 mm.

Common damper strokes range from 50 to 83 mm for single seaters, small sports cars and kit cars to 150 mm for touring cars. For rally and off-road use, even greater strokes are required.

The joints in the locating eyes should be spherical plain bearings. These have proven themselves and do not generate any additional (non-linear) elasticities, as is the case with the rubber bushings and pin joints commonly used in passenger cars. Rubber bushings can also be replaced with spherical plain bearings at a later date. The mounting holes fit common bearings (with 1/2″ or 15 mm inner diameter). Only two additional spacer sleeves need to be provided if the original bearing width is to be achieved, for example in order to use the existing mounting bracket.

Like hardly any other components on racing cars, you can't expect more for your money than you invest. The performance range is wide and so are the costs. Roughly, the ratios look like this. The basis for comparison is a non-adjustable damper of a standard car with the price *K*. Then these prices result [4]:

- Rebound/compression combined adjustable, take up in Silentblocs: 3 K.
- Rebound/compression combined adjustable, light alloy spring seat, mounting in spherical bearing, high quality: 6 K–7 K.
- Rebound and compression separately adjustable, medium quality: 12 K.
- Rebound and compression separately adjustable, dismountable, top quality: 15 K–20 K
- Rebound and compression separately adjustable, dismountable, separate reservoir, top quality: 23 K.

Multiple adjustable valves for different speed ranges, can be disassembled, separate expansion tank, top quality: up to 40 K.

Materials. Piston rods: Quenched and tempered steel C45E V (was Ck45 V) quenched and tempered to $R_m = 750–900\ N/mm^2$. Surface hardened to 58 + 2 HRC, additional hard chromium layer with superfinish treatment to a surface roughness $R_t = 0.2\ \mu m$ (for interaction with the rod seal of a monotube damper) [5].

Piston rod guide: Wrought aluminium alloy (AlMgSi 1 F 28) with hard anodised running surface [5].

Housing: Wrought aluminium alloy, extruded and hard anodised.

Vibration Absorber (*Mass Damper*)

An interesting alternative for reducing wheel load fluctuations (e.g. Fig. 3.23) is offered by an absorber mass, Fig. 3.44. The vibration absorber is mounted on the vehicle's centre plane on an axle in a frame-fixed manner so that the movable mass (1) can slide up and down. The heave motion of the vehicle causes the actual absorber mass to vibrate, which in turn acts back on the vibrations of the vehicle. The deflections of the vehicle vibrations (is a measure of the wheel load fluctuations) are greatly reduced in a certain frequency range by the absorber. Due to the mass, the spring and the enclosed air, which acts as a damping medium via holes, the absorber can be tuned to the vehicle. As a rule, its natural frequency is set slightly above that of the wheel (wheel vibration frequency). The effect is therefore limited to a selected range. For racing cars, which drive on circuits, this is no problem. After all, the vibration absorber can be tuned for any track with its ingrained unevenness. In Formula 1, some teams have successfully used absorbers. Renault, for example, drove with one absorber each on the front and rear axle in the order of 10 kg [22]. This extra weight of 20 kg may have raised the centre of gravity slightly (the total mass remained unaffected – there are, after all, considerably more ballast masses carried), but the balancing effect on

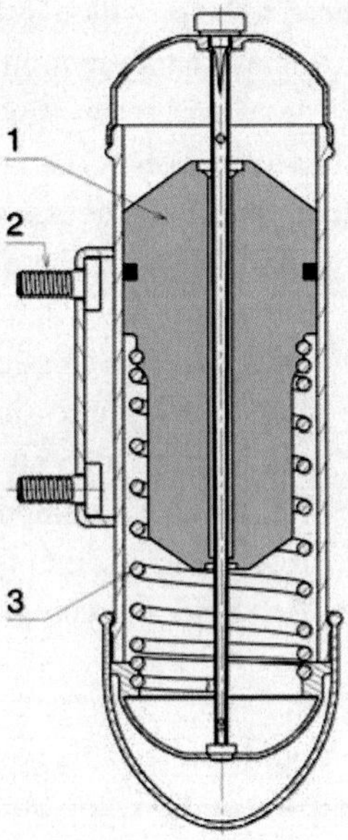

Fig. 3.44 Vibration damper, according to [21]. **1** Moving damper mass. **2** Fastening. **3** Spring. The housing is attached to the frame via the fastening (2). The absorber mass (1) slides up and down excited by the vehicle heave movements. A spring (3) and the enclosed air cause forces that influence the movement of the mass

wheel load fluctuations more than compensated for this. In the meantime, the use of such aids was probably prohibited by the regulations for fear of too large moving masses.

The use of a vibration damper is naturally limited for vehicles that travel on different surfaces. If the set-up is suitable on an asphalt track, the effect of the absorber is severely limited on gravel, when the wheel lift movements are much stronger.

Absorber masses are also cleverly used in series-produced passenger cars. For the torsionally soft bodies of convertibles, for example, the oscillation deflections are kept small by means of existing masses such as batteries and hydraulic pumps, but also by means of additional masses.

3.3 Stabilisers, Anti-Roll Bars, AE: Sway Bars

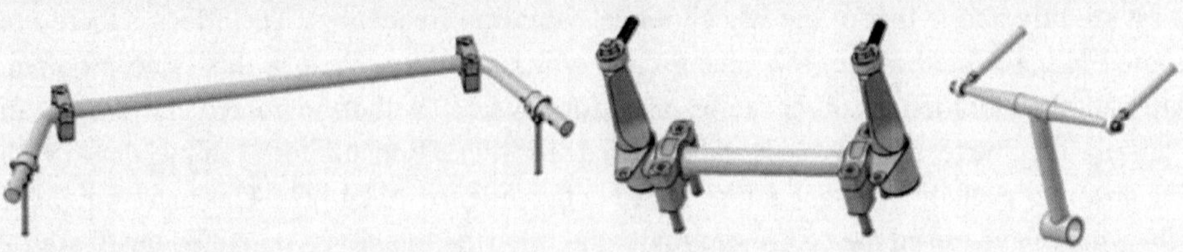

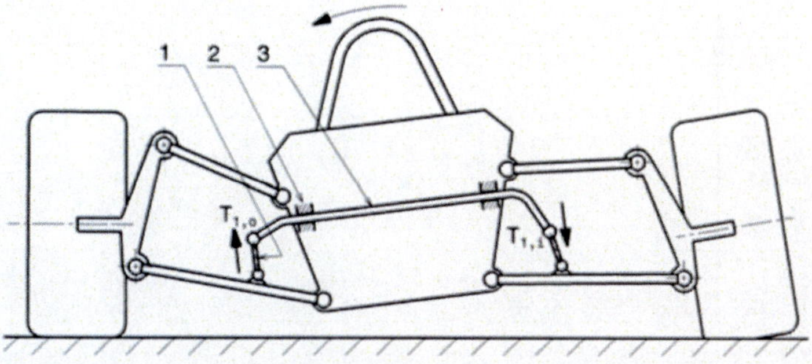

Fig. 3.45 Stabiliser with inclined body, schematic diagram. **1** Drop link. **2** Support. **3** Anti-roll bar. The stabilizer (3) is rotatably guided in the bearings (2) and connected at its ends in the joints T_1 via the two drop links (1) to the wheel suspension

Although the body springs absorb some of the roll moment that tilts the body during cornering, they are not designed to do so. The roll moment is balanced by the torsion stabiliser, which is bent and/or twisted in the case of alternating springing (Fig. 3.45). It is not active in the case of equal-sided springing. Because of this action of the stabilizer, the body spring need not be made unnecessarily stiff to reduce roll. Thus, the stabilizer serves to stabilize the position of the vehicle body parallel to the road. However, its name does not refer to its influence on driving stability, because other measures are more suitable for improving driving stability. For example, elastokinematic joints or electronically controlled individual wheel braking interventions are used on road vehicles. Stabilizers are indispensable for preventing the rolling tendency of high superstructures caused by transport tasks, such as container transport.

The anti-roll bar means a coupling of the two wheels of an axle and brings the independent suspension somewhat closer to the rigid axle in terms of vibration. On vehicles with independent suspension, the stabilizer should be avoided where possible. If there are obstacles in only one lane, the suspension acts harder due to the effect of the stabilizer. For this reason, the first step in vehicle development should be to try to use moderately all means which also counteract the tendency to roll. These are firstly the means which reduce the rolling moment itself, of which the reduction of the centre of gravity height is the most important. On the other hand, there are means on the vehicle which increase the roll stiffness of the axle: larger track width, suitable elastokinematics with an appropriately high rolling centre, taking into account the change in track width, or progressive springs. The most elegant way to prevent rolling is with the help of active suspension.

On racing cars, anti-roll bars are the means of adjusting the desired self-steering behaviour, which is not only predetermined by the concept, but also changes due to changing tyre behaviour, occurring drive/brake forces, weather conditions, etc. It may be sufficient for the balance of the car to fit an anti-roll bar to just one axle. The anti-roll bar is the most effective means of increasing the wheel load differences on an axle, i.e. its roll stiffness, when cornering. A larger wheel load difference between the inner and outer wheel

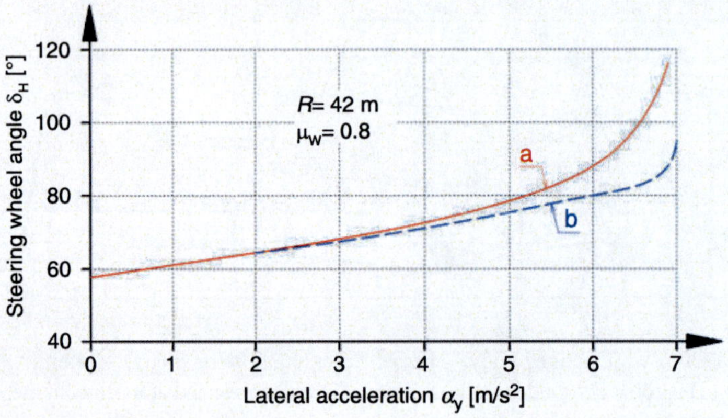

Fig. 3.46 Influence of a stabilizer on the driving behaviour [23]. The diagram shows the result of a driving test for a passenger car: Steady-state cornering with constant radius R (ISO 4138). The course of the steering wheel angle is shown for two different setups. The set-up with the stiffer front stabilizer (a, red) shows a significantly greater tendency to understeer (disproportionately greater steering angle with increasing lateral acceleration) and also does not achieve the maximum lateral acceleration permitted by the coefficient of friction $\mu_W = 0.8$. It can also be seen in passing in the diagram that stabilisers have no effect in the linear range of the tyre behaviour $F_{W,y} = f(\alpha)$. At low lateral accelerations (up to 2 m/s^2) and thus at low coefficients of friction, a change in the stabilizer stiffness has no effect on the driving behavior. Only at higher lateral accelerations does the greater wheel load shift at the front of tuning a become noticeable in the driving behaviour. Vehicle mass $m_{V,t} = 1544$ kg. (**a**) Stabilizer rates $c_{Ro,S,f} = 16.5$ N/mm, $c_{Ro,S,r} = 3.0$ N/mm, (**b**) Stabilizer rates $c_{Ro,S,f} = c_{Ro,S,r} = 9.75$ N/mm

leads to a larger tyre slip angle, because the tyre builds up lateral forces degressively, i.e. not linearly over the wheel load. If a disproportionately greater lateral force with a greater wheel load is to be generated at the decisive outer tyre, a greater slip angle is required for this, cf. Figs. 1.38 and 2.1. A stiffer stabiliser on the front axle thus generally increases the tendency to understeer (Fig. 3.46), for example, while greater stabilisation of the rear axle makes the front-wheel-driven car more neutral.

As the stabiliser increases the wheel load of the outer wheel on this axle during roll (this can go so far as to lift the inner wheel, Fig. 3.47), it also increases the wheel load of the diagonally opposite wheel, i.e. the "disadvantaged" inner wheel. At this axle, the wheel loads are thus somewhat balanced, which benefits the total transmissible lateral force at this end of the vehicle. In the same sense, the traction of this axle is also improved, but at the same time the traction of the other axle, where the stabilizer acts directly, is reduced.

A stiff vehicle frame is a prerequisite for the planned effect of an anti-roll bar. A frame that is (too) soft will be twisted by the stabilizer without the wheel loads changing significantly.

Stabilizers are sometimes omitted entirely on one axle. On the driven rear axle, traction can thus be increased when cornering, but at the expense of a more heavily loaded outer

Fig. 3.47 Effect of a stabilizer. With a front-wheel drive car, the tendency to understeer is naturally high. A stiff stabilizer bar on the rear axle helps to balance the wheel loads on the front axle when cornering, but at the expense of the wheel load shift on the rear axle. In a left-hand bend, the inner wheel of the rear axle visibly lifts off in the picture

front wheel [4]. The wheel load contribution by which the rear wheel on the inside of the bend is more heavily loaded comes from the diagonally opposite front wheel, whose spring has been hardened by the stabiliser.

Stabilizers should be mounted in such a way that adjustments or modifications can be made easily. The anti-roll bar ends require a range of movement during full wheel travel which must not lead to a collision even when the front wheels are turned in. The stabilizer back should be mounted so that it moves easily. As with the body springs (Sect. 3.3.1), a linear or progressive spring characteristic is also desirable for stabilizers. This behaviour can be achieved by arranging the coupling struts on the chassis accordingly. Degressive behavior of stabilizers must be avoided at all costs. Such a vehicle practically does not react to changes in stabilizer settings.

Stabilizers are also installed whose stiffness can be adjusted by the driver while driving. The design effort – provided the regulations allow it – is worthwhile in certain cases. During longer races, track conditions can change just as much as fuel tank contents and tyre condition. All this together influences the handling of the car, which thus moves away from the set desired condition. A skilled driver now has the possibility to trim the car back in the desired direction by means of such an adjustment lever and to continue to keep the lap times low.

These systems are infinitely variable, but this is due more to the mechanism than to any necessity. In fact, changes in vehicle set-up are made in noticeable leaps and bounds. In sprint-like races, adjustability via replaceable stabilizer elements of varying stiffness is sufficient. Adjustable anti-roll bars are usually found on the rear axle. For the simple reason that it is easier to implement because fewer components interfere with the accommodation.

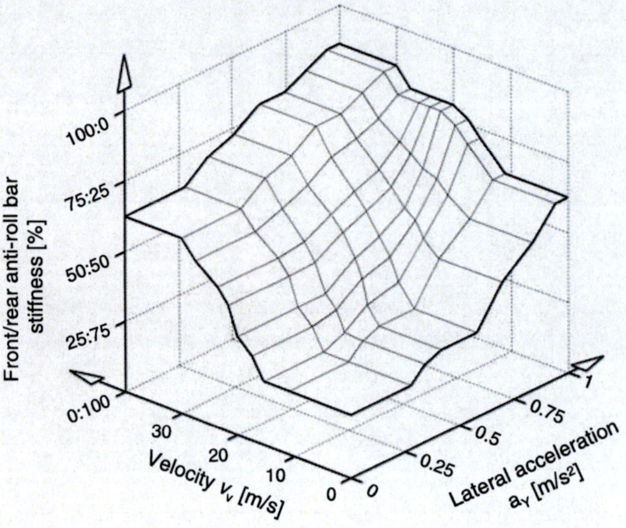

Fig. 3.48 Characteristic diagram of active stabilizers. The characteristic diagram represents the distribution of the front to rear stabilizer stiffnesses for a specific vehicle and a specific friction value

The latest developments in passenger cars are aimed at active stabilizers. In this case, the stabilizer torsioni bar is not in one piece, but separate, and the two ends are connected via a servomotor. By means of this servomotor, the two stabilizer arms can be deliberately twisted against each other, whereby the stiffness can be increased or decreased [24]. Figure 3.48 shows a map for a vehicle with active stabilizers on both axles on dry road ($\mu_W \approx 0.8$). The vehicle is a typical passenger car, i.e. with a slightly understeering basic design. At low driving speeds and low lateral accelerations, the rear stabilizer is stiffened by an on-board computer and the vehicle is thus neutral in self-steering behavior. In this state, the vehicle exhibits better handling characteristics and is significantly more willing to steer. The closer the driving condition approaches critical areas, the more the car's characteristics are set to understeer. At high speeds and/or high lateral accelerations, the front stabilizer bar is thus made stiffer and the rear one softer.

3.3.1 Calculation

In principle, a basic design will aim for a slightly understeering behaviour of the vehicle and thus select a slightly higher share of the front axle in the roll stiffness.[6] At this point, a more detailed consideration follows for determining a starting value for the stabilizer

[6]See Racing Car Technology Manual Vol. 5 *Data Analysis, Tuning and Development* Chap. 5 *Tuning*.

Table 3.7 Typical roll gradients of some vehicles, partly according to [26]

Vehicle type	Roll gradient, °/*g*	Vehicle type	Roll gradient, °/*g*
Mid-size car	5–6	High downforce racer	0.2–0.7
Sports car	3	Indy Car (2001)	0.1–0.2
Touring car with little downforce	1.0–1.8	Formula 1 Car (2002)	0.03–0.1

stiffness. As a starting point, the recommendation is not to exceed the maximum roll angle of 2.5° for touring cars and 1.5° for single-seaters at a lateral acceleration of 1 *g* [4]. In general, roll angles should remain below 4°. By way of comparison, series-produced cars roll by about 5–6° at 1 *g* lateral acceleration [25], if they are at all able to build up such tyre adhesion, cf. also Fig. 5.2. Table 3.7 provides a summary of typical roll gradients.

The quantities required for the calculation are compiled in (Fig. 3.49).

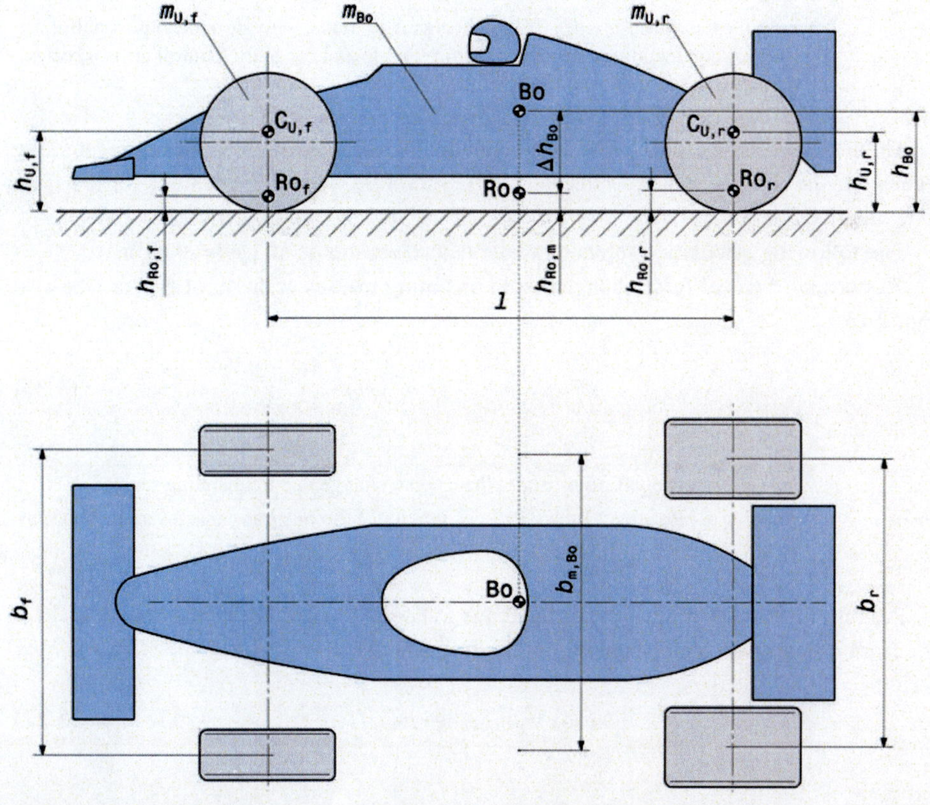

Fig. 3.49 Definitions of the distances required for the calculation. Indices: *f* front, *r* rear, Bo body center of gravity, C_U Centre of gravity of the unsprung masses. *Ro* Roll Center

Consideration of the masses involved:

$$m_{Bo} = m_{Bo,f} + m_{Bo,r} \tag{3.19}$$

m_{Bo}	Mass of the body (sprung mass), kg
$m_{Bo,f}$	Part of the body mass acting on the front axle, kg
$m_{Bo,r}$	Part of the body mass acting on the rear axle, kg

$$m_{Bo,f} = m_{V,f} - m_{U,f} \tag{3.20}$$

$m_{V,f}$ or $m_{V,r}$	Axle load front or rear, kg
$m_{U,f}$ or $m_{U,r}$	Unsprung masses front or rear, kg
	$m_{U,f}$ = front wheel masses (front wheel carrier, tyres, wheels, brakes) plus half the mass of connecting parts between front wheels and car body (control arms, springs, dampers, . . .)

The respective sizes for the rear axle result analogously by inserting the corresponding values for the rear axle (index *r* instead of *f*). Therefore, only the equations for the front axle are given below.

The following calculation refers to a lateral acceleration a_y of 1 g (= 9.81 m/s^2).

Proportion of wheel force change due to unsprung masses at the front or rear (see also Fig. 2.13):

$$\Delta F_{U,f} = m_{U,f} \cdot g \cdot h_{U,f} / b_f \tag{3.21}$$

$\Delta F_{U,f}$	Proportion of front wheel force change due to unsprung masses, N
$h_{U,f}$	Position of the Centre of gravity of the unsprung masses at the front, m
b_f	Track width, front, m

Proportion of the change in wheel force due to lateral acceleration of the body masses at the front or rear, which are supported in the roll centres:

$$\Delta F_{Bo,f} = m_{Bo,f} \cdot g \cdot h_{Ro,f} / b_f \tag{3.22}$$

$\Delta F_{Bo,f}$	Proportion of front wheel force change due to front body mass (geometric proportion), N
$h_{Ro,f}$	Height of front roll Centre, m

Proportion of the change in wheel force due to body roll about the roll axis caused by lateral acceleration:

$$\Phi_m = \frac{m_{Bo,r}}{m_{Bo}} \tag{3.23}$$

$$b_{m,Bo} = \left(\left(b_r - b_f\right) \cdot \Phi_m\right) + b_f \tag{3.24}$$

$$h_{Ro,m} = \left(\left(h_{Ro,r} - h_{Ro,f}\right) \cdot \Phi_m\right) + h_{Ro,f} \tag{3.25}$$

$$h_{Bo} = \left(h_{Bo,r} - h_{Bo,f}\right) \cdot \Phi_m + h_{Bo,f} \tag{3.26}$$

$$\Delta h_{Bo} = h_{Bo} - h_{Ro,m} \tag{3.27}$$

$$\Delta F_{Ro} = m_{Bo} \cdot g \cdot \Delta h_{Bo} / b_{m,Bo} \tag{3.28}$$

Φ_m	Ratio of the mass of the body acting on the rear axle to the total mass of the body, –
$b_{m,Bo}$	Mean track width under the Centre of gravity of the body, m
$h_{Ro,m}$	Mean roll Centre height under the Centre of gravity of the body, m
h_{Bo}	Height of the Centre of gravity, m
$h_{Bo,f}$ or $h_{Bo,r}$	Height of the Centre of gravity of the body section above the front axle or above the rear axle, m
Δh_{Bo}	Lever arm of the body around roll axis, m
ΔF_{Ro}	Proportion of the change in wheel force due to rolling of the body mass (elastic proportion), N

Total wheel force change due to lateral acceleration:

$$\Delta F_{W,Z} = \Delta F_{U,f} + \Delta F_{U,r} + \Delta F_{Bo,f} + \Delta F_{Bo,r} + \Delta F_{Ro} \tag{3.29}$$

$\Delta F_{W,Z}$	Total wheel force change, N

The "rigid" complete vehicle can be used for a rough check of the wheel force change:

$$\Delta F_{W,Z} = m_{V,t} \cdot a_y \cdot \frac{h_V}{b_m} \tag{3.30}$$

$m_{v,t}$	Total mass of the vehicle incl. driver, kg
a_y	Lateral acceleration. Here: $a_y = 9.81\ \mathrm{m/s^2}$
h_V	Height of the overall centre of gravity of the vehicle, m
b_m	Mean track width under centre of gravity, m

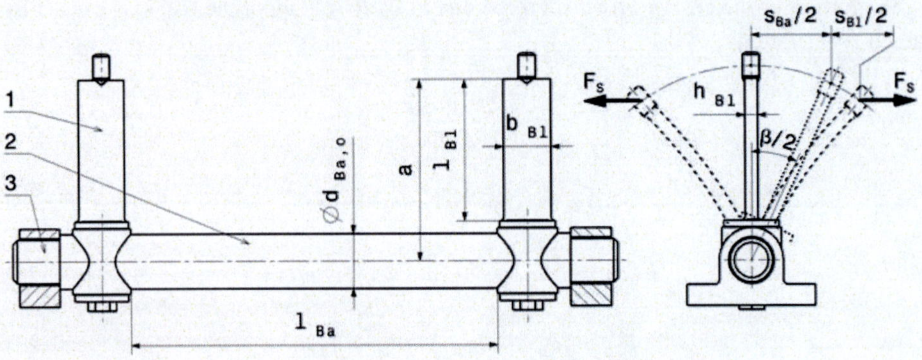

Fig. 3.50 Stabilizer calculation. **1** Blade. **2** Torsion bar. **3** Bearing

Roll resistance of the body due to the suspension springs:

$$c_{Ro,Sp,f} = \frac{c_{Sp,f}}{2 \cdot \left(s_f / s_{Sp,f}\right)^2} \cdot b_f^2 \cdot \frac{\pi}{180^\circ} \tag{3.31}$$

$c_{Ro,\mathrm{Sp,f}}$	Roll stiffness of the front body suspension related to the front axle, N m/°
$c_{\mathrm{Sp,f}}$	Rate of front body spring, N/m
$s_{\mathrm{SP,f}}$	Deflection of the body spring at the front with wheel travel s_{f}, m
s_{f}	Front wheel travel, m
Note: $s/s_{\mathrm{Sp}} = i_{\mathrm{Sp}}$	Spring ratio, see (3.1)

The quantities given in Fig. 3.50 are used to calculate the stabilizer stiffness. The action of the stabilizer shown is composed of three springs connected in series, namely the torsion of the bar and the bending of the two arms. A spring rate for the stabilizer is determined from the total deformation and the applied forces, according to the relationship: spring rate = total force/total deformation.

Area moments of inertia of stabilizer torsion bar and lever arm:

$$I_{p,Ba} = \frac{\pi \cdot \left(d_{Ba,o}^4 - d_{Ba,i}^4\right)}{32} \tag{3.32}$$

$I_{\mathrm{p,Ba}}$	Polar moment of inertia of the torsion bar, m^4
$d_{\mathrm{Ba,o}}$	Outer diameter of the torsion bar, m
$d_{\mathrm{Ba,i}}$	Inner diameter of the torsion bar (if hollow), m

$$I_{Bl} = \frac{b_{Bl} \cdot h_{Bl}^3}{12} \tag{3.33}$$

I_{Bl}	Axial moment of inertia of the lever arm, m^4
b_{Bl}	Cross-section width of the lever arm, m
h_{Bl}	Cross-sectional height of the lever arm, m

The stresses, and in particular the bending stresses, are lower compared to the body springs. For this reason, the frequently encountered solutions in passenger cars, in which the side arm of the stabilizer bar is used as a supplementary longitudinal linkage of the lower control arm in wheel suspensions with MacPherson strut, are also justifiable.

The deformations of the elastic stabilizer parts follow:

$$\beta = \frac{F_S \cdot a}{G \cdot I_{p,Ba}} \cdot l_{Ba} \tag{3.34}$$

β	Angle of twist of the torsion bar, rad. Value in °: $\beta^o = \beta \cdot 57,3^\circ$
F_S	Forces at the ends of the lever arms, N
a	Lever arm of the force F_S, m
G	Shear modulus of the torsion bar material, N/m^2. $G = 80 \cdot 10^9\ N/m^2$ for spring steel
l_{Ba}	Springy length of the torsion bar, m

$$s_{Ba} = a \cdot \beta \tag{3.35}$$

s_{Ba}	Displacement of the ends of the arms due to torsion of the bar, m

$$s_{Bl} = 2\frac{F_S \cdot l_{Bl}^3}{3E \cdot I_{Bl}} \tag{3.36}$$

s_{Bl}	Displacement of the ends of the arms due to bending of the arms, m
l_{Bl}	Resilient length of the arm, m
E	E-modulus of the arm material, N/m^2. $E = 206 \cdot 10^9\ N/m^2$ for spring steel

This results in the total deformation of the stabilizer to:

$$s_t = s_{Ba} + s_{Bl} \tag{3.37}$$

s_t	Total displacement of the ends of the arms, m

The spring constants for the stabilizer components thus increase:

$$c_{Ba} = \frac{F_S}{s_{Ba}} = \frac{G \cdot I_{p,Ba}}{a^2 \cdot l_{Ba}} \tag{3.38}$$

c_{Ba}	Linearized spring rate of a torsion bar spring related to the end point of a lever with length a, N/m

$$c_{Bl} = \frac{F_S}{s_{Bl}} = \frac{3}{2} \frac{E \cdot I_{Bl}}{l_{Bl}^3} \tag{3.39}$$

c_{Bl}	Linearized spring rate of the arms related to the articulation points of the forces F_S, N/m

The spring rate for the entire stabilizer follows from this to:

$$c_S = \frac{c_{Ba} \cdot c_{Bl}}{c_{Ba} + c_{Bl}} \tag{3.40}$$

c_S	Total, linearized spring rate of a stabilizer according to (Fig. 3.50) related to the two articulation points of the forces F_S, N/m

Roll resistance of the body due to stabilizers:

$$c_{Ro,S,f} = \frac{c_{S,f}}{\left(s_f / s_{S,f}\right)^2} \cdot b_f^2 \cdot \frac{\pi}{180^\circ} \tag{3.41}$$

$c_{Ro,S,f}$	Roll stiffness of the front anti-roll bar referred to the front axle, N m/°. Without term π/180° the unit is [N m/rad]
$c_{S,f}$	Linearized spring rate of the front stabilizer related to its lever end point, N/m
$s_{S,f}$	Travel of the stabilizer lever end (is joint T_1 in Fig. 3.45) of the front axle at travel of the front wheel s_f, m

Total body roll resistance due to body springs and stabilizers:

$$c_{Ro,f} = c_{Ro,Sp,f} + c_{Ro,S,f} \tag{3.42}$$

$c_{Ro,f}$	Total roll resistance of front axle, N m/°

$$\Phi_{c,Ro,f} = \frac{c_{Ro,f}}{c_{Ro,f} + c_{Ro,r}} \tag{3.43}$$

$$\Phi_{c,Ro,r} = 1 - \Phi_{c,Ro,f} \tag{3.44}$$

$\Phi_{c,Ro,f}$	Ratio of the roll stiffness of the front axle to the total roll stiffness, –
$\Phi_{c,Ro,r}$	Ratio of rear axle roll stiffness to total roll stiffness, –

Dynamic wheel load transfer of the decisive outer wheels:

$$\Delta F_{W,Z,f,o} = \Delta F_{Ro} \cdot \Phi_{c,Ro,f} + \Delta F_{Bo,f} + \Delta F_{U,f} \tag{3.45}$$

$\Delta_{FW,Z,f,o}$	Change in wheel force of the front outer wheel on the turn, N

Lateral transfer of wheel loads from one axle side to the other when cornering:

$$\Delta F_{W,Z,f} = \Delta F_{W,Z,f,o} - \Delta F_{W,Z} \cdot \frac{m_{V,f}}{m_{V,t}} \tag{3.46}$$

$\Delta F_{W,Z,f}$	Change in wheel force on front axle, N
$m_{V,t}$	Total mass of the vehicle incl. driver, kg
	$m_{V,t} = m_{V,f} + m_{V,r}$
	with: $m_{V,f}$ or $m_{V,r}$ axle load front or rear, kg

Resulting wheel contact forces due to wheel load displacement:

$$F_{W,Z,f,o} = \frac{m_{V,f}}{2} \cdot g + \Delta F_{W,Z,f,o} \tag{3.47}$$

$$F_{W,Z,f,i} = \frac{m_{V,f}}{2} \cdot g - \Delta F_{Z,W,f,o} \tag{3.48}$$

$F_{W,Z,f,o}$	Resulting wheel contact force at the front outer wheel of the turn, N
$F_{W,Z,f,i}$	Resulting wheel contact force at the front inner wheel of the turn, N

Roll angle of the body, (Fig. 3.51):

$$\varphi = \frac{m_{Bo} \cdot g \cdot \Delta h_{Bo}}{c_{Ro,f} + c_{Ro,r}} \tag{3.49}$$

$$\varphi = 57.3^{\circ} \cdot \frac{\Delta s_{1,f} + \Delta s_{2,f}}{b_f} \tag{3.50}$$

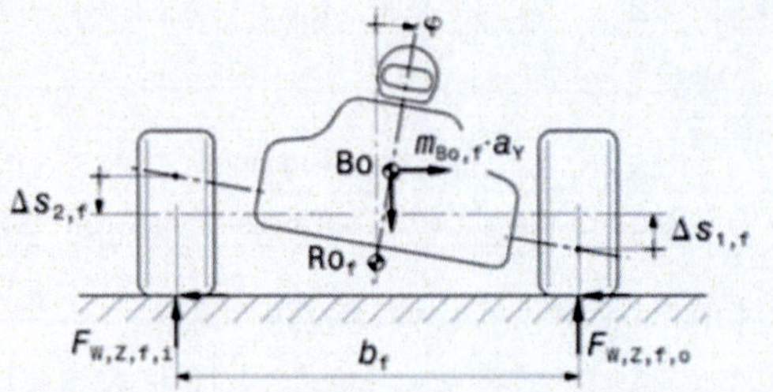

Fig. 3.51 For the definition of the roll angle φ. View of a vehicle from the front in a right-hand turn. The lateral tire forces accelerate the car to the right as action forces and thus cause the reaction force $m_{\mathrm{Bo,f}} \cdot a_{\mathrm{Y}}$ of the inert car body, which rolls around the roll center *Ro*

φ	roll angle of the body, °
$\Delta s_{1,f}$	Wheel travel of the outer front wheel during compression due to rolling, m
$\Delta s_{2,f}$	Wheel travel of the inside front wheel when deflecting due to rolling, m

3.3.2 Design of Anti-Roll Bars

Stabilizers are an additional spring in the suspension system that connects the wheels of an axle. The spring effect is generally achieved via torsion and bending. Accordingly, U-shaped and T-shaped stabilizers are found in use. However, in the case of high roll stiffness, disc spring assemblies are also used to reduce the roll angle, (Fig. 3.52).

U-Shaped Stabilizers

The closer the stabilizer is to the wheel, the smaller the forces are and it can be executed more easily. This principle can be used with U-shaped anti-roll bars. These are found mainly on production vehicles and on older racing cars. They usually consist of a piece of tubular or bar material, with coupling rods at the ends leading to the wheel-side mounts. In the case of a one-piece design, the stabilizer mounts must be separable for assembly.

The dimensions of the rod or tube naturally depend on the lever ratios and wheel loads, but the following range of values can serve as a rough comparison. In single-seaters, the stabilizer torsion bars have an outer diameter starting at 10 mm, in heavy touring cars this is hardly larger than 25–30 mm. The rod ends for the connections do not exceed the size with M8 thread.

Version Shaft with Lever

This version offers the possibility to mount the levers on the outside. In this way, shafts of different diameters can be installed without further modifications and thus easily different stiffnesses of the stabilizer can be realized.

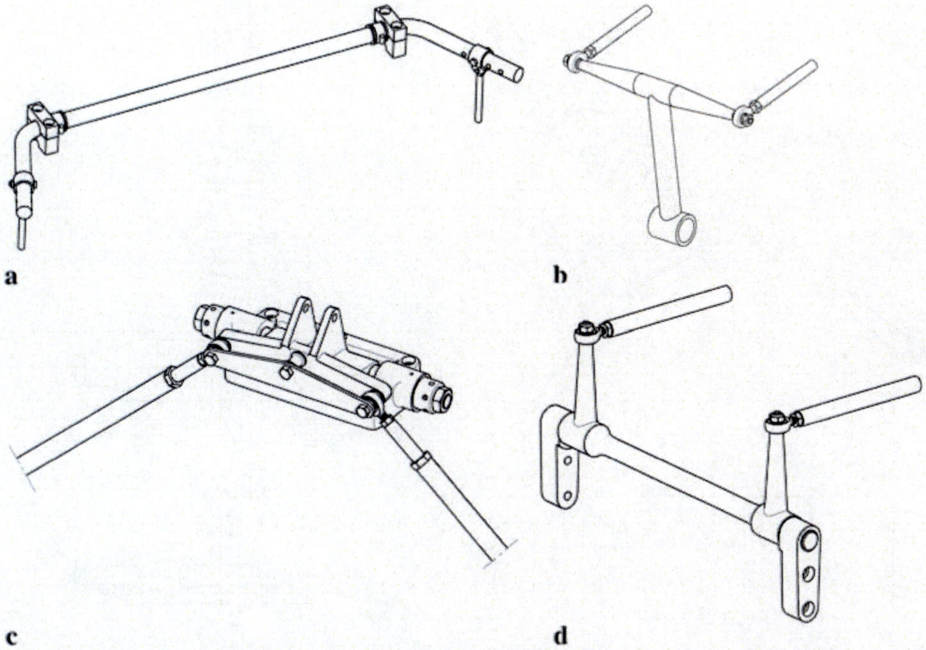

Fig. 3.52 Types of stabilizers. **(a)** U-shaped, bent from one piece. **(b)** T-shaped. **(c)** Disc springs with mono-damper arrangement. **(d)** U-shaped, multi-part (shaft with levers)

Stabilizer for Mono-Spring System

Such systems are found on rear or front axles. Design and function shown (Fig. 3.53).

The single spring strut is actuated via the bell crank (1). This is mounted on the axle (5) in such a way that it can rotate (= spring) and move along the axle (= roll). For this purpose, the lever bearing (2) has axial bearings (7) in addition to the two radial bearings (6). Belleville spring assemblies, which are preloaded with thrust bearings (4), are supported against these. Two spyholes (9) allow symmetrical adjustment. In the event of a rolling movement, the lever (1) is displaced along the axle (5) by a pressure rod (3). If this movement takes place, e.g. to the left, the right abutment (4, detail X) presses the cup springs against the thrust bearing (7). The stabiliser effect is adjusted by varying the arrangement of the cup springs. If they are installed in an O-shape (<><>), they have the softest effect. If they are arranged in the same direction (<<<<), they are stiffest. In between, combinations and different numbers lend themselves to: <<<>>>, <<>> < <, <><<<, <> <,. . . .

T-Shaped Stabilizers

These are used when the distance between the levers becomes too short for a "conventional" stabilizer torsion bar. The torsionally stressed part of the stabilizer is then arranged

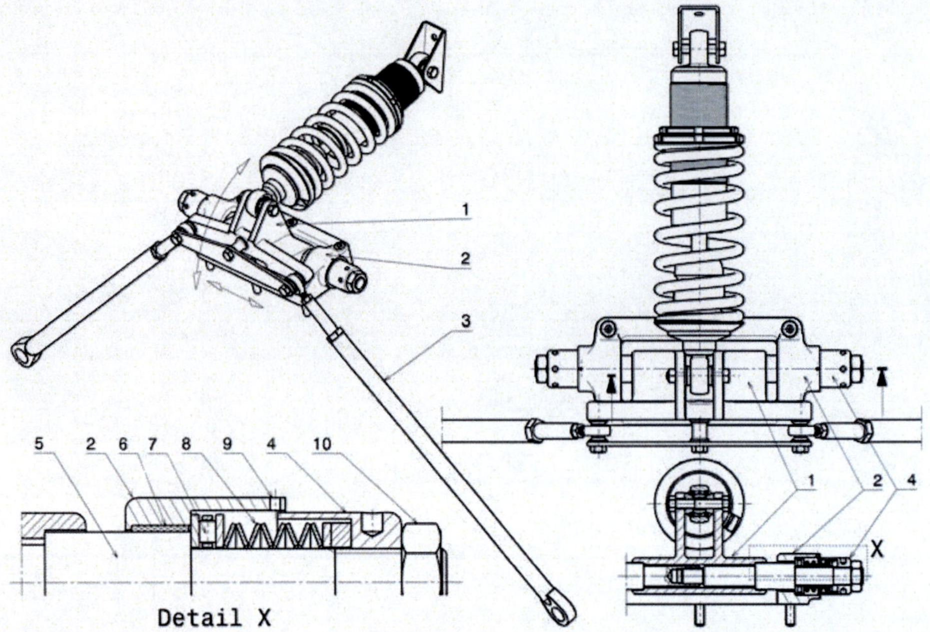

Fig. 3.53 Stabiliser for mono spring. **1** Rocker. **2** Lever bearing. **3** Pressure bar. **4** Thrust bearing. **5** Lever axle, 2-piece. **6** Radial bearing. **7** Axial bearing. **8** Belleville spring assembly. **9** Inspection hole. **10** Lock nut

almost vertically. With this type of anti-roll bar, a third spring can be added, which is only effective when the spring is on the same side. This variant is preferred for vehicles with high dynamics and aerodynamic downforce, (Fig. 3.54).

Cockpit-Adjustable Anti-Roll Bar

One way to change the stiffness of the stabilizer without disassembly is to use a multi-part design. If one or both levers are rotatably mounted, the moment of resistance of a sword-shaped lever can be varied by turning it. (Fig. 3.55) shows an example of such a lever mounting, and the operation is illustrated in (Fig. 3.56). The effect of rotation on the stiffness of the entire stabilizer can be taken as an example from (Fig. 3.57). The adjustment is made either by a push-pull cable, a linkage or by hydraulic systems. Of course, this adjustment option means further strain on the driver if it is to be exploited in the race while driving. Therefore, there are also teams that do without such a direct adjustment possibility of the anti-roll bar and "spare" their drivers – after all, there are already more than 15 adjustment knobs in the cockpit.[7] A possible further disadvantage can become unpleasantly noticeable. In the intermediate position between stiff and soft the blade can tilt, i.e. it

[7] See, e.g., Racing Car Technology Manual Vol. 2 *Complete Vehicle* Chap. 4, especially Fig. 4.25.

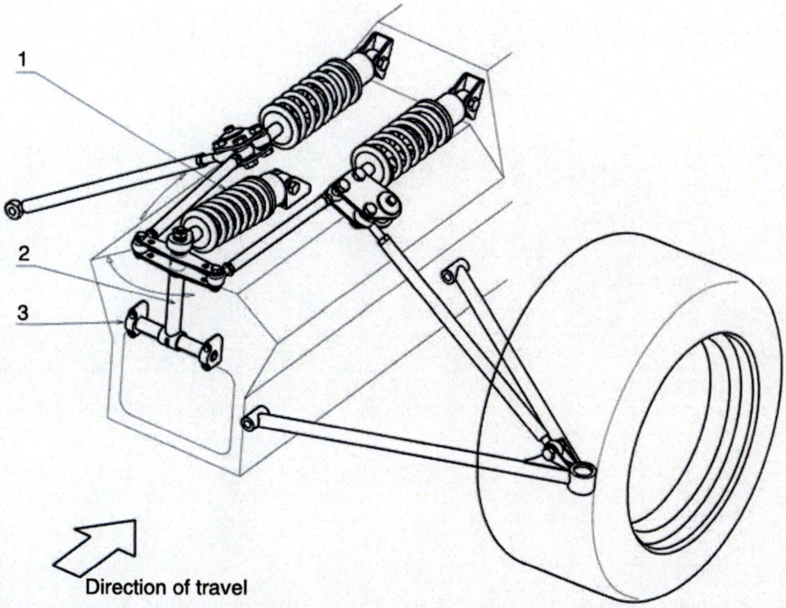

Fig. 3.54 Three-spring wheel suspension. The picture shows a rear suspension. **1** Auxiliary spring (third spring). **2** T-stabilizer. **3** Stabiliser bearing. The stabilizer (2) is twisted during alternate springing. The bearings (3) are positioned correspondingly far apart so that they introduce the torque into the chassis as small forces. In the case of equal-sided springing (heave) or at high driving speeds (downforce), the third spring (1) acts in addition to the two body springs. The stabilizer thereby oscillates around the axis of the bearings (3)

swerves sideways under the load. This is an unstable form of failure, like buckling of shells or buckling of push rods. It therefore occurs suddenly without noticeable notice. In this case, the stabilizer suddenly becomes softer during cornering. This phenomenon, also known as "*roll rock*", is particularly unpleasant when both levers are in the same intermediate position and thus fail simultaneously.

Material. U-shaped stabilizers bent from one piece are made of quenched and tempered steel, e.g. seamless drawn tube of 25CrMo4 (DIN EN 10083) for diameters up to 25 mm, 36CrNiMo4 (EN 10250) or 16CrMnMoV8–7–5-2 (1.7951) for larger diameters. Further materials can be found in Table 3.3 "Hot-rolled steels for heat treatable springs", as stabilisers are subjected to the same stresses as body springs.

In (Fig. 3.58), tube and rod material are contrasted as stabilizer backs, i.e. pure torsion spring. The pure torsion spring rate follows from Eqs. (3.34) above to $c_\beta{}^\circ = G - I_{p,Ba}/(57.3\ l_{Ba})$. If a back with 200 N m/° is required, a rod with a diameter of 19.5 mm can be used, which has a mass of 234 g. The same stiffness can be achieved with a tube. A tube with an outer diameter of 24.5 mm and a wall thickness of 1.5 mm has the same stiffness, but a

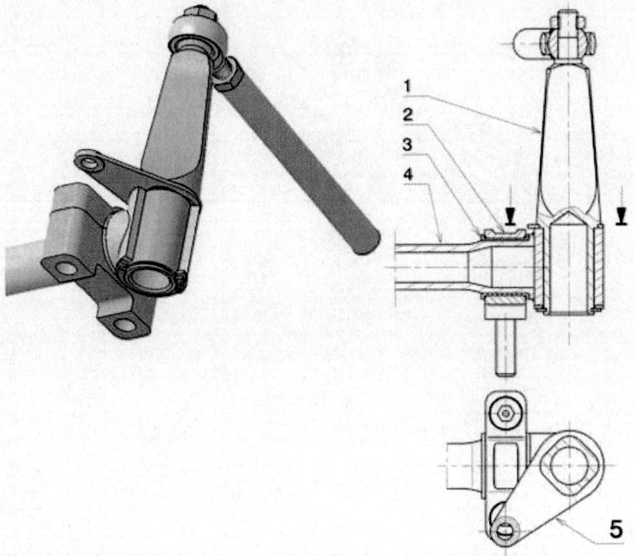

Fig. 3.55 Adjustable stabilizer, structure. **1** Lever arm. **2** Two-piece mount. **3** Two-piece bearing shell. **4** Stabiliser bar. **5** Adjusting plate

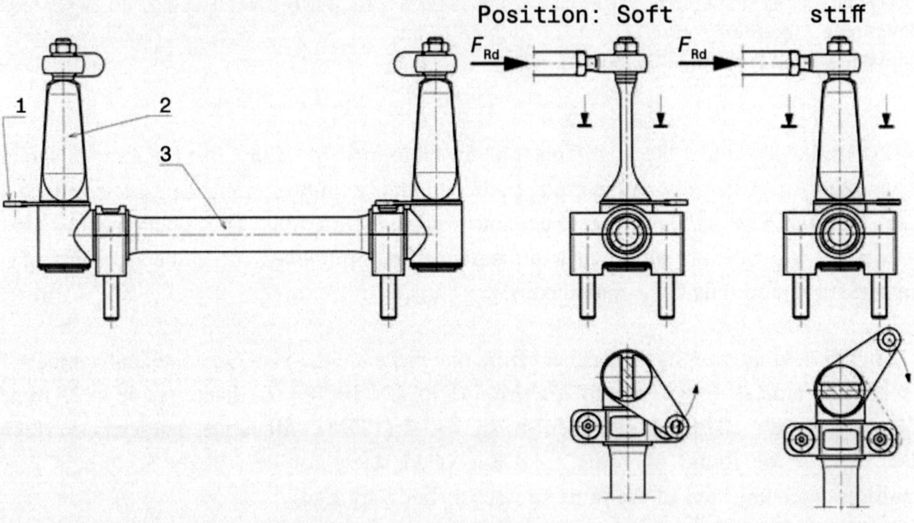

Fig. 3.56 Adjustable stabilizer, function. The force F_{Rd} of the strut bends the sword-shaped lever. This lever can be twisted in relation to the force. The two extreme positions, *soft* and *stiff*, are shown in plan and front view. The stiffnesses between the end positions lie within the extreme values. **1** Adjusting lug. **2** Arms. **3** Back

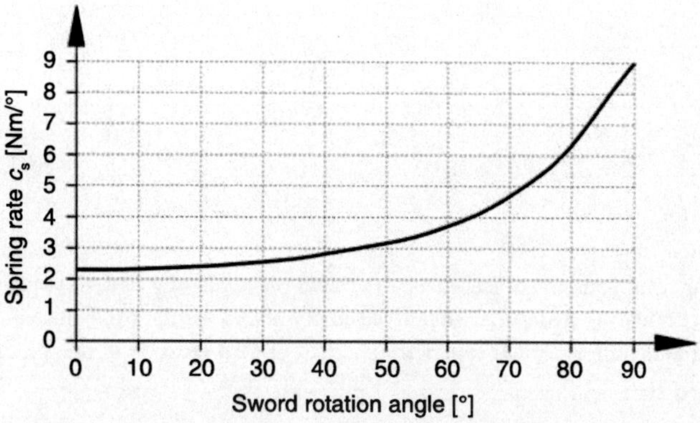

Fig. 3.57 Stiffness of an adjustable stabilizer versus the angle of twist of the arms. The stiffness of the entire stabilizer results from three springs connected in series (bar and 2 arms). The stiffness curve between the extreme positions of the arms – soft (0°) and stiff (90°) – is not linear

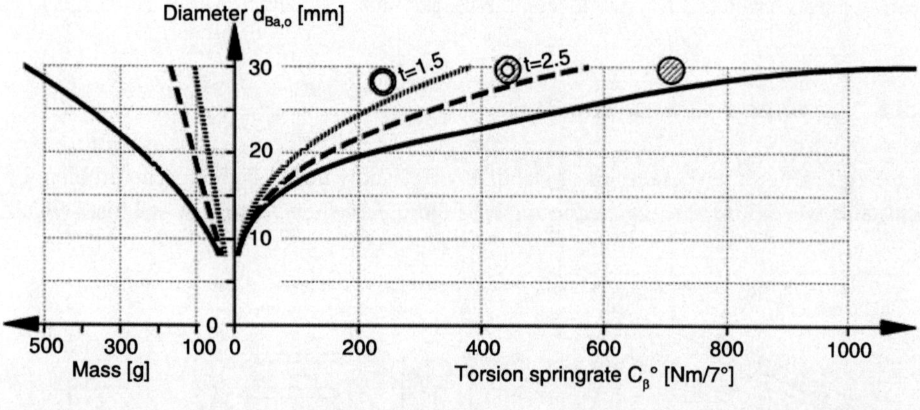

Fig. 3.58 Comparison of torsion springs made of steel. Three different versions of stabilizer backs are shown with a length of $l_{Ba} = 100$ mm, namely 2 tubes with wall thicknesses of 1.5 and 2.5 mm and a solid shaft

mass of only 85 g. If you choose a tube with a thickness of 2.5 mm, it must have a diameter of 21.8 mm and weighs 119 g.

Mount

Stabilizers are mounted directly in an aluminum block or a maintenance-free sliding bush (dry sliding bearing with PTFE sliding layer) with a thrust collar is additionally pressed into the bearing block. Rubber bushes are only used in production vehicles. A good compromise for racing vehicles with U-shaped stabilizers or if the back is bent during operation is

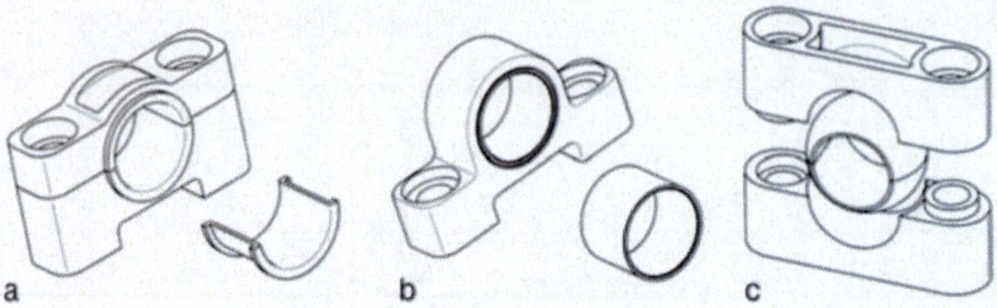

Fig. 3.59 Bearing arrangements of stabilizers. (**a**) Two-piece bearing pedestal with flanged bearing shells, (**b**) One-piece bearing pedestal with pressed-in dry plain bearing, (**c**) Adjustable plastic bearing with split housing and split spherical cap. The spherical cap can also be in one piece

to use a bushing in a spherical mount (adjustable bearing). The bearing should allow the necessary rotational movement with as little friction as possible. If the bearings are too preloaded, the body springs will respond poorly due to the coupling effect of the stabilizer when driving on undulating roads and the car will "copy" the course of the road.

The bearing blocks are one-piece if they can be attached to the stabilizer ends. If the bearing points are inside the two levers, the blocks must of course be split, (Fig. 3.59).

3.3.3 Examples of Anti-Roll Bars

In the following, a few examples of stabilizers will show how differently the problem of adjustable wheel load shifting can be solved. Figure 3.60 shows a narrow stabiliser which

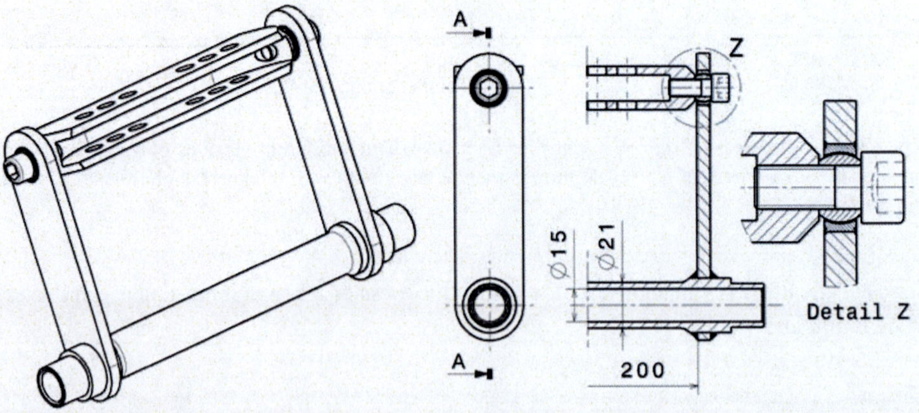

Fig. 3.60 Stabilizer on the front axle of a Formula BMW car. The stabilizer is mounted externally with the steering gear and is loaded via push rods from the bell crank via the U-shaped profile. The elasticity is practically only limited to the tube section. Several holes in the U-profile allow fine tuning with unchanged tube diameter

Fig. 3.61 U-shaped stabilizer on an older formula car (Lola T-328 SV, left front). The stabilizer is adjusted by moving the fastening clamp of the coupling rod (drop link)

Fig. 3.62 Front anti-roll bar (Formula 3000 Zytec), driving direction to the left. The stabilizer (1) is designed for extreme roll stiffness and functions similarly to that of (Fig. 3.53). The adjustment is made by different springs (2) and different pivot points (3) of the bellcranks. Travel sensors (4) are mounted on the suspension struts to measure the spring travel

sits on the nose of a formula car. A classic U-shaped design can be seen on (Fig. 3.61). On cars with pronounced aerodynamic downforce aids, extremely stiff stabilizers are used as in (Fig. 3.62). The stabilizer on (Fig. 3.63) can be adjusted from the driver's seat. A purely bending anti-roll bar is presented in (Fig. 3.64).

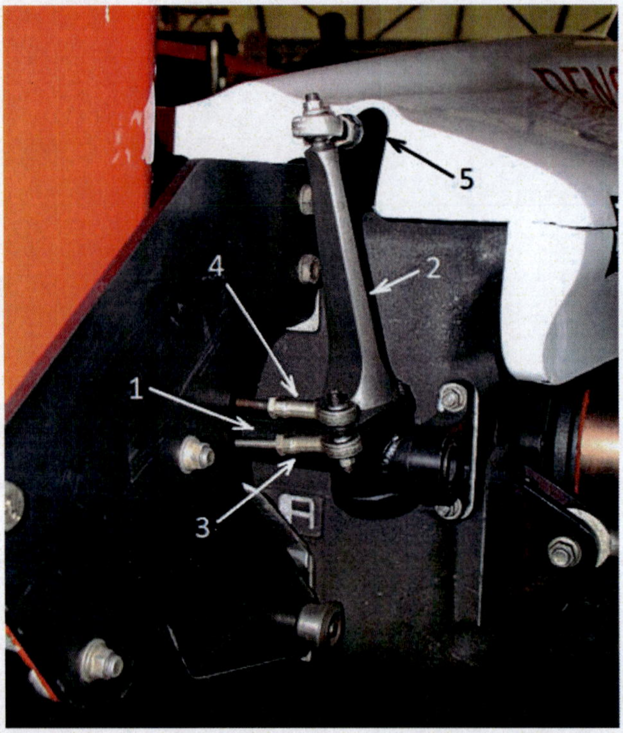

Fig. 3.63 Adjustable rear stabilizer of an Indy Car (Toyota RV8B). The section of the stabilizer shown is the arm that is subjected to bending. By rotating around the longitudinal axis, the governing section modulus and thus the behavior of the stabilizer can be changed. **1** Stabilizer torsion bar. **2** Stabilizer lever, right. **3** Push rod for turning the arm from the driver's position. **4** Coupling rod to stabilizer link on left. **5** Coupling rod to the bell crank between the push rod and the damper strut

Fig. 3.64 Rear stabilizer on a Formula Ford. With this stabilizer, practically only the bending of the two levers has an elastic effect. The torsion of the back does not come into play. Adjustment is made by changing the effective lever length by pushing one or both connecting sleeves (1) up or down. In the arrangement shown, the stabilizer is softest

References

1. Henker, E.: Fahrwerktechnik. Springer Vieweg, Wiesbaden (1993)
2. Richter, W.: Konstruieren neuartiger Produkte. Konstruktion Heft 47, pp. 268–276. Springer, Düsseldorf (1995)
3. Crahan, T.C.: Modelling steady-state suspension kinematics and vehicle dynamics of road racing cars, part II. SAE paper 942506. In: Smith, C. (ed.) Racing Chassis and Suspension Design. SAE International, Warrendale (2004)
4. Staniforth, A.: Competition Car Suspension, 3rd edn. Haynes, Sparkford (1999)
5. Reimpell, J., Betzler, J.: Fahrwerktechnik: Grundlagen, 4th edn. Vogel, Würzburg (2000)
6. Anderson, D.: Quicker corners with bump stops. In: Speedway Illustrated, pp. 68–71 (2008)
7. Smith, C.: Tune to Win. Aero Publishers, Fallbrook (1978)
8. Incandela, S.: The Anatomy & Development of the Formula One Racing Car from 1975, 2. Haynes, Sparkford (1984)
9. Muhs, D., et al.: Roloff/Matek Maschinenelemente, 17th edn. Springer Vieweg, Wiesbaden (2005)
10. Heißing, B.: Moderne Fahrwerksauslegung, Vortrag im Rahmen der ÖVK-Vortragsreihe, Graz 12. Mai. (2004)
11. Milliken, W., Milliken, D.: Race Car Vehicle Dynamics, 1st edn. SAE International, Warrendale (1995)

12. Rouelle, C.: Race Car Vehicle Dynamics and Data Acquisition. Seminar Binder, OptimumG (2012)
13. Haney, P.: The Racing & High-Performance Tire, 1st edn. SAE International, Warrendale (2003)
14. Braess, H.-H., Seiffert, U.: Vieweg Handbuch Kraftfahrzeugtechnik, 4th edn. Vieweg, Wiesbaden (2005)
15. Matschinsky, W.: Radführungen der Straßenfahrzeuge, 2nd edn. Springer, Berlin (1998)
16. Krimmel, H., et al.: Elektronische Vernetzung von Antriebsstrang und Fahrwerk. ATZ. **5**, 368–375 (2006)
17. Thomas, C.: Off-road suspension. High flying birds, in Professional Motorsport World Heft Jul.–Aug. 2017, pp. 36–39. UKi Media & Events, Dorking (2017)
18. Kasprzak, J.: KAZ Technologies. Damping Calculations Seminar, Michigan (2011)
19. Fuhr, F., et al.: Optimizing vehicle dynamics by in-house function development. In: Pfeffer, P. (ed.) 7th International Munich Chassis Symposium, pp. 143–158. Springer Vieweg, Wiesbaden (2016)
20. N.N.: Formula Renault 2000 Manual. Renault Sport Promotion Sportive (2001)
21. Milliken, W.F.: Chassis Design: Principles and Analysis. Society of Automotive Engineers, Warrendale (2002)
22. Formula One Technology: Sonderheft von Race Engine Technology. High Power Media, Somerset (2007)
23. Zomotor, A.: Fahrwerktechnik: Fahrverhalten. In: Reimpell, J. (ed.) Vogel Buchverlag, Würzburg, 2nd edn. Vogel Fachbuch, Kraftfahrzeugtechnik (1991)
24. Öttgen, O., Bertram, T.: Modellgestützte Entwicklung eines mechatronischen Fahrwerkregelungssystems. ATZ. **6**, 546–553 (2007)
25. Der neue BMW 5er, Sonderausgabe ATZ MTZ. Springer Vieweg, Wiesbaden (2003)
26. Segers, J.: Analysis Techniques for Racecar Data Acquisition, 1st edn. SAE International, Warrendale (2008)

4 Types of Suspensions

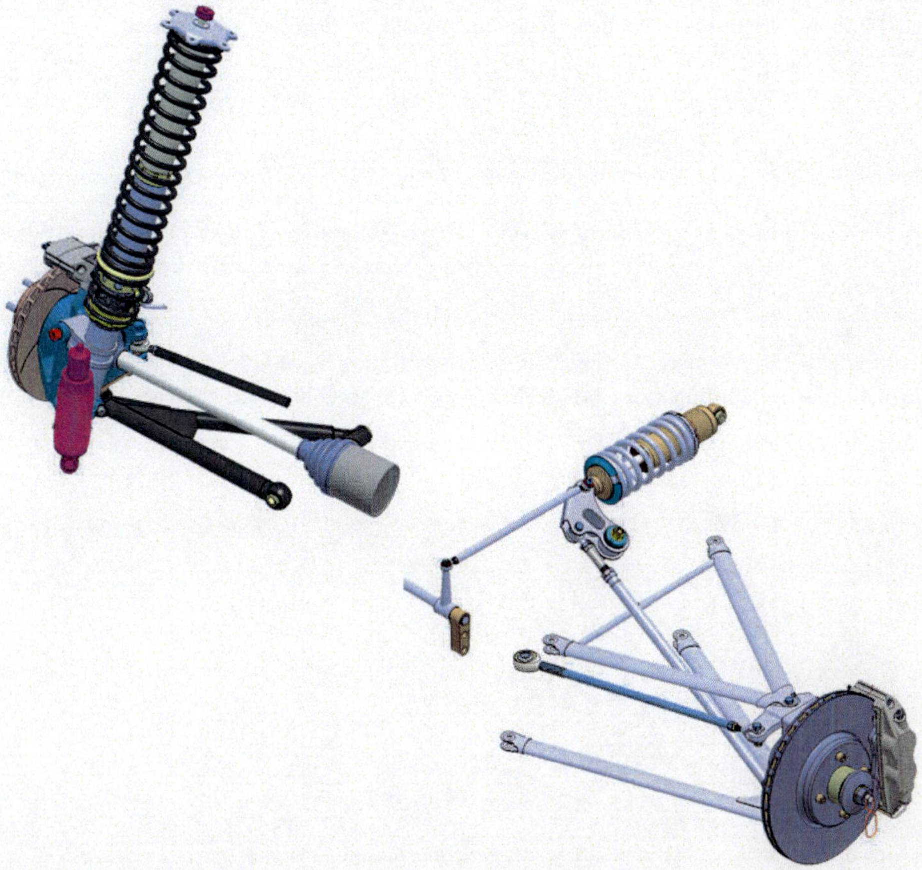

M. Trzesniowski, *Suspension System*,
https://doi.org/10.1007/978-3-658-39847-7_4

In most racing vehicles, the double-wishbone axle for front and rear has become established. It offers the greatest freedom in design for the designer, but also subsequently in the setup for the race engineer. The moving masses can be kept low even with large track widths. In addition, this design can be used for both driven and non-driven axles. Rally vehicles use wheel-guiding suspension struts at the front and rear.

4.1 Suspensions of Passenger Cars

Depending on the requirements or regulations, various designs have become established. In the case of passenger cars, apart from comfort, it is important that there is room for the engine (which is generally housed at the front) and that the front wheels have sufficient turn-in. The rear axle should be as flat as possible for large trunks and simple fuel tank shapes. Figures 4.1 and 4.2 show designs that meet these requirements and are therefore widely used.

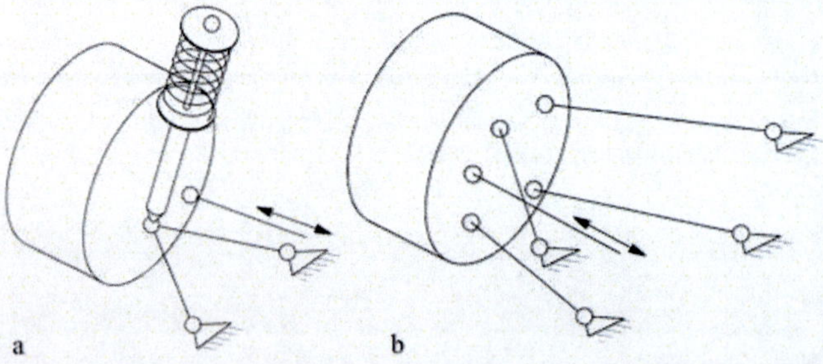

Fig. 4.1 Passenger car front axles. (**a**) McPherson axle. (**b**) Multi-link axle, has max. 4 links

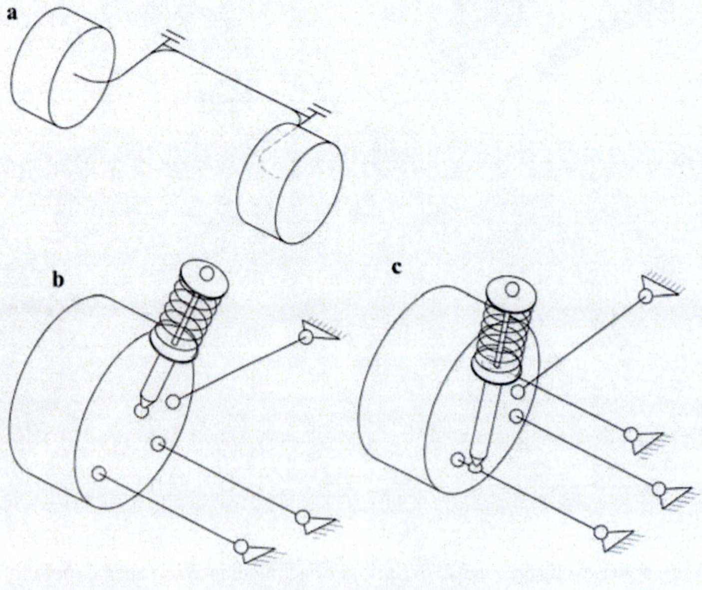

Fig. 4.2 Passenger car rear axles. (**a**) Twist beam axle. (**b**) McPherson axle. (**c**) Multi-link axle, has max. 5 links

Table 4.1 Total chassis of passenger cars, according to [1]

Class	Drive	Front axle	Rear axle	Comment
Compact class	Front	McPherson	Composite links	
	All Wheel	McPherson	Multi-link	
			McPherson	
Middle class	Front	McPherson	Composite links	
		Multi-link		Floating subframe at the front
	All Wheel	McPherson	Multi-link	
		Multi-link	McPherson	Floating front subframe
	Rear	McPherson	Multi-link	
			McPherson	Floating subframe rear
Upper	Front	McPherson	Multi-link	Floating subframe rear
Middle class		Multi-link		Floating front subframe
	Rear	McPherson	Multi-link	
			McPherson	Floating subframe rear
Upper class	Front	Multi-link	Multi-link	Floating front subframe
	Rear	McPherson	Multi-link	Floating subframe rear
		Multi-link		

In addition to their technical characteristics, these designs also differ according to economic aspects. For this reason, there are designs that are only used on more sophisticated vehicles. Table 4.1 provides an overview of combinations of axles currently found on the market.

4.2 Racecar Suspensions

4.2.1 Double Wishbone Suspension

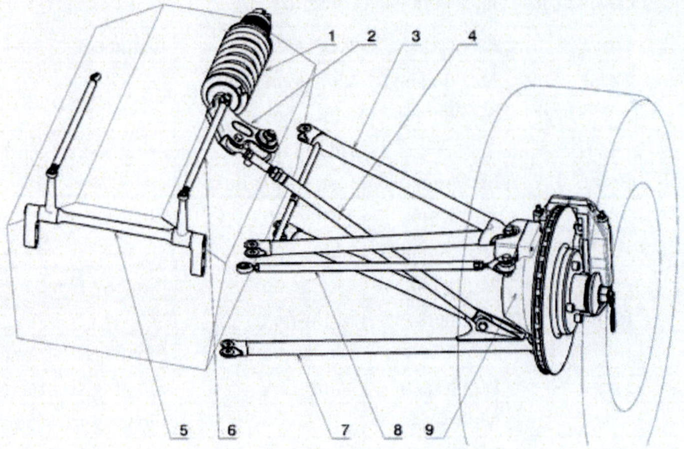

Fig. 4.3 Parts of a double wishbone axle. The suspension of one wheel is shown. **1** Coil-over spring damper unit. **2** Rocker. **3** Upper wishbone. **4** Push rod. **5** Anti-roll bar. **6** Linkage. **7** Lower wishbone. **8** Tie rod. **9** Wheel carrier *(upright)*

The double wishbone suspension is the dominant design above all in formula cars and production sports cars, but it can also be found in all other racing vehicles such as touring cars or raid vehicles. It allows great variability in the course of the most important geometric variables defining the wheel position and comes close to the ideal of combining distant wheels (the rubber tyre is only able to generate relatively small forces) with a compact body with as little mass as possible. Driven and non-driven axles can be represented and the brake can be mounted on the wheel side or on the fuselage side. Their design-related disadvantage, namely taking up a lot of space, does not come into play in racing vehicles. In production vehicles, where passengers and luggage space have to be accommodated, the assessment is different.

In the basic construction only the steering (i.e. tie rod laterally movable) distinguishes the front from the rear axle, Fig. 4.3. The picture shows the currently common suspension of a wheel, which can basically be installed on the left or right and front or rear. The wheel carrier (9) is connected to the car body via two wishbones (3) and (7) and a track rod (8). The track rod can be connected to a steering gear and thus the axle can be steered. If the axle is to be driven, a side shaft must also be routed from the final drive to the wheel hub. The vertical forces coming from the tyre are introduced into the frame via the pressure rod (4) through the bell crank (2) via the spring (1). A stabilizer (5) elastically connects both wheels of an axle via its bell crank.

Starting from the basic design with two triangular wishbones shown in Fig. 4.3, there are numerous variations for passing on the wheel load and for integrating the stabilizer.

The effective distance c of the two links (Fig. 4.4) should be as large as possible in order to obtain small forces in the body and link bearings. This also keeps resulting deformations in the components small and play in the joints has less effect on the wheel. Some vehicles

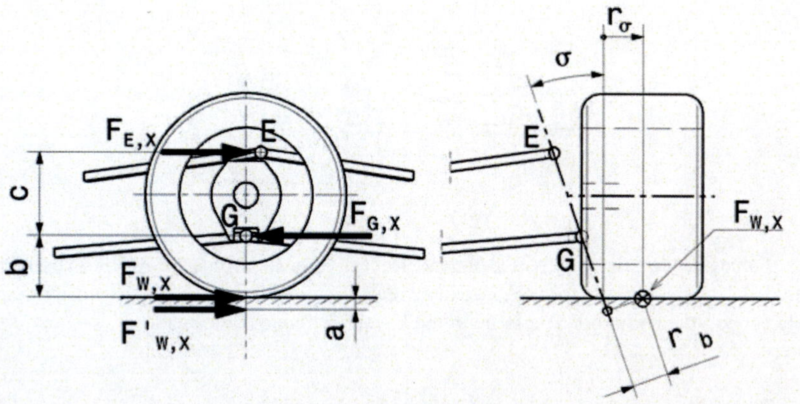

Fig. 4.4 Forces at the points of articulation in the case of circumferential forces. Effect of a circumferential braking force $F_{W,X}$ on the forces in the joints *E* and *G* on the wheel carrier. The scrub radius r_σ is positive

deviate strongly from this ideal in order not to influence the air flow negatively. The effect of aerodynamics is therefore given priority over these strength considerations.

Some basic considerations show that the lower wishbone is more heavily loaded. A circumferential force $F_{W,X}$, e.g. a braking force, must be balanced by the two joints *E* and *G* on the wheel carrier, Fig. 4.4. This braking force acts on the steering axis *EG* with the lever arm r_b. In order for the magnitudes of the joint forces $F_{E,X}$ and $F_{G,X}$ to be determined, $F_{W,X}$ must be shifted perpendicular to the steering axis (applies to the common case of the brake lying in the wheel). As a result, this comes to lie under the roadway as $F'_{W,X}$. In the side view, the displacement path *a* results in:

$$a = r_b \cdot \sin\sigma = +r_\sigma \cdot \cos\sigma \cdot \sin\sigma$$

If the scrub radius r_σ is negative, $F'_{W,X}$ is above the road surface.

From the side view the forces follow to:

$F_{W,X} = F_{G,X} - F_{E,X}$	$F_{G,X} = F'_{W,X} \cdot \frac{a+b+c}{c}$
	$F_{E,X} = F'_{W,X} \cdot \frac{a+b}{c}$

For small reaction forces in the joints *E* and *G*, (*a* + *b*) should therefore be small and *c* as large as possible.

A negative scrub radius therefore has advantages in this respect, because the length *a* is negative here.

A basic consideration of the lateral force influence shows that here too the lower wishbone transmits the greater forces and the effective distance *c* should be large,

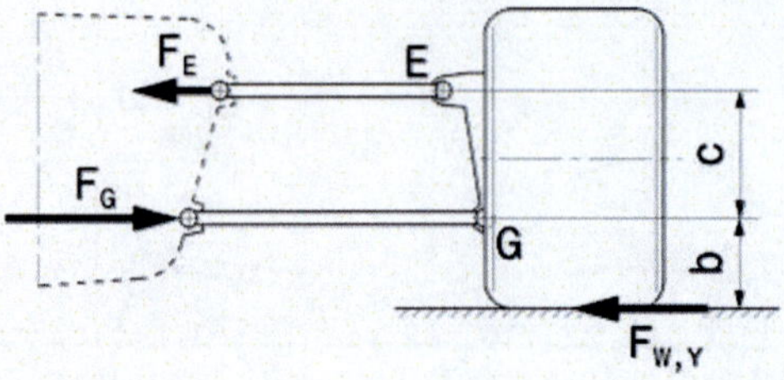

Fig. 4.5 Forces at the joint points with lateral forces. Only the influence of the lateral forces is considered. A lateral cornering force $F_{W,Y}$ causes reaction forces F_E and F_G in the wishbones, which adversely support the body roll. The force in the lower wishbone is greatest

Fig. 4.5. The track distance *b* of the lower joint *G* should in turn be as small as possible. For the case of horizontally lying control arms applies:

$F_{W,Y} = F_G - F_E$	$F_G = F_{W,Y} \cdot \frac{b+c}{c}$
	$F_E = F_{W,Y} \cdot \frac{b}{c}$

The link lengths should also be kept large. This keeps the angular deflections small during compression and rebound and thus the changes in the wheel position geometry as well. A disadvantage that must not be forgotten here is the greater buckling sensitivity with longer, slim control arms.

A wishbone can be simplified as two bars that only transmit tension/compression forces, which is ideal in terms of lightweight construction. In fact, the conditions are generally more complex because the control arms are rigidly connected to each other, at least on the wheel carrier side, and this point can therefore also transmit moments. Now the wishbones must be able to absorb different amounts of longitudinal and lateral forces. The largest and most frequent forces are important for the design. On a non-driven front axle, large longitudinal forces mainly occur against the direction of travel when braking, and when cornering, high lateral forces occur on the outer wheel, i.e. directed towards the car. On a driven rear axle, the driving forces and the lateral forces towards the fuselage are significant. In order to ensure that all forces originating from the tyre can be absorbed by the two arms of a wishbone plane, a compromise will therefore be necessary in which the angle of the control arms to the forces and parameters such as the resulting force, length of the control arm (mass, buckling, kinematics) and installation space must be taken into account, Fig. 4.6.

A detailed consideration of the bar forces caused by the forces acting on the tyre is given in Sect. 2.3.5 *Calculation*.

On the steered front axle there is an additional limiting consideration, namely the maximum turn of the wheels, Fig. 4.7.

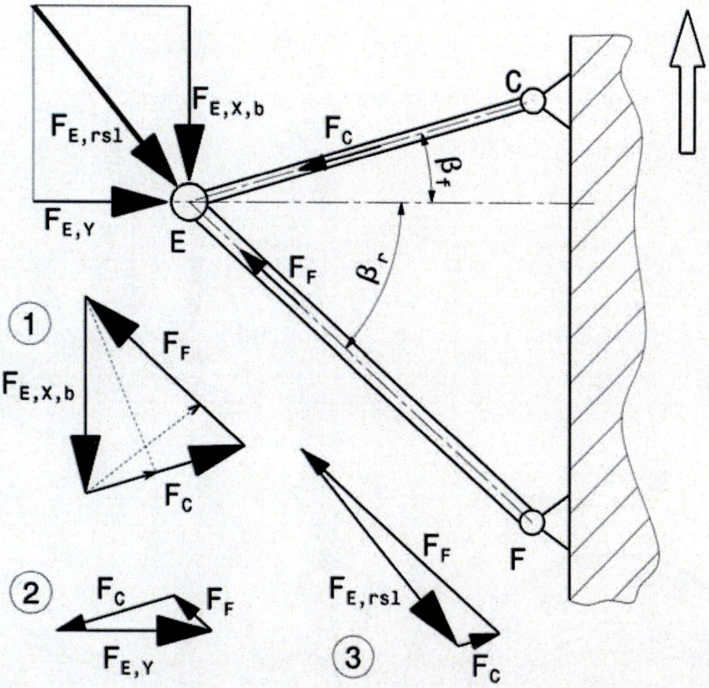

Fig. 4.6 Wishbone design according to forces. A wishbone plane on the left side of the vehicle is considered. This is a front axle, so the braking forces dominate the longitudinal forces. When considering the forces in the link arms, 3 cases are considered individually. **Case 1**: Only the part $F_{E,X,b}$ of the braking force acts on the joint E. Arm EF absorbs a greater force F_F than arm EC (F_C), whose angle β_f is smaller. Arm EF also experiences a compressive force, while arm EC experiences a tensile force. From the triangle of forces it can be seen: If the force F_F is to become smaller, either the angle β_f must become larger (dashed), or the angle β_r is chosen to be larger. With these measures, F_C would also become smaller, i.e. the changes would be favourable from this point of view. **Case 2**: Only the lateral force F acts on the joint $E_{E,Y}$ alone. **Case 3**: Braking and lateral forces act on the joint E as the resultant $F_{E,rsl}$. In the example, the maximum forces in the longitudinal and transverse directions are considered to act simultaneously, which is not possible in reality due to the tyre characteristics, i.e. the combined forces are actually smaller than the maximum forces in the longitudinal direction

Variation *Options*

Starting from the basic arrangement of parallel links of equal length, three basic variations with different driving behaviour can be derived, Fig. 4.8.

Figure 4.9 shows the changes in some variables that are important for the driving behaviour during equilateral springing and roll of the car body for the three basic variants of double wishbone arrangements.

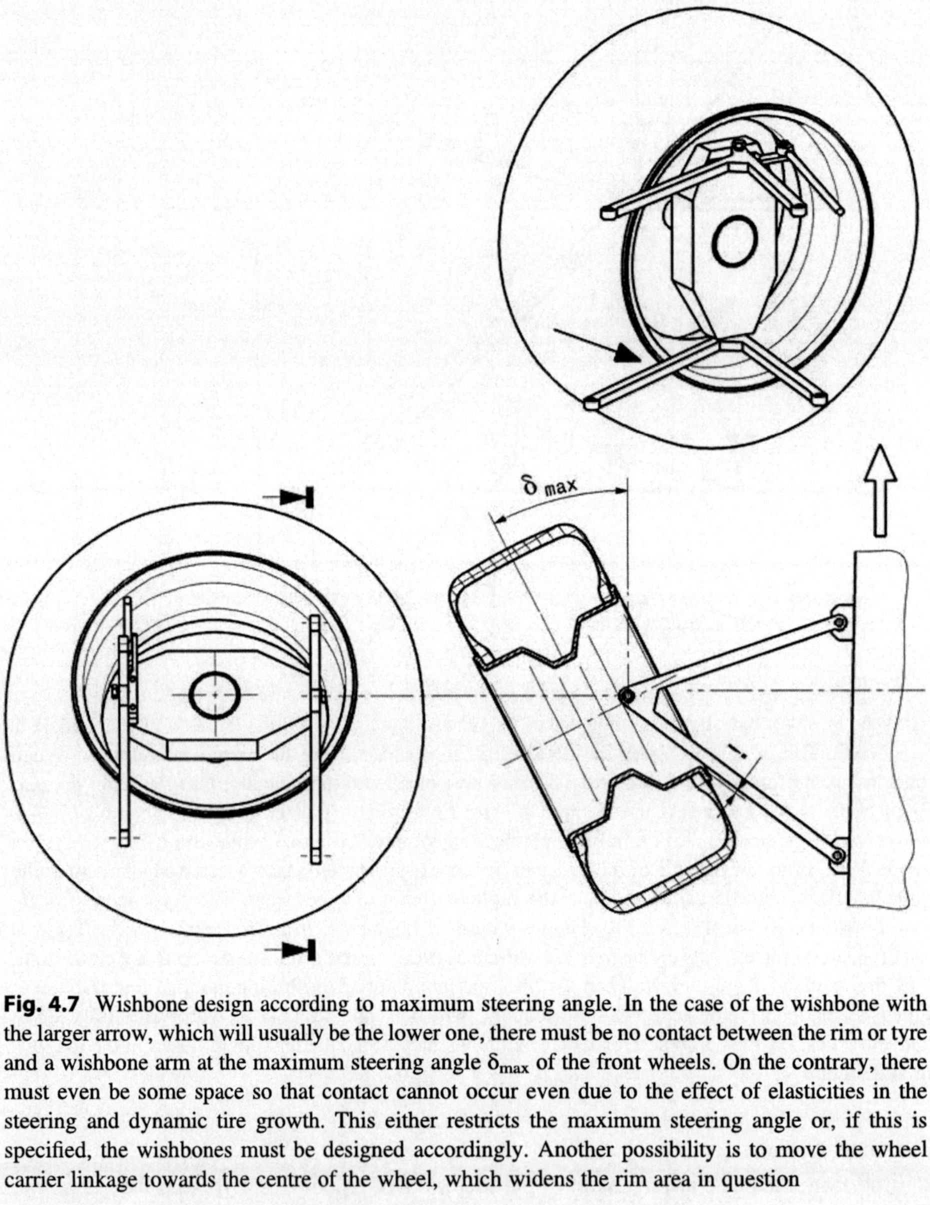

Fig. 4.7 Wishbone design according to maximum steering angle. In the case of the wishbone with the larger arrow, which will usually be the lower one, there must be no contact between the rim or tyre and a wishbone arm at the maximum steering angle δ_{max} of the front wheels. On the contrary, there must even be some space so that contact cannot occur even due to the effect of elasticities in the steering and dynamic tire growth. This either restricts the maximum steering angle or, if this is specified, the wishbones must be designed accordingly. Another possibility is to move the wheel carrier linkage towards the centre of the wheel, which widens the rim area in question

Parallel Links with Equal Length

Link and wheel carrier form a parallelogram. The cross instant center is at infinity and the roll center is exactly on the road. Consequently, the camber angle does not change at all during heave, but the track width varies greatly (scrub). Because the control arms are horizontal in design position, the track width decreases during bump or rebound. Larger

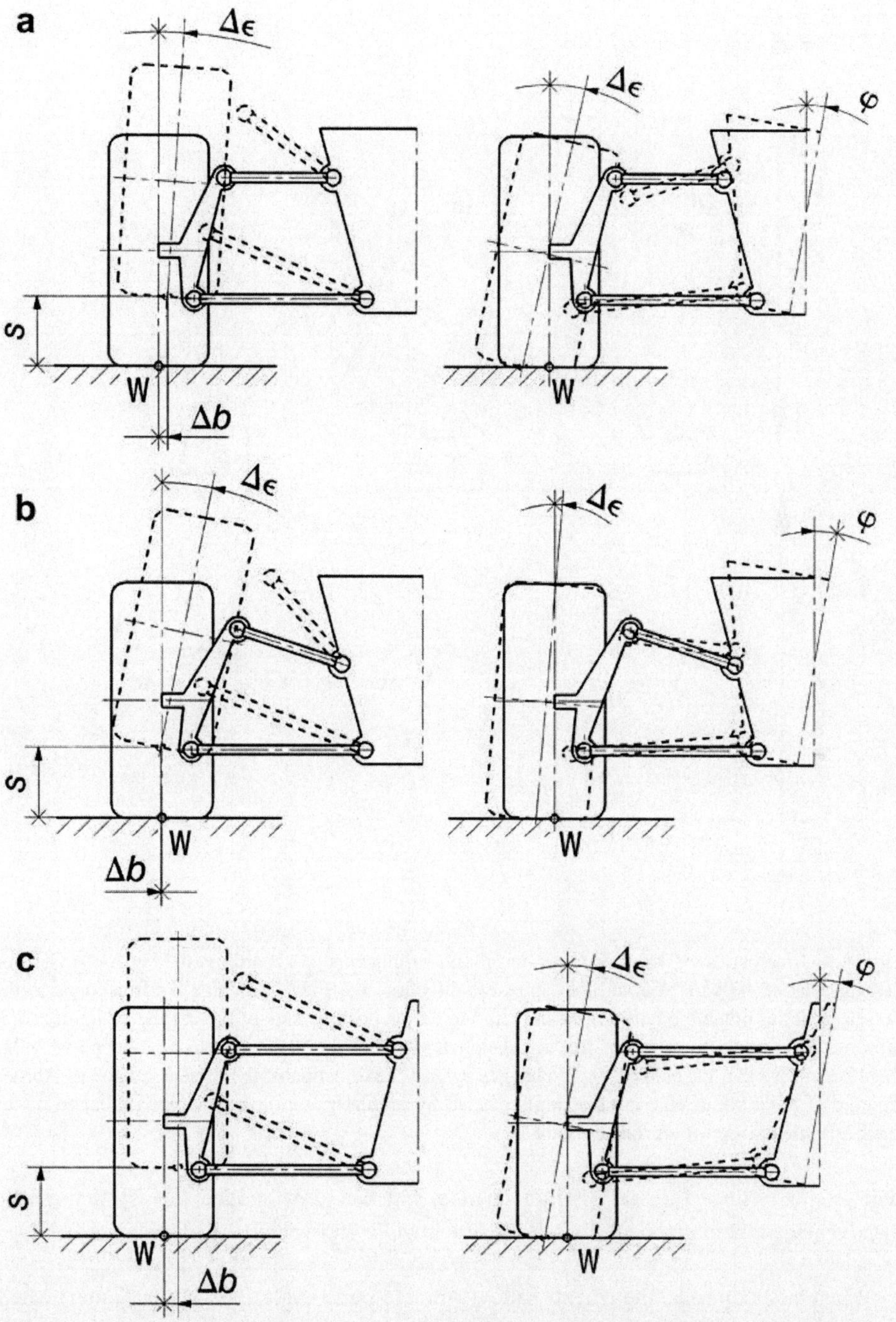

Fig. 4.8 Principle design options for double wishbone axles. For comparison purposes, the lower control arm is in the design position parallel to the road surface for all variants. The axles are shown in the design position (solid lines) as well as sprung and rolled (dashed). In design position, all wheels have camber angle 0° for comparison purposes. *S* wheel travel, φ Roll angle of the frame, *W* Wheel contact point, Δb Change of track (scrub), $\Delta \varepsilon$ Change of the camber angle. (**a**) Parallel links of equal length. (**b**) Parallel links of unequal length. (**c**) Non-parallel, unequal-length links

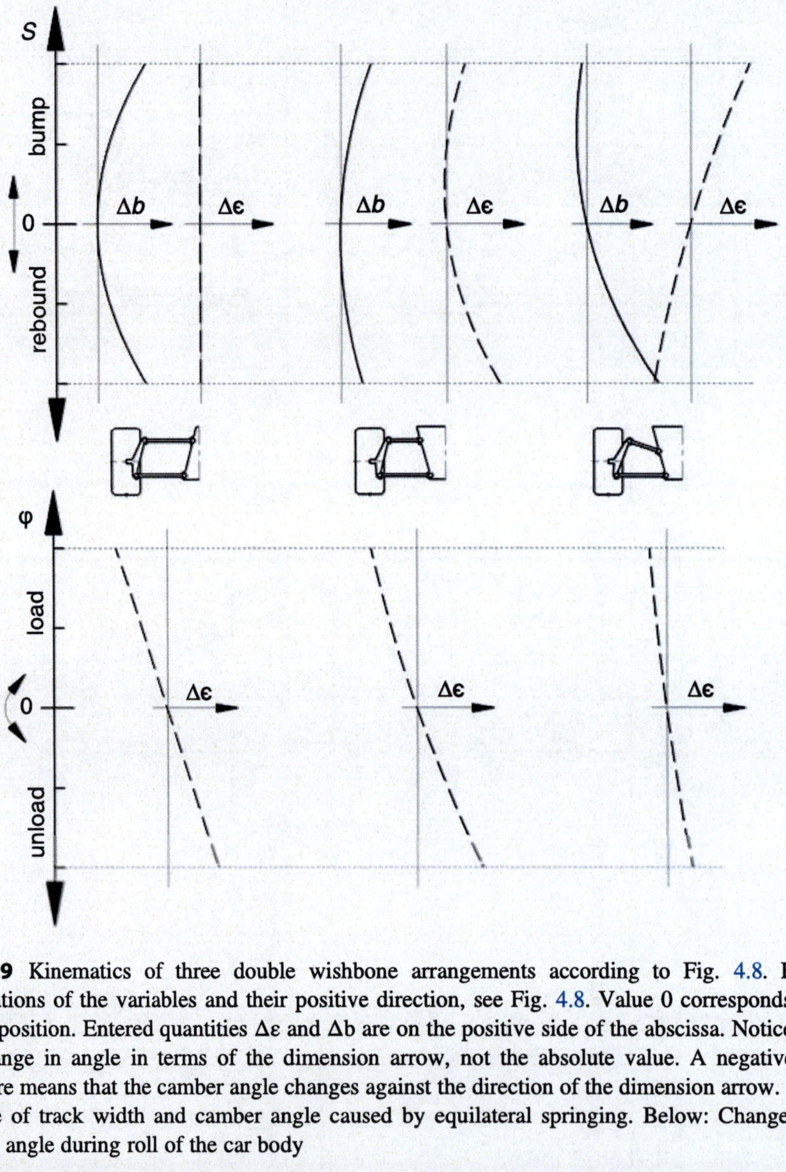

Fig. 4.9 Kinematics of three double wishbone arrangements according to Fig. 4.8. For the designations of the variables and their positive direction, see Fig. 4.8. Value 0 corresponds to the design position. Entered quantities Δε and Δb are on the positive side of the abscissa. Notice: Δε is the change in angle in terms of the dimension arrow, not the absolute value. A negative value therefore means that the camber angle changes against the direction of the dimension arrow. Above: Change of track width and camber angle caused by equilateral springing. Below: Change of the camber angle during roll of the car body

link lengths reduce the track width change, but can only reduce this system-related disadvantage due to practical limitations (track width and vehicle width restricted).

When the body rolls, the wheels follow with the same angle. The loaded (outer) wheel can (depending on the camber angle in the design position) thereby go into unfavourable positive camber.

Parallel, Unequal Links
If the upper link is made shorter than the lower link, the kinematic behaviour changes considerably compared to the above initial arrangement. The camber angle changes in a favourable manner when the wheel compresses and decompresses in the direction of a negative angle. By this change of angles of wheel, also change of track width is influenced favourably. This is greatly reduced and can be kept to a useful minimum by clever choice of link lengths and joint arrangements. If the upper link is chosen too short, it will come into a stretched position during compression, which must be avoided at all costs. This must be taken into account especially if large suspension travel is desired, e.g. for off-road use.

When the frame rolls, the loaded wheel moves towards negative camber, but the unloaded wheel moves disadvantageously towards positive camber values.

Non Parallel, Unequal Links
If the link lengths and their position are chosen arbitrarily, the transverse instant centers can be placed arbitrarily in the design position and thus the kinematic behaviour of selected variables can be designed arbitrarily, as it were. In this way, the camber change during rolling can be reduced compared to the arrangement with parallel control arms. The change in track width during compression can also be kept small. In the design shown, however, the track width changes considerably when the suspension is released.

In order to keep the camber change low during roll or to steer in the direction of negative values at the outer wheel of the turn, unequal-length and non-parallel wishbones are used. By corresponding inclination of the wishbone axis of rotation on the car side (in the side view), additional braking or acceleration pitch compensation can be created.

The kinematic effects of geometric changes of a typical suspension according to Fig. 4.10 are described in Table 4.2. In individual cases, the effects of changes must of course be examined in more detail, but the tendential effects can be used as a rough guideline for optimization. Lowering the pivot point *C*, for example, leads to a large increase in the roll centre height, a very large increase in the camber angle change during bouncing and a large reduction in the camber angle variation during roll.

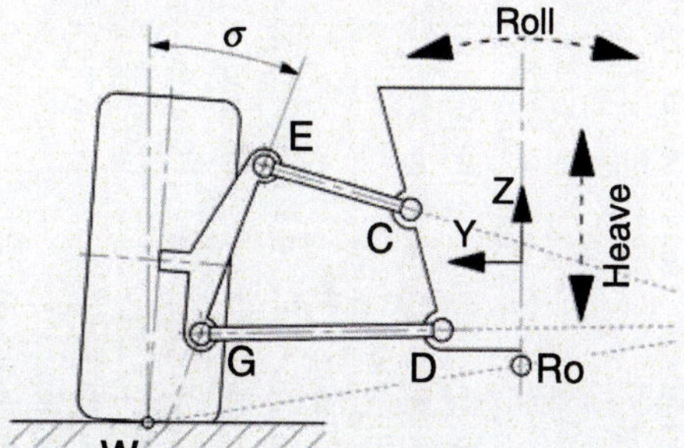

Fig. 4.10 Designation of the articulation points of a double wishbone axle with unequal-length, non-parallel control arms for Table 4.2

Table 4.2 Trend changes of kinematic quantities with variation of the axle geometry, [2]

		Change of location of the roll center					Changing the camber angle			
		With (equilateral) springing			On roll		When springing		On roll	
Modifications of the initial design by moving the pick-up points in 25 mm steps in *Y*- or *Z*-direction		Height, static	Descent on compression	Raise on rebound	Lateral movement	Vertical movement	On compression	On rebound	Leftwards	Right
Changing the inclination of the upper link *EC*	*C* downwards	>>	>	≤	<<	>>	>>>	>>>	<<	<<
	C upwards	<<	<	>	>>>	<<<	<<<	<<<	>>	>>
Modification of the length of the upper link *EC* (E remains unchanged)	*C* to the right	=	≥	>	≥	≤	≤	=	=	=
	C to the left	=	≤	<	≤	≥	≥	=	=	=
Change of the kingpin inclination angle (*G* and *C* remain unchanged)	Link *EC* longer	=	≥	≥	≥	≤	≤	=	=	=
	Link *EC* shorter	=	≤	<	≥	≥	≥	=	=	=
Changing the inclination of the lower link *GD*	*D* downwards	<<<	<<<	<<<	<<	<<<	>>>	<<<	>>	>>
	D upwards	>>>	>>>	>	>>	>>>	<<<	>>>	<<	<<
Modification of the length of the lower link *GD* (*G* remains unchanged)	*D* to the right	≤	<<	<<	<	>	=	=	=	=
	D to the left	≥	>>	>>>	>	<	=	=	=	=
Change of the kingpin	Link *GD* longer	≤	=	<<	≤	>	≥	=	=	=

inclination angle (*E* and *D* remain unchanged)	Link *GD* shorter	≥	=	>>	≥	<	≤	=	=	=
Simultaneously changing the steering inclinations	*C* + *D* upwards	<<<	<<<	<	<<	<<	>	>	<	<
	C + *D* down	>>>	>>>	>	>>	>>	<	<	>	>

Legend:

< Reduction in roll center height, lateral movement, or camber change

> Increase in roll centre height, lateral movement or camber change

Extent of change: = none / ≤, ≥ very small / <, > small / <<, >> large / <<<, >>> very large

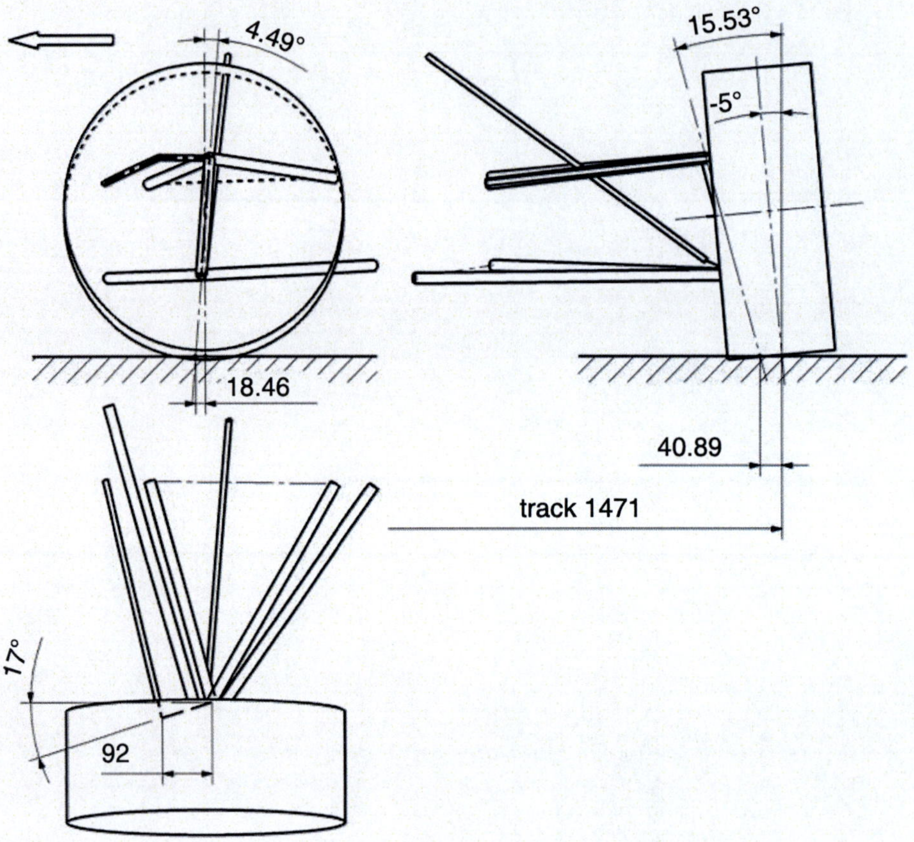

Fig. 4.11 Suspension design front axle (Formula Renault 2000). The left front wheel is shown. This axle is not driven. The wheel load is transmitted to the mono-spring by a push rod

Figures 4.11 and 4.12 show examples of the design of a front and a rear axle of the same vehicle.

4.2.2 Parts of Double Wishbone Axles

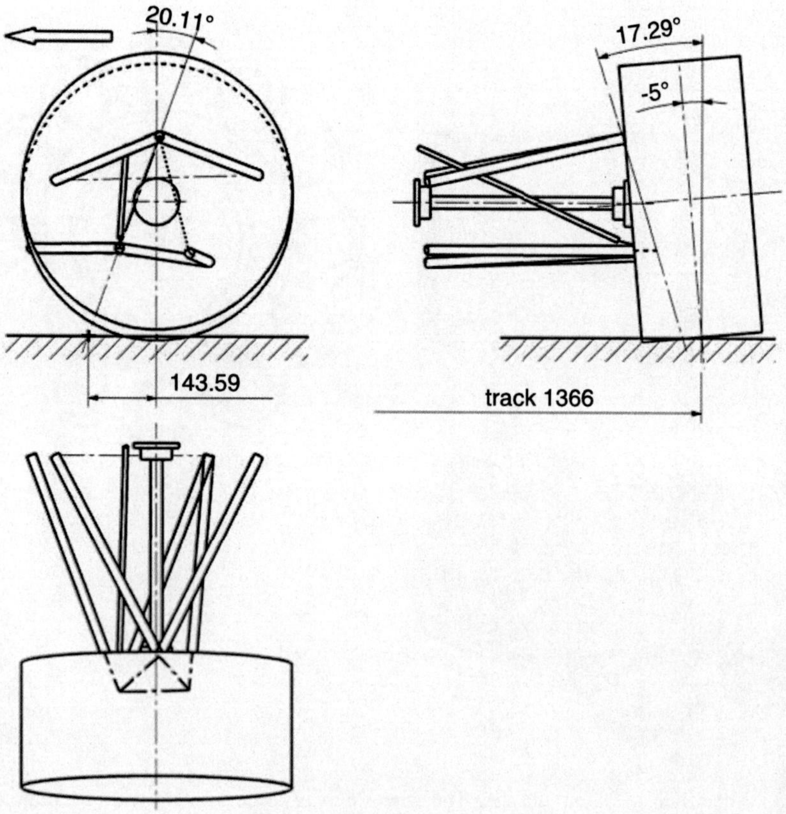

Fig. 4.12 Suspension design rear axle (Formula Renault 2000). The left rear wheel is shown. The axle is driven via two side shafts. The dampers/struts are actuated by a bell crank via the push rod. The wheel carrier is only connected to the wishbone at the top by means of a ball joint. At the bottom, the wheel carrier has two joints. A wishbone and a tie rod located at the rear in the direction of travel absorb horizontal forces. The reason for this arrangement is the greater forces on the lower control arms. The push bar is attached to the lower wishbone in such a way that there is sufficient space for the drive shaft

Hub

The wheel hub receives the wheel, i.e. it centres the wheel and establishes a torque connection between the brake disc and, in the case of a driven axle, between the drive shaft and the wheel. In addition, it must also absorb the bending caused by the cornering forces of the tyres and transmit this to the wheel bearings. The hub is rotatably connected to the wheel carrier via the wheel bearing. The inner rings of the wheel bearings therefore sit directly on the hub. Hub and wheel-bearing thus practically will be one unit, together with bearings, spacer-sleeves and seals. Here, however, these parts are to be considered separately for reasons of clarity.

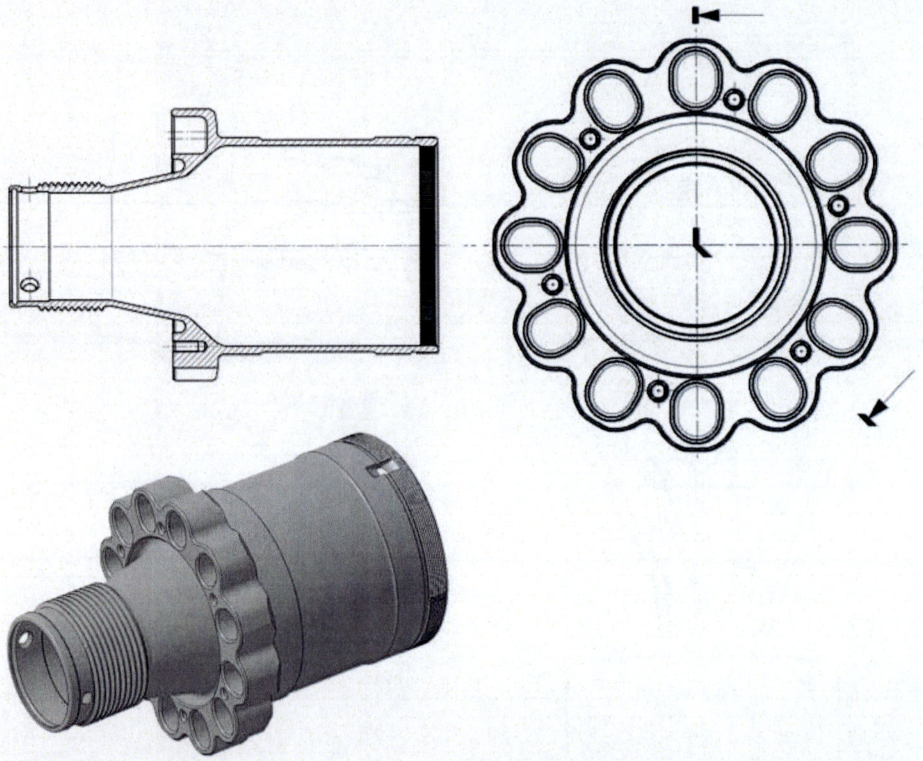

Fig. 4.13 Wheel hub of a formula car. The wheel hub is made of steel with a minimum wall thickness of 2 mm. The wheel is held in place by a central lock. Part of the torque connection is provided by journals which are pressed into the wheel. The corresponding holes in the hub are highly oval so that the position of the wheel is not overdetermined by four or more pivots. So that the brake disc, which is also carried along by the pivots, does not fall off when the wheel is changed, it is bolted to the flange with a few small bolts
The hollow hub accommodates the automatic locking unit for the central nut on the inside

To ensure that the hub is flexurally rigid and yet light, it is designed with the largest possible diameter and hollow, Fig. 4.13 Wheel hub of a formula car. The diameter of the hub is limited by the bearing friction, which increases with the diameter of the wheel bearings, and by the space left by the rim for the brake disc and caliper.

A central bolting allows quick wheel changes. The thread pitches should be relatively coarse so that it is easy to fit the central nut in the required haste. The wheel mounting is described in more detail in Sect. 1.4 *Wheel Mounting*.

As stiff as the hub must be designed against bending, the neck area between the thread and the wheel contact is the "shank" of the bolt in a central bolt connection. In order for this bolted joint to have the desired high working capacity, the shank must not be designed too stiff in the direction of tension, i.e. in the axial direction. The elastic compliance of a tension member with length l and cross-section A is characterized, as is well known, by

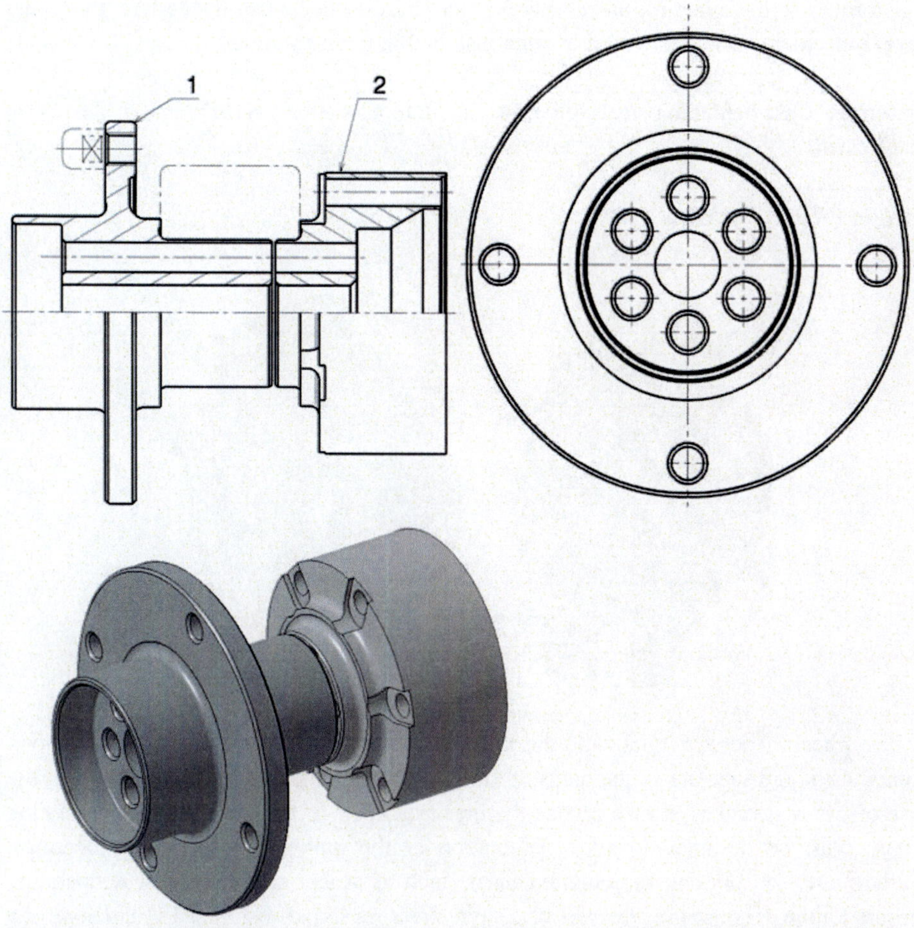

Fig. 4.14 Wheel hub for compact bearing (Formula Renault rear axle). The wheel hub is in two parts and is installed on the driven rear axle. **1** Hub flange. **2** Flange for axle shaft. The moment connection to the wheel is partly ensured by pins which are screwed into the wheel flange (1) (shown in dashed lines). The wheel is centred by the inner collar of the wheel flange. The inner ring of the compact bearing (dashed) is clamped between the two parts of the hub. The drive shaft is screwed directly to the flange (2)

l/(EA) (*E* modulus of elasticity of the material). Thus, a large length and a small cross-section are helpful. The central nut can also contribute to the elasticity if it is designed accordingly.

Hubs can also be designed in several parts to accommodate compact bearings, Fig. 4.14 Wheel hub for compact bearing (Formula Renault rear axle).

Wheel hubs of driven wheels can directly accommodate the counter track for tripod joints to reduce weight. A tripod enables the transmission of a uniform rotary motion with high torques at the same time and allows the side shaft the necessary freedom for the wheel travel movement, which is why it is often installed in racing vehicles.[1]

Materials Case hardening steels, quenched and tempered steels. 41SiNiCrMoV7 6 (DIN) heat treated.

***Upright* (Wheel Carrier)**

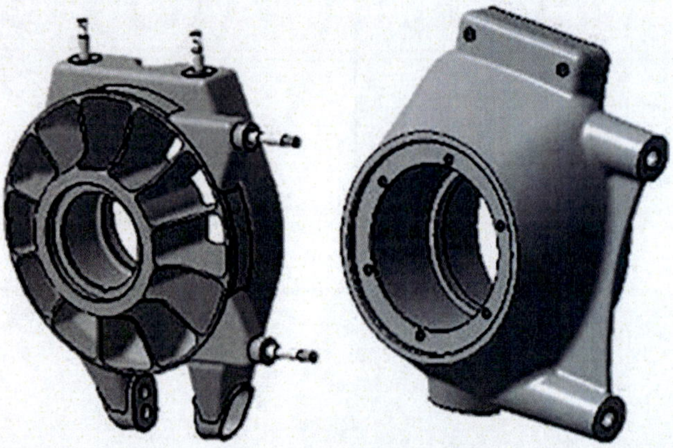

The wheel carrier accommodates the wheel bearing and all wheel-side suspension pivot points. Although separate components, it forms a close functional unit with the hub. In the case of external brakes, it also carries the brake calliper. In this case, the wheel bearing mount must be designed to allow ventilation of the brake disc and wheel bearings. Furthermore, sensors can be mounted on it, such as wheel speed sensors, temperature sensors for the brake or the receivers of the tyre pressure signal. Figure 4.15 Functions of a wheel carrier shows an example of a wheel carrier, the chassis points it provides and its other functions. The size of the wheel carrier results from the following considerations. The mounting points of the wishbones *E* and *G* should be as far apart as possible (cf. Figures 4.4 and 4.5). On the other hand, the wheel carrier must fit into the rim and it must be possible to turn the steered wheel by the desired steering angle without the wishbones or track rods touching rotating wheel parts.

The method shown In Fig. 4.16 is frequently used, in which the upper wishbone connection E and the mounting of the track rod U are made in a bracket and this is only bolted directly to the wheel carrier. The advantage of this arrangement is the ease of camber adjustment. If the camber angle is adjusted, it is only necessary to replace the intermediate plates between the bracket and the wheel carrier, which ensure the desired distance

[1]For more details see Racing Car Technology Manual Vol. *3 Powertrain*, Sect. 5.5.3 *Shaft Joints.*

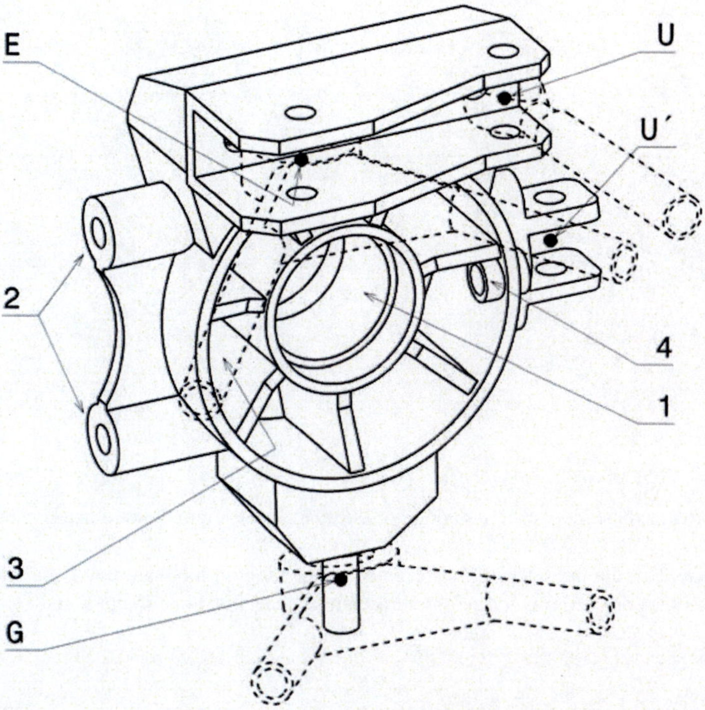

Fig. 4.15 Functions of a wheel carrier. A wheel carrier provides the following chassis points: *E* Upper pivot point for wishbone, *G* Lower pivot point for wishbone, Axis *EG*: Kingpin inclination axis on front axles, *U* Mounting tie rod, *U′* alternative mounting of the tie rod, Other functions: **1** Wheel bearing mounting. **2** Mounting of the brake calliper. **3** Cooling air for wheel bearing and brake. 4 Mounting of sensors, e.g. wheel speed

(Fig. 4.17). If the wishbone and track rod are attached separately to the wheel carrier, the toe-in must also be readjusted when the camber is changed. However, a disadvantage of this high-lying track rod is that the distance between points *E* and *U*, i.e. the length of the track lever, is limited by the rim. For this reason, alternative arrangements of the track rod attachment (point *U′*) are also found at the height of the wheel centre. This position ensures the largest possible track lever within a given rim size. Such an arrangement is particularly suitable on the non-steered rear axle, where the lateral forces are also generally greater.

The wheel carrier can also directly accommodate the compression or tension struts that are usually attached to a wishbone. This means that the wishbones do not experience any bending moment due to the wheel load and no wishbone bearing has to transmit the wheel load. The dimensioning of these parts can therefore be correspondingly lighter. Particularly on the steered front axle, however, there is the additional requirement that the joint center of the push/pull rod lies exactly on the kingpin inclination axis. Otherwise the wheel load

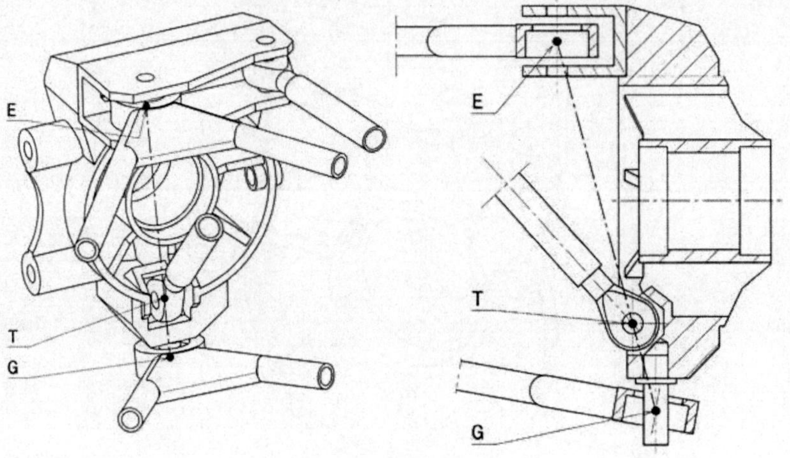

Fig. 4.16 Direct connection of a compression strut to the upright (axonometric view and step section)
The joint centre *T* of the pressure bar lies exactly on the kingpin inclination axis *EG*. The kingpin inclination axis (steering axis) is formed by the centres of the wishbone joints *E* and *G*

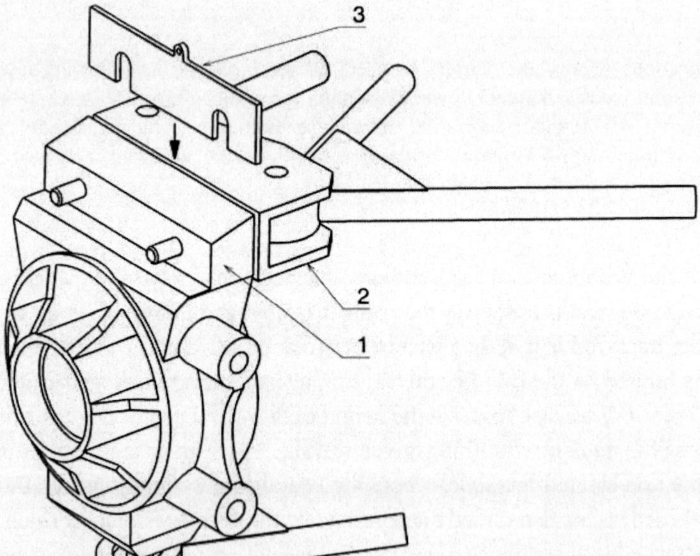

Fig. 4.17 Camber adjustment with adjustment plates. **1** Upright. **2** Bracket. **3** Adjusting shim. The wishbone bearing and the track rod are accommodated in a common bracket (2). This is screwed to the wheel carrier (1) with two screws. Adjustment plates (3) of different thicknesses are placed between them. This allows the camber angle to be adjusted specifically. So that the retaining screws only have to be loosened, the adjustment plates have U-shaped recesses instead of the screw holes. They also have a tab on the top for easy removal when dismantling

distribution will change unfavourably during steering. Figure 4.16 shows the possibility of direct connection to the wheel carrier for a push rod.

In practice, the experience is quite different. In Formula 1, Ferrari used this solution on the front axle in the 2000 season, which had already been introduced by BAR in 1999. In a Le Mans race car, however, exactly this arrangement was changed back to the common one, i.e. thrust strut support at the lower wishbone, after problems [3].

The wheel support also accommodates the wheel bearing. The surroundings of the outer ring should support it as uniformly as possible in the radial direction in order to avoid out-of-roundness of the bearing raceways under load. This results in uneven rolling path stress and increased bearing temperature (friction, life). The thinner the bearing outer ring (thin section bearings!), the more important it is to have a sufficiently thick bearing support. This is particularly important for materials with a lower compressive strength and lower modulus of elasticity. A compromise solution for light alloy wheel carriers is provided by steel rings that form the bearing environment, i.e. are themselves pressed into a larger locating bore.

The camber angle setting should be easily adjustable at the race track. Adjustment is made possible by rod ends with threaded inserts on the wishbone (see Fig. 4.48), adjustment plates under the car side brackets (Fig. 4.49) or direct adjustment options on the wheel carrier, Figs. 4.17 and 4.18.

Figure 4.18 shows an alternative adjustment possibility when the wishbone eye and the tie rod connection are mounted together on the wheel carrier.

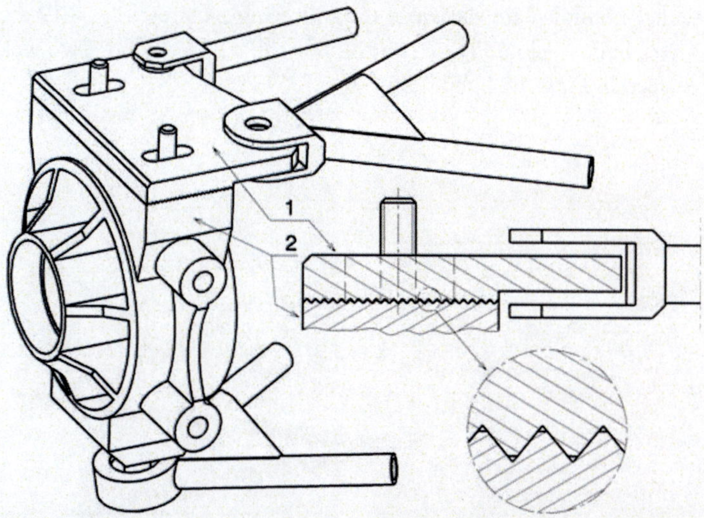

Fig. 4.18 Camber adjustment with grid plate. **1** Serrated plate. **2** Upright. The wishbone and track rod are mounted on a plate (1). This plate is bolted to the top of the wheel carrier (2). Wedge-shaped grooves run transversely to the tyre side forces, creating a positive fit between the grid plate and the wheel carrier. After loosening the two fastening nuts, the grid plate can be adjusted in locking steps. It has corresponding slotted holes for this purpose

Materials Cast from magnesium, aluminium or titanium alloy.

Welded from steel, titanium alloy (e.g. TiAl6V4 (DIN17851)) or aluminium alloy parts.

Special casting process in which light metal is reinforced with fibres (MMC – Metal Matrix Composite).

Made from solid aluminium, titanium or steel alloys.

Wheel carriers are also used as a built variant (differential design). Trapezoidal plates are bolted to a tube for wheel bearing mounting, which accommodate the wishbones and brake calipers.

Cast or forged parts made from aluminium alloys or steel are used in production vehicles. Light metal-based wheel carriers have a greater coefficient of thermal expansion than those made of steel, so a greater overlap is required for the interference fit of the bearing outer ring to maintain an interference fit even at high temperatures. At room temperature, this increases the back pressure of the bearing seating, which the material must be able to withstand. Steel wheel carriers do not have this problem. However, the disadvantages of light alloy wheel carriers only come into play when the required service life of sports cars is high (100,000 to 150,000 km for wheel bearings).

A titanium Formula 1 wheel carrier weighs about 1.8 kg for the rear axle and 1.5 kg for the front axle.

Welded steel wheel carriers are the first choice for heavy vehicles because they are the most rigid solution. Castings also have long delivery times and changes cannot be implemented as quickly as in milled and welded designs.

Some executed examples are shown in the following pictures: Fig. 4.19 a cast wheel carrier, Fig. 4.20 a built version, Fig. 4.21 a solution milled from solid. Two welded wheel carriers can be seen in Figs. 4.22 and 4.23.

Wishbone

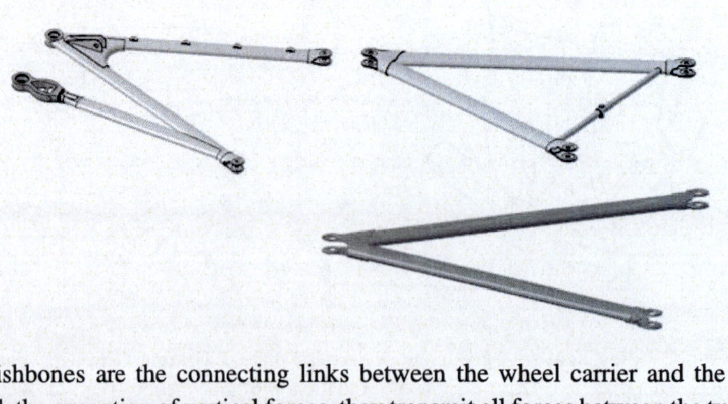

The wishbones are the connecting links between the wheel carrier and the frame or body. With the exception of vertical forces, they transmit all forces between the tyre and the vehicle body. Some wishbones also transmit the wheel loads via the compression or pull

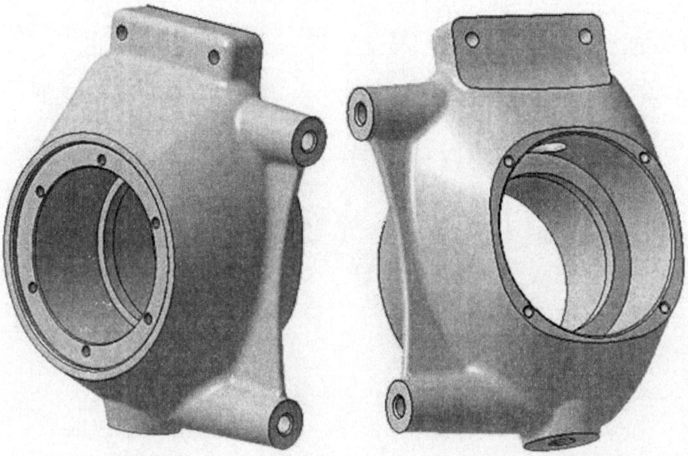

Fig. 4.19 Cast wheel carrier (Formula Renault). The hollow cast light alloy wheel carrier is installed on the front axle and accommodates a compact wheel bearing. The two slugs for screwing the brake caliper to the front and rear are present on the blank. By milling away one side each, the wheel carriers for the left and the right side become from it. At the top a bracket (steering bracket) is fastened with two bolts, which represents an axle joint and the track lever. The camber is adjusted by placing small shims in between. Below the connection to a spherical bearing in the lower wishbone is made by a stud bolt

rod connection. Apart from these designs, therefore, practically no bending moments occur in the transverse control arms because they are articulated to the adjacent parts and can thus only transmit longitudinal forces.

The wishbones are considered by many teams to be typical wear parts that are routinely replaced after each race. So they are supposed to be relatively inexpensive. As important as their strength is in operation, however, in the event of an accident they are supposed to be the first parts to break away without deforming the mounting points of the frame or chassis. This prevents major consequential damage to the supporting structure. In Formula 1 and Formula 3, retaining cables are prescribed for the wheel carrier to prevent wheels from detaching from the vehicle in such an event.

The links are usually assembled from several semi-finished products, welded from two half-shells, cast or milled from the solid. The lowest mass can be achieved with a composite of fiber-reinforced plastic and metal ends. In series production vehicles, control arms are forged, cast or welded from semi-finished products.

The ends of the control arms are connected to neighbouring parts such as wheel carriers, chassis, pull/push rods, etc. via joints. The most common joints are spherical plain bearings, to which the following section is devoted.

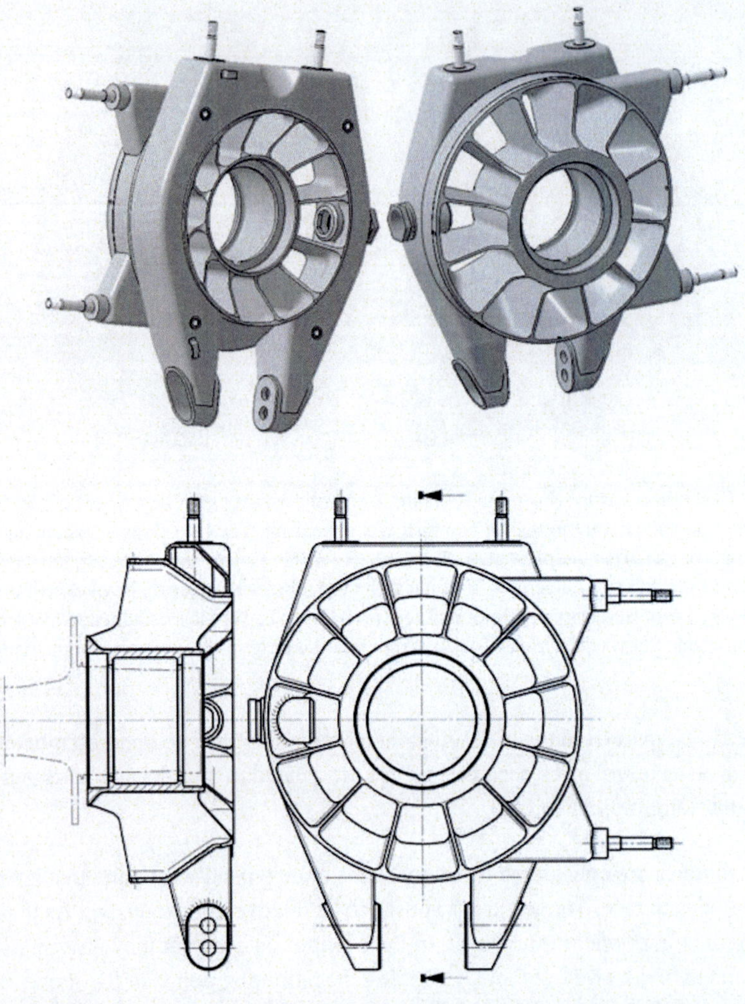

Fig. 4.20 Composite wheel carrier (Formula 1). The wheel carrier is welded together from several steel parts. The inner part, which holds the wheel bearings, is machined from solid. The gaps left by the 2.5 mm thick ribs are wire-cut. In the sectional view, the wheel hub is shown with a dotted line

In principle, there are two possibilities of joint design, namely the eye or the fork at the wishbone end. Figures 4.24 and 4.25 give some examples of the design of control arm ends.

In the case of circuit vehicles with extremely small suspension travel and production to the tightest tolerances (i.e. Formula 1), there is also a clever solution for wishbone connections on the car side which works completely without play, namely flexible metal links (solid-state joint, material-locking joint) (*flexure pivot*), Fig. 4.26. However, the replacement of such joints depending on the operating hours is becoming a vital necessity.

Fig. 4.21 Front left wheel carrier, Formula BMW. You can see the upper and lower wishbones, whose articulation points form the steering axis, the tie rod connection and the brake caliper as well as the wheel bearing. The lower wishbone bearing absorbs the greater vertical forces because the pressure bar acts on the lower wishbone. For this reason, the spherical plain bearing is installed rotated by 90° in relation to the upper bearing

Fig. 4.22 Front right wheel carrier, Formula 3000 (Reynard D94). The wheel bearing area is closed with a cover. A central cooling air supply to the brake is not necessary for the vehicle, because it is built for a hill climb and the brake must come up to temperature quickly

Fig. 4.23 Right front wheel carrier of a historic vehicle (Hawke DL 2b). The wheel carrier is welded from sheet metal parts

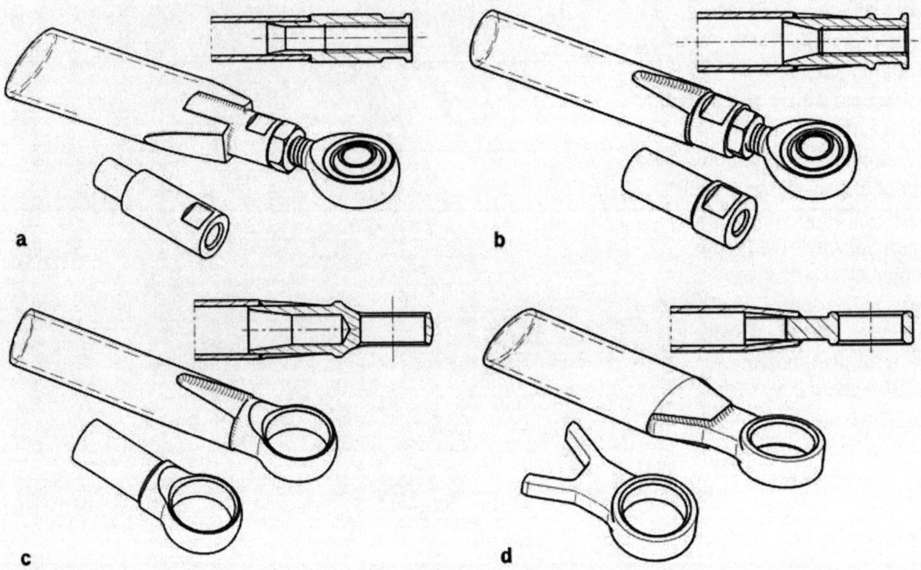

Fig. 4.24 Designs of link ends with eyelet. Various mounting parts are welded to the link: (**a**) Threaded sleeve in flat link. (**b**) Threaded sleeve in tubular link. (**c**) Eyelet in tubular link. (**d**) Flat link with eyelet

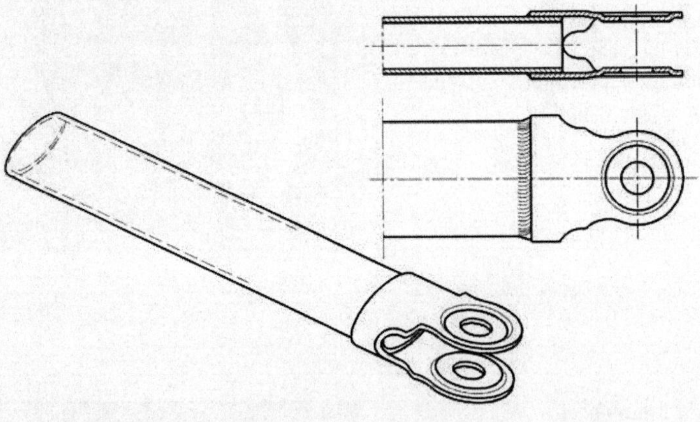

Fig. 4.25 Types of link ends with fork (clevis). The end of the link is welded with a fork-shaped sleeve

Such solid-state joints also do not exhibit any hysteresis as is unavoidable with classic joints with (adhesive) friction. The successful LMP1 cars from Audi, which have won the endurance classic at Le Mans many times, show that reliability can be controlled. When arranging the joints, it is important that they are not subjected to large lateral forces. Ideally, the axis of the tension/compression bar acting on this control arm passes through the joint

Fig. 4.26 Flexible titanium links as wishbone connection (Formula 1, Ferrari). This extremely simple-looking solution for a joint is in fact only possible with small spring deflections and the greatest precision of the parts to be connected, including the chassis. Otherwise, these backlash-free connections merely produce a hardening of the chassis due to tensioning of the components involved

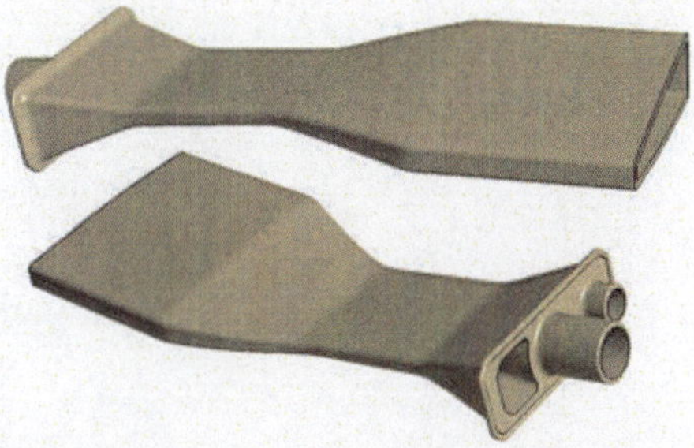

Fig. 4.27 Solid state joint made of TiAl6V4. The part is milled from the solid and serves as a car-side wishbone mount for a Formula 1 vehicle

point on the wheel side and the control arm does not transmit any (lateral force) bending moment.

If money is no object, the material TiAl3V8Cr6Mo4Zr4 (Beta C/ Grade 19) is the first choice for a solid-state joint. TiAl6V4 (DIN 17860) is a comparatively inexpensive alternative. The material is easy to machine and has a high fatigue strength. The modulus of elasticity of $1.07 \cdot 10^5$ N/mm^2 is about half that of steel.

Figure 4.27 shows such a joint worked from solid.

The control arms have predetermined breaking points at the articulation points to the frame. If a wheel collides with an obstacle, the control arms deform and break away without damaging the frame. In Formula 1 and Formula 3, tether ropes are prescribed to connect the wheel carrier to the chassis in order to prevent consequential damage caused by detached parts.

The joining of assembled wishbone arms is designed similarly to the joining of individual ends. In addition, gusset plates can be welded in for angular stiffening, Figs. 4.28, 4.29, and 4.30.

Wishbones that also transmit wheel loads have the tension/compression bar connection as close as possible to the bearing point to the wheel carrier where the vertical force is introduced (Fig. 4.31). This keeps the bending moment in the control arms small. The bearing to the wheel carrier must of course be able to absorb the entire vertical force of this

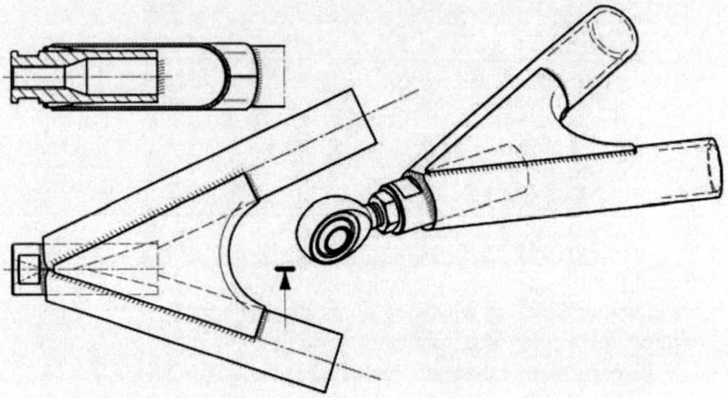

Fig. 4.28 Joining with threaded sleeve. The two link arms are welded together with a threaded sleeve into which a rod end is screwed. A gusset plate stiffens the connection

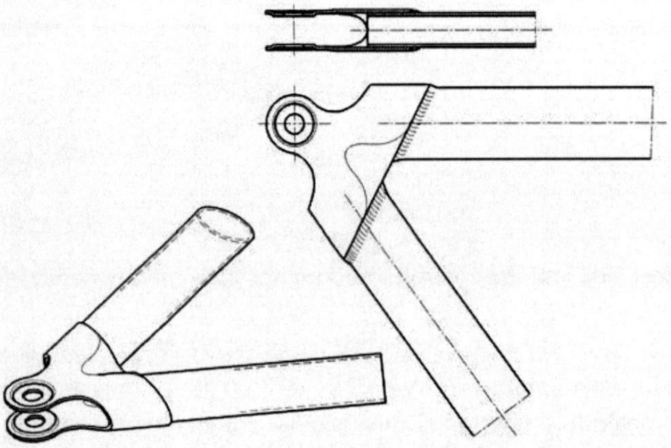

Fig. 4.29 Joining with forked sleeve. The two link arms are welded together in a sleeve that supports a fork for the joint mount

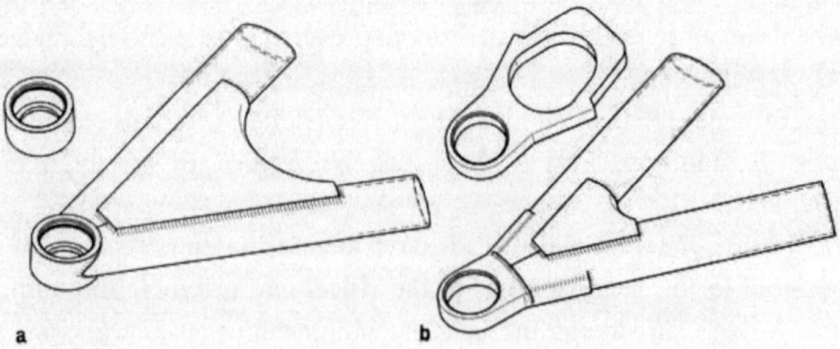

Fig. 4.30 Merging to joint eye. The two link arms are welded with an eye to the joint admission (**a**) Ring piece. (**b**) Plate with eye

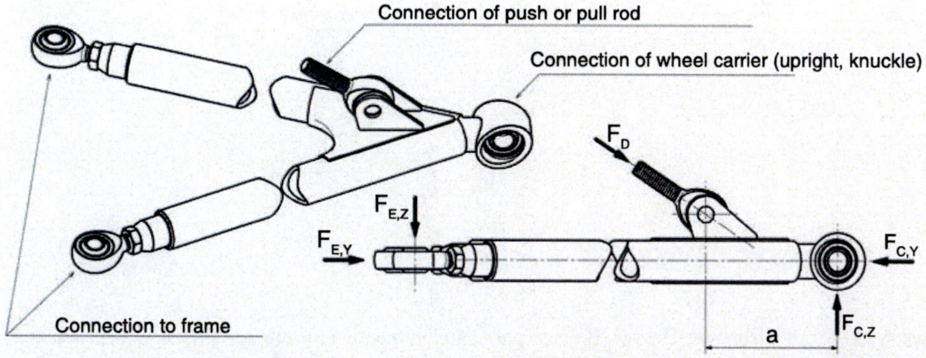

Fig. 4.31 Vertical force application in wishbones. The spherical plain bearing to the wheel support is installed in an upright position so that it can transmit the vertical force $F_{C,Z}$ safely. Its distance a to the compression member connection should be as small as possible

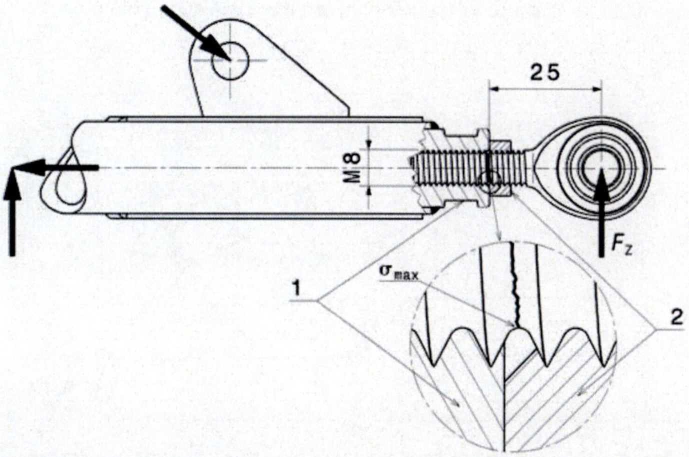

Fig. 4.32 Bending moment in rod end shank. In such an installation situation, a bending moment acts on the rod end under the load F_Z shown. The notch effect of the thread is aggravated by the pretension of the lock nut (2) and leads to the highest tensile stress σ_{max} in the first thread turn in the receiving piece (1) on the tensile side of the bend. In addition, typical dimensions are entered

wheel. Therefore, for large forces, these bearings are arranged so that the vertical wheel load is the radial load of the bearing (eye positioned).

Rod ends with an externally threaded shaft are poorly suited at this point because the alternating bearing forces cause an alternating bending moment in the threaded shaft of the rod end. In addition, the notch effect of the thread has a particularly damaging effect under alternating stress and leads to fractures of the threaded shaft far below the permissible bearing load, Fig. 4.32.

Figures 4.33, 4.34, and 4.35 present some executed examples of wishbones.

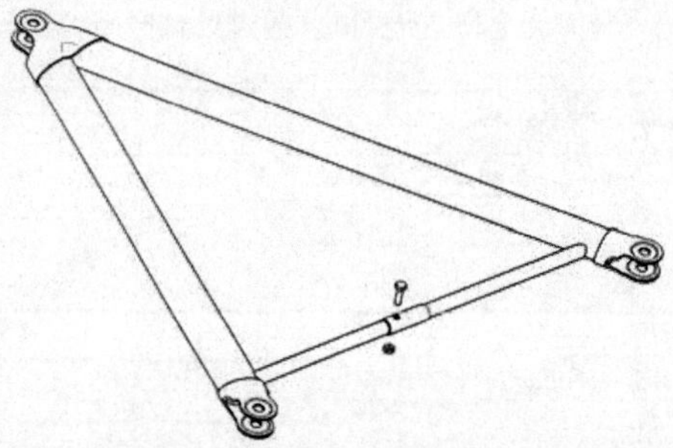

Fig. 4.33 Front wishbone. The two linkage points to the monocoque are designed as predetermined breaking points. In the event of a collision between the wheel and an obstacle, the links break away. To prevent the released link ends from breaking through the monocoque wall and injuring the driver's legs, they are connected behind the predetermined breaking point with a longitudinal strut (*anti-intrusion bar*). The strut itself is interrupted by a socket in the middle. During assembly, the articulation points are first bolted to the vehicle and then the sleeve is drilled through transversely and bolted. However, this bolting is not carried out by all teams, because this makes it easier for the bar ends to compensate for distance tolerances between the articulation points

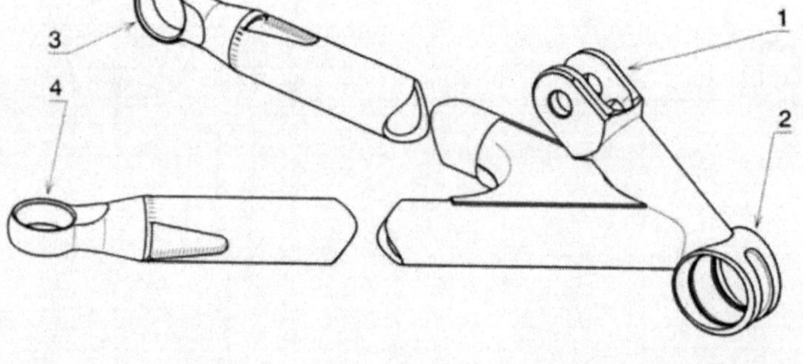

Fig. 4.34 Front lower wishbone. This wishbone accommodates a spherical bearing (2) which, on the one hand, allows the wheel carrier to rotate for steering and, on the other hand, absorbs the wheel load. An intermediate piece guides the wheel load directly into a fork (1), which is bolted to a rod end of the compression bar. The intermediate piece is aligned in such a way that the bending by the compression bar remains in a wishbone arm. The bearing (3) at the end of this arm is therefore installed in an upright position so that it advantageously transmits the reaction force as a radial load. The second arm with bearing (4) mainly transmits tension/compression forces

Materials Steel (e.g. 15 CDV6 (1.7734.5)), aluminium and titanium alloys, fibre-reinforced plastics with laminated metal ends or bushings.

For some vehicle types, the regulations require that the control arms be made of homogeneous material. Chrome plating of wishbones is also prohibited by some regulations. Metal control arms are usually kephos-coated.

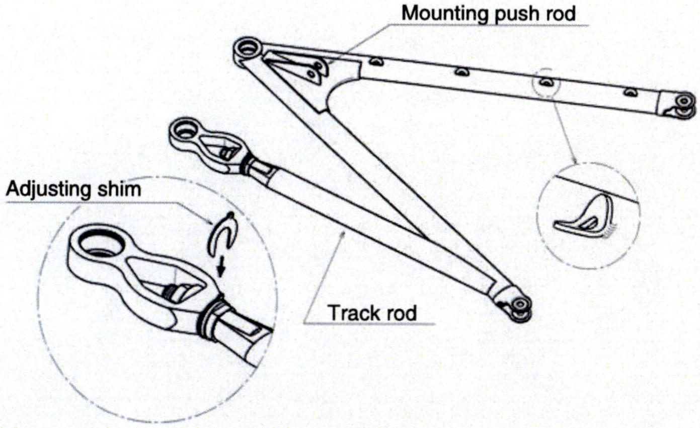

Fig. 4.35 Rear wishbone (Formula Renault 2000). This control arm also transmits vertical forces from the tire to the pressure bar. In addition, the track rod is integrated. The screw connection between the bearing and the track rod (see enlargement) allows the toe-in to be adjusted. For this purpose, adjusting washers of different thicknesses are placed between as required before tightening the nut. Another detail shows how a simple mounting base for the brake line was welded on with bent washers

Cross Sectional Shapes

Ideal from strength considerations (if no bending moment is introduced, pure tension/compression forces remain) is the circular ring cross-section; for aerodynamic reasons, flat elliptical or wing profiles are usually used. Figure 4.36 shows elliptical cross-section shapes. If larger forces are to be transmitted, a circular tube with slot welds can be inserted within an elliptical profile for reinforcement.

Spherical Bearing (Rose Joint)

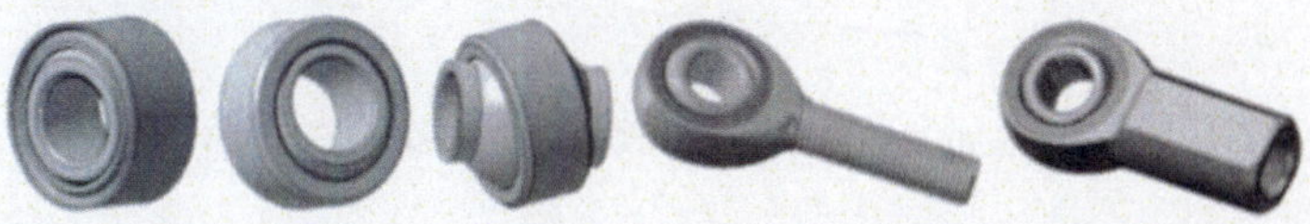

Rod end bearings and spherical bearings have become established as connecting joints between parts that move in relation to each other, such as wishbones and wheel carriers, thrust *rods* and bell cranks, etc. A rod end is basically a bolt-on mount with a spherical bearing rolled into it.

The spherical plain bearings (also known as uniball joints) transmit radial loads without clearance, but also absorb axial forces. They are mounted in brackets or appropriately designed mounts using fitting bolts. The spherical plain bearing is pressed into a bore and secured with retaining rings (Figs. 4.37 and 4.38) or caulked in. For the latter type of

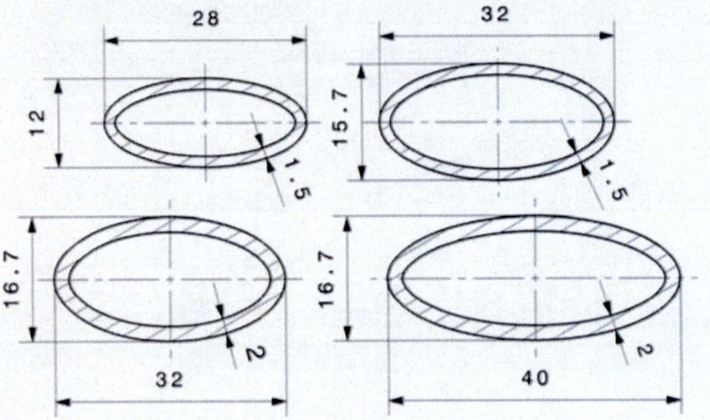

Fig. 4.36 Elliptical cross-sectional shapes of steel control arms. These cross-sections are often used for aerodynamic advantages for control arms in the air flow

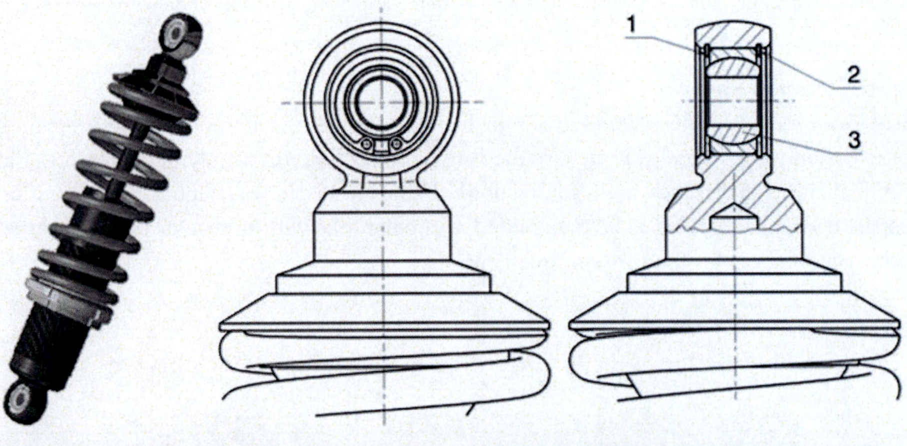

Fig. 4.37 Spherical bearings in the mounting eyes of a spring/damper strut. **1**, **2** Circlip. **3** Spherical bearing

securing, however, a separate design with radial grooves is required, Fig. 4.39. The permissible swivel range depends on the design, see Fig. 4.40. Maintenance-free bearings with a PTFE sliding layer on the outer ring are recommended in the chassis area. However, some designs with a sliding layer tend to have a greater breakaway torque than their two-piece counterparts, which is why the simpler bearings without an intermediate track are preferred in the strut eye area. If even lower friction and clamping tendency are desired, only rolling bearings (e.g. needle bearings) remain, but with the restriction that the movement is limited to one degree of freedom. The rotary axes must therefore be exactly aligned for this application, otherwise the effect of friction reduction is negated.

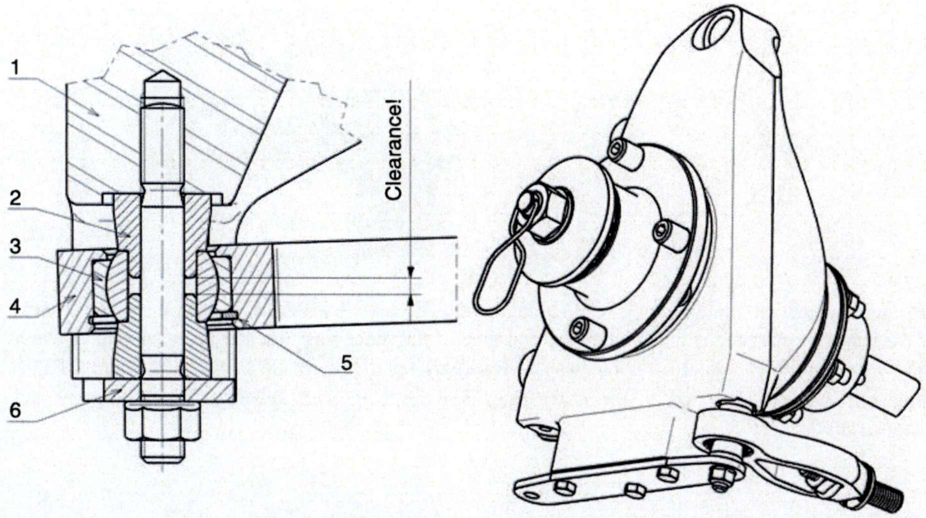

Fig. 4.38 Bolting of a spherical plain bearing. **1** Upright or inboard housing, e.g. transmission. **2** Conical spacer. **3** Spherical bearing. **4** Wishbone end or spherical bearing mounting. **5** Circlip. **6** Sheet metal shackle. The axial distance between the two tapered sleeves ensures clear contact with the end faces of the spherical plain bearing and facilitates removal of the sleeves

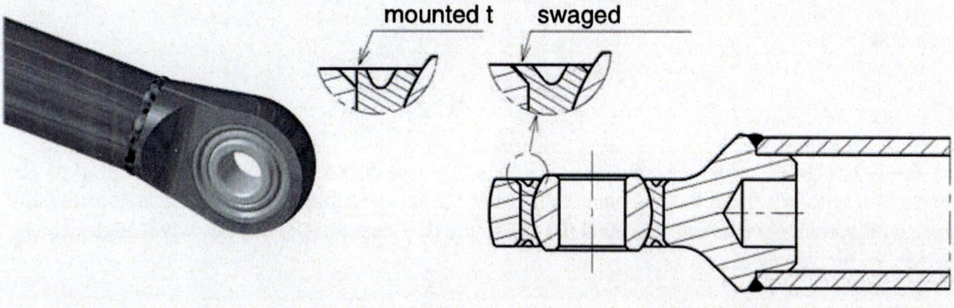

Fig. 4.39 Spherical plain bearing swaged into a weld-on end of a compression strut. As delivered, the spherical plain bearing has grooves on both sides and is inserted into the locating bore with a light press fit. The two edge webs are then pressed outwards against the chamfers of the locating bore using a suitable tubular tool

The operating temperature range of spherical plain bearings extends from −50 to 120 and, in some designs, up to 160 °C.

The largest possible swing angle when installed is further limited by the design of the adjacent parts, Fig. 4.41. For particularly large swing angles, special designs are offered, such as those used as axle bearings, see Fig. 5.76.

The tolerances of the locating holes for spherical plain bearings are based on the manufacturer's recommendations. The following can be used as a rough guideline:

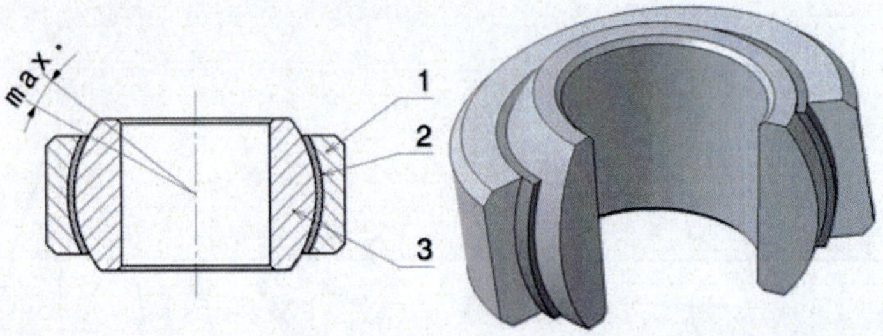

Fig. 4.40 Spherical plain bearing with sliding layer. The max. Swivel angle from the point of view of the bearing surface is reached when the end face of the inner ring touches the outer ring. **1** Outer ring, with hollow spherical inner raceway. **2** Raceway layer (e.g. made from Teflon fabric, PTFE composite material) bonded to the outer ring. **3** Inner ring with spherical outer raceway and cylindrical bore

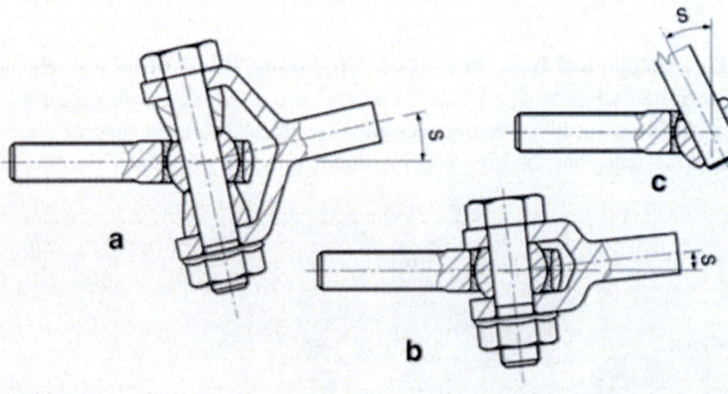

Fig. 4.41 Design of joints with spherical plain bearings. (**a**) Fork with additional spacer sleeves. (**b**) Connection with simple fork. (**c**) Connection with pin. The possible swivel angle *S* is considerably larger with variant (**a**) than with variants (**b**) due to the two sleeves. The largest swivel angles can be realized with a simple pin

- ∅H7: For low forces and when axial displacement of the bearing is desired.
- ∅M7: Results in an interference fit for combined loads and impact-type forces.
- ∅N7: For joint mountings made of light metal.

Tighter interference fits should only be used if shock loads are present, the seating is strong enough and the increased breakaway torque of the bearing due to the interference fit does not represent a disadvantage.

In the case of large spherical plain bearings, conical sleeves with an interference fit can be inserted in the bore and thus compensate for the difference in diameter to the fitting screw without play, Fig. 4.38. In addition, the two sleeves enable the full swivel angle range to be utilizedd. The bolted connection shown is *double shear*, but still easy to fit because of the separately bolted sheet metal bracket. The bearing mounting in the wishbone with pressure rod connection is designed in such a way that the circlip (5) is located on the side that only transmits vertical forces during deflection. The opposite shoulder transmits the greater forces during compression.

The problem of length tolerances in double-shear joints can be circumvented, see Fig. 4.42.

Single shear connections are possible, but should be avoided in the chassis area for safety reasons. If a single shear connection is chosen, many regulations require, for good reason, the use of a washer at the free end with an outside diameter larger than the diameter of the locating bore of the bearing, Fig. 4.43.

Rod ends are spherical plain bearings that are already pressed and caulked into a mounting with an internal or external thread. There is a wide range of rod ends to choose from. Versions are available with different threads (metric or inch, internal or external thread, left-hand or right-hand), made of stainless steels, high-strength materials, aluminium mountings, with lubrication grooves, etc. Table 4.3 gives a rough overview of data of designs with usable thread sizes.

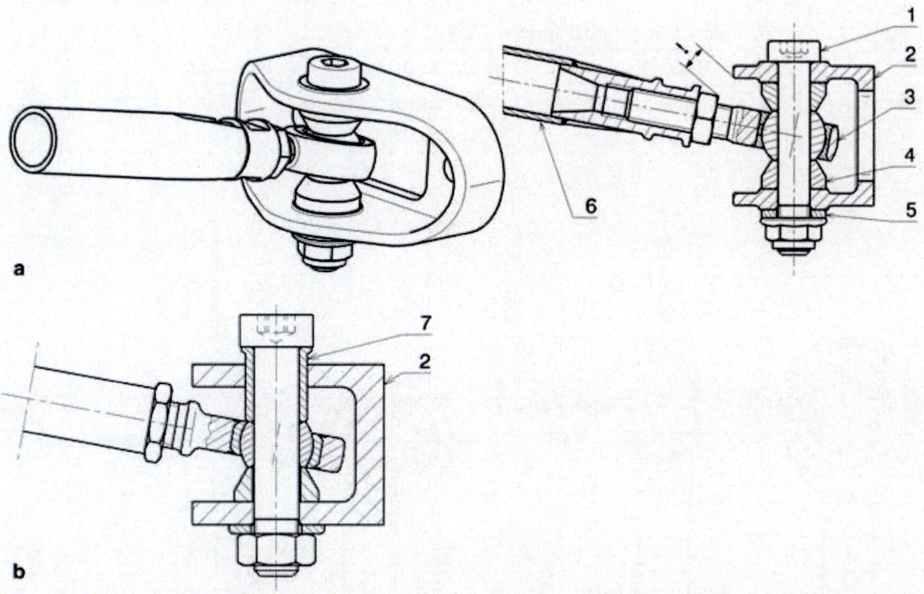

a

b

Fig. 4.42 Rod end fitting in a bracket. (**a**) Conventional bolt connection, (**b**) Backlash-free bolt connection. (**a**) A fitting screw (1) takes the ball segment of the rod end (3) and braces it between two spacer cone sleeves (4) with the bracket (2). The bracket is in turn attached to the vehicle frame or to the wheel carrier. It can be seen that, for reasons of space alone, the conical sleeves require a certain length so that the permissible swivel angle of the joint can be utilised by the transverse link (6). The washer (5) prevents the nut from reaching the end of the bolt thread. (**b**) A fitting sleeve or floating bush (7) provides length compensation when pretensioning the bolt (1) and thus avoids the disadvantage of variant (**a**), in which the bracket and all intermediate parts must be machined to the exact length so that the bracket is not bent by the bolt force

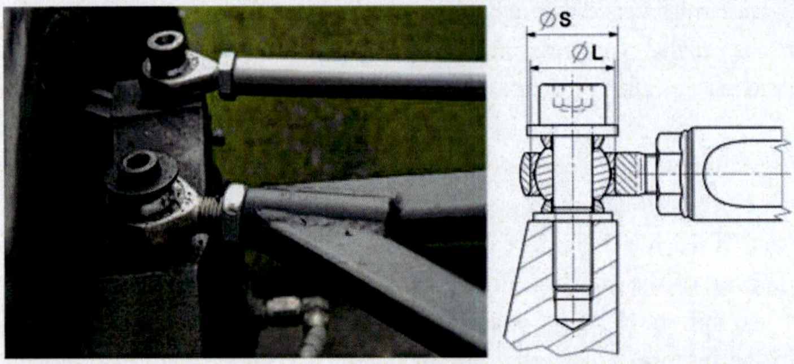

Fig. 4.43 Single-shear screw connection of a rod end. The diameter *S* of the washer under the bolt head is so large that the link remains mechanically connected to the wheel carrier even if the spherical plain bearing breaks out of the bore with ∅ *L*

Table 4.3 Values of rod ends made of steel, for designations see Fig. 4.44 [4, 5]

	B		Perm. stat. axial force	Max. articulation angle
Thread *G*	mm	Perm. stat. radial force	kN	°
M6	6	7.3 to 23.4	0.1 to 0.15 F_{rad}	7.5 to 8
M8	8	13.6 to 36.9	0.1 to 0.15 F_{rad}	7 to 9
M10	10	20.6 to 48.7	0.1 to 0.15 F_{rad}	6 to 8.5

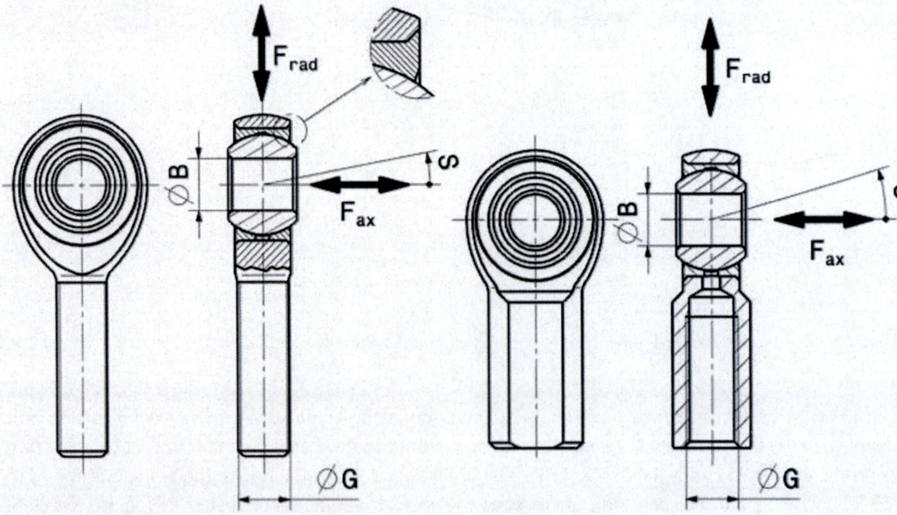

Fig. 4.44 Connection dimensions of rod ends. *B* Eye diameter, *G* Thread diameter, *S* Articulation angle, F_{ax} Axial force, F_{rad} Radial force

For comparison, similar inch threads: ¼″ = 6.35 mm; 5/16″ = 7.94 mm; 3/8″ = 9.525 mm.

In the case of triangular wishbones that transmit large forces, spherical plain bearings are also installed as pure revolut joints. The two joints on the car side are on one axle and ensure the connection to one wishbone arm each. Figure 4.45 shows an example of a heavy version of a Dakar Rally chassis. Locating sleeves (7) are welded into the adjacent frame uprights (1) to define the bearing axis. The actual spherical plain bearing (5) is sealed by O-rings (6) in the receptacle (4). The tolerance compensation of this double-section connection is provided by the adjusting screw (3), which is locked with the nut (2). The

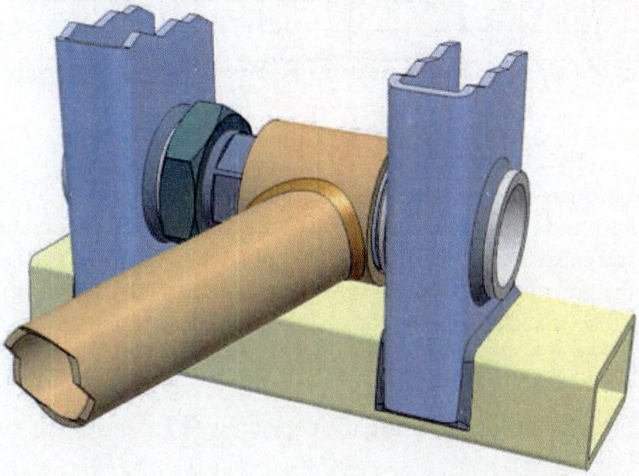

a

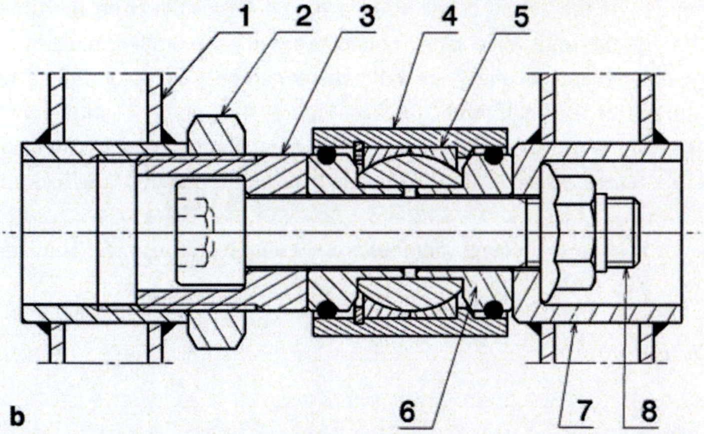

b

Fig. 4.45 Wishbone bearing on the car side of a Dakar Rally car, after [6]. (**a**) Axonometric representation of the spherical plain bearing in the vehicle frame. (**b**) Section through the bearing assembly. 1 Frame, 2 Lock nut, 3 Spacer screw, 4 Wishbone pick up, 5 Spherical bearing, 6 Spacer with O ring, 7 Weld-in sleeve, 8 Clamping bolt

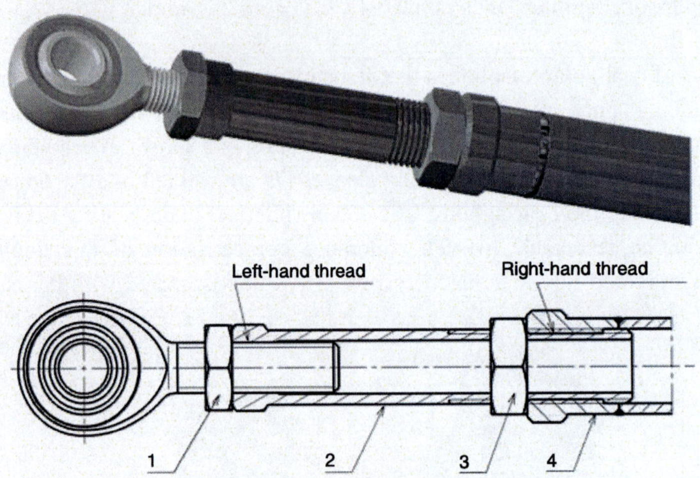

Fig. 4.46 Length adjustment for a pressure rod with rod end. **1, 3** Lock nuts. **2** Compensating sleeve (threaded sleeve). **4** Push rod. The screw connection of the rod end with the pressure rod is designed in several parts. The rod end has a left-hand thread, the compensating sleeve a right-hand thread. This allows the length of the pressure rod to be changed continuously without dismantling by turning the compensating sleeve. The rod itself and the rod end are not turned in the process. The two nuts are used for countering

entire bearing assembly is preloaded using the cap screw (8) and a collar nut with clamping piece.

A disadvantage of the screw connection in Fig. 4.42 results from the limited length adjustment. Firstly, the transverse screw connection with the bracket must be completely removed for a change and secondly, the adjustment can only be made in half turns of the rod end, i.e. only in steps of half thread pitches. Figures 4.46 and 4.47 show how a possible improvement in the sense of a simple fine adjustment looks. As a disadvantage of this arrangement, the increased stress on the threaded shaft in the area of the lock nut must be mentioned.

In the case of wishbone linkages, another type of stepless adjustment is available in the assembled state, Fig. 4.48.

Brackets (Clevis)

On the car side, wishbones are connected in an articulated manner with brackets. The brackets are bolted directly to the monocoque (or chassis), the frame or the transmission

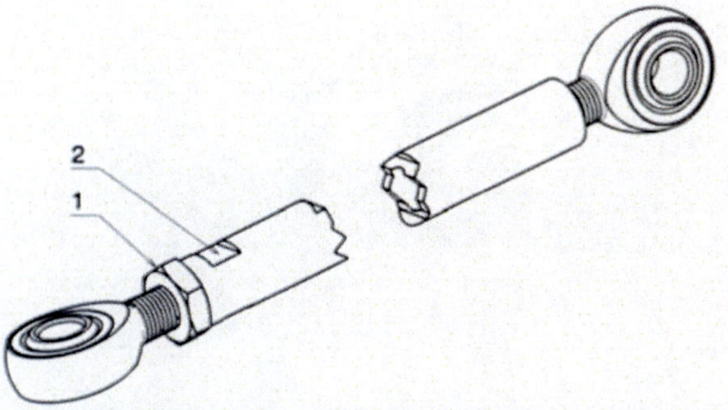

Fig. 4.47 Stepless length adjustment for one rod. **1** Lock nut. **2** Rod with driving feature. The two rod ends each have left-hand and right-hand threads. By turning the rod, the distance between the bearings can thus be varied without disassembly. The rod has a bicone at one end for open-end wrench engagement. The setting is secured with the lock nut

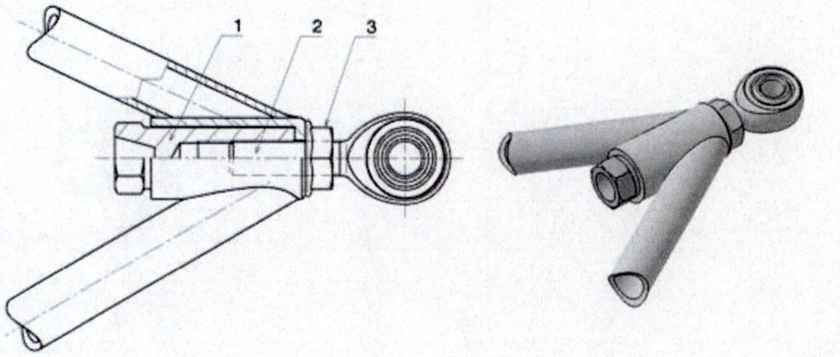

Fig. 4.48 Length adjustment for a wishbone connection. **1** Adjusting nut. **2** Rod end. **3** Lock nut

housing. If several screw positions are provided, the behaviour of the suspension can be changed by changing the height of the joints, e.g. the position of the pitch instant centers against brake compression or starting pitching or the position of the roll centre, Fig. 4.49. By inserting plates of different thicknesses, the lateral position of the joint and thus the camber, caster, etc. can be easily and repeatedly adjusted by moving one joint (of two of a triangular wishbone).

For safety reasons, the bolts that make the connection between the link arm and the console should be installed inserted from above. Figures 4.50, 4.51, and 4.52 show some examples of brackets.

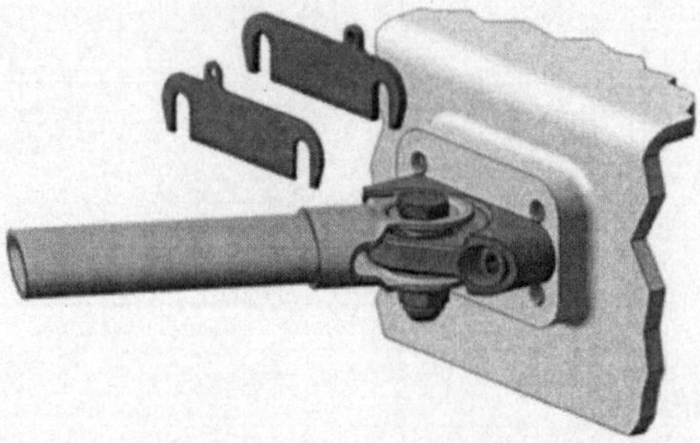

Fig. 4.49 Bracket for a wishbone with fork end (clevis). The gearbox housing has several threaded holes for mounting the bracket. This allows the height position of this pivot point to be adjusted, which influences the position of the pitch instant centers and the roll centers. In addition, plates of different thicknesses can be provided for interleaving

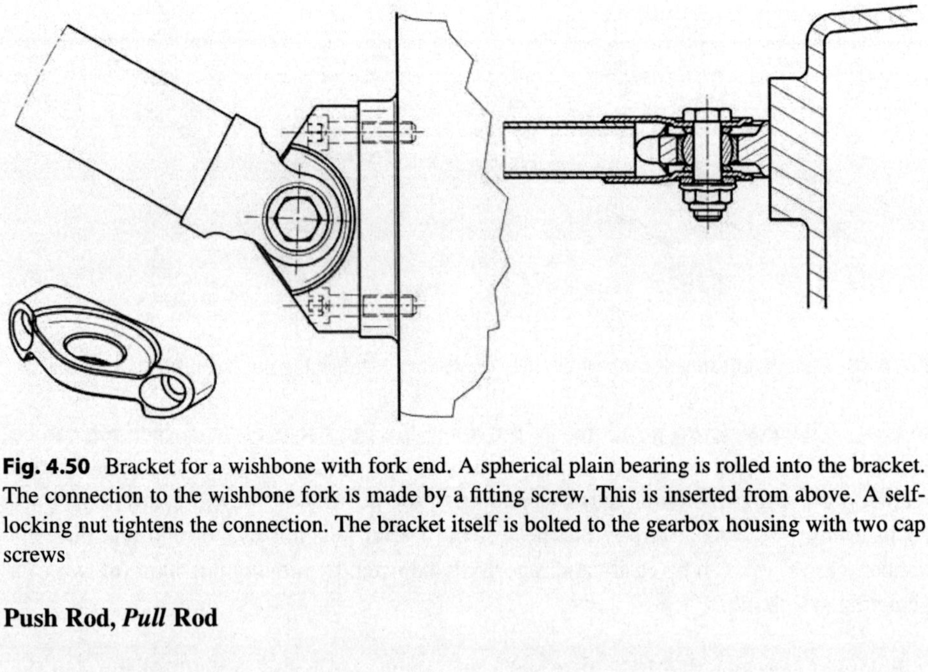

Fig. 4.50 Bracket for a wishbone with fork end. A spherical plain bearing is rolled into the bracket. The connection to the wishbone fork is made by a fitting screw. This is inserted from above. A self-locking nut tightens the connection. The bracket itself is bolted to the gearbox housing with two cap screws

Push Rod, *Pull* Rod

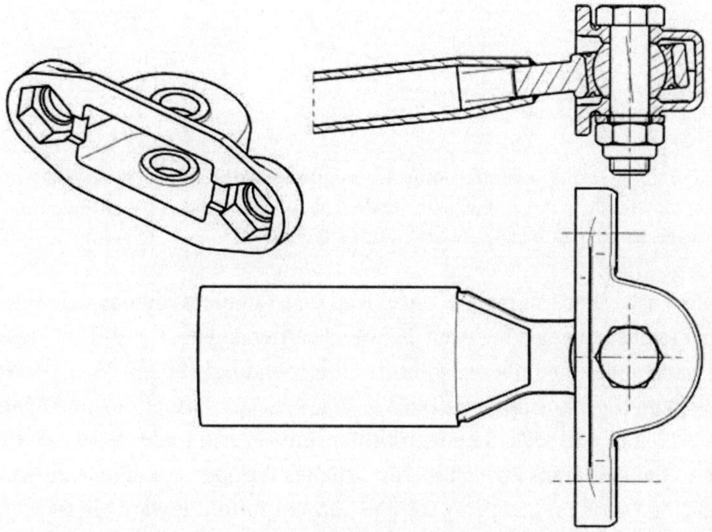

Fig. 4.51 Recessed bracket for a wishbone with joint eye. The console, an investment casting, is countersunk flush into the chassis and fixed with two hexagonal bolts. On the inside, the bracket has a collar at the top and bottom, which is milled to the desired distance for the spherical plain bearing

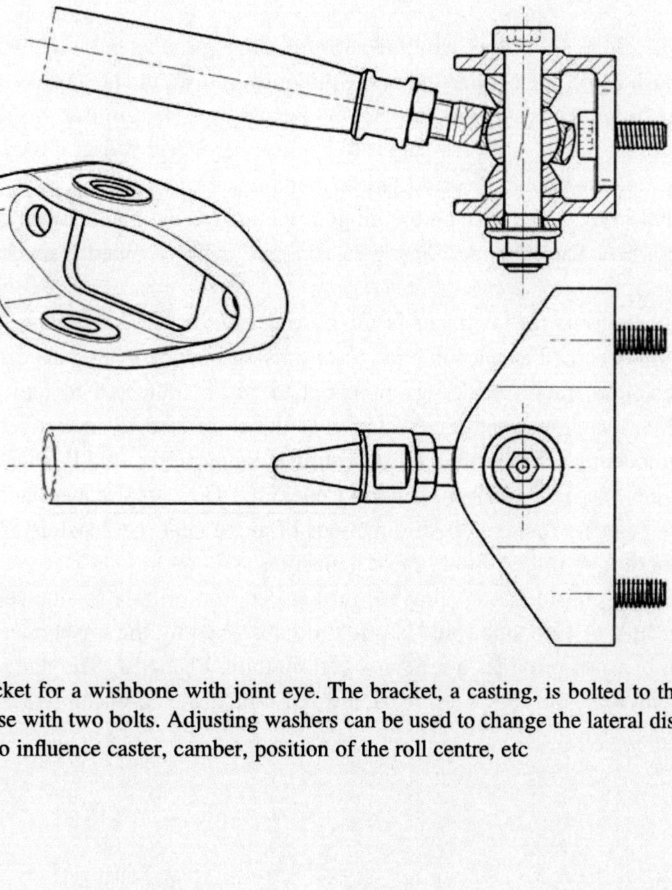

Fig. 4.52 Bracket for a wishbone with joint eye. The bracket, a casting, is bolted to the chassis or transmission case with two bolts. Adjusting washers can be used to change the lateral distance of the joints in order to influence caster, camber, position of the roll centre, etc

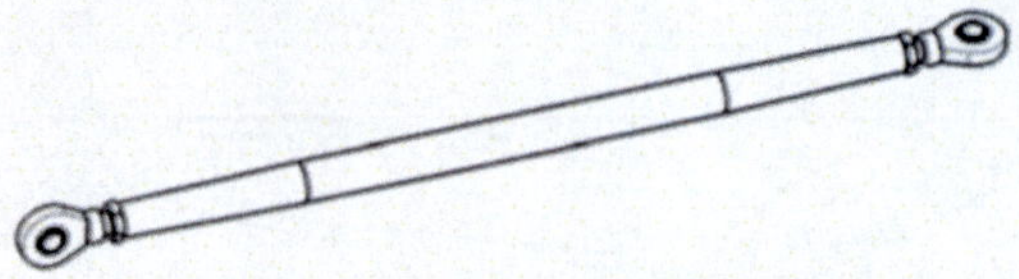

Fig. 4.53 Push rod. The compression member is designed as a part subject to buckling stress, i.e. the cross-section in the middle is larger than at the ends. The rod ends each have left-hand and right-hand threads, so that the length can be adjusted by turning the rod

The simplest way of absorbing the wheel load on a double wishbone axle is to arrange a spring damper strut at an angle between the wheel carrier and the body. This requires only a few moving parts and the number of joints is correspondingly small. Many vehicles drove and still drive with this constellation (see e.g. Fig. 4.74). A further development led to the use of compression or pull rods. Slender rods transmit vertical forces from the wheel carrier or wishbone to the bell crank, which in turn actuates the spring. Large, heavy parts of the wheel suspension (springs, dampers) are thus located further inwards in the car, which is favourable for the mass moment of inertia and air resistance of the overall vehicle. The rods are designed so that their length can be changed by means of threaded pieces, which can be used, for example, to change the distance from the ground or the static wheel load distribution in a targeted manner, Fig. 4.53. Some possibilities for suitable articulated fittings can be found in the section above on spherical plain bearings.

In principle, there are two possibilities for transferring the wheel load, namely via pull rods or via push rods. The connection of the pull/push rod should be as close as possible to the wishbone to wheel carrier joint so that the bending moment in this wishbone remains small. The ideal solution in this respect is to connect the wheel carrier directly to the strut. In this way, no wishbone is subjected to bending. However, the ball centre of the connecting link must be exactly on the kingpin inclination axis, at least on the front axle, if additional wheel load changes during steering are to be avoided (see also Sect. 4.2.2 *Wheel Carrier*).

In general, elements for the transmission of forces in the sense of lightweight construction are ideally designed as tension bars. Compression members have the disadvantage of the unstable failure mode buckling compared to struts subjected to tensile loads (see appendix). Mass-saving strength considerations therefore lead to such a design for compression members that the member cross-section is larger in the middle than at both ends, because the member is actually bent during buckling. However, it must not be forgotten that dampers generate forces in both directions of movement, i.e. tension struts must also transmit compressive forces during rapid deflection.

The following considerations provide further decision criteria for the selection of the transmission link. A tyre side force is primarily absorbed by the wishbones. However, a compression bar also provides a relieving contribution, Fig. 4.54. The flatter the angle α, the more lateral force the compression bar absorbs with its reaction component $F_1 \cdot \cos(\alpha)$.

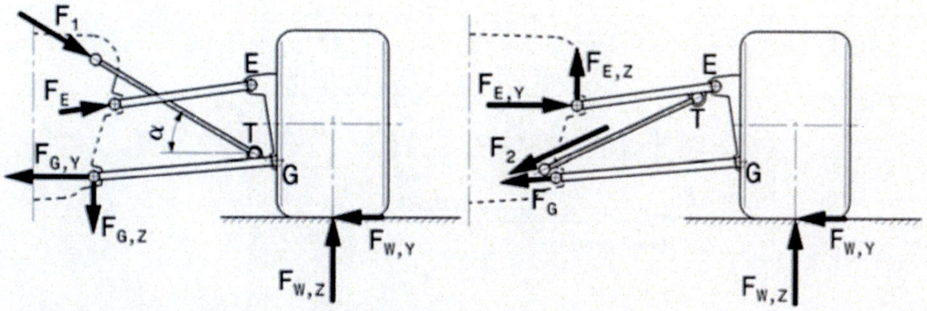

Fig. 4.54 Forces in control arms and struts during cornering. Left push rod, right pull rod. The reaction forces in the tension or compression struts have different effects

For a (usual) angle of 30°, this is 87% of F_1 . Another advantage is the lower load on the upper wishbone and thus its support on the car side. On front axles, the mounting points are usually located in an area that is difficult to stiffen because the steering, driver's feet or engine require space. Tie rods are therefore disadvantageous in this respect, because they put a lot of pressure on the upper wishbones in particular. Tie rods thus cause greater compliance in this area. In addition, when the outer wheel is displaced, the lateral force of the tyre and the reaction force of a pull rod also lead to compressive forces on the lower wishbone.

The position of the pull rod in relation to a lateral force leads to (undesirable) lifting of the car body in the turn. In this respect, the arrangement of the push rod is more advantageous.

An argument in favour of the pull rod arises from the arrangement of the spring/damper units. These can be located low down in the car and thus secure their share of a low centre of gravity. On vehicles with a high nose, this arrangement is prohibited for geometric reasons alone. In this case, only push rods come into question.

The lower wishbone is generally more heavily loaded, so the position at the upper wishbone is more favourable for the connection point T (i.e. for a pull rod) and the joint G *of* the lower wishbone is not even more heavily loaded by an additional application of force.

In vehicles with extreme aerodynamic emphasis (Formula 1, Indycar), the reduced obstruction of the flow around the transmission area (*coke-bottle effect*) and thus more favorable flow to the lower rear wing speaks for a traction detachment of the rear axle.

Weight Optimisation

Compression bars (push rods) also buckle under pure compressive force if they are slender (i.e. very long and thin). The limiting criterion for their design is therefore the buckling stress σ_{buckle}. For the bar supported in Fig. 4.55 applies:

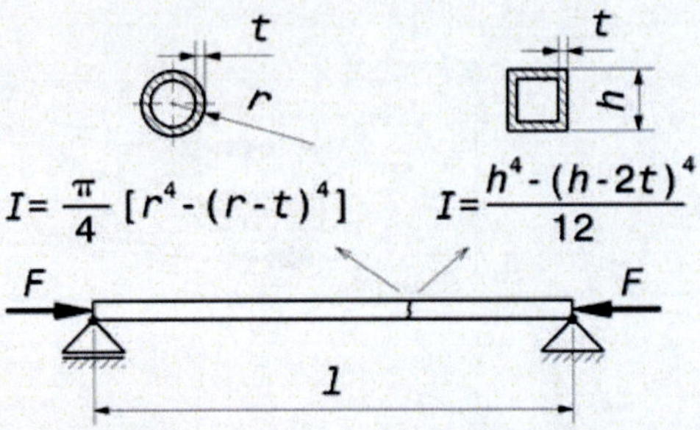

Fig. 4.55 Compression strut with circular ring or square profile. The rod is articulated on both sides and is loaded with a longitudinal force *F*
t Wall thickness of the profile, mm
r, h Cross-sectional dimensions, mm
l Bar length, mm

$$\sigma_{buckle} = \pi^2 \cdot E \cdot \frac{I}{l^2 A}$$

E	E-modulus, N/mm²
I	Area moment of inertia, mm⁴
A	Cross-sectional area, mm²

For weight-optimal compression members, i.e. loaded to this limit, with a given ratio of the profile *r/t* or *h/t, the* following applies, according to [8]:

For ring cross section	For square cross section
$r = \left[\frac{1}{\pi^3}\left(\frac{r}{t}\right)\frac{F \cdot l^2}{E}\right]^{\frac{1}{4}}$	$h = \left[\frac{3}{2\pi^2}\left(\frac{h}{t}\right)\frac{F \cdot l^2}{E}\right]^{\frac{1}{4}}$

From this, for a given material, i.e. *E*, and the bar length *l*, the cross-sectional dimensions can be determined. For this load, it is found that the thin-walled square tube is slightly better than the circular tube because more material is arranged in the outer area.

A basic comparison of materials can be made for compression members according to Fig. 4.55 according to the following relationship [9]:

$$m_{Rd} \propto \frac{\rho}{\sqrt{E}}$$

m_{Rd}	Bar mass, kg
ρ	Density of the material, kg/m³

Table 4.4 Comparison of some materials for compression members

Material	Density, kg/m^3	E-modulus, N/m^2	Characteristic value $\rho/E^{1/2}$, kg/m^2 N$^{1/2}$
Quenched and tempered steel	$7.85 \cdot 10^3$	$2.06 \cdot 10^{11}$	$1.73 \cdot 10^{-2}$
Titanium TiAl6V4 F 89	$4.43 \cdot 10^3$	$1.16 \cdot 10^{11}$	$1.30 \cdot 10^{-2}$
Aluminium AlMgSi1 F 32	$2.70 \cdot 10^3$	$0.70 \cdot 10^{11}$	$1.02 \cdot 10^{-2}$

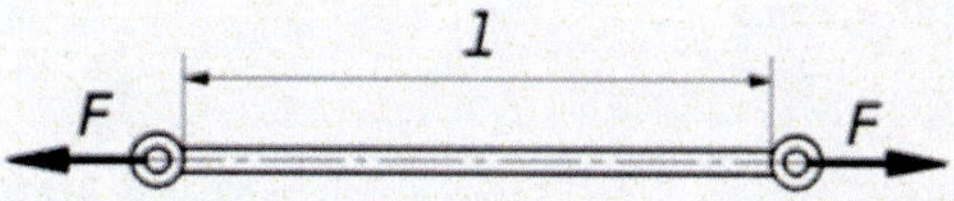

Fig. 4.56 Pull rod. l Rod length

For light compression bars, the value $\rho/E^{1/2}$ must therefore be as small as possible, e.g. Table 4.4.

Aluminium compression bars can therefore be made with a lower mass than those made of titanium or even steel, although aluminium has the lowest modulus of elasticity of the three materials.

Pull rods only know one type of failure and that is the fracture when the material strength is exceeded. If it is assumed that the bar must not permanently deform under load, the yield strength or the yield point $R_{p0.2}$ forms the design criterion. For a given tensile force F, this results in a minimum bar mass m_{Rd} of (Fig. 4.56):

$$m_{Rd} \geq \frac{\rho}{R_{p0.2}} F \cdot l$$

m_{Rd}	Bar mass, kg
ρ	Density of the material, kg/m^3
$R_{p0.2}$	Yield strength of the material, N/m^2

It can be seen that only the characteristic value $\rho/R_{p0.2}$ is decisive for a weight comparison of materials. The smaller this value is for a material, the lighter a pull rod can be made with it, Table 4.5.

Comparable pull rods made of titanium are therefore lighter than those made of steel or aluminium.

Materials Steel tube, titanium tube, fibre-reinforced plastics (CFRP).

Typical diameters are in the range of 20 mm with lengths of about 270 to 600 mm.

Table 4.5 Comparison of materials for minimum weight tension rods (pull rods)

Material	Density, kg/m^3	Yield strength, N/m^2	$\rho/R_{p0.2}$, kg/(N m)
Quenched and tempered steel 42CrMo4	$7.85 \cdot 10^3$	$765 \cdot 10^6$	$1.026 \cdot 10^{-5}$
Aluminium AlCuMg 1 F40	$2.75 \cdot 10^3$	$265 \cdot 10^6$	$1.038 \cdot 10^{-5}$
Titanium TiAl6V4 F 89	$4.50 \cdot 10^3$	$820 \cdot 10^6$	$0.549 \cdot 10^{-5}$

Bell Crank

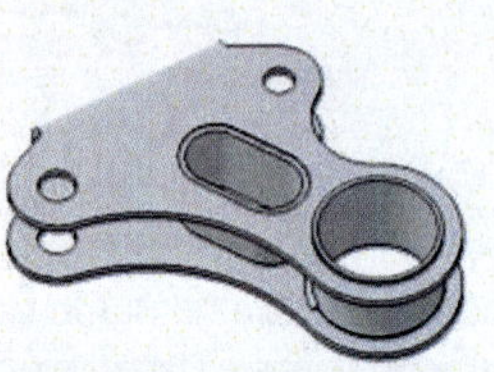

Bell cranks transfer forces and movements from pull or push rods to springs and dampers. Although a lever represents additional weight, it enables (almost) any position of springs and dampers in the vehicle. At the same time, a ratio can be set between the forces. This ratio is not constant, but changes with the angular position of the lever. By clever arrangement of the lever, a desired progressivity in the wheel-related spring characteristic can be achieved. This alone justifies the additional effort that levers entail.

Effect

The lever achieves the highest efficiency when the transmission link encloses a right angle with the lever arm (connection of the pivot points), (Fig. 4.57). However, this position occurs theoretically only twice when the lever rotates around its bearing. In all other positions, the effective lever arm is smaller.

The movement of the rod and the spring/damper strut when the suspension is sprung is therefore worth a closer look (Fig. 4.58). When the lever is rotated about its bearing, the relative positions of the push/pull rod and the spring/damper change. For isolated consideration of the action on one side of the lever, we specify a constant moment $F_{Rd} \cdot r_{Rd}$, i.e. only the link on the other side moves when the lever is sprung. It can be seen that a deviation of $\gamma = \pm$ 10 to 15° of the connecting link from the ideal position has hardly any noticeable effect.

However, if one wants to achieve a steady progressive spring rate, the ratios must not reverse and the path ratio between rod and spring r_{Rd}/r must decrease (cf. Fig. 3.9). This means that the angle γ must move towards the 90° position (i.e. $\gamma = 0°$) during compression, whether it comes from the positive or negative side. Thus, in any case, the 90° position is not reached until the compression stop. Or described differently: The angle

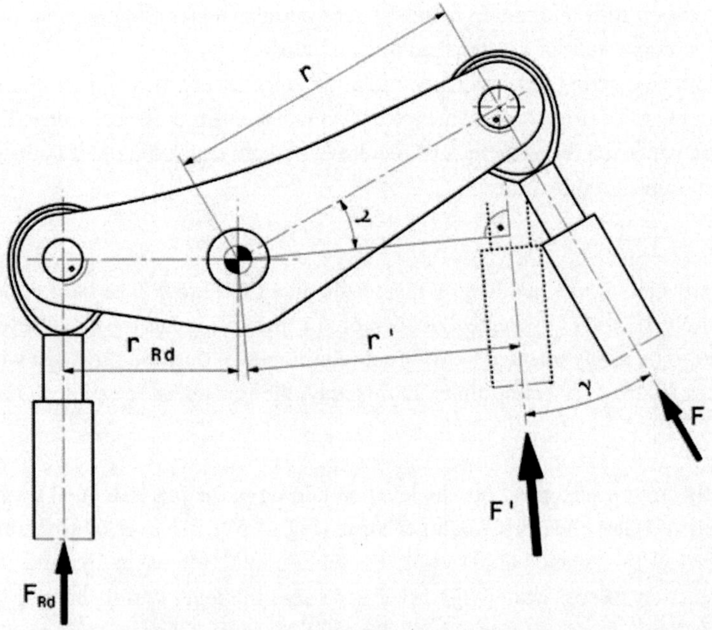

Fig. 4.57 Leverage. Direct transmission occurs when the transmission links (pull/push rods) are perpendicular to the lever arms. Then the following applies: $F = F_{Rd}\frac{r_{Rd}}{r}$ with the lever ratio r_{Rd}/r. However, if the rod deviates from this ideal position by angle γ, the effective lever arm is reduced to $r' = r\text{-}cos(\gamma)$ and it becomes: $F' = F_{Rd}\frac{r_{Rd}}{r\cos(\gamma)}$

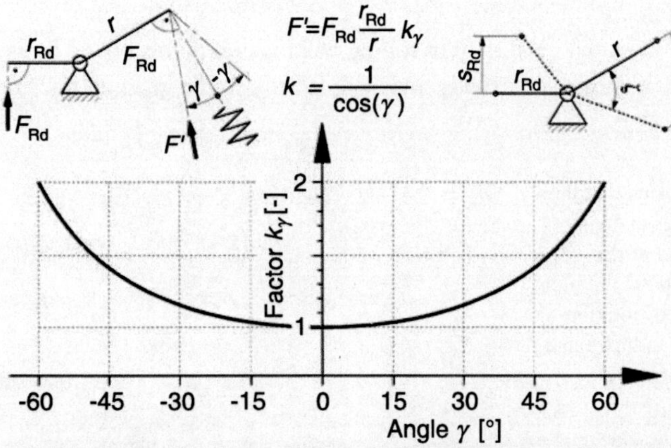

Fig. 4.58 Effects of a lever movement. γ Angular deviation of a connecting link from the ideal position, °. k_γ Factor describing the change in the leverage ratio due to γ, –. The lever ratio r_{Rd}/r is increased by the factor k_γ. The same applies to the ratio of the swivel paths, which is equal to the lever ratio: $\frac{s_{Rd}}{s} = \frac{r_{Rd}\cdot\sin\zeta}{r\cdot\sin\zeta} = \frac{r_{Rd}}{r}$

between the shock strut and the lever must either initially be less than 90° and open when the spring is compressed, or greater than 90° and close.

In fact, both connecting links will move relative to the lever when springing, i.e. for both sides the statements of (Fig. 4.58) come into effect and superimpose each other. The change in the total transmission ratio can thus be increased or even cancelled out if both connecting links work in opposite directions.

Leverage

If the lever arms are of unequal length, this results in a ratio r_{Rd}/r. The ratio of push rod to spring is usually 1:2 to 1:3 in the design position, i.e. the spring has the larger lever arm. To enable levers to be easily adapted to different requirements, they are provided with several locating holes for the connecting links. In this way, the transmission ratio can be varied in steps.

In the fully compressed position, the lever should lie in the plane defined by the axes of the pull/push rod and the spring/damper strut (Fig. 4.59). Its axis of rotation must be perpendicular to this plane. Otherwise, the longitudinal forces in the rod will cause additional reaction forces, which will adversely stress the lever and its bearing and cause additional frictional forces, and the shock strut will be twisted during springing.

Even if the arrangement of spring/damper strut and compression bar is as in Fig. 4.59, the parts need a spatially articulated connection, Fig. 4.60. The spring/damper axis lies in the plane of the lever and therefore only moves in this plane during compression. The connecting eye on the frame side must allow the angle β_{Sp} to gimbal if its axis of rotation is not parallel to the lever axis (which is usually the case, because this makes the mounting bracket for this eye simpler). The push rod swivels by the angle β_{Rd} into the lever plane during compression.

The levers are mounted on sliding or rolling bearings on pivots bolted or welded to the frame. The rear bell cranks are usually attached directly to the gearbox housing, to which

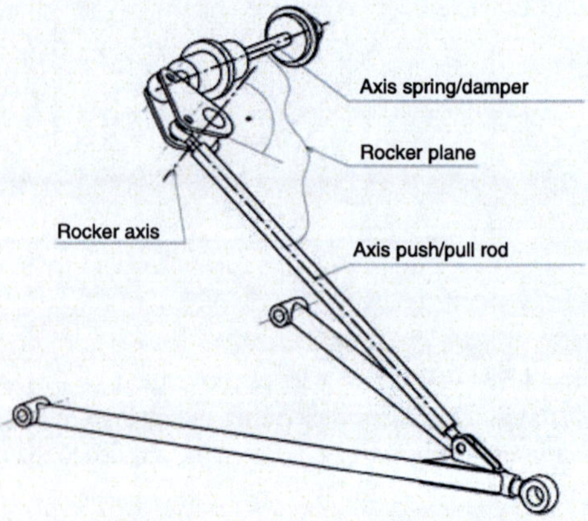

Fig. 4.59 Ideal position of the lever axis. The lever plane is spanned by the axes of the spring/damper strut and the pull/push rod. The axis of rotation of the lever is normal to this plane in the design position

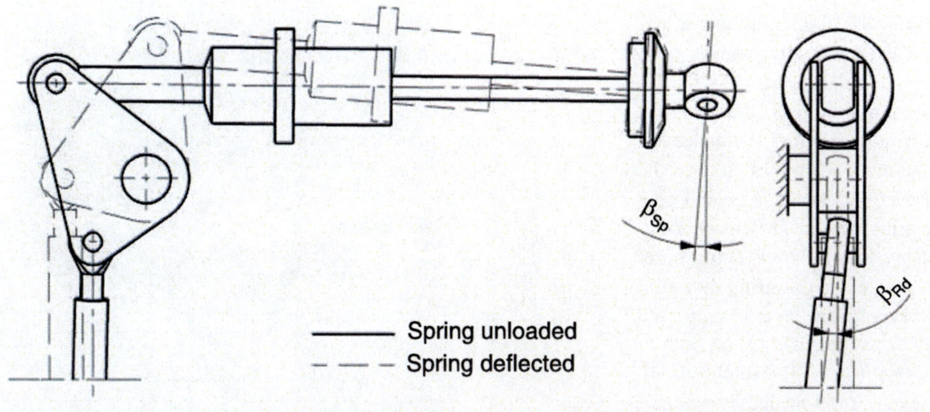

Fig. 4.60 Movements of the lever connecting links. β_{Sp} Movement of the damper strut. β_{Rd} Movement of the push rod. Top view of the lever plane (left) and corresponding view from the front (right)

corresponding slugs or lugs are cast (Fig. 4.61). In the case of box frames, where it is necessary to apply force over a large area, bearing journals with a foot are used, (Fig. 4.62). This foot can be connected to the frame using suitable connection techniques (rivets, screws, laminating).

Combined needle roller bearings offer an interesting alternative to the arrangement of several rolling bearings. These have a thrust bearing in addition to the needle bearing, which can only support about ¼ of the radial load, but this is easily sufficient for the usual lever bearing arrangement. Figure 4.63 shows an example of a double direction needle roller angular contact ball bearing.

Choice of Material

For basic material selection, the following consideration will be helpful. A bell crank basically consists of two beams which are bent. For a beam subjected to bending according to (Fig. 4.64), the following results for its mass:

$$m = 6F\frac{\rho}{\sigma_{b,\max}} \cdot \frac{l^2}{h} \cdot 10^{-9}$$

ρ	Density of the material, kg/m^3
$\sigma_{b,max}$	Permissible bending stress of the material, N/mm^2

With otherwise identical dimensions, the material with the smallest value of $\rho/\sigma_{b,max}$ is therefore the most favourable in terms of lightweight construction. A corresponding comparison of materials is carried out in Table 4.6.

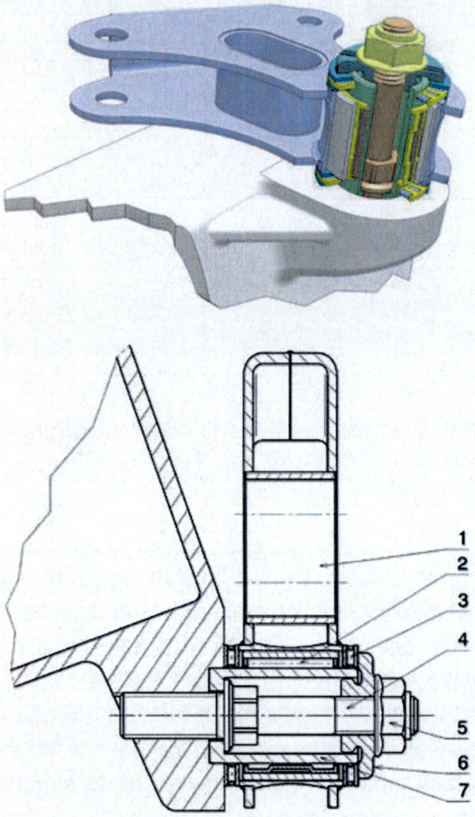

Fig. 4.61 Bell crank mounting. **1** Rocker. **2** Radial bearing (needle roller bearing). **3** Thrust bearing. **4** Stud. **5** Lock nut. **6** Adjusting nut. **7** Pivot sleeve. The lever (1) rotates on a needle bearing (2) which is guided on a bearing sleeve (7). This bearing sleeve is held in a locating bore in the gearbox housing by a stud screw (4). The lever is supported by thrust bearings (3) on both edge surfaces. The running clearance is adjusted by means of an adjusting nut (6) and secured by means of the lock nut (5)

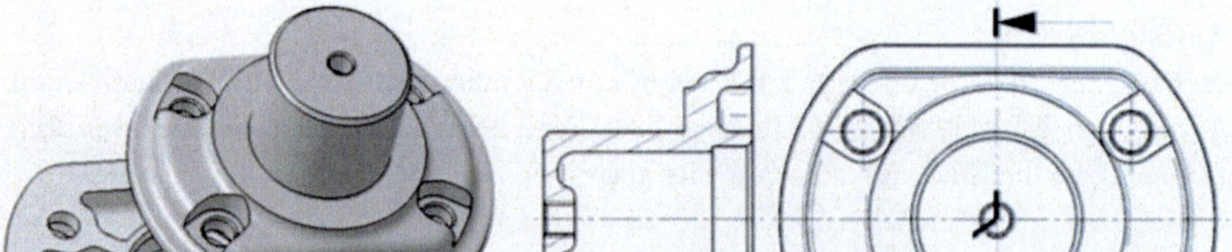

Fig. 4.62 Bolt on type rocker post. The trunnion is cast (e.g. from unalloyed cast steel GS 52 (1.0552) to DIN) and machined on the running surfaces

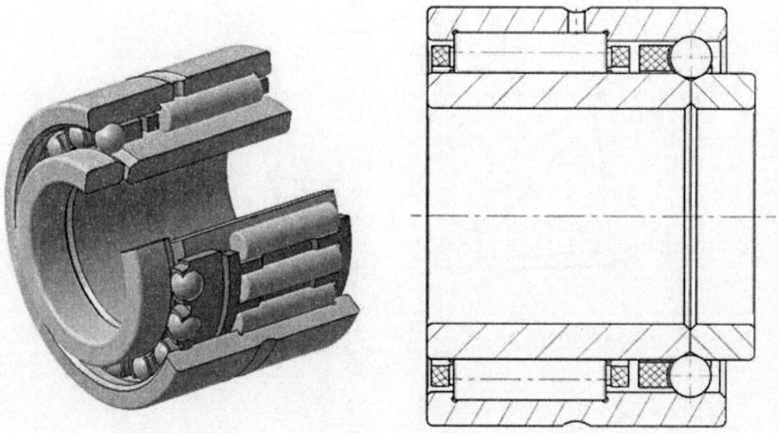

Fig. 4.63 Needle angular contact ball bearing. The bearing shown is double direction acting

Fig. 4.64 Bending beam under concentrated load. *w, h, l* Dimensions, mm, *F* Force, N

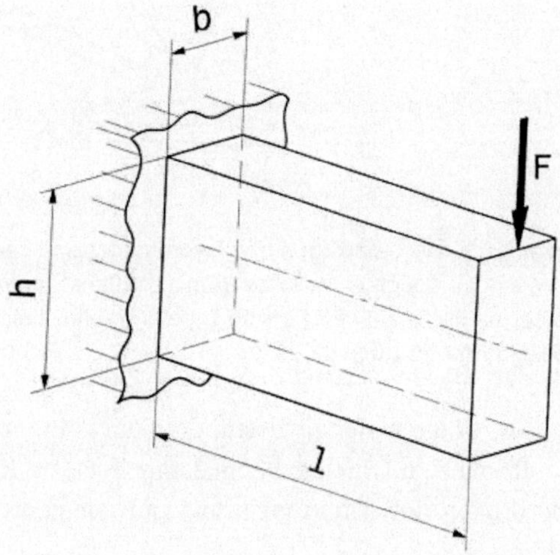

Table 4.6 Comparison of materials for bending beams

Material	Density kg/m^3	Permissible alternating bending stress N/mm^2	Characteristic value $\rho/\sigma_{b,max}$ kg/Nm
Quenched and tempered steel 42CrMo4	$7.85 \cdot 10^3$	480	$1.64 \cdot 10^{-5}$
Aluminium AlZnMgCu1,5 F53	$2.70 \cdot 10^3$	140	$1.93 \cdot 10^{-5}$
Titanium TiAl6V4 F89	$4.43 \cdot 10^3$	560	$0.79 \cdot 10^{-5}$

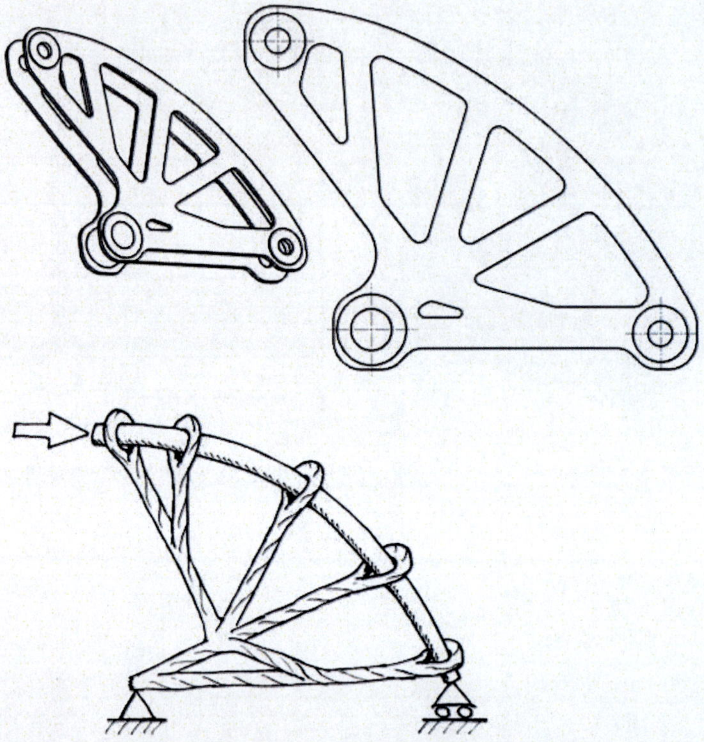

Fig. 4.65 Decomposition of a lever into tension and compression elements, after [10]. Based on the conceptual design of the lever from a compression bar, which is already preformed and whose further deformation is prevented with ropes (graphic below), a lightweight construction proposal for its design follows (above)

So, of the materials listed, titanium is the most suitable for a light bending beam.

Rockers in which the connecting eyes are located on one side of the lever axis can also be broken down into push and pull elements in the sense of direct load transmission,[2] (Fig. 4.65).

Fabrication

Rockers are milled from the solid or designed as a sheet metal bending/welding construction.

Surface Finish

Steel parts can be nickel-plated. Smaller aluminium parts are anodised.

Wishbones and push rods are often black nickel-plated or kephos-coated.

[2]See Racing Car Technology Manual Vol. 2 *Complete Vehicle*, Section 2.5 *Design Principles.*

Fig. 4.66 Right hand bell crank of a front suspension (Formula 3000, Reynard D94). The lever is mounted on needle bearings and has axial thrust bearing washers at the pivot. It transmits the wheel contact force via the pressure rod (1) to the suspension strut (2) and to the stabiliser (3)

Fig. 4.67 Rear suspension bell crank (Renault 2000 formula). The lever is a sheet metal bending and welding construction (see Fig. 4.68)

Chroming of steel chassis parts is prohibited by FIA regulations.

The (Figs. 4.66, 4.67, 4.68, and 4.69) show some executed examples of bell cranks together with the interacting connecting parts.

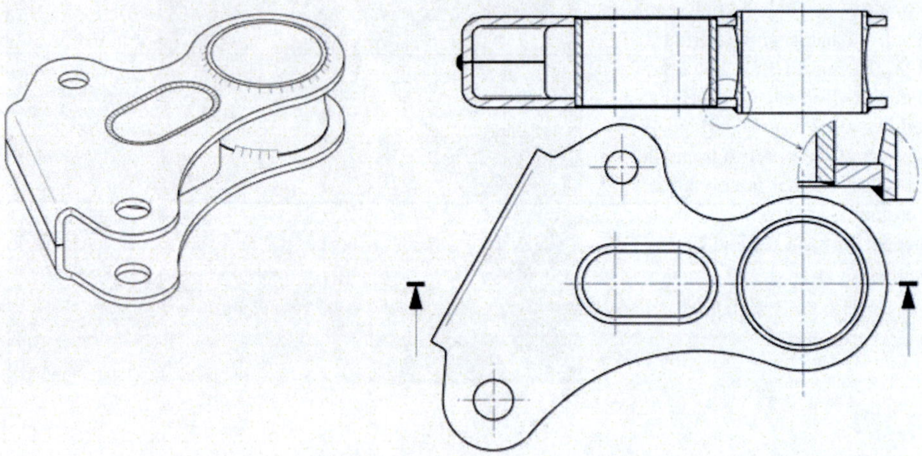

Fig. 4.68 Sheet metal lever. The lever is constructed in several parts. Stamped sheet metal parts are folded and welded together via sleeves

Fig. 4.69 Rocker of the rear suspension of a formula car (Reynard Zytec, driving direction to the left). The lever is mounted directly on the gearbox housing on which brackets for the wishbone connection are also cast. The lever also operates a T-shaped stabilizer bar

4.2.3 McPherson Axle

Due to its advantages in terms of installation space, the wheel-guiding suspension strut axle is found on the majority of front-wheel drive passenger cars. Figure 4.70 shows a typical example. Not least because of their proximity to the passenger car, rally vehicles also use this type of suspension on the front and rear axles. This allows a four-wheel drive to be represented and at the same time, if the wheel carriers are cleverly designed, the same component can be installed in all four positions. This means that no mirrored component is required for the left and right. Rally teams are particularly keen to simplify logistics in this way, as it greatly simplifies the supply of spare parts in a competition.

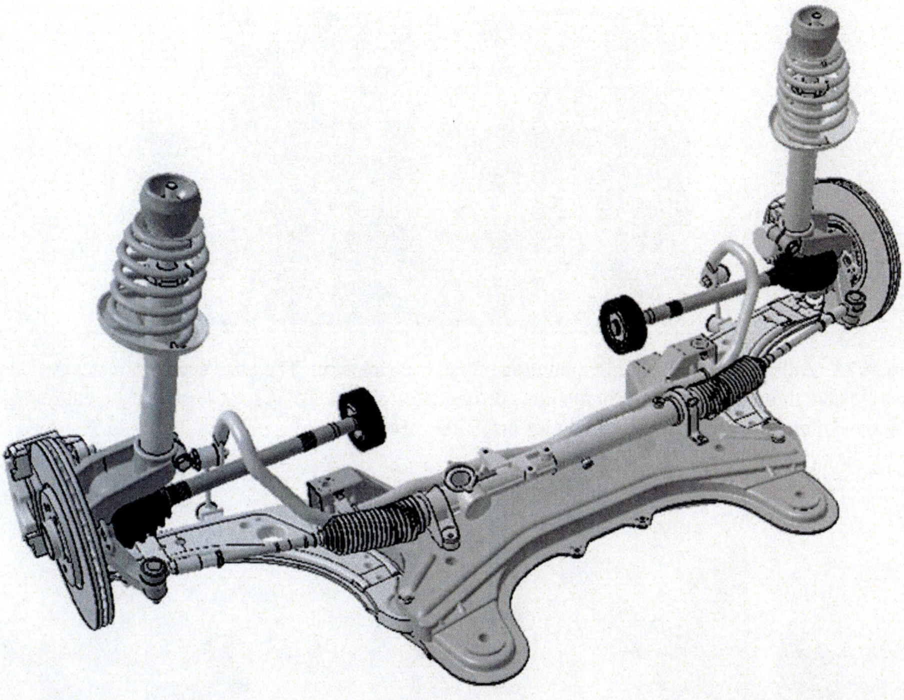

Fig. 4.70 Wheel-guiding strut axle of a passenger car as front axle. The wheel carrier (also referred to here as a pivot knuckle) accommodates the suspension strut at the top and is connected to the wishbone at the bottom via a wheel joint. The two wishbones are mounted via rubber bushes on a subframe, which also accommodates the steering gear. The subframe itself is elastically decoupled from the body. The axle is driven. The stabilizer, which is connected to the wishbones via coupling rods (drop links), deflects in an arc to the cardan shafts

The adjustment of the formative wheel position variables such as kingpin inclination and kinematic caster is carried out in passenger cars by means of slotted holes in the lower wishbone. However, the adjustment range is very small, as it is only necessary to be able to compensate for manufacturing tolerances. In racing vehicles with the most varied requirements, plates for the strut head bearing have proven their worth. These are screwed into the suspension strut dome of the body and enable the strut to be swivelled in the transverse direction (kinpin inclination) and the plate together with the head bearing to be moved in the longitudinal direction (caster) via slotted holes (Fig. 4.71).

4.2.4 Beam Axle (Rigid Axle)

The rigid axle may be considered quite outdated for use in competition cars, but there are actually still racing series in which the beam axle is standard due to regulations. First and

Fig. 4.71 Adjustable head bearing mounting of a suspension strut. The plate accommodates the head bearing and allows pivoting within the large central slotted hole. To adjust the caster, the entire plate is moved along three narrow oblong holes inside the strut dome

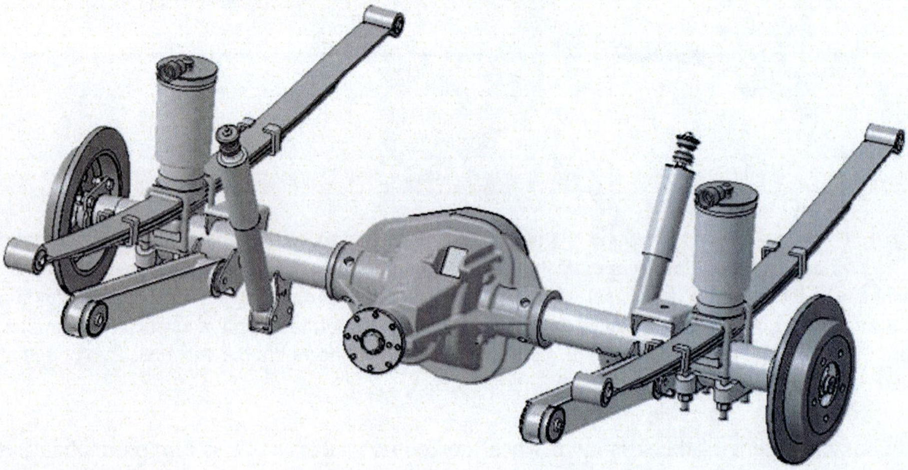

Fig. 4.72 Driven beam axle of a passenger car. The two axle journals are directly connected to each other via two tubular pieces that are welded into the housing of the final drive. The axle is guided in the longitudinal direction by two trailing arms. In the lateral direction, the stiffness of the leaf spring is sufficient to absorb the lateral forces. Air suspension elements help to keep the level constant with different loads

foremost is the North American NASCAR series. Figure 4.72 shows an example of a passenger car axle.

The *twist beam axle* can be regarded as a variation of a rigid axle. Although the two wheels of an axle are directly connected to each other by an axle beam, this element is designed with an open and thus torsionally soft profile (Fig. 4.73). This transverse connection therefore also acts as a stabiliser. The wheels have a certain independence similar to independent wheel suspension via the torsion of the cross member, but part of the wheel load is transferred between the wheels.

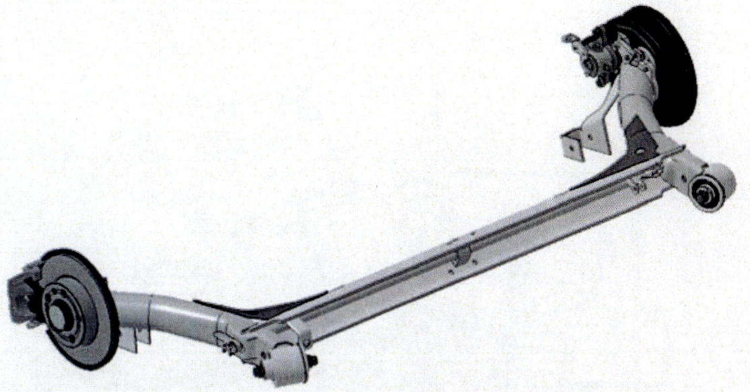

Fig. 4.73 Non-driven twist-beam axle of a passenger car

4.3 Examples of Race Car Suspensions

Finally, a few pictures of executed wheel suspensions should illustrate the overall system. The (Figs. 4.74, 4.75, 4.76, 4.77, 4.78, 4.79, 4.80, 4.81, 4.82, and 4.83) show wheel suspensions of various racing vehicle classes from different eras.

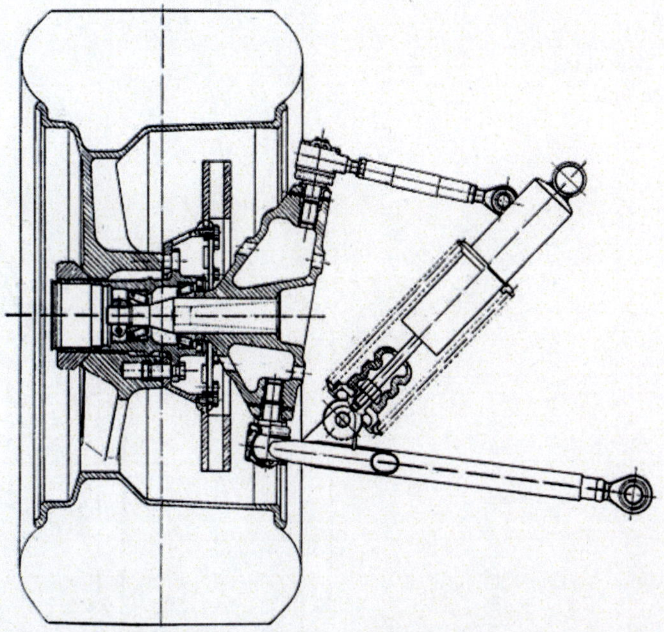

Fig. 4.74 Front axle of an older racing car (Porsche 917) [7]

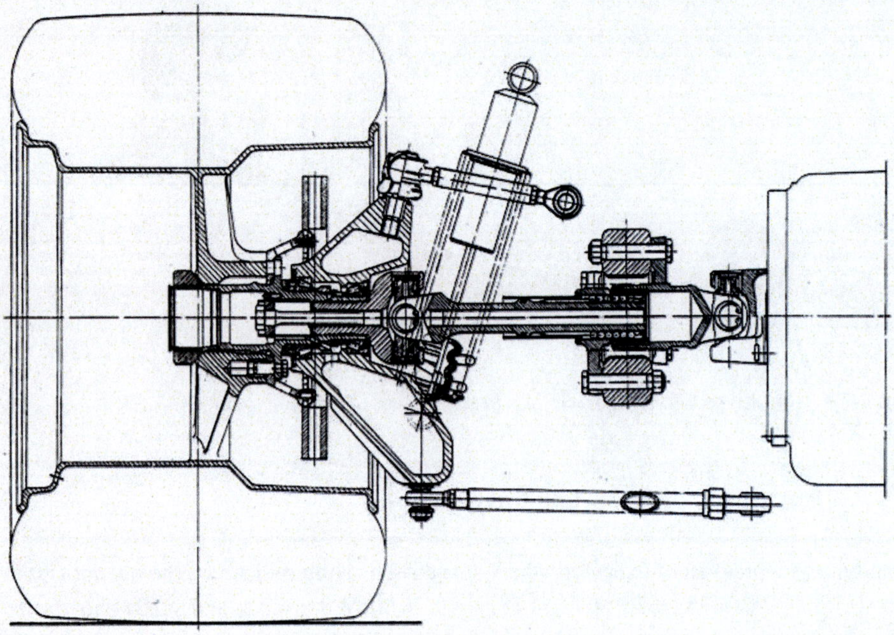

Fig. 4.75 Rear axle of an older racing car (Porsche 917) [7]

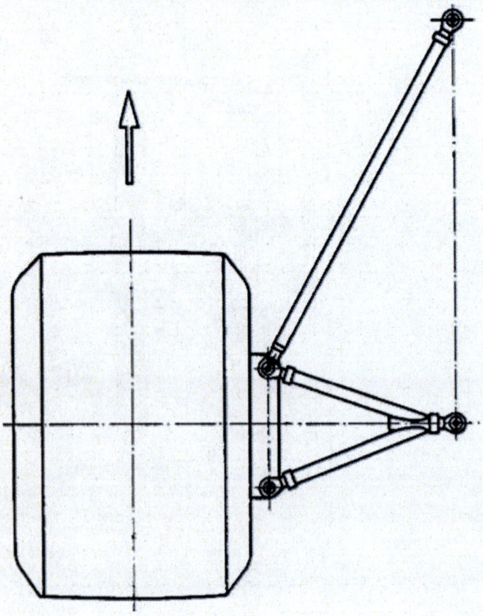

Fig. 4.76 Top view of the wheel suspension from (Fig. 4.75) [7]

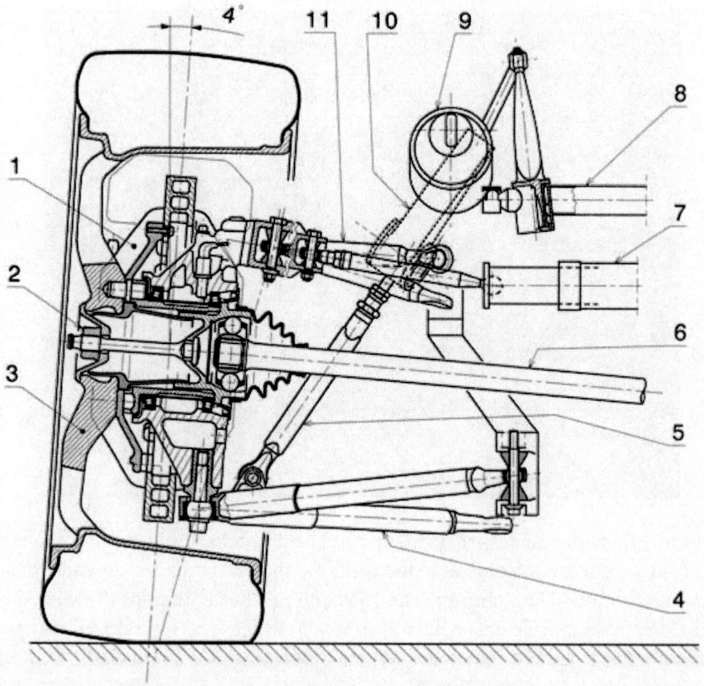

Fig. 4.77 Front axle of a touring car (Opel Calibra ITC 96) [11]. 1 8-piston caliper, 2 Central fastening, 3 Magnesium wheel, 4 Lower wishbone, 5 Push rod, 6 Drive shaft, 7 Hydraulic steering, 8 Variable anti-roll bar, 9 Coil-over-damper unit, 10 Rocker, 11 Upper wishbone. The vehicle has four-wheel drive (See e.g. Racing Car Technology Manual Vol. 3 Powertrain, Section 5.6.3 Designs). The stabilizers are adjusted electrohydraulically

The view shows the lower wishbone plane with a wide, frame-side base for absorbing the braking forces. The wishbone is thus arranged upside down compared to usual wishbones.

Fig. 4.78 Formula Renault rear suspension bell crank and stabiliser, direction of travel to the right. The push rods (1) transmit the wheel load to the struts via the bell crank. At the same time, they twist one arm of the stabilizer (2). The stabilizer arms move up to adjustable stops (3) when the wheels are displacet. If a different stabilizer characteristic is required, the entire stabilizer is replaced. Three different diameter variants are offered by the vehicle manufacturer. The brackets (4) of the stabilizer linkage can be adjusted 3 or 4 times. This allows the adjustment of the roll center and the anti squat feature

Fig. 4.79 Formula BMW rear suspension, driving direction upwards. In the case of an equilateral wheel travel movement, the force is transmitted to the mono-spring via the two push rods (1) and the angle lever (2). During alternating stroke movement, the bell-crank lever is displaced along its axis of rotation by a thrust strut (1) and pressed against a disc spring assembly (3) (acting as a roll stabiliser). The preload and central position of the cup springs can be changed via screw stops on both sides

Fig. 4.80 Front suspension of a Formula Renault car. The monoshock is actuated by two thrust rods (1, 2) and a bell-crank lever (3). The bell-crank lever can be moved along its axis of rotation against Belleville washers and thus acts as a roll stabiliser

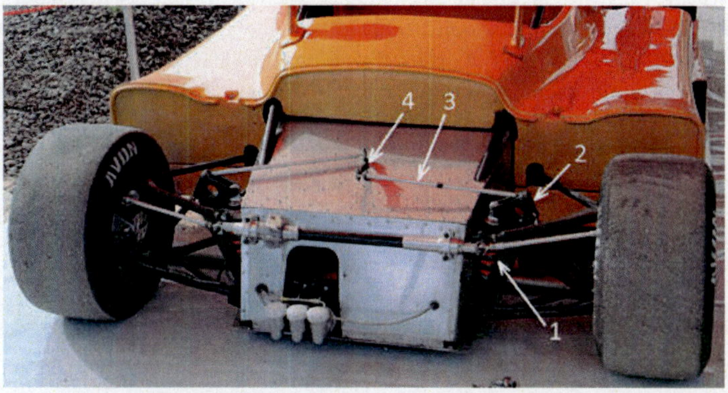

Fig. 4.81 Front suspension of a production sports car. The spring struts (1) are approximately vertical and are actuated via the bell crank (2). The two bell cranks are connected to each other via linkages (3) and intermediate levers (4) in such a way that wheel loads are transmitted when the springs are alternated. This system therefore acts as a roll stabiliser

Fig. 4.82 Rear suspension of a production sports car (Norma N20). The illustration shows the basic relationship of racing cars. Removing the bodywork reveals the drivetrain, which is flanged directly to the engine. The rear suspension is mounted on the gearbox housing and transfers the wheel load via push rods and bell cranks to the struts, which sit longitudinally on the gearbox. The upper wishbone supports the wheel carrier at two points via a connecting bracket. This allows the camber to be adjusted without changing toe-in. The short U-stabilizer is mounted on the flange between the gearbox and the clutch bell housing

4.4 Data

Tables 4.7 and 4.8 provide some data for comparison purposes and as starting values for the design, respectively.[3]

[3] A compilation of typical starting values for the development of race cars can be found in the Racing Car Technology Manual Vol. 5 *Data Analysis, Tuning and Development*, Chap. 5 *Tuning* Table 5.3.

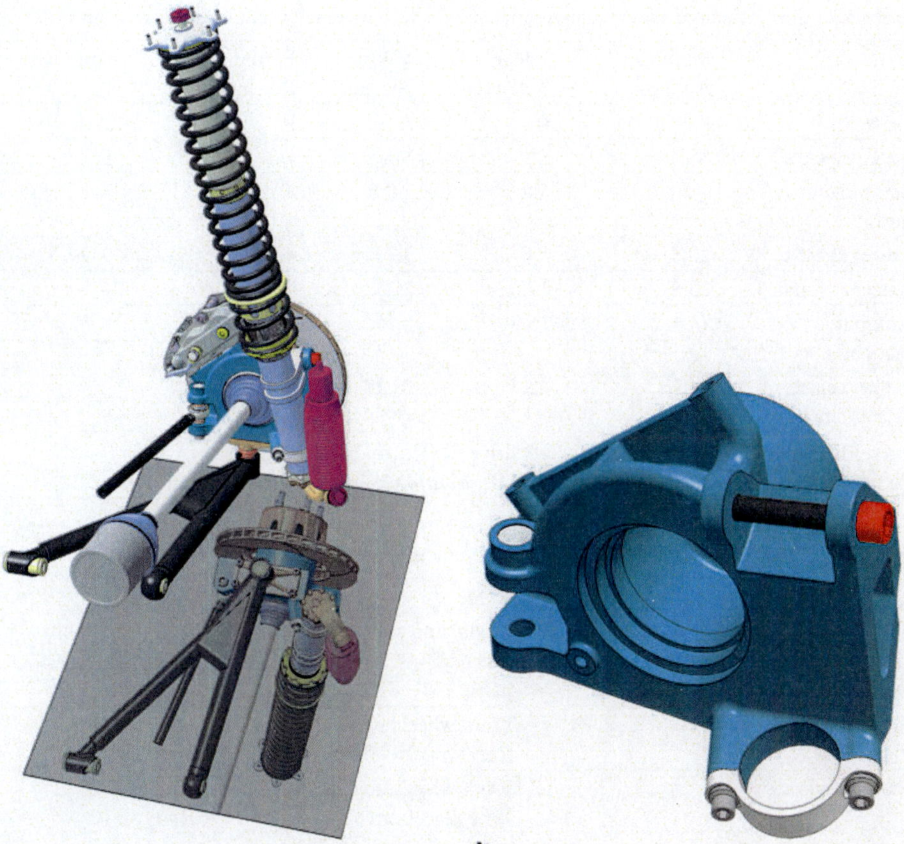

a b

Fig. 4.83 Wheel suspension of a rally car with four-wheel drive (Ford Fiesta RS WRC). (**a**) Left front suspension, (**b**) Wheel carrier for left side of vehicle. The aluminium wheel carrier is milled from the solid and can be used on one side of the vehicle at the front and rear. The lower bracket is exchanged: On the rear axle, it accommodates not only the wheel-side joint of the lower wishbone but also that of the track rod. The suspension strut is bolted to the wheel carrier and has a rearward tilt on the front and rear axles. The integrated damper carries an external gas reservoir and its rebound and compression damping can be adjusted. The body spring is supported at the bottom by a short helper spring. The built steel wishbone is supported on the car side by a subframe, which is also a tube-welded construction. The brake discs are internally ventilated and are held in place by a pot that bolts to the wheel. On asphalt, 355-mm-diameter brake discs and 8 × 18-inch wheels are used; on gravel and snow, 300-mm discs are paired with 7 × 15-inch wheels

Table 4.7 Comparison of kinematic parameters of selected vehicles, passenger cars from [12]

	Mercedes A-Class	BMW 3 Series	VW phaeton	Formula Renault [13]	Audi R8 (LMP)
Drive[a]	F	R	A	R	R
Axle principle[b]	SS	SS	ML	D	D
Scrub radius r_σ, mm	−20.7	4.8	−0.5	40.89	–
Caster angle τ, °	2.83	5.8	3.71	4.49	9–12.5
Caster $r_{\tau,k}$, mm	13.8	17.0	26.12	18.46	–
Kingpin inclination σ, °	14.1	15.36	5.15	15.53	8.7
Wheel center offset, mm	44.12	83.3	22.19	–	

[a]Drive: F: Front-wheel drive, A: All-wheel drive, R: Rear-wheel drive
[b]Axle principle: SS: Suspension strut axle, ML: Multi-link axle, D: Double wishbone axle

Table 4.8 Comparison of kinematic ranges of vehicle types

Designation	Value Production vehicles Passenger cars partly [14]	 Race cars partly [15]
Camber ε, °	Front wheels –1 to +0.5[a] Rear wheels −0.5 to −1.67[a]	–0.5 to −6
Caster angle τ, °	Front wheel drive −2 to +8 Rear wheel drive +4 to +11	+2.5 to +4.5 Rally: >10 (on sand), 2.5 (gravel)
Caster $r_{\tau,k}$, mm	0 to 39.5 [16]	18
Kingpin inclination σ, °	3 to 12	5 to 8
Scrub radius r_σ, mm	−30 to +75	2 to 4
toe (toe-in/toe-out)	Front wheels 0° ± 20′ (−3 to +3 mm) Rear wheels 0° ± 60′ (−5 to +5 mm)	A little more than series. Front axle: 0.2° to –1.5°. Rear axle 0° to 1
Wheel vertical movement, mm	−80 to +120	−20 to +30 (formula car) −150 to +350 (rally car)
Max. body roll angle at 1 *g* lateral acceleration, °	8	0.8

[a]Max. ±4° according to [16]

References

1. Heißing, B.: Moderne Fahrwerksauslegung. Vortrag im Rahmen der ÖVK-Vortragsreihe, Graz 12. Mai (2004)
2. Crahan, T.C.: Modelling Steady-State Suspension Kinematics and Vehicle Dynamics of Road Racing Cars, Part II. SAE Paper 942506. In: Smith, C. (Hrsg.) Racing Chassis and Suspension Design. SAE International, Warrendale (2004)
3. Lis, A.: Bericht über 2KQ Rennfahrzeug von Reynard. Racecar eng. **27**, 34 ff (2000)
4. N.N.: Gelenklager, Gleitbuchsen, Gelenkköpfe: Katalog 236 Fa. INA, Helmut ELGES GmbH, Steinhagen (1999)
5. Aurora Bearing Company catalogue No. 503. Aurora, Illinois/Montgomery (2019)
6. Schweighart, T.: Das optimale Dakar-Rallye-Fahrwerk. Diplomarbeit an der FH JOANNEUM, Graz (2006)
7. Reimpell, J.: Fahrwerktechnik, 5. Aufl. Vogel, Würzburg (1982)
8. Klein, B.: Leichtbau-Konstruktion, 5. Aufl. Vieweg, Wiesbaden (2001)
9. Hintzen, H. et al.: Konstruieren und Gestalten, 3. Aufl. Vieweg, Braunschweig (1989)
10. Mattheck, C.: Verborgene Gestaltgesetze der Natur. Optimalformen ohne Computer, 1. Aufl. Forschungszentrum Karlsruhe GmbH, Karlsruhe (2006)
11. Indra, F.: Grande complication, der Opel Calibra der ITC-Saison 1996. Automobil Revue N. 50 (1996)
12. diverse Sonderausgaben ATZ MTZ für Audi A4. Springer Vieweg, Wiesbaden (2002–2005)
13. N.N.: Formula Renault 2000 manual. Renault Sport Promotion Sportive (2001)
14. Braess, H.-H., Seiffert, U.: Vieweg Handbuch Kraftfahrzeugtechnik, 4. Aufl. Springer Vieweg, Wiesbaden (2005)
15. McBeath, S.: Competition Car Preparation, 1. Aufl. Haynes, Sparkford (1999)
16. Henker, E.: Fahrwerktechnik. Springer Vieweg, Wiesbaden (1993)

Steering

5

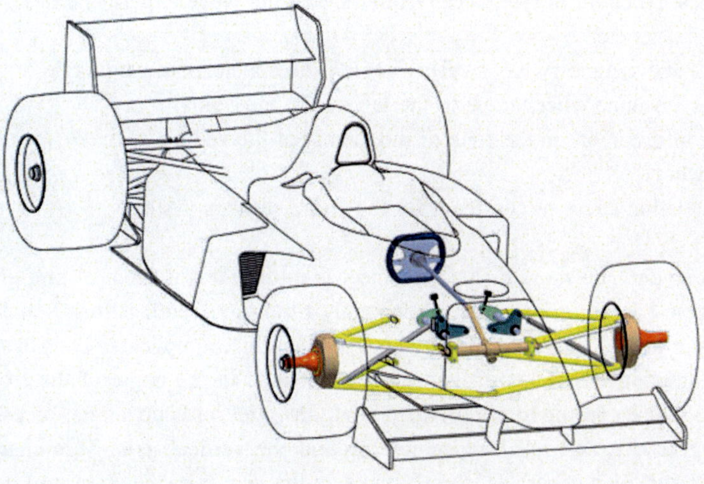

The driver of a multi-track vehicle actually has only a few levers at his disposal to influence the behaviour of the car in the desired way. An important system in this respect is the steering. At the same time, it provides the driver with important information about the tire-road contact, but at the same time it should filter out disturbing influences from the road.

M. Trzesniowski, *Suspension System*,
https://doi.org/10.1007/978-3-658-39847-7_5

5.1 Requirements

The steering serves to maintain the target course of the vehicle via steering wheel rotation by the driver. To do this, the driver relies not only on information provided by the eyes, but also on other important sensory impressions, e.g. the body roll, the lateral forces acting on the driver via the seat, the tire return torque felt via the steering wheel, and the change in the vehicle's sideslip angle, which is a measure of the change in the slip angle of the rear axle.

The following requirements are of particular importance:

- Wheel steering angle: Maximum steering angle for the tightest corner of the track and for countersteering in case of tail swerve (rear-end break-away)
- Ease of movement
- Zero backlash
- Return to the middle position (straight ahead)
- Security
- Kinematics: Isolation of tyre forces from the steering wheel where practical, i.e. do not transmit shocks etc.
- Exactness and symmetry, i.e. steering angle causes a clear steering angle of the wheels as well as the same wheel angle to the left and to the right.
- reporting information on the state of movement of the vehicle to the driver through the steering wheel[1]
- Low installation space requirement for the entire steering system

These requirements are more difficult to meet than they might seem at first glance. The steering system cannot be considered separately from the overall human-vehicle system. On the contrary, it is an actuator and feedback element in a complicated control system. By turning the steering wheel, the driver causes a change in the angle of the front wheels (either directly or according to a predefined law). The tyres build up lateral forces due to the side slipping caused and turn the vehicle around the vertical axis. This change in the direction of travel also generates lateral forces on the rear axle, which in turn can only be generated by side slipping. The driver reacts to this change of the vehicle's position by steering. Figure 5.1 shows the course of slip angles during an abrupt turn (steering angle jump, J-turn).[2] In sequence, a slip angle is first imposed on the steered front wheels and only after the vehicle has begun to turn is a slip angle imposed on the rear wheels. The car needs a few seconds to settle into a steady state.

For comparison, Fig. 5.2 shows the target ranges in passenger-car chassis development. These values concern a steady-state cornering with a radius of 100 m.

[1] The haptic sensory channel is the fastest in humans. Reaction times of approx. 0.1 s can be addressed by the driver [1].

[2] For more details on this test, see Racing Car Technology Manual, Vol. 5 *Data Analysis, Tuning and Development*, Chap. 5 *Tuning*.

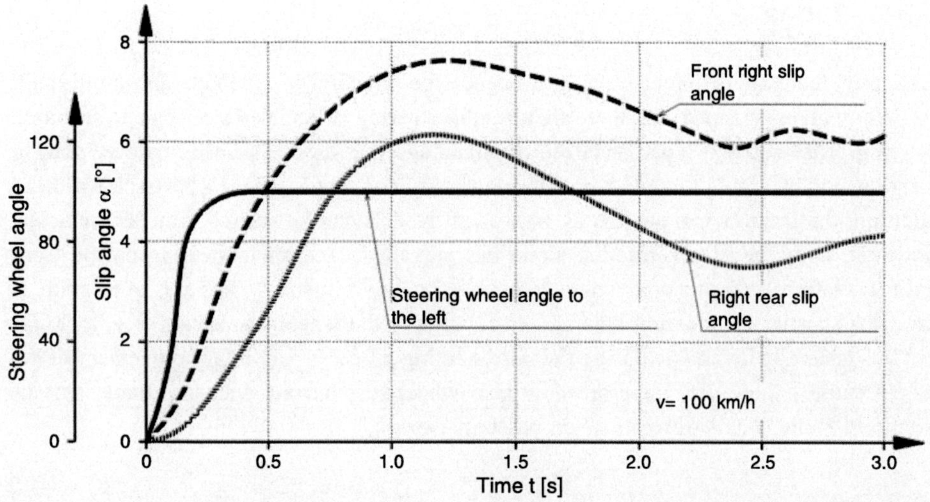

Fig. 5.1 Steering angle jump on a passenger car, according to [2]. The steering wheel is turned by 100° in 0.2 s during this test. In the case of the outer front tyre considered, the slip angle increases to over 7° as a result. The slip angle of the corresponding rear tyre is always smaller during the time under consideration. The vehicle therefore understeers and the slip angle decreases over time

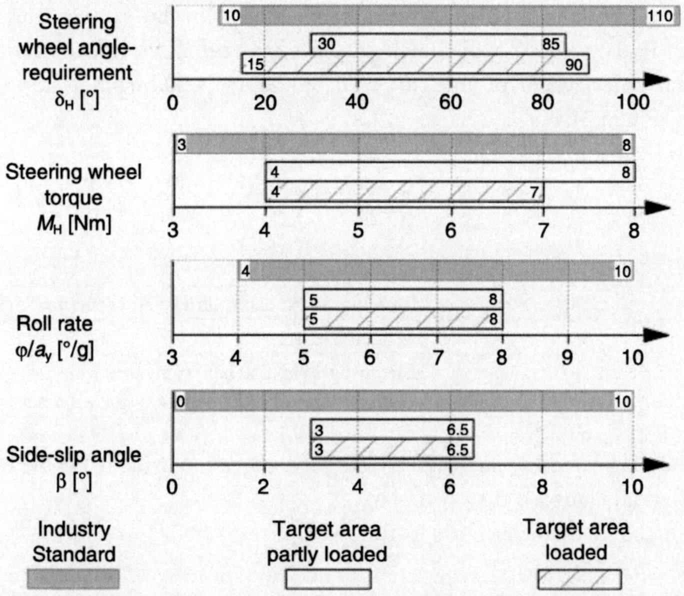

Fig. 5.2 Passenger car target values for cornering, according to [3]. Steady-state cornering with 100 m radius

5.2 Terms

Basically, there are different ways to influence the direction of a vehicle. On multi-track vehicles with pneumatic tyres, there are turntable steering, articulated steering and stub axle steering. The first two types have the disadvantage that the footprint is reduced during steering and disturbing forces act on a lever arm that corresponds to half the track width. In addition, the front or rear wheels as well as all wheels can be steered. For fast vehicles, however, one type of construction alone has prevailed: Ackerman steering on the front axle. Therefore, only this type will be considered in the following.[3] The axis of rotation of the wheel carrier or "steering knuckle" in relation to the wheel suspension (e.g. a "king-pin") is generally invariable during the pure steering process (pure rotary movement of the wheel carrier); however, there are now also wheel suspensions with a variable axis of rotation ("virtual" axis of rotation) on passenger cars.

5.2.1 Steer Angle, Ackermann Angle

The main chassis dimensions wheelbase and track width have a direct influence on the steering angle requirement if the turning circle diameter is desired.[4]

If a vehicle drives a corner very slowly, i.e. strictly speaking without lateral force, then all wheels must be tangential to circular paths that have the same center point. This consideration leads to the so-called Ackermann condition. These ideal steering angles at the inner and outer wheel of the curve according to Ackermann follow the relation according to (Fig. 5.3):

$$\cot(\delta_{A,o}) = \cot(\delta_i + \frac{j}{l}) \tag{5.1}$$

δ_i, $\delta_{A,o}$	Steering angle according to Ackermann, °. (See Fig. 5.3)
L	Wheelbase, mm
$j = b_f - 2r_\sigma$	Steaaring axis distance, mm. (see Fig. 5.4) If the scrub radius is negative, the sign becomes positive

The difference in steering angles between inside (i) and outside (o) is the toe-out angle (steering differential angle): $\Delta\delta_A = \delta_i - \delta_{A,o}$.

The turning circle diameter D_S is as follows:

[3] For a discussion of all-wheel steering, see Sect. 5.7.

[4] See also Racing Car Technology Manual, Vol. 2 *Complete Vehicle*, Chap. 2 *Concept* especially Fig. 2.7.

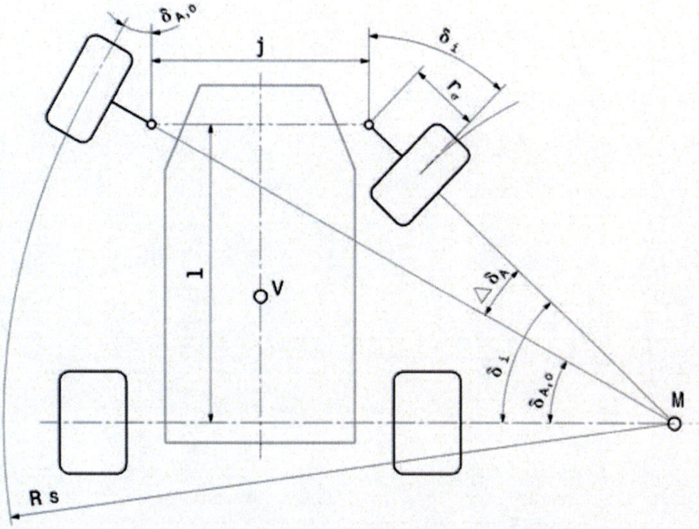

Fig. 5.3 Kinematic relationships during cornering (Ackermann). Vehicle behaviour with pure rolling around the instant center M. *j* distance of steering axes on road, (see Fig. 5.4), $\delta_{A,o}$ Steering angle outer wheel, δ_i Steering angle inner wheel, $\Delta\delta_A$ Ackermann angle, R_S turning circle radius l Wheelbase, V vehicle centre of gravity

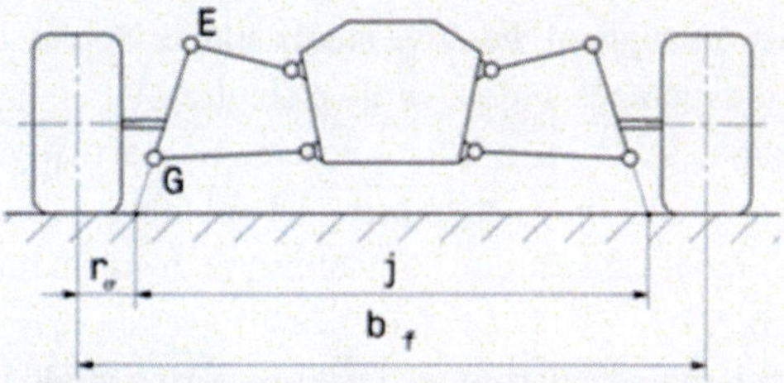

Fig. 5.4 Route designations for (Fig. 5.3). EG Kingpin inclination axis (steering axis), *j* Steering axis distance on the roadway, b_f Front track width, r_σ Scrub radius, in the representation positive scrub radius

$$D_S = 2R_S = 2\left(\frac{l}{\sin(\delta_{A,o,\max})} + r_\sigma\right) \tag{5.2}$$

D_S	Turning circle diameter, mm
$\delta A_{,o,max}$	Largest steering angle of the outer wheel, mm

It can be seen that for manoeuvrable vehicles the wheelbase must be small and the steering angle large. The turning angles are limited by the design of the wishbones (Sect. 4.

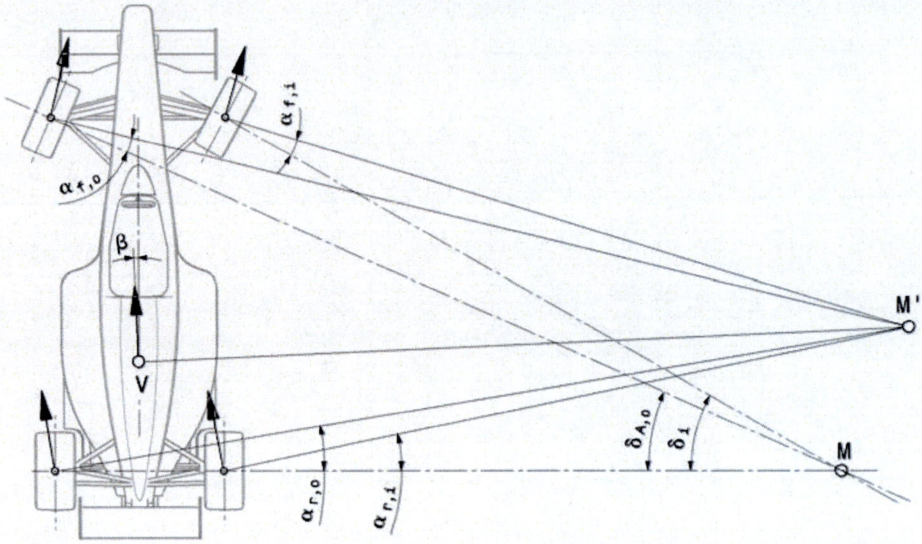

Fig. 5.5 Cornering under lateral acceleration. $\alpha_{f,r}$ Front or rear slip angle, inside or outside corner, β attitude angle, M instantaneous centre acc. to Ackermann, M′ actual instant center

2.1 *Double Wishbone Axle*), by the space required for compression/rebound with closed wheels and by the deflection angle of the drive shafts with driven axles. A short wheelbase has driving dynamic disadvantages and is hardly a solution.

When a vehicle negotiates a corner, lateral acceleration actually occurs and the tyres must build up a slip angle so that they can transmit lateral forces. The instant center around which the vehicle rotates moves forward in comparison to the ideal instant center according to Ackermann, (Fig. 5.5).

In the case of powerful engines, part of the lack of steering angle is compensated for by the accelerator pedal, i.e. by large slip angles on the rear axle (power oversteer, load oversteer). However, a certain steering angle is needed for vehicles where swerving cars have to be counter-steered, such as rally and raid vehicles. In this case, variable steering ratios also pay off, allowing the driver to achieve the required large wheel angle more quickly (i.e. without having to change the steering).

Further considerations for maximum turn-in result from the actual behavior of the tires when cornering. In fast turns, an Ackermann design leads to the outer slip angles being smaller than the inner ones. This means that the lateral force potential is not fully exploited, because the higher wheel loads on the outside of the turn would allow more lateral force. If the outer wheel receives more steering input than the inner wheel (Fig. 5.6), the steering responds more quickly and forces a larger slip angle on the more heavily loaded outer wheel. The lateral control of the front axle can be increased in this way (cf. Sect. 2.2.2 *Development Objectives*). However, this advantage only has an effect in fast bends; in tight bends there is hardly any possibility of fully exploiting the lateral control force of the tyres.

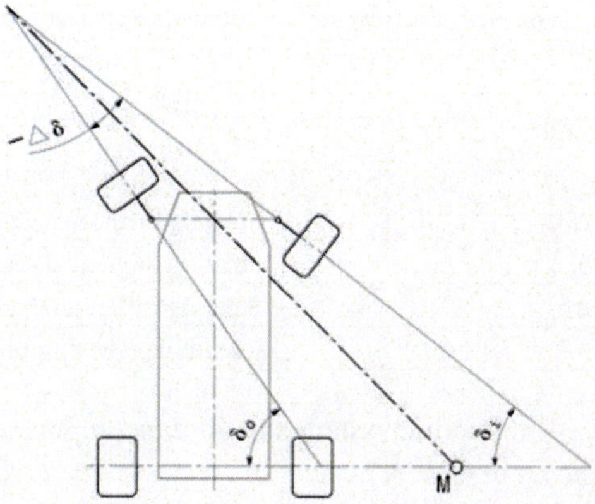

Fig. 5.6 Increasing the lateral guidance force at the front in wide bends, i.e. at low steering angles. The wheel on the outside of the corner is turned in more than the wheel on the inside of the corner. The steering differential angle Δδ is therefore negative

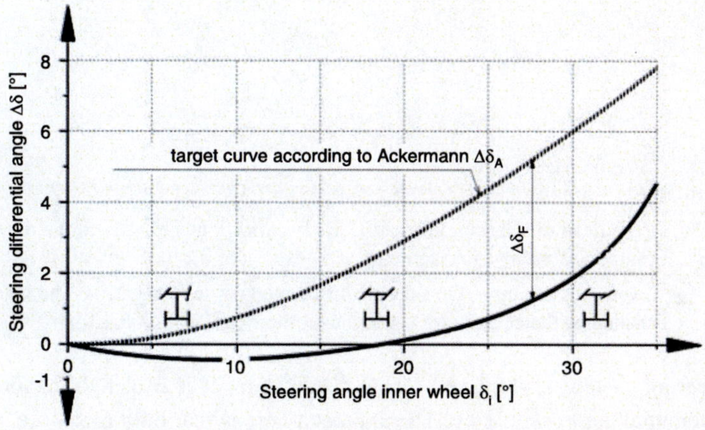

Fig. 5.7 Course of an ideal track difference angle Δδ. The diagram shows the course of the toe-out angle plotted against the steering angle of the inside wheel δ_i . In addition, the course of the toe-out angle $\Delta\delta_A$ according to Ackermann is plotted. The resulting steering deviation $\Delta\delta_F$ is plotted for a steering angle of. At low steering angles, the outside wheel turns in more than the inside wheel. In between there is a transition area where the wheels are parallel and at large steer angles the course approaches up to about half of the Ackermann curve

This effect should therefore only have an influence up to corner radii of $R = 20$ m [2], which corresponds to a steering angle of 5–10° depending on the vehicle. For larger steer angles, the actual curve of the toe-out angle should approach the target curve (Ackermann angle) again, so that tyre wear and rolling resistance in tight turns are reduced. Figure 5.7 shows such a curve of the steering differential angle.

The steering angle can therefore be the same at low steering angles on the inside and outside wheels (parallel steering, i.e. toe-out angle is zero). This allows better use of the available space. Parallel steering can be a useful compromise for road courses [4].

The deviation from the ideal Ackermann angle is called steering error:

$$\Delta\delta_F = \delta_o - \delta_{A,o} = \Delta\delta_A - \Delta\delta \tag{5.3}$$

$\Delta\delta_F$	Steering error, better: intentional steering deviation, °
$\delta_{i\ resp.\ o}$	Steering angle inside or outside
$\Delta\delta_A$	Steering differential angle according to Ackermann, °. $\Delta\delta_A = \delta_i - \delta_{A,o}$
$\Delta\delta$	Intentional steering differential angle, °. $\Delta\delta = \delta_i - \delta_o$

The steering system should thus produce a course of the steering angle difference as shown in (Fig. 5.7).

Sometimes the deviation from the Ackermann interpretation is also expressed as a percentage:

$$\text{Percent Ackermann} = \frac{\Delta\delta}{\Delta\delta_A} \cdot 100\% \tag{5.4}$$

0% Ackermann:	Parallel steer wheels
100% Ackermann:	The desired steering differential angle corresponds exactly to the differential angle according to Ackermann
$-x$ % Ackermann:	Anti-Ackermann – the steering differential angle is negative. The wheel on the inside of the corner turns in less than the wheel on the outside

If the steering system is designed for a specific tyre, it is easier to decide what the steering differential angle should be. The approach here is that both tires – i.e. inside and outside the turn – ideally build up their maximum lateral guidance force $F_{W,Y}$ at the same time when cornering in order to achieve the greatest lateral acceleration. Depending on the tyre design, the maximum of $F_{W,Y}$ shifts towards larger or smaller slip angles with increasing wheel load $F_{W,Z}$ (Fig. 5.8). In order for both tyres to reach the maximum of $F_{W,Y}$ at the same time despite the wheel load shift, the steering must compensate for this influence (Fig. 5.9).

In the case of tyre A (red), the inside wheel reaches the maximum lateral force $F_{W,Y}$ at a larger slip angle ($A_i = 7.7°$) than the outside wheel ($A_o = 6.6°$). A slight Ackermann design can now ensure that the inside cornering wheel has a larger slip angle. This allows both tires to build up their maximum lateral force at the apex of the corner. The opposite is true for tire B (blue). Ideal design of steering is, corner inner wheel will hit less than outer wheel (anti-Ackermann). Then at inner wheel results smaller slip-angle than at outer wheel. With this constellation, both tires of the steered axle can reach their maximum lateral guidance at the same time. The actual conditions are more complicated because, among other things, the camber is neglected in this consideration.

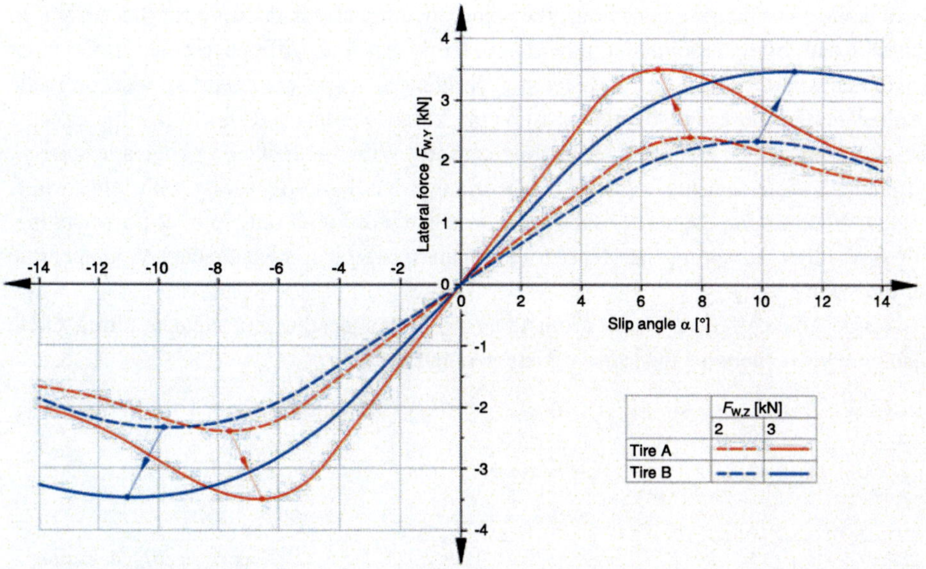

Fig. 5.8 Comparison of the lateral force characteristics of two tyres. For tire A (red), the slip angles of the maxima shift from $F_{W,Y}$ to lower values with increasing wheel load $F_{W,Z}$. Tire B (blue) shows the opposite behavior

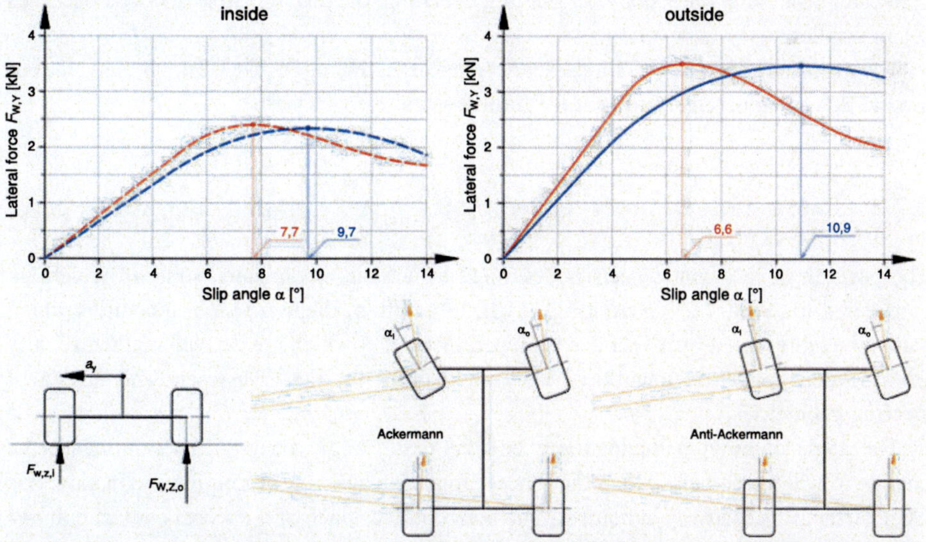

Fig. 5.9 Influence of the tyre on the choice of steering design. The tyre characteristics are taken from (Fig. 5.8). For reasons of clarity, only those parts of the tyre curves are shown that are needed. Due to the wheel load shift during cornering, the wheel loads on the inner and outer wheel are different. Therefore, the curves for lower wheel loads are shown for the inner wheels, and those for higher contact forces are shown for the outer wheels

In addition to the tyre behaviour, the wheel load transfer is decisive for this design. In vehicles with high aerodynamic downforce, the wheel load difference between the inner and outer wheel is smaller. The steering differential angle can therefore also be made smaller. Furthermore, a distinction must be made between fast and slow turns. In fast turns the steering angle is smaller than in slow corners. With small steering angles, a steering differential angle that increases with wheel angle has less effect than with large wheel angles. When tuning the car for such turns, you can work with static pre- or post-tracking. For slow turns, an appropriate (anti-)Ackermann design can advantageously compensate for the static setting.

A side effect of a deviation from the Ackermann design is a smaller turnig circle diameter with otherwise the same vehicle geometry [2]:

$$D_S = 2\left(\frac{l}{\sin(\delta_{A,o,\max})} + r_\sigma\right) - 0.1 \cdot \Delta\delta_F \tag{5.5}$$

$\Delta\delta_F$	Deliberate steering deviation, °
D_S	Turning circle diameter, m
r_σ	Scrub radius, m
l	Wheelbase, m

It can be seen that for every 1° of steering deviation, the turning circle diameter becomes 0.1 m smaller.

The maximum steer angles for passenger cars are in the range of 45–50°, while values of around 20° are generally sufficient for formula cars.

5.2.2 Parameters of Steering Geometry

The position of the front wheels is described by various parameters, some of which are considered in (Sect. 2.2 *Terms*) (Fig. 5.10). In addition, there are other quantities which cannot be represented by a visible design parameter, but which are derived mathematically from several geometric quantities. They are helpful for the assessment and design of steering geometries.

The angle of kingpin inclination σ and the caster angle τ significantly influence the change in wheel camber ε when the wheel turns. The steering or kingpin inclination axis EG intersects the roadway at point A. The horizontal distance of the wheel contact point W from A in the rear view is called the scrub radius r_σ, although the wheel contact point W generally does not revolve with this radius during steering because the actual, spatial distance between the two points is greater. In the side view, the caster becomes visible. The axis of kingpin inclination is inclined with the caster angle τ against the vertical. The wheel center does not have to be on the projection of the steering axis, but can be shifted

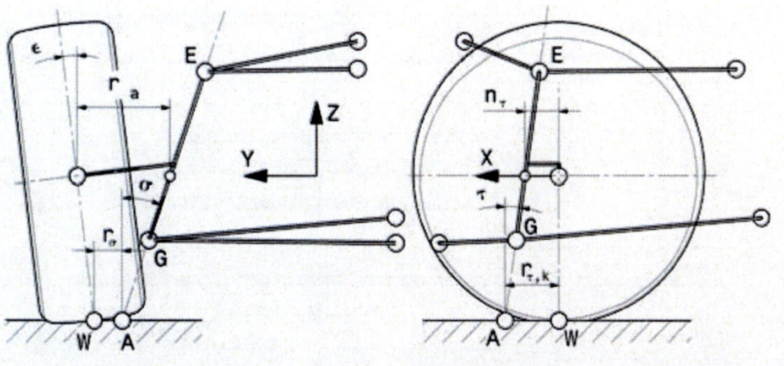

Fig. 5.10 Characteristics of steering geometries, according to [5]. View from behind (left, transverse plane YZ) and view from the side (right, longitudinal plane XZ)). σ king pin inclination angle, τ Caster angle, r_σ scrub radius, steering offset at ground, kingpin offset at ground, r_a kingpin offset at hub, wheel centre offset, $r_{\tau,k}$ Trail, castor offset at ground, n_τ castor trail at wheel centre

forward (positive *X direction*) or backward (negative *X direction*) by a caster offset n_τ . The distance between points W and A in the side view is called caster distance $r_{\tau,k}$. If a negative caster offset n_τ (i.e. in the negative *X direction*) is provided, the caster distance $r_{\tau,k}$ is reduced by this amount and the camber change during steering becomes more favourable.

Corresponding to the caster offset, a wheel-center offset *r* is designated$_a$. This is the horizontal distance of the wheel centre from the kingpin inclination axis as viewed from the rear (in the *X direction*). This distance is also referred to in the literature[5] as the longitudinal, interference or drive force lever arm, because all forces coming from the tyre in the case of a purely rolling wheel are transmitted via the wheel bearings at the wheel centre to the wheel carrier and thus also to the steering.

Figure 5.11 shows target values for some parameters from passenger car development.

With the above geometric quantities, which describe the wheel position, the effects of tyre forces can be calculated, Fig. 5.12.

If a braking force $F_{W,X,b}$ acts on the tyre in the X-direction, this together with the scrub radius provides a spatial moment about the Z-axis [5]:

$$M_{A,Z,b} = F_{W,X,b} \cdot r_\sigma \tag{5.6}$$

$M_{A,Z,b}$	Moment of braking force about the intersection point A of the steering axis, N m
	This moment turns around the Z-axis and not around the steering axis
$F_{W,X,b}$	Braking force of a tyre, N
r_σ	Scrub radius, m

[5] However, it is not used universally: sometimes it also refers to the actual distance between the centre of the wheel and the steering axle, as in this book, see Sect. 2.2.1 *Terms*.

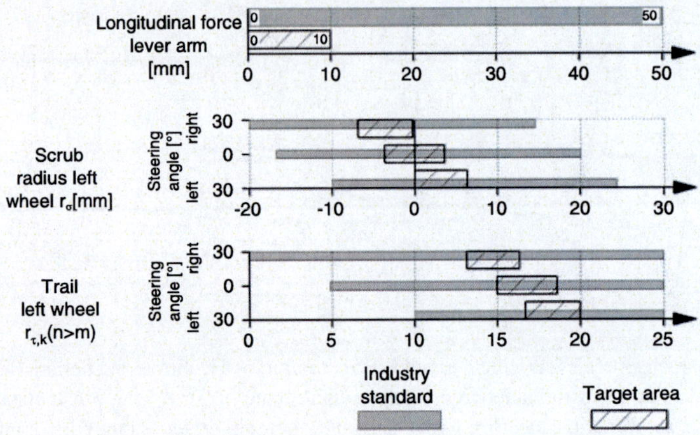

Fig. 5.11 Target ranges of characteristics of passenger car front axles, according to [3]

Fig. 5.12 Forces on the front tyre. Shown is a left front tire. *W* Wheel contact point, *A* Intersection of the kingpin inclination axis with the roadway, *EG* Steering or kingpin inclination axle, δ Steering angle, $F_{W,X,b}$ Braking force, $F_{W,Y}$ Lateral force, $F_{W,Z}$ Wheel load, $r_{\tau,k}$ castor trail in wheel centre, $r_{\tau,T}$ pneumatic trail

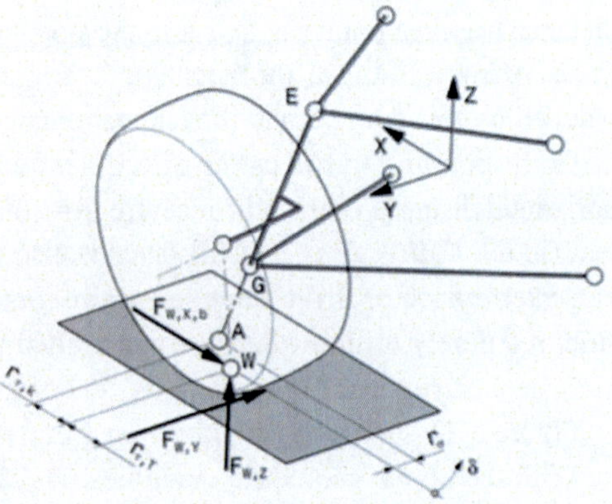

The moment about the kingpin inclinatioin axis results from the projection of the moment vector onto it:

$$M_{A,b} = F_{W,X,b} \cdot r_\sigma \cdot \cos\sigma \cdot \cos\tau \tag{5.7}$$

$M_{A,b}$	Moment of braking force about the steering axis, N m

It can be seen that as the scrub radius r_σ increases, the moment of the braking force increases. For this reason, the scrub radius should be as small as possible. In this way, different friction conditions do not have such a strong effect on the steering during braking. In passenger cars, negative scrub radii (i.e. r_σ points outwards from the wheel contact point W) are also used with the aim of generating a stabilizing countersteering effect in the case of brakes acting on unequal sides. Geometrically, a negative scrub radius means that the pivot points E and G of the wheel carrier must move further towards the centre of the wheel and thus the brake disc must be moved further outwards. With narrower drop-centre rims, this can result in a loss of brake disc diameter of around 25 mm if the rim diameter remains the same.

Corresponding to the braking force, a lateral force $F_{W,Y}$ acts over the longitudinal distance between the force and the point of impact A of the steering axis with the road surface. The total distance results from the sum of the kinematic trail (caster trail distance $r_{\tau,k}$) and the pneumatic trail $r_{\tau,T}$.

Because the steering axis is not perpendicular to the road, the distance between the centre of the wheel and the road inevitably changes. This causes the front end to rise or lower when steering. The wheel load therefore has an influence on the steering torque that must be applied by the driver. This phenomenon is measured mathematically by the wheel-load lever arm q. The wheel-load lever arm referred to the *Z-axis* (at point A) is:

$$q = r_\sigma \cdot \tan\tau + r_{\tau,k} \cdot \tan\sigma \tag{5.8}$$

q	Wheel load lever arm, mm
Angles and lengths: (see Figs. 5.10 and 5.12)	

The moment of the wheel load around the *Z-axis* results directly from this to:

$$M_{A,z} = qF_{W,z} \tag{5.9}$$

$M_{A,z}$	Moment about *Z-axis*, N mm

Due to the spatial inclination, the moment $M_{A,EG}$ of the wheel load about the steering axis EG differs from $M_{A,z}$ due to the included angle:

$$M_{A,EG} = q\cos(\beta)F_{W,z} \tag{5.10}$$

$M_{A,EG}$	Moment about the steering axis EG, N mm
β	Angle between steering axis and Z-axis, ° $\tan(\beta) = \sqrt{\tan(\sigma)^2 + \tan(\tau)^2}$

A wheel load does not exert a moment about the steering axis if the steering axis is vertical or the vertical force intersects the steering axis.

The wheel load lever arm q is defined positively if the moment generated by the wheel load has a restoring effect, i.e. it attempts to reduce the amount of the steering angle δ. Because the wheels are moved to the straight-ahead position by the wheel load in that case, the steering is also said to be weight-resetting. This definition of a positive wheel load lever arm means that the wheel load has a restoring effect (weight jacking) as soon as q and δ have the same sign.

The wheel load lever arm can also be viewed as changing the elevation of the front end above the steering angle:

$$q = -\frac{dz}{d\delta} \tag{5.11}$$

Thus, when the wheel load lever arm q *is* positive, a steering movement with a positive steering angle δ (which is the inside turn of the wheel) results in a lift of the car body.

The wheel load lever arm should be as small as possible so that fluctuations in the wheel load do not have a disturbing effect on the steering.

However, the weight jacking is practically only significant in the straight-ahead position. When driving fast and with larger steering angles, the self-aligning moments of the lateral forces are considerably larger, Fig. 5.13.

In general, the steering angles are smaller at high speeds than in slow corners. Therefore, in this context, it does not matter what the toe-out angle of the front wheels is, e.g. Ackermann design or parallel steering. Due to the side slipping of all wheels, the instant centre M moves forward. The torque felt on the steering wheel is primarily generated by the lateral forces $F_{W,Y,f}$ from side slipping. The restoring effect of the wheel loads, on the other hand, is small. In contrast to the lateral camber forces $F_{T,Y,\varepsilon}$, the lateral slip forces act around the tyre trail $r_{\tau,T}$ behind the wheel contact point W and are greater on the outside of the corner because of the wheel load shift to the outer wheels. If the front axle is driven, the driving forces are added to the forces in Fig. 5.13, which act via the deflection force lever arm (kingpin offset) r_a to rotate around the steering axis. The self-aligning torque on the steering gear is calculated with all forces [5]:

$$M_{Sg} = \left[F_{W,Yf,o} \cdot \left(r_{\tau,kf,o} + r_{\tau,Tf,o}\right) - F_{T,Y,\varepsilon f,o} \cdot r_{\tau,kf,o} - F_{W,Zf,o} \cdot q_{f,o} - F_{W,X,af,o} \cdot r_{af,o}\right] / i_{T,o} + \left[F_{W,Yf,i} \cdot \left(r_{\tau,kf,i} + r_{\tau,Tf,i}\right) + F_{T,Y,\varepsilon f,i} \cdot r_{\tau,kf,i} + F_{W,Zf,i} \cdot q_{f,i} + F_{W,X,af,i} \cdot r_{af,i}\right] / i_{T,i}$$

$$M_H = M_{Sg} / i_{Sg} \tag{5.12}$$

M_{Sg}	Torque at steering gear with rotary movement, N mm
$r_{\tau,k}$	Kinematic caster, mm
$r_{\tau,T}$	Pneumatic trail (tyre caster), mm
$F_{W,Y}$	Lateral forces due to tyre side slip, N
$F_{T,Y,\varepsilon}$	Lateral forces due to tyre camber, N
$F_{W,Z}$	Wheel loads, N
$F_{W,X,a}$	Driving force at one wheel, N
r_a	Driving force lever arm, mm. (see Fig. 5.10)

(continued)

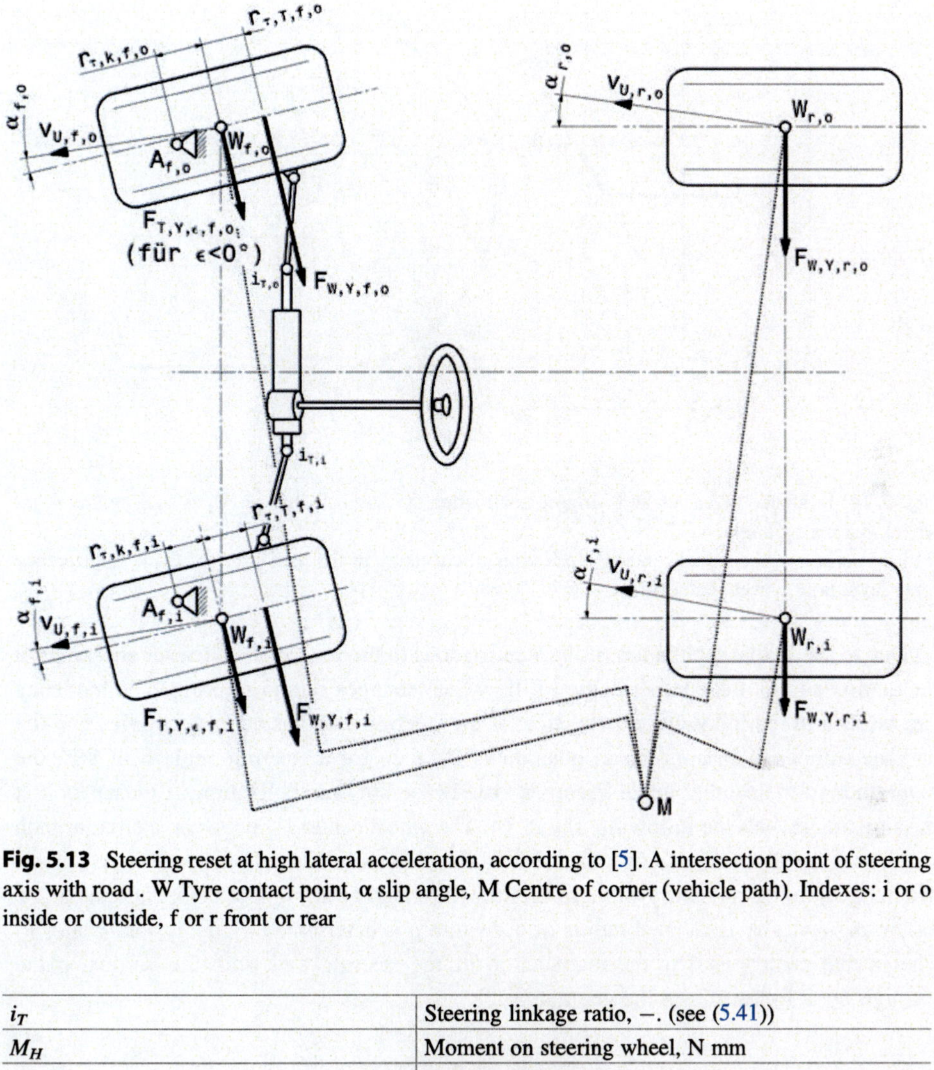

Fig. 5.13 Steering reset at high lateral acceleration, according to [5]. A intersection point of steering axis with road . W Tyre contact point, α slip angle, M Centre of corner (vehicle path). Indexes: i or o inside or outside, f or r front or rear

i_T	Steering linkage ratio, –. (see (5.41))
M_H	Moment on steering wheel, N mm
i_{Sg}	Steering gear ratio, –

For most wheel suspensions, the kinematic caster $r_{\tau,k}$ increases above the turn-inside steering angle δ and decreases above the turn-outside steering angle. As lateral acceleration increases, the pneumatic trail (tyre caster) $r_{\tau,T}$ decreases. As a result, the influence of the turn-outside lateral force $F_{W,Y,f,o}$ becomes smaller and smaller, and it may even eventually have the opposite effect, i.e. the lateral force wants to increase the magnitude of the steering angle. However, this generally does not have the same effect because the steering linkage ratio increases when the wheel is turned on the outside of the bend as opposed to the inside of the turn. Thus, the contribution to steering torque of the outside cornering tire decreases relative to the inside cornering tire.

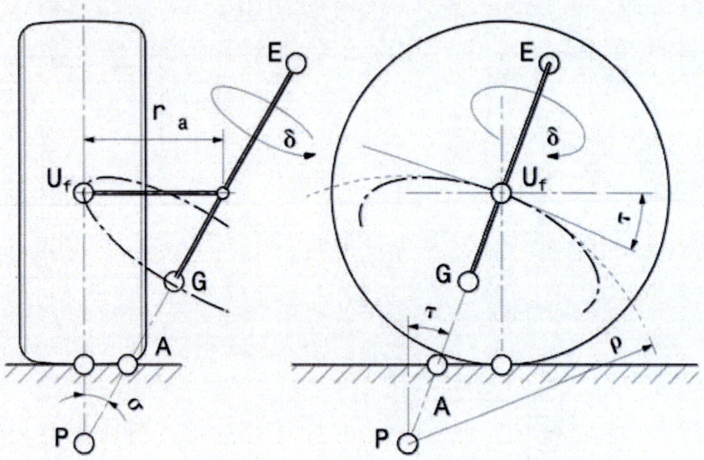

Fig. 5.14 Influence of caster and kingpin inclination on camber change. U_f Wheel centre Front wheel, δ Steering angle
P Instant center of the wheel center U_f, ρ Radius of curvature of the path curve of U_f, r_a Interference force lever arm, wheel center offset

Due to the spatial inclination of the steering axis to the road surface (caster and kingpin inclination angle), the camber angle of the wheel changes during steering. The tendency can be seen in the following observation. If the steering axis has no caster angle ($\tau = 0°$) and the camber in straight-ahead position is 0°, then for a steering angle δ of 90°, the magnitude of the camber angle becomes exactly the kingpin inclination. A closer look at the situation reveals the following, Fig. 5.14. The wheel center U_f moves on a circular path when steering about the steering axis EG. In the side view this path appears as an ellipse. The tangent to this trajectory curve in the straight-ahead position ($\delta = 0°$) is inclined at the caster angle τ. The associated radius of curvature ρ is determined by the instant center *P*. The instant center itself is the intersection of the steering axis with the vertical plane through U_f. It follows from the geometry:

$$\rho = r_\sigma / (\tan \sigma \cdot \cos \tau)$$

The curvature of the curve $\varepsilon(\delta)$ is therefore proportional to the kingpin inclination angle σ. A positive kingpin inclination bends the curve towards positive camber angles when steering is applied.

The rate of change of camber versus steering angle depends on the caster and kingpin inclination angles [5]:

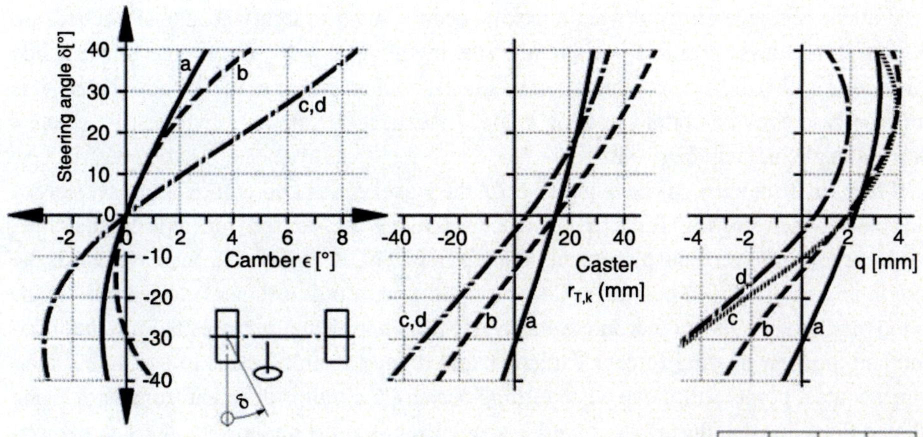

	σ	τ	r_σ	$r_{\tau,k}$
a	5°	3°	50	16
b	12°	3°	0	16
c	12°	9°	15	5
d	12°	9°	0	5

ε(δ=0°)= 0°

Tyre radius 300 mm

Fig. 5.15 Course of camber, caster and wheel load lever arm q during steering, according to [5]

$$\frac{d\varepsilon}{d\delta} = \frac{\tan\tau \cdot \cos\delta + \tan\sigma \cdot \sin\delta}{\tan\varepsilon \cdot (\tan\tau \cdot \sin\delta - \tan\sigma \cdot \cos\delta) + 1} \tag{5.13}$$

ε	Camber angle, °
δ	Steering angle, °

Figure 5.15 compares the effects of several different steering geometry designs.

Version a has a low kingpin inclination and low caster and therefore has larger distances to the wheel contact point, i.e. scrub radius r_σ and kinematic caster $r_{\tau,k}$. Variants c and d have the largest angles for kingpin inclination and caster and differ only in the scrub radius.

The tangent of the camber curve at steering angle = 0° of the versions c and d is much flatter than that of the other curves. The inclination of the tangent of the curves a and b is about three times larger, which corresponds to the ratio of the caster angles τ. The larger kingpin inclination of variant b curves the curve much more than that of variant a. This leads to the fact that from a steering angle of −30° (i.e. outside the turn) the camber becomes positive. The caster becomes larger on the inside of the corner for all versions. Likewise, the caster decreases on the outside of the corner to negative values, thus achieving a sort of negative caster trail.

In all variants, the effect of weight jacking occurs in the straight-ahead position because the wheel load lever arm q is positive at a steering angle $\delta = 0°$. In the case of the inside wheel, this is also the case at other steering angles, but in the case of the outside wheels, it is only the case above a certain steering angle. As expected, version a requires the greatest steering angle, namely $\delta = -30°$.

If the trail distance $r_{\tau,k}$ is not equal to zero, the wheel contact point moves transversely to the vehicle when steering. If the trail distances of both wheels of the front axle are the same, this leads to a transverse displacement of the front end of the vehicle. In any case, this is the case in all straight-ahead positions. Conversely, different trail distances cause both wheels to move relative to each other in the transverse direction. This increases the tire deformation and thus the steering forces. Variants b and d have a scrub radius of 0 mm, but their wheel contact points still move when turning because the trail is different from zero. If one wanted to realize that during steering the wheel actually turns on the spot, the steering axis would have to intersect the road at the wheel contact point W. Then both the scrub radius and the caster would be zero.

5.2.3 Steering Ratio

Once the required maximum steering angle has been determined, the necessary transmission ratio between the steering wheel and the front wheels must be determined. Laws for road vehicles and regulations for racing vehicles require (to this day) a permanent mechanical connection between the steering wheel and the steered wheels. The steering movement is transmitted from the steering wheel to the vehicle wheels via the steering linkage (tie rods, steering push rods, etc.), which is actuated by a steering gear. The latter has an internal gear ratio i_{Sg}, in order to reduce the steering wheel forces. The steering linkage also has a ratio i_T between the steering gear and the vehicle wheels that generally varies with the steering angle.

The kinematic steering ratio i_S from the steering (hand) wheel to the vehicle wheels is calculated from the steering wheel angle δ_H and the wheel lock angles to, Fig. 5.16:

$$i_S = \delta_H / \delta_m \tag{5.14}$$

i_S	kinematic steering ratio, –.
δ_H	Steering wheel angle, °.
δ_m	Mean steering angle of the wheels, °. $\delta_m = (\delta_o + \delta_i)/2$

The gear ratio will generally not be constant over the entire steering range. Therefore, the above equation only applies to angular ranges of steering wheel and steering angle. The overall steering ratio is limited at the lower end by the directness of the steering response at high driving speeds; here values below 14 are rare for passenger cars. The upper limit

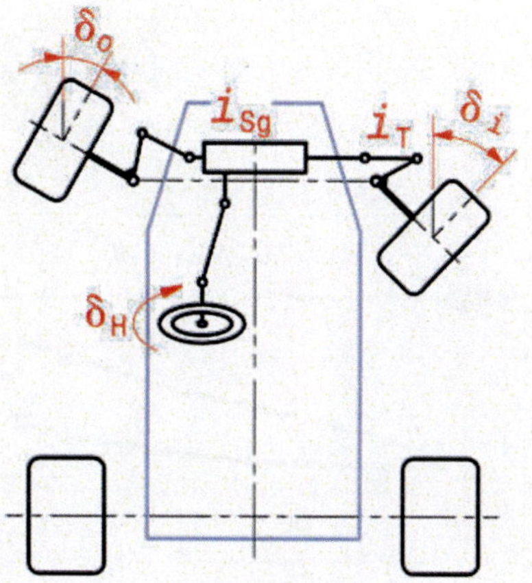

Fig. 5.16 System overview Ackerman steering. The steering-wheel angle δ_H at the steering wheel is converted into steering angle δ_o or δ_i of the wheels outside or inside via the steering gear with the inner transmission ratio i_{Sg} and via linkage with transmission ratio i_T

results from the amount of steering effort that can reasonably be expected in the parking area; this is thus directly related to the presence of power assistance and hardly ever exceeds a ratio of 20. In terms of design, the total steering ratio corresponds to the product of the steering linkage ratio and the steering gear ratio:

$$i_S = i_T \cdot i_{Sg}$$

i_T	Steering linkage ratio, (see Sect. 5.5 *Transmission*)
i_{Sg}	Steering gear ratio, (see Sect. 5.4 *Steering Gear*)

For the steering linkage ratio, the mean value of the outer and inner steering angle must be taken into account. With known effective track levers (projection of the levers in a plane normal to the kingpin inclination axis), it can be determined from the ratio of track lever to pitman arm.

The kinematic steering ratio differs from the real conditions in that all transmission links have elasticities and clearances. This means that the steering wheel can be turned without the wheels moving. How large such deviations can be when stationary is illustrated by a measurement on a passenger car with rack-and-pinion steering, Fig. 5.17.

The actual effective steering ratio, which the driver notices when driving, is the so-called dynamic steering ratio and results from the kinematic ratio by superimposing the compliance of the transmission links:

$$i_{dyn} = i_s + \frac{\Delta\delta_{H,e}}{\Delta\delta_H} \tag{5.15}$$

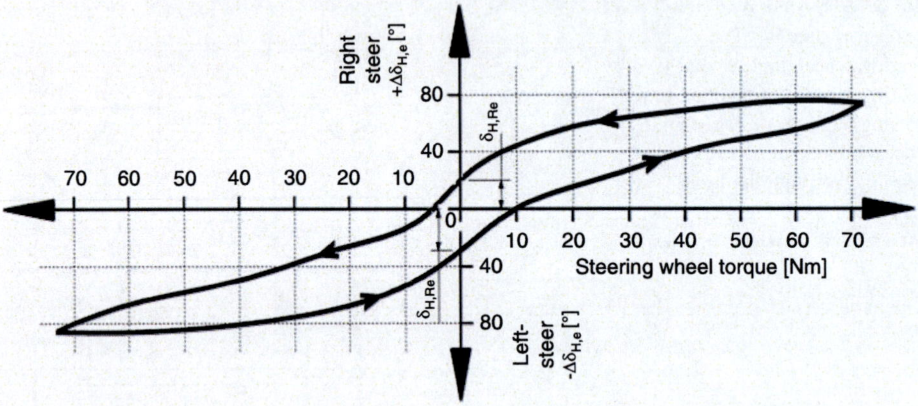

Fig. 5.17 Elasticity measurement on a car steering system at standstill, according to [2]. The wheels were held in place during the measurement and a moment was applied to the steering wheel. As expected, the compliance $\Delta\delta_{H,e}$ at the steering wheel increases with increasing torque. However, the resistance, i.e. the stiffness of the steering increases and the curve becomes flatter. The steering wheel is turned to the right and left. Hysteresis occurs in this process and residual angles $\Delta\delta_{H,Re}$

i_{dyn}	Dynamic steering ratio, –
$\Delta\delta_{H,e}$	Elastic compliance on the steering wheel, °
$\Delta\delta_H$	Steering angle range at the steering wheel at which $\Delta\delta_{H,e}$ occurs, °

So what the driver feels is an increase in steering ratio due to elasticities in the steering system. The steering wheel actually has to be turned further for a certain steering angle of the wheels than would theoretically be necessary. As the steering torque increases – e.g. due to aerodynamic downforce or off-road – the amount of compliance increases.

The interaction of several link chains in passenger car steering systems with rack-and-pinion transmissions also leads to unwanted changes in the kinematic steering ratio over the steering angle range. Naturally, front-wheel drive vehicles, with their restricted space due to the engine and transmission, perform worse here than vehicles with standard drive. For example, the ratio drop from the straight-ahead position to the full steering angle of one side is between 17 and 30% for front-wheel drive vehicles, while the drop is only 5–15% for rear-wheel drive vehicles [2].

Table 5.1 lists typical steering ratios.

In general, the basic design of the steering ratio is based on the specification of how the vehicle should react to steering inputs around the centre position. In the road test, the

Table 5.1 Steering ratios of different vehicles

Touring car 1998 [6]	13:1 to 16:1	Kurek GT 6, self-built sports car	0.75 revolutions from lock to lock
Formula 1 (max. approx. 20° steering angle at the wheel), depending on track and driver's preference	Unsupported 9 to 12:1 [7]. 14:1, i.e. 1.5 turns from lock to lock [8]. 5.4:1 i.e. approx. 0.6 revolutions from lock to lock [9].	Mercedes CLK-C297 Class [10]	12:1 with support, i.e. 1.5 turns from lock to lock
Formula Renault [11] (max. 16° steering angle at the wheel)	approx. 10:1 i.e. 0.9 turns from lock to lock	Mercedes C-Class (DTM '94) [10]	1.5 turns from lock to lock
Passenger car [12]	14 to 20:1; i.e. 4 to 5 turns from lock to lock	Ford Focus WRC06	1.5 turns from lock to lock

steady-state yaw gain[6] (yaw rate gain) is determined as the characteristic variable for this. This value can be represented mathematically with the aid of the single-track model[7]:

$$\frac{\dot{\psi}}{\delta_H} = \frac{1}{i_S} \cdot \frac{v}{l + G_U v^2} \tag{5.16}$$

$\dot{\psi}$	Yaw rate, yaw speed, °/s. Value range racing cars and passenger cars ±50 °/s. Unit of yaw gain factor s^{-1}.
l	Wheelbase, m
G_U	Self-steering gradient, rad s^2/m
v	Driving speed, m/s

The optimum yaw gain range for passenger cars is 0.3 s^{-1} at 80 km/h [1]. Today's passenger cars are designed in such a way that the yaw amplification reaches its maximum between 100 and 120 km/h. The understeer gradient characterizes the self-steering behavior. The understeer gradient characterizes the self-steering behavior and is determined on the circular track. The decisive influencing variables can be identified from the

[6]See Racing Car Technology Manual Vol. 5 *Data Analysis, Tuning and Development* Chap. 6 *Development.*

[7]A single-track model is created from a dual-track model by combining the two tires of an axle into one. Roll is neglected in this model. See e.g. Racing Car Technology Manual Vol. 5 *Data Analysis, Tuning and Development* chap. *tuning*.

observations on the single-track model. The relative position of the yaw moment-generating tires to the vehicle's center of gravity essentially dictates the driving behavior:

$$G_U = m_{V,t} \frac{c_{\alpha,r,eff} l_r - c_{\alpha f,eff} l_f}{c_{\alpha,r,eff} c_{\alpha f,eff} l} \tag{5.17}$$

$c_{\alpha,f,eff}$, $c_{\alpha,r,eff}$	Effective cornering stiffness of the front or rear axle, N/rad
$m_{V,t}$	Total vehicle mass, kg

The effective cornering stiffnesses of the axles can be determined approximately for the design [1]:

$$c_{\alpha f,eff} \approx \frac{c_{\alpha f}}{1 + (r_{\tau,T} + r_{\tau,k}) \frac{c_{\alpha f}}{c_S}}, c_{\alpha,r,eff} \approx c_{\alpha,r} \tag{5.18}$$

c_S	Steering stiffness from tyres to steering wheel, N m/rad
$c_{\alpha,f}$, $c_{\alpha,r}$	Angular stiffness of the front or rear axle, N/rad
$r_{\tau,T}$, $r_{\tau,k}$	Caster trail distances due to tyres and steering geometry, (see (5.12))

The lateral force behaviour of an axle is composed of the behaviour of both tyres. At high lateral accelerations and high centre of gravity the result is not simply the double of a single tyre because of the degressive tyre behaviour, nevertheless for racing vehicles – low centre of gravity and consideration of the linkage behaviour – the following simplification can be made:

$$c_{\alpha f} \approx 2c_{\alpha,Tf}, c_{\alpha,r} \approx 2c_{\alpha,T,r} \tag{5.19}$$

$c_{\alpha,T,f}$, $c_{\alpha,T,r}$	Lateral stiffness (lateral force coefficient) of a tyre on the front or rear axle, N/rad
	Typical values of passenger car tyres are between 1250 and 2500 N/° [1] or 71,620 and 143,240 N/rad
	Values of racing tires are around 1700 to 3900 N/° or 97,400 and 223,450 N/rad

High steering stiffness is critical for precise steering feel. In passenger cars, however, excessive stiffness is avoided because of the resulting bumpiness. The steering stiffness c_S related to the steering shaft results from the series connection of the stiffnesses of the front axle and the steering shaft or the torsion bar in power-assisted steering systems [1]:

$$\frac{1}{c_S} = \frac{1}{c_f} + \frac{1}{c_{ts} i_S^2 A_S} \tag{5.20}$$

c_f	Stiffness of front axle suspension due to elasticities of track rods, track levers and joints, N m/rad
c_{ts}	Torsional stiffness of the steering shaft or torsion bar, N m/rad. A typical torsional stiffness for passenger cars is 2 N m/° or 115 N m/rad
A_S	Steering assistance, ratio *of* a power steering system, –. Without power steering, $A_S = 1$

An electric or hydraulic steering booster reduces the torque to be applied by the driver to the steering wheel M_H, which is caused by the lateral forces on the front tyres:

$$A_S = \frac{1}{i_S} \cdot \frac{M_S}{M_H}, \text{mit } M_S = (F_{W,Y,l} + F_{W,Y,rs})(r_{\tau,T} + r_{\tau,k}) \tag{5.21}$$

M_S	Steering torque caused by lateral forces $F_{W,Y}$ on the front tyres, N m
M_H	Hand torque on steering wheel, N m

Another parameter can be used to check the selected steering ratio: The lateral acceleration achieved in response to a steering input, the steering sensitivity:

$$\frac{a_y}{\delta_H} = \frac{\dot{\psi}}{\delta_H} v \text{ in } (\text{m/s}^2)/\text{rad or } S_H = \frac{a_y}{\delta_H} \cdot \frac{\pi}{180^\circ} \text{ in } (\text{m/s}^2)/^\circ \tag{5.22}$$

S_H	Steering sensitivity, (m/s^2)/° or S_H/g in g/°
a_y	Lateral acceleration, m/s^2

Figure 5.18 plots the calculated yaw gain and steering sensitivity against speed for a Formula 3 car and a passenger car. For the Formula 3 car, this has been done for two steering ratios. Compared with the passenger car, the Formula 3 car exhibits the significantly stronger yaw movement at the same steering wheel angle and correspondingly also the stronger steering sensitivity. The vehicle is delivered with the steering ratio $i_S = 13$.

Steering ratios and largest steering angles of racing vehicles are adapted to extreme conditions depending on the track. In general, the higher the driving speed, the larger the steering ratio. Indy cars with their enormously high continuous speeds have ratios of around 12.5–16, while the value of Formula 3 cars is between 11 and 14.

For the narrow street circuit in Monaco, Formula 1 cars have a maximum steering angle of about 22° [13].

In this context, it is interesting to compare the typical steering movements of a normal driver and a professional rally driver, Fig. 5.19. As expected, both achieve the largest steering angles at low frequencies. From 0.5 Hz, the steering movements of the normal

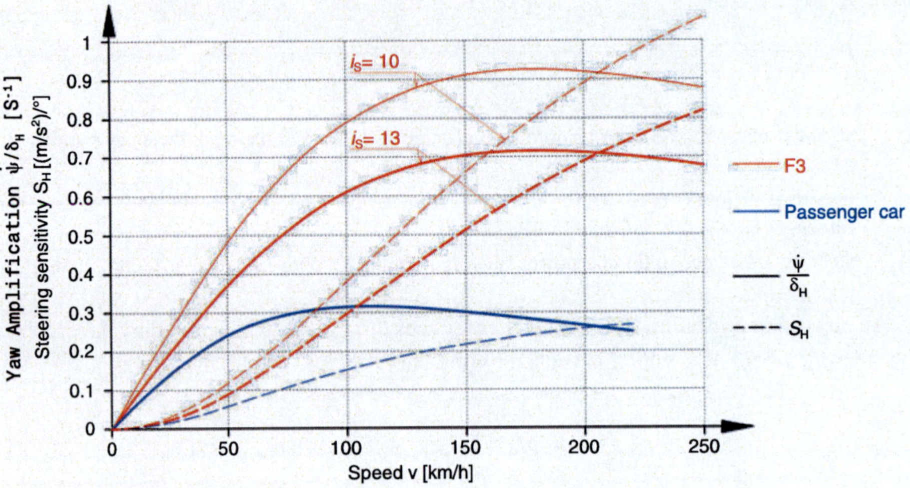

Fig. 5.18 Static yaw gain and steering sensitivity for two vehicles. For a Formula 3 car (red) and a passenger car (blue), the course of these variables is plotted against the driving speed. The car has power steering. In the case of the Formula 3 car, the effects of two steering ratios i_S can be seen. Values Formula 3: $m_{V,t} = 550$ kg, $l = 2.675$ m, axle loads f/r 48:52%, $c_{\alpha,\,T,\,f} = 100{,}268$ N/rad, $c_{\alpha,\,T,\,r} = 157{,}563$ N/rad, $c_f = 30{,}000$ N m/rad, $c_{ts} = 15{,}000$ N m/rad, $A_S = 1$. Car values: $m_{V,t} = 1550$ kg, $l = 2.8$ m, axle loads *f*/*r* 55:45%, $c_{\alpha,\,T,\,f} = 75{,}000$ N/rad, $c_{\alpha,\,T,\,r} = 75{,}000$ N/rad, $c_f = 40{,}000$ N m/rad, $c_{ts} = 115$ N m/rad, $A_S = 2$, $i_S = 17$

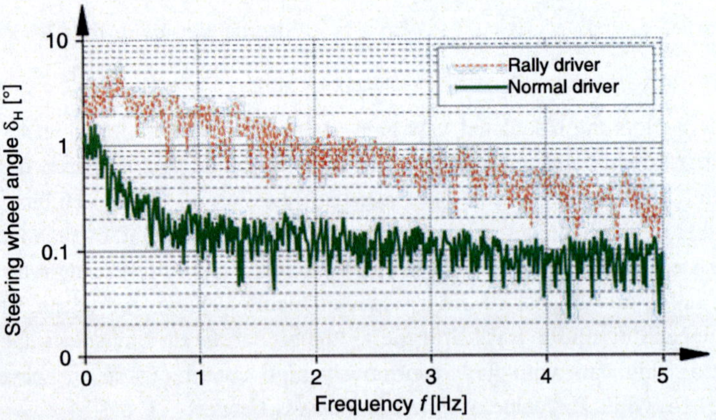

Fig. 5.19 Comparison of steering angle spectra, after [14]. The rally driver is Petter Solberg during the Argentina Rally in the 2002 World Championship

driver are in the range of the background noise. It is noticeable that the rally driver steers about three times as much as the normal driver over the entire frequency range. In the lower frequency range up to approx. 1 Hz the steering wheel angles hardly change, towards higher frequencies the steering inputs decrease until at 5 Hz they almost reach the magnitude of the road driver. In the range around 0.5 Hz, the professional windss the steering wheel seven to nine times further than the passenger-car driver. It is important to mention that the rally vehicle has a steering ratio that is at least twice as fast as that of the passenger car.

Typical steering wheel rotation speeds of normal drivers show maximum values of 160 °/s. In emergency situations, 300–450 °/s are reached [15].

One way of avoiding the compromise of a certain steering ratio that is nevertheless necessary is to use steering gears with variable ratios.

In passenger car construction, these have been standard for a long time, but in racing they have only recently been introduced. However, the development goals are not the same either. In a racing vehicle, the gear ratio must change during a relatively small steering angle, and the gear ratio must be large for small steering angles. In fast-paced corners, i.e. above 200 km/h, the steering angles are small, but the large aerodynamic downforce forces allow for high lateral accelerations, which correspondingly increase the steering forces for the driver at constant steering ratios. Conversely, in slower turns, i.e. around 100 km/h, the steering angles are considerably larger and the downforce smaller [17].

A variable steering ratio thus provides the driver with ergonomic relief, which is particularly noticeable on long routes, Figs. 5.20 and 5.21. However, the behaviour of such a steering system also takes some getting used to for drivers who are trained to use a constant steering ratio. The steering-wheel degree of lock in fast corners is larger compared to conventional steering and can be misinterpreted by the driver as understeer. Likewise, the steering torque to be applied by the driver is lower in fast corners, which can be mistaken by the driver for the vehicle's behaviour with lower tyre grip.

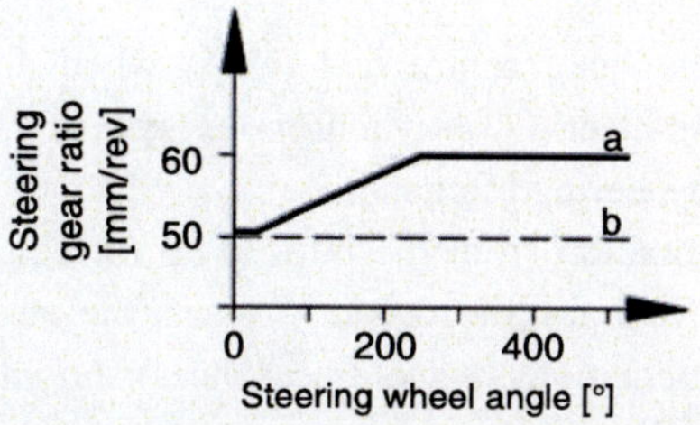

Fig. 5.20 Variable steering ratio of a hydraulically assisted rack-and-pinion steering system on a sporty passenger car (Porsche 911 Carrera), after [16]. (**a**) Model year 2005. At small steering wheel angles, the ratio is similar to that of the predecessor model at 17.1:1. At steering wheel angles of more than 30°, the steering ratio becomes increasingly direct, down to 13.8:1. (**b**) Predecessor model with constant ratio

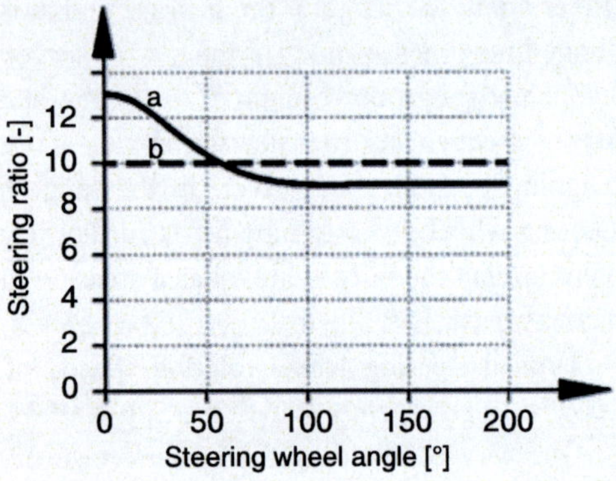

Fig. 5.21 Course of a variable steering ratio over the steering wheel angle for a Formula 1 car, after [17]. (**a**) Variable steering ratio. (**b**) Constant steering ratio

The driver's arms are hardly crossed during the race in standard formula cars and the driver does not have to reach around when steering on standard tracks, i.e. the maximum, ergonomically sensible steering wheel angle results from the two wrists touching.

5.2.4 Steering Assistance

For ergonomic reasons, the maximum hand force must not exceed 250–400 N [18]. In passenger cars, hand force support has long been standard in all segments. This is not the case on racing vehicles. Passenger car steering systems are designed with comfort in mind, and the aim is to achieve a high degree of filtering of disturbing influences (changes in rolling resistance due to road unevenness, fluctuations in braking force, wheel imbalance, etc.) from the tyres. Isolating the steering from road contact, on the other hand, is completely undesirable in competition vehicles, since the feedback behaviour is essential in detecting the grip limit of the front tyres. In addition to the requirements mentioned in the introduction to this chapter, there are the following [1], which primarily concern the control quality of the amplification unit:

- Steering precision: pronounced centre feel (helps when driving straight ahead and steering – without "over-steering"), synchronous system reactions to driver inputs, spontaneous response to steering inputs.
- Steering feedback: Information from the bottom up (driving condition, road surface) should – as far as relevant for the driver – reach the steering wheel. Disturbance variables (brake force fluctuations, wheel imbalances) should be filtered out.
- Steering dynamics: Important for fast steering manoeuvres (slalom, chicane, swerving)
- Steering feel: Harmonious and not synthetic. This is the case when the steering gain increases linearly over lateral acceleration [1].

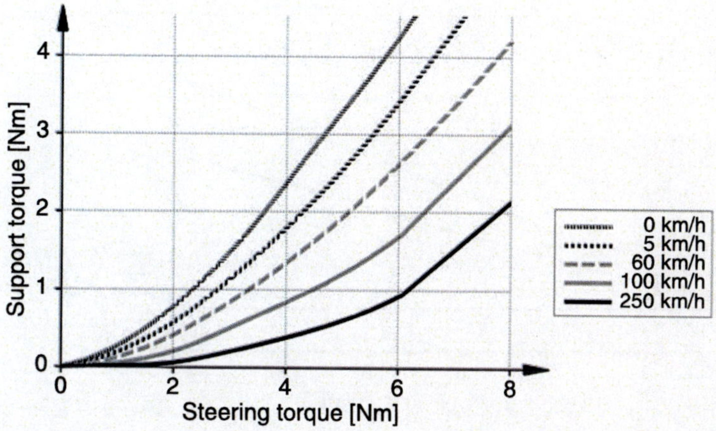

Fig. 5.22 Support characteristics of an electromechanical passenger car steering system, after [19]. The assisting torque of an electric motor depends on the applied steering torque of the driver and the driving speed

Power-assisted steering devices are permitted on some racing vehicles. However, as with road vehicles, only in such a way that the wheels can still be steered by hand even if the assistance fails.

In Formula 1 cars, approx. 30% auxiliary power of a servo assistance proves to be the upper limit in order to maintain the necessary road feel for the driver [8].

As an example of what is technically feasible, Fig. 5.22 shows the assistance characteristics of an electromechanical steering system for a passenger car. In addition to the pinion from the steering shaft, a second pinion, which is driven by an electric motor, engages on the rack. The assistance provided by the electric motor is set by a control unit as a function of the driving speed and the steering torque applied. In addition, active return assistance to the center position of the rack allows precise tuning of the steering.

The assistance effect is mathematically described in (5.21) as steering gain ratio A_S. Without steering assistance $A_S = 1$ and the gradient hand torque on the steering wheel over lateral acceleration is constant [1]:

$$\frac{dM_H}{da_y} = \frac{m_{V,f}(r_{\tau,T} + r_{\tau,k})}{i_S} = C_A \tag{5.23}$$

dM_H / da_y	Steering wheel torque gradient, N m/(m/s^2). For optimum steering behaviour, the value for passenger cars should be 2.5 to 3.5 N m at $a_y = 1$ m/s^2 [1]
$m_{V,f}$	Proportion of total vehicle mass above front axle (axle load), kg
C_A	Restoring factor, N m/(m/s^2). Represents a measure for the restoring (self-aligning) capacity and thus for the centering

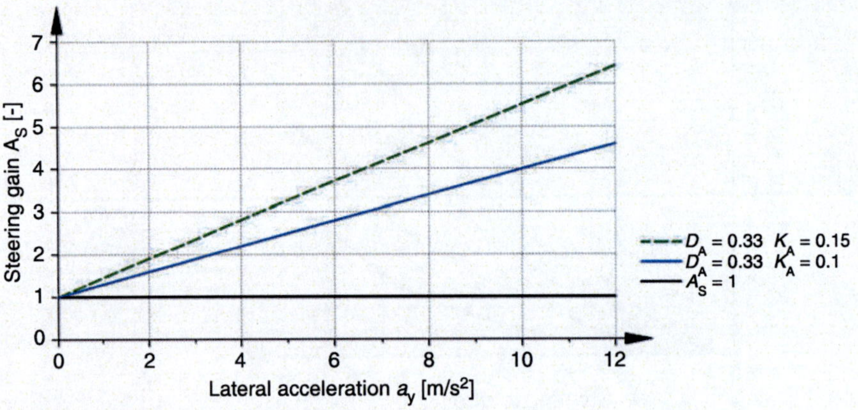

Fig. 5.23 Steering gain ratio versus lateral acceleration. The two curves with the increasing steering gain belong to a passenger car. Without steering assistance, $A_S = 1$. Car values: $m_{Vf} = 852.5$ kg, $r_{\tau,T} + r_{\tau,k} = 0.06$ m, $i_S = 17$. The values of the parameters for (5.24) are given

Vehicles whose steering wheel torque curve is subjectively perceived as harmonic exhibit a distinctly linear curve of steering gain A_S over lateral acceleration [1]. A linear curve can be represented in this way:

$$A_S = C_A\left(D_A + K_A a_y\right) \tag{5.24}$$

D_A	Gradient factor, (m/s²)/(N m). $D_A = 1\ /C_A$, is therefore a measure of the basic support
K_A	Degressivity factor, 1/(N m). Determines how much the increase in steering wheel torque flattens out over lateral acceleration, cf. Figure 5.24. Without steering assistance ($A_S = 1$), $K_A = 0$

Figure 5.23 shows 2 curves of steering gains A_S over lateral acceleration for a passenger car. For a vehicle without power steering, the value is a constant 1.

With the known quantities, the steering wheel torque curve can also be expressed in terms of lateral acceleration:

$$M_H = \frac{1}{\frac{D_A}{a_Y} + K_A} \tag{5.25}$$

The steering wheel torque for passenger cars around the centre position at a lateral acceleration of 1 m/s² should be 2.5–3.5 N m [1]. Some resulting steering wheel torque curves versus lateral acceleration can be seen in Fig. 5.24. In addition to curves for passenger cars, the torque curve for a Formula 3 car is plotted. It has no steering assistance ($K_A = 0$) and therefore the basic assistance D_A shapes the steering torque. Ultimately, the

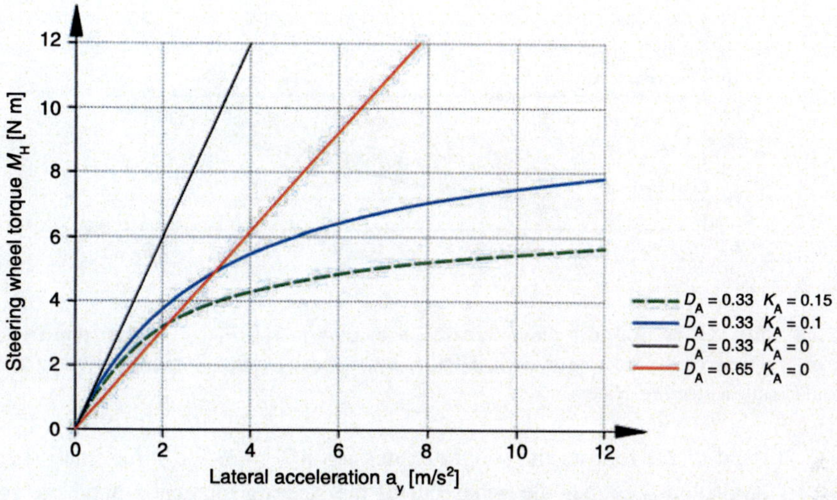

Fig. 5.24 Steering wheel torque versus lateral acceleration. The values of the decisive parameters correspond to those of Fig. 5.23. In addition, a Formula 3 car (red) is entered here. Formula 3 values: $m_{V,f} = 264$ kg, $r_{\tau,T} + r_{\tau,k} = 0.0755$ m, $i_S = 13$

steering feel is essentially determined by the chassis geometry (trail distances $r_{\tau,T}$, $r_{\tau,k}$) and the steering ratio i_S.

The gradient of the steering torque from the central position serves as a further objective parameter for assessing a steering system. For the analytical representation, the steering moment is required as a function of the steering angle. The steering moment over the lateral acceleration is (cf. (5.23)):

$$M_H = \frac{m_{V,f}(r_{\tau,T} + r_{\tau,k})}{i_S A_S} a_y = \frac{C_A}{A_S} a_y \tag{5.26}$$

The relationship between steering wheel angle and lateral acceleration for steering movements around the centre position (small steering wheel angles $\delta_{H,0}$) is provided by the definition of the self-steering gradient[8] for very large corner radii R:

$$\delta_{H,0} = \lim_{R \to \infty} \left(i_S \frac{l}{R} + i_S G_U a_y \right) = i_S G_U a_y \tag{5.27}$$

[8] See, e.g., Racing Car Technology Manual Vol. 5 *Data Analysis, Tuning and Development*, Chap. 6 *Development*.

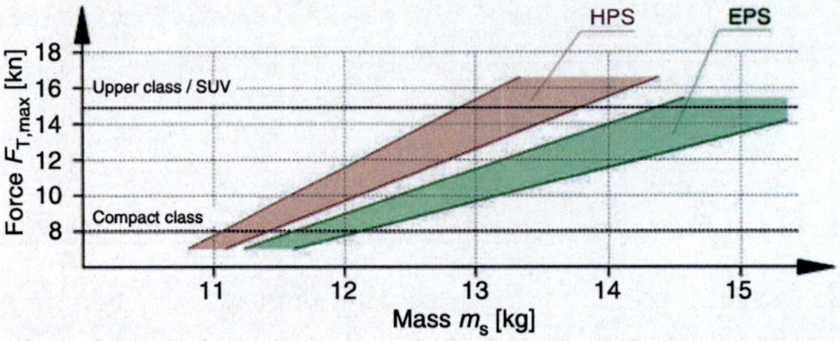

Fig. 5.25 Mass comparison of steering systems, according to [1]. $F_{T,max}$ ...maximum track rod force, m_S ...mass of the steering system, HPS... hydraulically assisted steering system, EPS... electromechanical steering system

With (5.26) and (5.27) follows the sought expression $M_H = M_H(\delta_H)$ for small steering angles. Its derivation describes the behaviour of the steering moment around the centre position:

$$\left.\frac{dM_H}{d\delta_H}\right|_{\delta_H=0} = \frac{C_A}{i_S A_S G_U} \text{ in N m/rad or } \frac{C_A}{i_S A_S G_U} \cdot \frac{\pi}{180^\circ} \text{ in N m/}^\circ \qquad (5.28)$$

For optimum steering behaviour, a value range of 0.3–0.5 N m/° is recommended for passenger cars [1]. For racing vehicles without power steering, values of around three to four times this value are achieved, depending on the driving speed and aerodynamic downforce.

In rally vehicles, electromechanical power steering systems (EPS), which are currently displacing hydraulically assisted steering systems (HPS) in passenger cars, are proving to be unusable. Average steering angle speeds of 700 °/s and peak values of 1500 °/s cannot be achieved with such systems [20].

Another principle-related disadvantage of EPS steering systems is that they are heavier than HPS systems, and the more mechanical power is required, Fig. 5.25. In the small car segment, both systems still have about the same mass [1].

At this point, it is important to remember that the steering system not only transmits the driver's inputs from top to bottom to the tires, but that useful and disturbing information is also transmitted from the road to the driver. Ideally, filtering takes place so that only driving-relevant feedback is perceived by the driver. As unfavorable as friction is for humans when performing a control task, it has a helpful effect in this context. Shocks and periodic excitations such as wheel imbalance, brake force fluctuation, etc. are damped by increased friction in the transmission system [1]. The steering system as a spring-mass system capable of oscillation naturally exhibits a frequency-dependent behavior. A comparison of the transmission behaviour of different steering systems is provided by the Bode diagram Fig. 5.26. The general limit of the frequency for driver-relevant feedback signals can be seen in the range of 20–30 Hz [1]. The inertia of a steering system increases with its

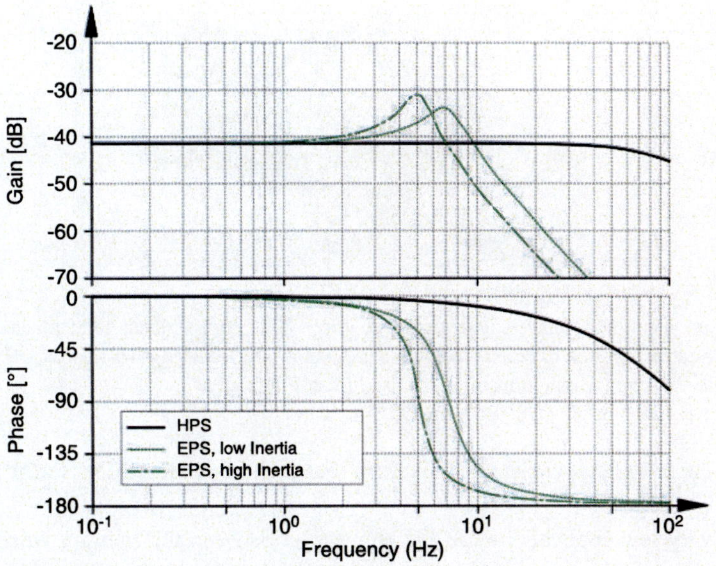

Fig. 5.26 Bode diagram of various steering systems, after [1]. A Bode diagram represents a transmission response according to magnitude (amplitude response) and phase (phase response). Magnitude refers to the value of the transfer function between the response, displacement of the rack or rotation of the torsion bar in the steering valve, and the excitation, force on the rack, whereby only the magnitude of the variables under consideration is used. Phase describes the phase shift between excitation and response as a function of the excitation frequency. HPS... hydraulically assisted power steering, EPS... low and high inertia electromechanical steering. In contrast to the HPS, EPS show pronounced resonance phenomena around 6 Hz. In this case, the system response lags behind the excitation by 90°. A system with greater inertia exhibits resonance at a lower frequency with greater signal overshoot. It can be seen that the HPS exhibits the more constant transfer behavior over the entire relevant frequency range, and as frequencies increase, the gain from excitation to impact becomes weaker. Such behavior acts as a low-pass filter and thus supports the requirement that (high-frequency) interference signals are not transmitted unchanged from bottom to top to the steering wheel

moving masses. For this purpose, the following parts were considered for the transmission behavior in Fig. 5.26: The rack, for HPS the hydraulic piston, for EPS the rotor of the electric motor together with the belt drive and the recirculating ball nut.

5.3 Steering Shaft

The steering shaft transmits the driver's steering movement to the steering gear. The shaft diameter for steel is about 15–22 mm, depending on the steering wheel diameter and shaft length. Hollow and solid shafts are used. The design criterion is generally not the maximum

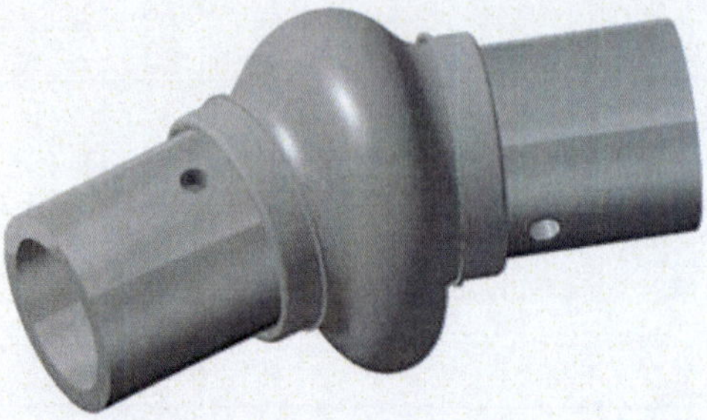

Fig. 5.27 Compact shaft joint (sealed universal joint). This encapsulated joint made of steel is welded into the steering shaft or connected with a cross bolt. The total length is approx. 70 mm with a mounting hole ∅16 mm. Such joints are also used in shift linkages

torque that the driver can create with his arms, but the (torsional) stiffness of the steering system.

Only in very few applications can the connection between the steering wheel and the steering gear be realized by a simple and straight connecting shaft (steering spindle). Steering spindles are often designed with one or two angle joints. Cardan joints are common, e.g. Figure 5.27. Constant velocity joints have not proved successful for this application. If such joints are designed without play, the friction will be inappropriately high. Conversely, if the frictional behaviour is acceptable, the clearance between the balls and the raceway becomes too large. However, shaft joints are not only present for reasons of space, but also have the additional advantage that in the event of an accident the steering shaft can yield in the longitudinal direction. However, the steering play increases with the number of joints. These joints should therefore be as torsionally stiff as possible. In order to reduce the effective play, it pays to select a larger diameter for the forks than is required by the forces, Fig. 5.28.

Steering shafts located between the driver's feet must be installed laterally offset with three pedals (between clutch and brake pedal) and therefore require two shaft joints. However, there are also vehicles which, in such a case, nevertheless have only one joint, but instead have a laterally pivoted steering wheel [21].

For small deflection angles β (up to approx. 5°), the universal joints can be replaced by joint discs. Steering columns with a universal joint produce a degree of non-uniformity *U* that is noticeable to the driver at deflection angles >15°:

$$U = \frac{\omega_{2,\max} - \omega_{2,\min}}{\omega_1} = \tan\beta \cdot \sin\beta \tag{5.29}$$

U	Degree of non-uniformity, –
ω_1 *or* ω_2	Angular velocity of the interconnected shafts 1 and 2, rad/s
β	Deflection angle between the shafts, °

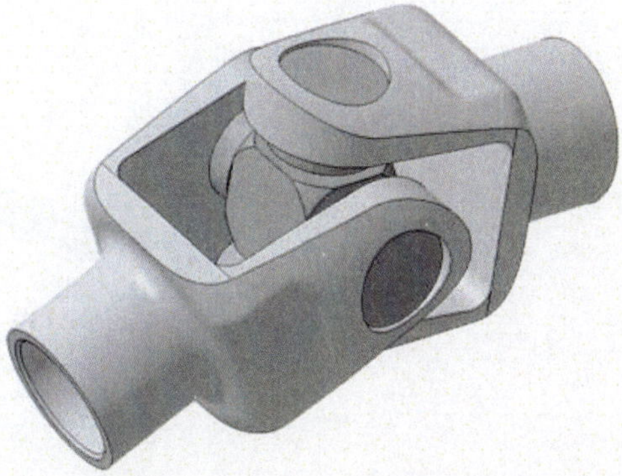

Fig. 5.28 Hooke's universal joint. This shaft joint is dimensioned according to its deformation, i.e. larger than the transmittable steering torque would require. With a steering shaft diameter of 18 mm, the forks have a maximum width of 43 mm

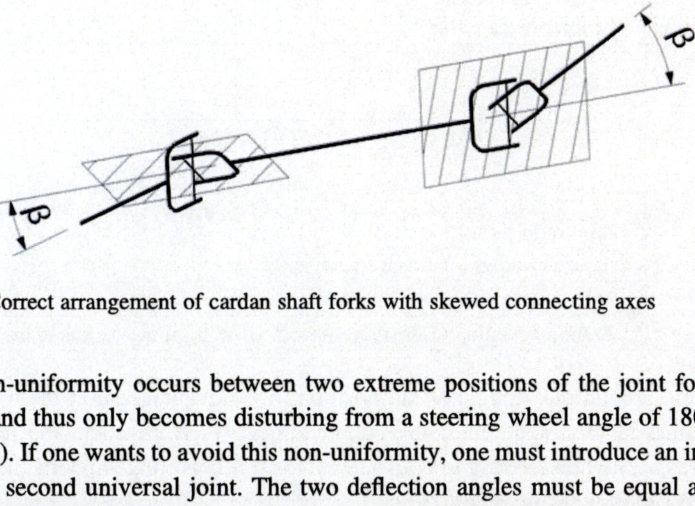

Fig. 5.29 Correct arrangement of cardan shaft forks with skewed connecting axes

This non-uniformity occurs between two extreme positions of the joint fork (cf. also Fig. 5.32) and thus only becomes disturbing from a steering wheel angle of 180° ($\delta_H = -$ 90° to +90°). If one wants to avoid this non-uniformity, one must introduce an intermediate shaft and a second universal joint. The two deflection angles must be equal and the two forks of the intermediate shaft must lie simultaneously in their planes formed by input and output shafts (see Figs. 5.29 and 5.33).

In any case, the position of the fork in relation to the joint plane is important for the steering feel. The general case – skew axes with 2 unequal deflection angles – also offers the possibility of setting a symmetrical hand torque curve. The manipulated variable is the only freely selectable geometric parameter here: The offset angle γ between the forks of the intermediate shaft, Fig. 5.30.

Initially, only one joint, the first joint, is considered. From the point of view of a universal joint, the transmission ratio between the input and output shafts is decisive for the steering feel. When the input fork is in the joint plane ($\varphi_1 = 0^\circ$), it is at a maximum. A quarter turn later – the fork is perpendicular to the joint plane ($\varphi_1 = 90^\circ$) – it is minimum. The progression repeats periodically over 180°. It holds:

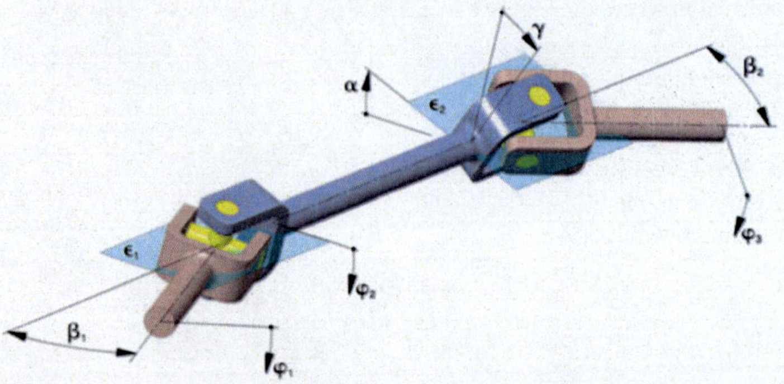

Fig. 5.30 Cardan shaft with intermediate shaft in general arrangement. β_1, β_2 Deflection angle between input and output shafts. ε_1, ε_2 Joint plane; these are spanned by the respective input and output axes. α Angle between joint planes. φ_1, φ_2, φ_3 Angle of rotation of each shaft. γ Angle of offset between the forks of the intermediate shaft

$$i = \frac{\omega_2}{\omega_1} = \frac{M_1}{M_2} = \frac{2\cos(\beta_1)}{2 - \sin^2(\beta_1)[1 + \cos(2\varphi_1)]} \tag{5.30}$$

i	Joint ratio, –
ω_1 or ω_2	Input or output rotational speed, rad/s
M_1 or M_2	Input or output torque, N m
φ_1	Angle of rotation of the drive fork, °. $\varphi_1 = 0^\circ$, if the fork is in the joint plane

For steering around the straight-ahead position, it is advantageous if the transmission ratio is minimal in this position: If the output torque M_2 is constant, the hand torque $M_1 = iM_2$ increases when steering to both sides. This self-centering supports the driver via a pronounced center feel (ditch effect).

Like the rotational speed, the cardan error $\Delta\varphi$ (= $\varphi_2 - \varphi_1$, i.e. the difference between the output and input angle) changes periodically. The largest advance occurs at about 45° and the largest retard occurs at about 135° input rotation angle. For the output angle φ_2 of the first joint applies:

$$\tan(\varphi_2) = \frac{1}{\cos(\beta_1)}\tan(\varphi_1) \text{ or } \varphi_2 = \arctan\left(\frac{1}{\cos(\beta_1)}\tan(\varphi_1)\right) \tag{5.31}$$

The time derivative of the angle of rotation curve provides the angular velocity curve and thus the relationship between the cardan error and the transmission ratio.

For the second joint, the relationships (5.30) and (5.31) basically apply in the same way. However, if the behaviour of the entire shaft train is to be determined, attention must be paid to the relative position of the forks to each other. If the first input fork is in its joint plane $\varepsilon_1(\varphi_1 = \varphi_2 = 0^\circ)$, the input fork of the second joint is 90° to its reference plane, the joint plane ε_2. For the correct application of (5.31), this must be taken into account. In addition, this fork may be rotated by the offset angle γ with respect to its original position (which is the first fork of the intermediate shaft). Similarly, the reference plane for the output angle φ_3 (= joint plane ε_2) may be rotated by angle α about the intermediate shaft axis. Therefore, for the input angle $\overline{\varphi}_2$ to calculate the output angle φ_3 of joint 2 must be set in (5.31):

$$\overline{\varphi}_2 = \varphi_2 + 90^\circ + \gamma - \alpha \tag{5.32}$$

The result $\overline{\varphi}_3$ must be corrected by the added 90° fork offset (γ and α are not affected – the references are actually twisted by these amounts):

$$\varphi_3 = \overline{\varphi}_3 - 90^\circ$$

A simpler way is to replace (5.31) with a relationship that is true for the fork position 90° to the joint plane:

$$\tan(\varphi_3) = \cos(\beta_2)\tan(\varphi_2) \tag{5.33}$$

If a fork misalignment γ and a joint plane misalignment α also occur, this must be taken into account analogously to (5.32). This and the combination (5.33) with (5.31) provide a relationship for the behaviour of the entire propshaft train:

$$\tan(\varphi_3) = \cos(\beta_2)\tan\left[arctan\left(\frac{1}{\cos(\beta_1)}\tan(\varphi_1)\right) + \gamma - \alpha\right] \tag{5.34}$$

For $\gamma = \alpha$, the total cardan error $\Delta\varphi_t = \Delta\varphi_1 + \Delta\varphi_2$ becomes minimal and (5.34) becomes:

$$\tan(\varphi_3) = \frac{\cos(\beta_2)}{\cos(\beta_1)}\tan(\varphi_1) \tag{5.35}$$

Using (5.32) inserted in (5.30), the transmission ratio i_2 of the second joint can also be determined. The total transmission ratio i_t follows from the product of the individual transmission ratios:

$$i_t = i_1 \cdot i_2 = \frac{\omega_2}{\omega_1} \cdot \frac{\omega_3}{\omega_2} = \frac{\omega_3}{\omega_1} = \frac{M_1}{M_3}$$

i_t	Total ratio, –
$\omega_1, \omega_2, \omega_3$	Angular velocities of the input, intermediate and output shafts, rad/s

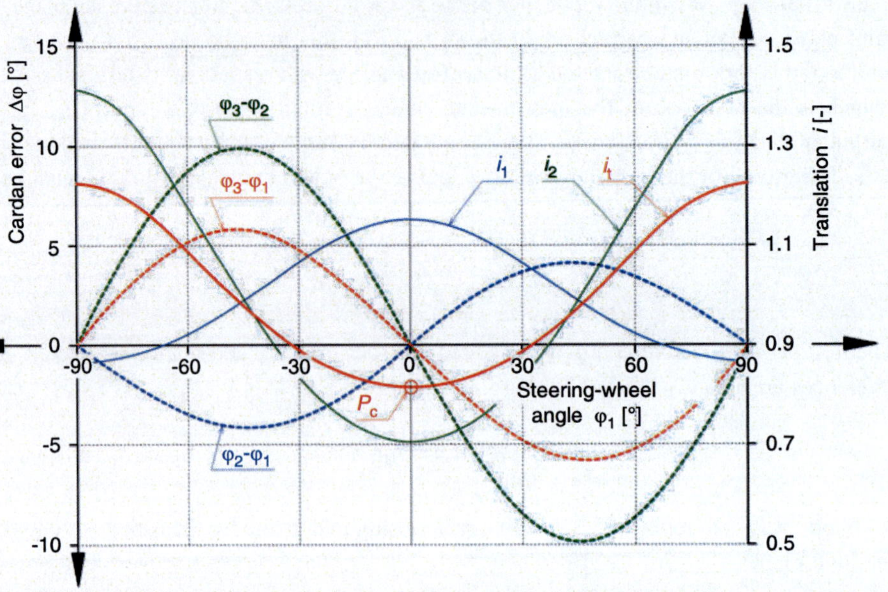

Fig. 5.31 Course of cardan error and transmission ratio for a double-joint shaft with phase correction. The two joints have different deflection angles β, therefore their cardan errors are out of phase. The fork offset angle γ of the intermediate shaft was adjusted to the relative position of the joint planes to each other (angle α). Therefore, the minimum of the total gear ratio coincides with the straight-ahead position of the steering wheel ($\varphi_1 = 0^{\circ}$). When steering, the driver feels an increase in the hand torque in both directions (ditch effect). $\beta_1 = 30°$, $\beta_2 = 45°$, $\gamma = \alpha = 30°$. P_c Center Point (minimum transmission ratio)

Figure 5.31 shows the course of the cardan errors and transmission ratios for a spatially arranged cardan shaft train with 2 joints. The offset angle of the intermediate shaft forks was adapted to the position of the joint planes ($\gamma = \alpha$). The total transmission ratio i_t thus ideally shows its minimum in the middle position of the steering wheel ($\varphi_1 = 0^{\circ}$).

Joint flexion angles above 30° should be avoided because the steering shafts are also subjected to bending due to additional bearing forces originating from the universal joints. The moments fluctuate between two extreme values in a similar way to the angular velocities (see also Fig. 5.32):

$$\begin{aligned} M_{S,2,\min} &= M_{S,1} \cdot \cos(\beta) \\ M_{S,2,\max} &= M_{S,1} / \cos(\beta) \\ M_{b,S,\min} &= M_{S,1} \cdot \sin(\beta) \\ M_{b,S,\max} &= M_{S,1} \cdot \tan(\beta) \end{aligned} \tag{5.36}$$

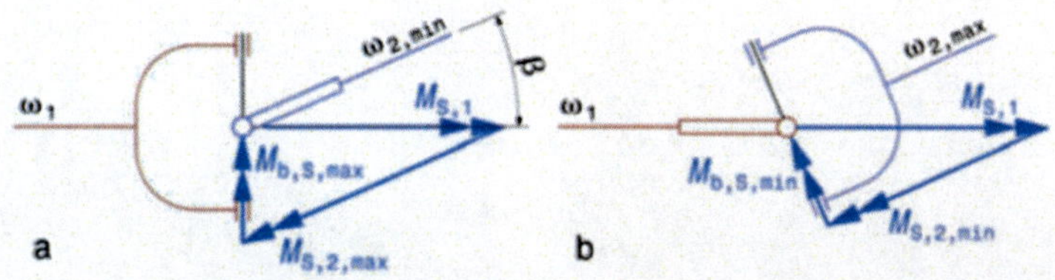

Fig. 5.32 Extreme universal joint positions. The largest and smallest values of the angular velocities and the torques result for the positions *a* and *b* with the fork of shaft 1 in the image plane and normal to it

Fig. 5.33 Compensation of a deflection angle of the steering shaft > 15° [28]. In order to avoid noticeable irregularities in the steering movement for the driver, two flexing joints are installed in a W arrangement and the three shaft axes are in one plane

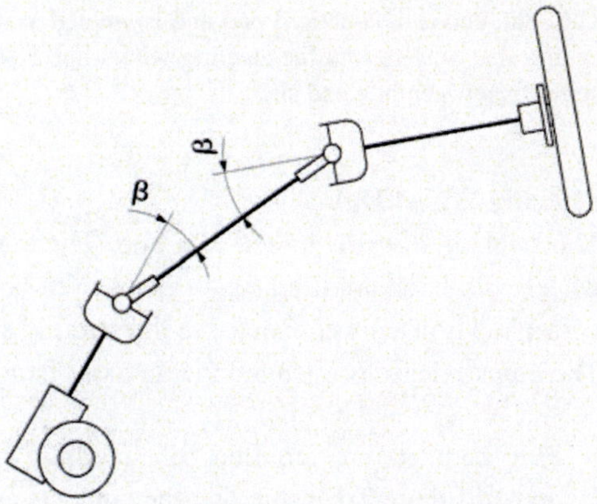

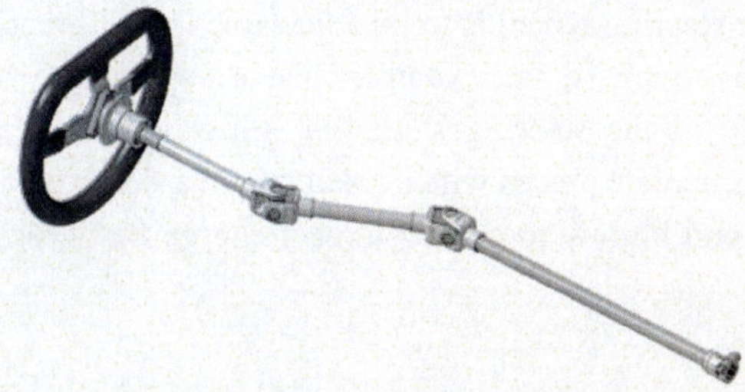

Fig. 5.34 Steering shaft of a Formula Renault car. The shaft has two flexing joints in a Z arrangement. This allows the steering wheel to deflect axially during the impact test in accordance with FIA F.32000 (see Racing Car Technology Manual, Vol. 2 *Complete Vehicle*, Chap. 3 *Safety*)

$M_{S,1}$ or $M_{S,2}$	Torsional moment on steering shaft 1 or 2, N m
$M_{b,S}$	Bending moment in the steering shaft due to universal joint, N m

Figure 5.33 shows a possibility to overcome larger deflection angles. Flexion joints in the steering shaft increase safety for the driver in the event of a frontal collision because the steering shaft cannot transmit longitudinal forces and buckles, Fig. 5.34.

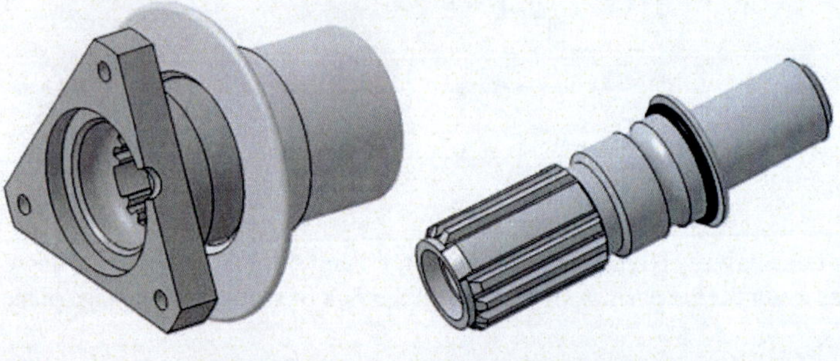

Fig. 5.35 Steering wheel hub with weld-on end for steering shafts (with weld-on splined section). The shaft end is a purchased part and is welded to the individual steering shaft. It has asymmetrical splines that only fit into the steering wheel hub in one position. This is necessary if there are plug contacts between hub and shaft

Steering Wheel Hub

The steering wheel is bolted to a hub. Three M6 bolts are sufficient for formula steering wheels, six bolts are used for vehicles with larger steering wheel diameters. The steering wheel hub is usually connected to the steering shaft (due to regulations) via a quick release. The quick release is operated via an axial flange ring, which must be yellow, Fig. 5.35.

Plug contacts are required for steering wheels with additional electronic functions (switch, display). To ensure that the contacts are aligned when the steering wheel is fitted, the shaft-hub connection must be asymmetrical (see Figs. 5.35 and 5.36).

If the position of the steering wheel is to be adjustable in the direction of the axis to suit the driver's preferences, there is, for example, a telescopic piece between the steering wheel hub and the end of the steering shaft that allows the shaft to be lengthened or shortened in steps. Intermediate pieces with a fixed length are also used. These spacers may also be basket-shaped and hollow to absorb impact energy by deformation to protect the driver, Fig. 5.37.

Bearing

The steering shaft or the steering shaft parts must be mounted so that they can rotate relative to the frame, Fig. 5.38. The tasks of such steering bearings are as follows:

- support the steering spindle without play
- effectively dampen vibrations and noise
- exhibit high stiffness
- be as low-friction as possible.

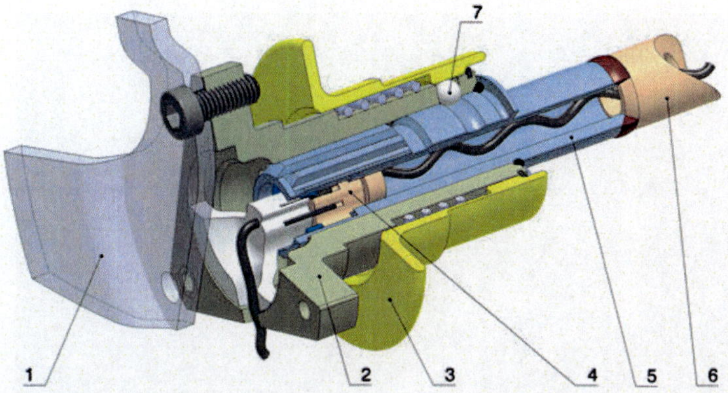

Fig. 5.36 Attached steering wheel hub with quick-release fastener (partially cut open). The steering wheel (1) is screwed onto the hub flange (2) with three bolts. The hub itself transmits the steering torque to the shaft end (5) and thus to the welded-on steering shaft (6) by means of splines. Axial locking is provided by locking balls (7) which engage in the groove at the shaft end. To release the steering wheel, the yellow flange ring (3) is pressed towards the steering wheel, thus releasing the balls. In addition, this version has electrical plug contacts (4)

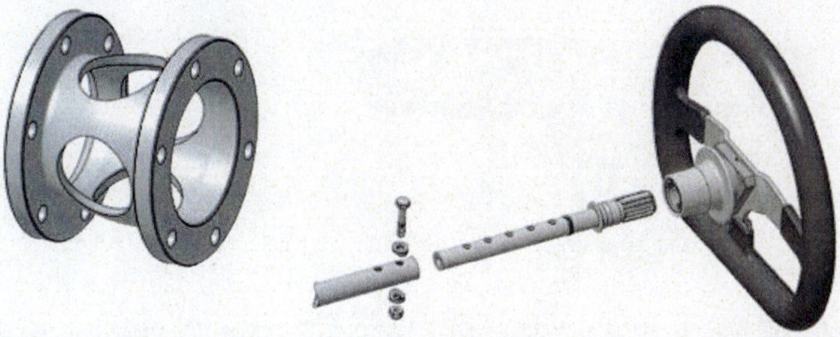

Fig. 5.37 Steering wheel spacers. *Left*: Fixed length. For six-hole bolt connection. *Right*: Telescopic piece that can be screwed in stages. Note the two saddle washers, which ensure sufficient pressure distribution on the tubular shaft part under the hexagonal heads

Spherical plain bearings (Fig. 5.39), plastic plain bearings but also rolling bearings are used. If spherical plain bearings are used, designs without an intermediate sliding layer (see Sect. 4.2.2 *Spherical Plain Bearings*) should be used. These show less tendency to stick-slip effects, which can be troublesome especially at the steering in the case of bearings with large diameters.

As a simple but meaningful test of the friction conditions, the following test can be carried out on the fully assembled steering: Operate the steering wheel with the last limb of the little finger with the front wheels jacked up. It must be possible to turn the wheel easily from lock to lock, otherwise the steering is too sluggish [11].

The steering bearings may also be fitted in such a way that the height of the steering wheel can be adjusted, Figs. 5.40 and 5.41.

Fig. 5.38 Cockpit-side steering shaft bearing in a monoposto (Dallara Formula 3)
In the unfinished vehicle assembled for inspection purposes, the bracket for the steering shaft bearing on the transverse frame can be clearly seen

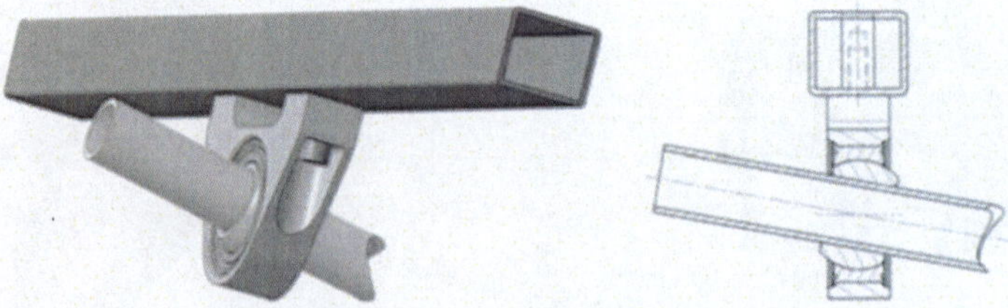

Fig. 5.39 Spherical plain bearing as steering shaft bearing. The spherical plain bearing adjusts itself advantageously to the required angle of the steering shaft without causing additional friction

In passenger cars, special needle roller bearings with a rubber tolerance ring are used to compensate for tolerances in the steering column's outer tubes. There are also full complement or cage-guided angular contact ball bearings that accommodate the steering shaft in the usual two-bearing arrangement. Four-point ball bearings are also used in longitudinally adjustable steering columns.

The steering shaft is connected to the steering gear via a shaft-hub connection that is specified by the steering gear. Square section and splined section shafts are commonly used. Figure 5.42 shows a connection piece for splines which is welded to the steering shaft. To transmit the steering movement without backlash, the connection has a clamping screw. This is arranged in such a way that at the same time it represents a positive locking device to prevent it being pulled off the stub shaft.

An example of a complete steering shaft is shown in Fig. 5.43.

Fig. 5.40 Steering shaft bearing of a production sports car. The steering bearing is housed in an aluminum housing that is attached to the frame with two bolts on tabs. Several holes in these lugs allow the height of the steering wheel to be adjusted to suit the driver. The longitudinal adjustment allows the sleeve in the steering shaft (cf. Fig. 5.37)

Fig. 5.41 Steering shaft in a touring car (DTM Mercedes AMG C-Coupé). The steering shaft bearing is attached to the cockpit crossmember via two cantilevers. Oblong holes allow a height adjustment of the steering shaft

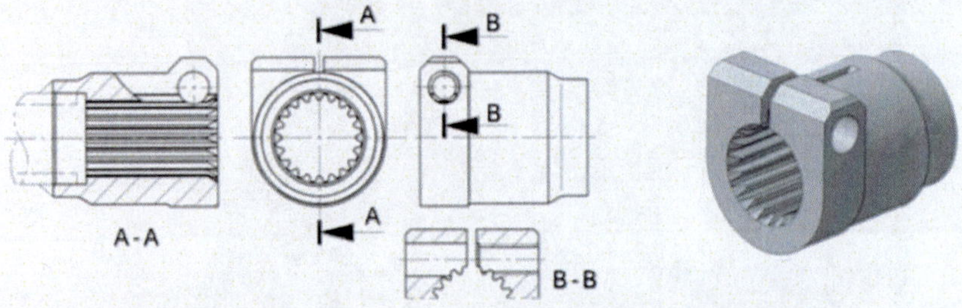

Fig. 5.42 Connecting piece to the steering gear. The part is welded to the steering shaft (section A-A, dashed)

Fig. 5.43 Steering shaft in a production sports car (Osella PA 205). The steering shaft runs stretched from the steering wheel (which is removed in the picture) to the steering gear connection (far left in the picture). Nevertheless, an encapsulated flexure joint is inserted in the shaft. The two-piece radial bearing next to this joint is made entirely of polyamide

For the design of the parts of a steering system one will start on the one hand from the steering torques to be expected, which are dictated by the tyres, their coefficient of friction and the wheel load (see (5.37)) and on the other hand from the maximum values which a human driver is at all capable of applying. The maximum steering torque can be assumed to be between 100 and 135 N m for light formula cars and between 175 and 240 N m for touring cars. Furthermore, the steering wheel and steering shaft should be able to absorb radial forces of 700 N without noticeable or even permanent deformation. However, the maximum steering wheel torque should not exceed 10 N m during driving, otherwise the driver will tire too quickly.

Table 5.2 Criteria for verifying the static strength of car steering columns [1]

Load situation (moment at the steering wheel)	Criterion
Operating load: 150 N m	No functional impairments No plastic deformations No loss of preload force of the screws No cracks or fractures
Misuse load[a]: 250 N m	No safety-relevant functional impairments No safety-relevant plastic deformations No inadmissible drop in preload force of the screws No cracks or fractures
Fracture behaviour	No deformation-free fracture Test termination at 350 N m without breakage

[a]e.g. when breaking open the steering wheel lock, pushing a wheel off the kerb

The following empirical equation can be used for the pivoting moment (bore torque) of a tyre when stationary [22]:

$$M_{W,S} \approx \mu_W \frac{F_{W,z}^{1.5}}{\sqrt{p_T}} \tag{5.37}$$

$M_{W,S}$	Swing moment (bore torque) of a tyre, N m
μ_W	Coefficient of friction between Tyre and road surface, –
$F_{W,z}$	Tyre contact force, N
p_T	Tyre inflation pressure, Pa = N/m^2

Table 5.2 can be helpful for dimensioning the entire transmission line, i.e. steering wheel – shaft – flexure joint – pinion connection. This table lists the criteria for verifying the static strength of standard passenger car steering columns.

5.4 Steering Gear

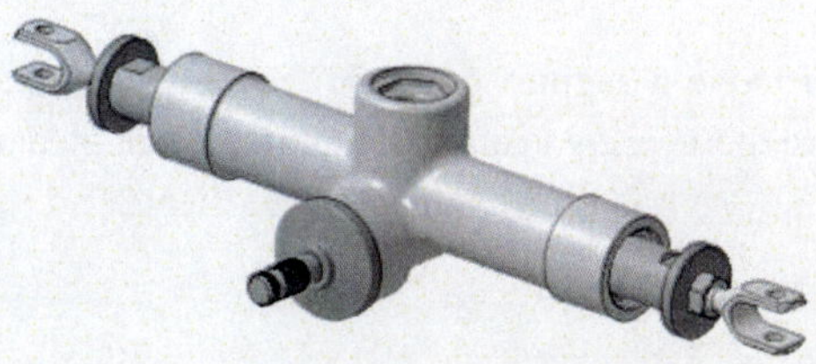

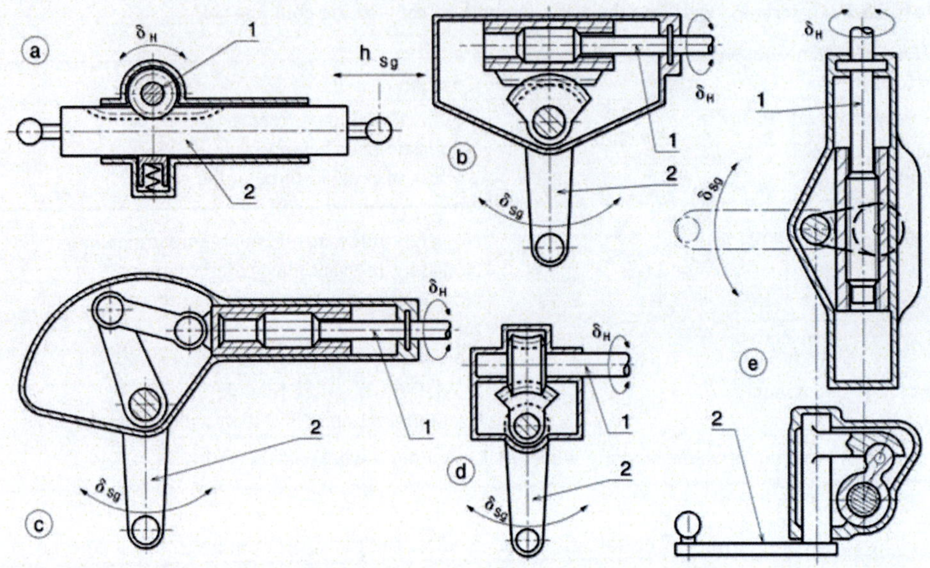

Fig. 5.44 Types of steering gear, according to [5]. (**a**) Rack and pinion steering: 1 Steering shaft with pinion, 2 Racks. (**b**) Steering nut steering: 1 Threaded steering shaft, 2 Steering column levers with gear segment. (**c**) Crank drive steering: 1 Steering shaft with thread, 2 Steering column levers. (**d**) Globoidal worm steering (Gemmer steering): 1 Steering shaft with globoidal worm, 2 Steering column levers with worm gear segment. (**e**) Steering nut steering: 1 Steering shaft with thread, 2 Steering column levers with ball socket

The steering gear converts the steering angle δ_H generated by the driver at the steering wheel into an adjustment of the steering linkage, e.g. the angle of rotation δ_{Sg} of a pitman arm, which sits on the pitman arm shaft – this is the output shaft of a steering gear, Fig. 5.44b–e. The steering gear ratio is then:

$$i_{Sg} = \frac{d\delta_H}{d\delta_{Sg}} \tag{5.38}$$

i_{Sg}	Steering gear ratio, –
δ_H	Steering angle at steering wheel, °
δ_{Sg}	Angle of rotation of the steering column lever, °

The simplest way to move a support point on the steering linkage is rack and pinion steering, Fig. 5.44a. Here the steering gear ratio can only be defined as the ratio of the angle δ_H and the rack stroke h_{Sg}, so it is dimensionally dependent:

$$i_{Sg,Rack} = \frac{d\delta_H}{dh_{Sg}} \tag{5.39}$$

$i_{Sg,Rack}$	Steering gear ratio of the rack, °/mm
δ_H	Steering angle at steering wheel, °
h_{Sg}	Rack stroke, mm

If the gear ratio of the rack is constant, the total travels can be put into ratio and it is valid:

$$i_{Sg,Rack} = \delta_{H,t}/h_{Sg,t} = 2\pi \cdot j_H/h_{Sg,t} \tag{5.40}$$

$i_{Sg,Rack}$	Steering rack ratio, wheel/mm
$h_{Sg,t}$	Total rack stroke, mm
$\delta_{H,\,t}$	Steering wheel rotation angle for total rack stroke, rad
j_H	Number of steering wheel revolutions at total stroke $h_{Sg,t}$, –

By intervening in the gear geometry, variable gear ratios can also be achieved with rack-and-pinion steering systems (see Fig. 5.51). The main advantage of rack-and-pinion steering is its simple design and the small space requirement, but not necessarily the relatively stiff direct conversion of the steering wheel rotation into a track rod displacement without the intervention of intermediate levers. The straight-line form of movement also restricts the design possibilities when designing the steering geometry, which is generally effective in three dimensions.

Equally common on passenger cars is a steering system with a "steering nut" connected to the steering shaft by a thread (now a recirculating ball screw to reduce friction), Fig. 5.44b. In the illustration, the steering nut carries a rack segment that meshes with a gear segment on the pitman arm shaft. A crank gear may also be substituted for the gearing, Fig. 5.44c, in which case the steering nut assumes the role of the piston. In this design, the steering gear ratio is variable.

Figure 5.44d shows a steering gear with a globoid worm and a worm wheel, whereby the steering wheel angle δ_H with constant transmission ratio i_{Sg} is converted into a rotation angle δ_{Sg} of the pitman arm. Newer designs of this type have a bearing clevis on the pitman arm shaft, in which there is a profiled roller representing two teeth of the worm wheel to reduce friction ("worm-roller steering"). The worm steering is sensitive to shocks, can be designed without play and with a progressive transmission ratio.

The designs shown in Fig. 5.44b, c are, like the rack and pinion steering a, particularly suitable for superimposing a hydraulic servo assistance (piston and cylinder) because of the straight-line stroke movement of their output element.

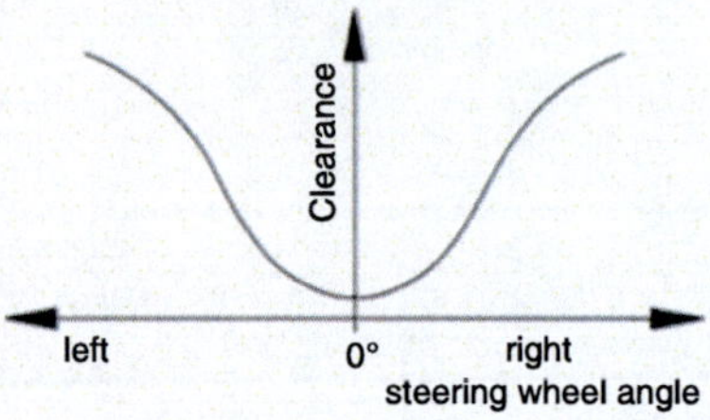

Fig. 5.45 Permissible course of the steering play over the steering wheel angle (schematic). In the straight-ahead position, play is most noticeable because the wheels – apart from toe-in, caster, etc. – are not guided and therefore run unstably. When cornering, lateral forces occur and push the wheels to a stop

A variant of the steering gear with steering nut is shown in Fig. 5.44e. The steering movement moves a steering nut via a thread. This guides a spherical cap on the steering column shaft with an attachment. As the nut is axially displaced, the pitman arm shaft rotates and at the same time tilts the nut about its axis because the ball cup moves along an arc of a circle specified by the pitman arm shaft. This pivoting movement of the nut is always directed towards the pitman arm shaft, as a result of which the effective lever arm around the pitman arm shaft decreases as the steering angle increases, i.e. the gear ratio i_{Sg} decreases (this can also be reversed by further intervention in the kinematics). The steering nut thus swivels once in the direction of rotation of the steering spindle and the other time in the opposite direction, so that the relative angle of rotation and thus the feed is reduced once and increased the other time. This results in an asymmetry in the transmission course and a slightly unequal number of steering wheel rotations to the left and right.

All versions with a single steering nut have a high sliding friction, which the driver perceives as sluggish, and cannot be adjusted (thread play).

The play between the steering wheel and wheel swivel movement should be smallest around the straight-ahead position ("pressure point", center point). Wear occurs primarily in the straight-ahead position and moderate play at large steering angles is insignificant for driving, Fig. 5.45.

Position of Steering Gear

First of all, the articulation points of the chassis are determined according to the driving performance (see Chap. 2 *Chassis*) and only then a suitable position for the steering gear and the linkage of the track rods is found.

The position of the steering gear or, more precisely, the tie rod connection in relation to the joints of the wheel suspension is decisive as to whether an (undesirable) self-steering behaviour occurs when the wheel compresses (bump). How the position of the transmission elements is determined is described in Sect. 5.5. The steering gear is aligned with the track rod when its position is known. In principle, the steering gear can be positioned in front of or behind the axle. In the latter position, the length of the steering shaft is shorter, but the

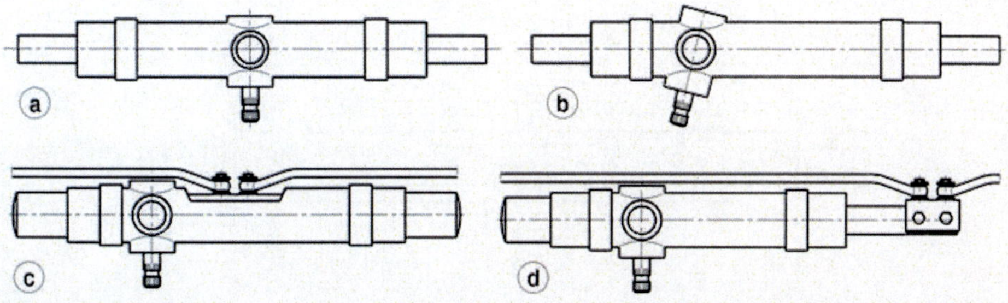

Fig. 5.46 Designs of rack and pinion steering gears. (**a**) Steering shaft in the middle, side output. (**b**) Steering shaft on the side, swivelled connection. (**c**) Centre output. (**d**) Single-sided output

steering gear can restrict the area for the driver's feet or lower legs if it is at the level of the upper wishbones. A shorter steering shaft is also not necessarily lighter. As soon as a shaft joint is required between the steering wheel axis and the steering gear pinion, the mass of the joint together with the necessary intermediate bearing with its mountings is added.

5.4.1 Rack-and-Pinion Gear

Rack and pinion steering gears are almost universally found on racing vehicles. The steering gear ratio $i_{Sg,\mathrm{Rack}}$ (see Fig. 5.44a) is roughly selected in the range 6.5°/mm (360 °/55 mm) to 9 °/mm (360 °/40 mm). Thus, the total steering ratio i_S is about 7:1 to 8:1 [21]. Basically, there are several places where the pinion can be placed on or under the steering rack. The output to the tie rods does not have to be lateral either, Fig. 5.46 shows principle design possibilities.

If the teeth of the rack point upwards in the installed position, sealing of the rack passages on both sides is required. This prevents deposits (chips, small stones,. . .) from remaining between the tooth flanks and binding or lockinging-up the steering.

If self-steering behaviour is to be avoided when the wheels are sprung, the length of the steering gear must be smaller than the distance between the wishbone axes at whose height the steering gear is fitted. In this way, the tie rod joint points can be shifted via screwed-in rod ends until they lie on the wishbone axes. The minimum length of the rack is determined by the desired stroke, Fig. 5.47:

$$l = 2\left(3\frac{h}{2} + a + b\right) = 3h + 2(a + b).$$

In addition, there are the axial installation spaces of seals.

At the maximum possible steering angle, neither the tyres nor the wheels must collide with any part of the chassis. In addition, the steering linkage must not be punctured or

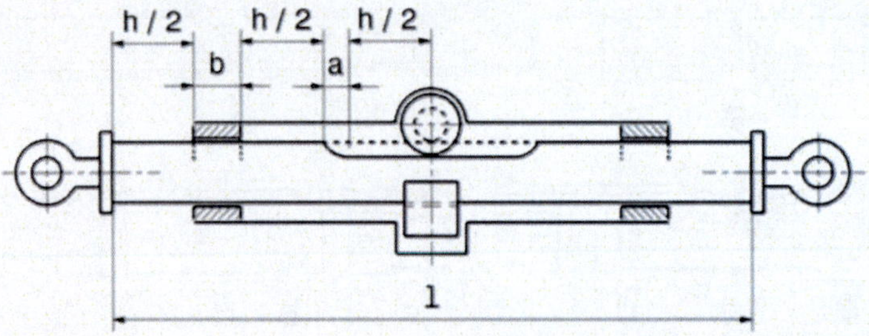

Fig. 5.47 Minimum length *l of* a toothed rack with laterally mounted tie rods. *h* Rack stroke, *a* Tooth runout, *b* Guide length

Table 5.3 Rack and pinion forces in series production vehicles, according to [23]

Axle load of the front axle, kg	Max. rack forces, N
550	6500
650	8000
850	8000
950	9000

become unstable. Therefore, a defined, mechanical steering stop (positive stop) is required. This is usually implemented using washers that are screwed onto the ends of the steering rack and run against the guide housing.

The rack diameter results from the largest axial force. This occurs at the highest steering wheel torque. On production vehicles, the so-called curb-stone push off test is carried out. In this test, no permanent deformations may occur on the steering rack when the front wheel is locked with the permissible front axle load and with full servo assistance with a steering wheel torque of 80 N m. The same considerations lead to the design of the tie rods for corresponding buckling stiffness.

The values in Table 5.3 can be used as reference values for the forces that occur.

A complete rack and pinion steering system of a passenger car is shown in Fig. 5.48.

The structure of a rack and pinion steering gear is shown in Figs. 5.49 and 5.50. The steering shaft is connected directly to the end of the pinion shaft (6). Common sizes of spline for connecting the steering shaft are in the range 12 × 14 to 17 × 20 (DIN 5481). The steering movement is transmitted to the rack (8) via the pinion. The pinion is held by two rolling bearings (4 and 7), whereby the deep groove ball bearing (4) acts as a locating bearing. A drawn cup needle roller bearing is located on the non-locating bearing side. The chosen design allows a space-saving construction of the bearing arrangement with low coefficients of friction. Both bearings are protected by the adjacent construction. The bearings are greased during mounting and are therefore maintenance-free for the operating life.

The spring-loaded thrust piece (9) ensures that the rack is pressed against the pinion without play. However, to prevent the rack from being pushed away by the tooth force to such an extent that the tooth meshing decreases impermissibly or even the pinion jumps

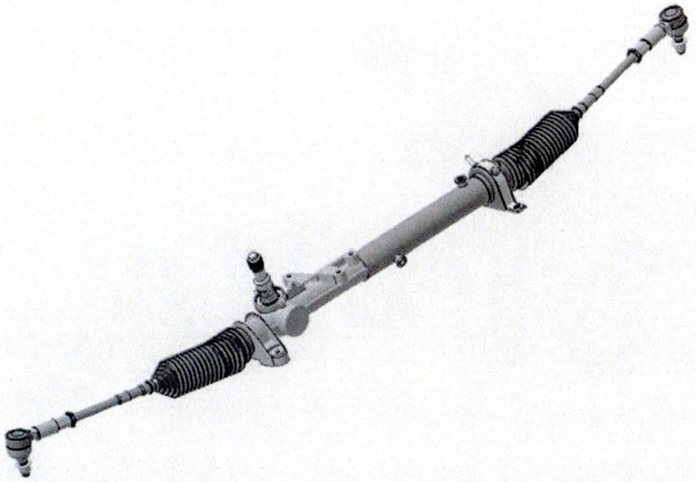

Fig. 5.48 Rack and pinion steering of a passenger car. The steering has a hydraulic assistance that acts on the rack. The left bracket of the steering gear (next to the steering shaft connection) is firmly bolted to the axle stub and at the right end an elastomer ring is held by a sheet metal bracket as a floating bearing. The two tie rods are attached to the side of the rack and the ball joints and the rack outlet are protected by bellows

over in the event of an overload, the maximum travel of the thrust piece is limited to the value s by the adjusting ring. The clearance s is adjusted with the spacer ring (12). The screw plug (13) is pulled against the ring (12). The O-ring (11) in the thrust piece acts as a damper and prevents rattling noises. The coil spring (10) has a preload force of 0.6 to 1.0 kN, depending on the steering size. The surface of both meshing partners should have a Rockwell hardness of at least 55 HRC. The frictional forces between the thrust piece and the back of the rack should be as low as possible for good response and feedback behaviour of the steering.

The gearing shown is a spur gearing. Helical gearing is also used for large gear ratios. The tooth engagement becomes softer due to the step overlap.

The rack does not have to have a constant pitch, but can have a variable pitch, Fig. 5.51. This allows a variable ratio to be represented in the steering gear itself.

Materials

Steering gear housing: Al sand casting EN AC-AlSi7Mg0.3 T6 (DIN EN 1706) with 230–310 N/mm^2 tensile strength, 190–240 N/mm^2 0.2% yield strength and 75–110 HB hardness.

Rack: Induction hardenable heat-treatable steels, e.g. Cf 53, 41 Cr4 (DIN EN 10083), C40 (DIN EN 10027–1) with 640 N/mm^2 tensile strength and 440 N/mm^2 yield strength,

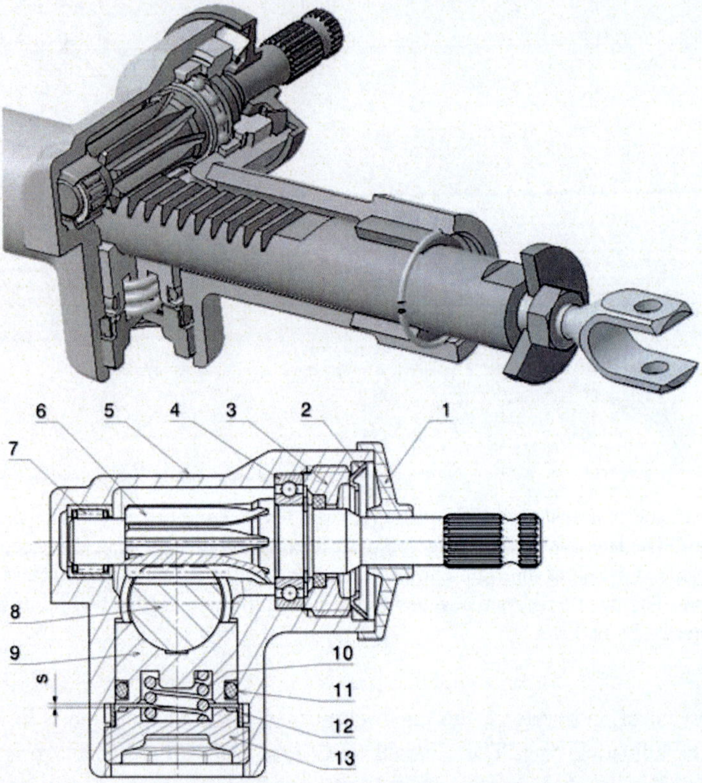

Fig. 5.49 Rack and pinion steering gear, section through pinion, after [2]. **1** Cover cap. **2** Locking plate. **3** Screw connection with sealing ring. **4** Deep groove ball bearing. **5** Steering gear housing. **6** Pinion. **7** Needle bearing. **8** Rack. **9** Thrust piece. **10** Spring. **11** O-ring. **12** Adjusting ring. **13** Screw plug
s max. 0.12 mm

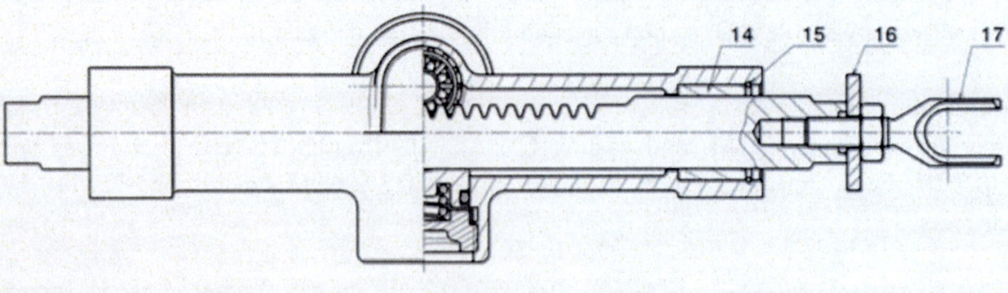

Fig. 5.50 Rack and pinion steering gear with side output, longitudinal section of Fig. 5.49. **14** Guide bush. **15** Snap ring. **16** Washer. **17** Fork (clevis)

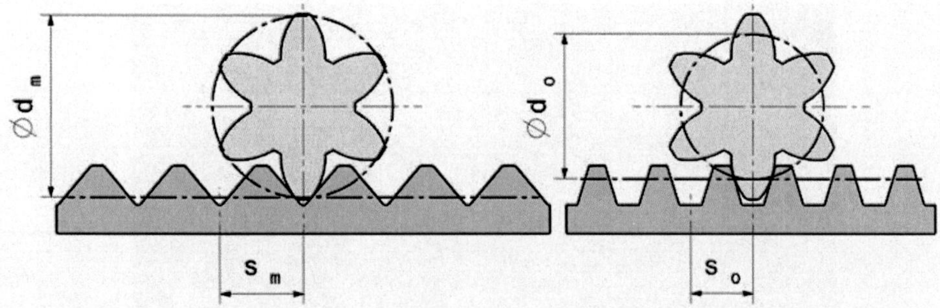

Fig. 5.51 Rack with variable pitch. In the middle of the rack, the pinion contacts the tooth flank at a larger pitch circle diameter d_m than on the outside, where the diameter d_o is effective. The displacement path of the rack thus decreases with increasing steering angle from s_m to s_o

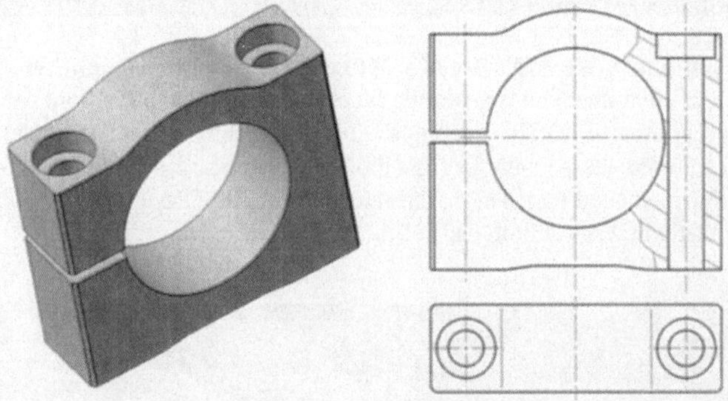

Fig. 5.52 Support bracket for rack-and-pinion steering gear. The steering gear housing is bolted to such a bearing block at both ends. The separating slot on one side enables targeted clamping of the tubular housing. The housing can be axially displaced within its bearing width, e.g. to compensate for manufacturing tolerances of the frame and chassis

37CrS4 as well as 41CrS4 (DIN EN 10027–1) with 775 N/mm^2 tensile strength and 620 N/mm^2 yield strength; titanium alloy PVD coated.

Pinion: Case-hardened steels, e.g. 20 MnCr5, 20MoCr4 (DIN EN 10008); titanium alloy PVD-coated.

Mount

Rack and pinion steering gears are firmly bolted to the frame with cast-on brackets or with separate support blocks (Fig. 5.52). On standard vehicles, a rubber mount is used with 2 bearings for reasons of comfort and tolerance.

Fig. 5.53 Rack and pinion steering gear of a formula car (Formel König). The steering gear is mounted at the end of the frame with two bearing blocks. (The nose with the front wing and the crash element is mounted in front of it.) The steering shaft opens slightly to the left so that the accelerator and brake pedals can be depressed with the right foot. The clutch pedal is depressed on the other side of the steering shaft with the left foot. The tie rods are in the plane of the upper wishbone. The steering stops are implemented via discs at both ends of the steering rack

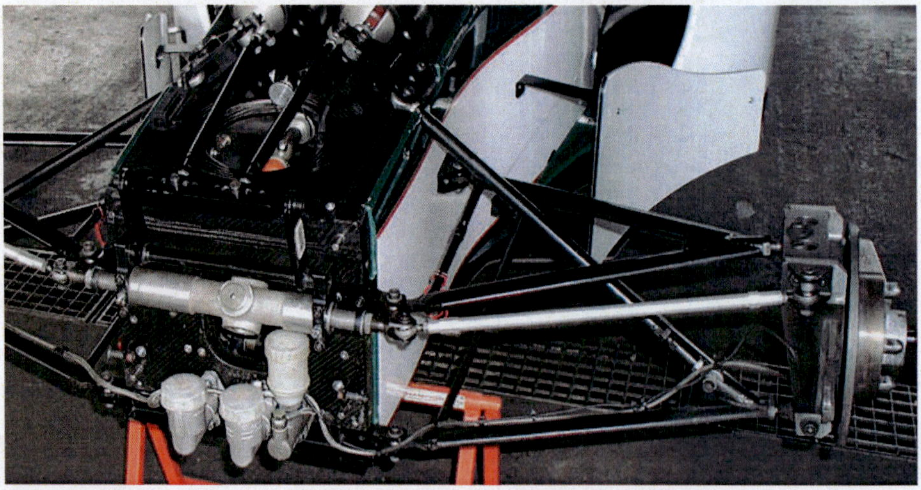

Fig. 5.54 Rack and pinion steering gear of a formula car (Formula BMW). The steering gear has a symmetrical design with lateral output. The tie rods lie in the planes of the upper wishbones

Examples

Figures 5.53, 5.54, and 5.55 show examples of rack-and-pinion steering gear designs and provide an illustrative system overview.

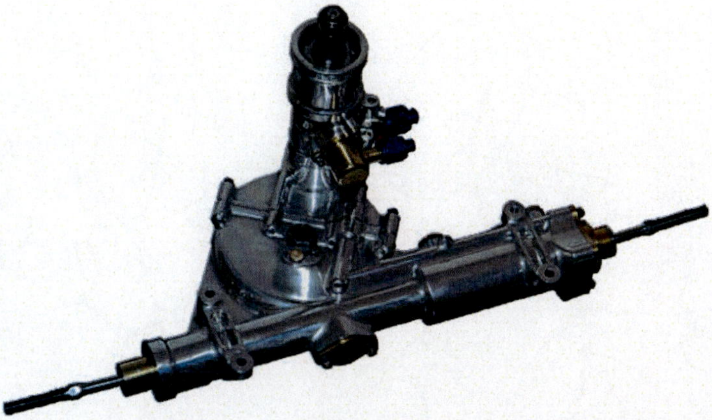

Fig. 5.55 Rack and pinion steering gear of a Formula 1 car (Ferrari). The steering gear is designed precisely for a particular car and features hydraulic support. It is screwed directly to the foremost bulkhead of the monocoque by means of the four integrated screw bosses. An intermediate gear causes a height offset between the steering shaft connection and the tie rod connections. The large intermediate gear is mounted eccentrically. By turning its bearing, the tooth backlash is adjusted at both meshing points

5.5 Transmitting Device and Axle Bearing

Regardless of whether rack-and-pinion steering or another steering gear is used, the movement must be transmitted from the steering gear fixed to the frame to the track levers on the wheel carrier side. In the case of independent wheel suspension, this is best done by means of linkages (movably coupled linkages) which, in addition to the steering movement, also have to accommodate the different wheel-travel movements during heave. Figure 5.56 shows some possible arrangements of linkages. In version a the axes of rotation of the steering gear and the opposite guide lever are parallel. The steering column lever forms with the guide lever and the middle part of the three-part track rod a flat articulated quadrilateral (more exactly parallelogram). The two levers actuate the outer track rod parts. The disadvantage of this design is friction. All six joints of the linkage make almost the full steering angle when steering. In addition, there is the undesirable influence of the play in the joints, which adds up. Version b corresponds largely to version a, except that here the axes of the steering gear and the guide lever are adapted to the inclination of the steering axis EG. This is particularly necessary with large kingpin inclination angles, because otherwise the self-steering behaviour becomes too great when the wheels are sprung. In version c, the middle section of the track rod is mounted in ball joints. This gives this rod a further degree of freedom, namely rotation about its rod axis. Therefore, the joint centres of the two outer track rods must lie on this axis of the middle track rod so that they cannot perform any undesired rotation. In variant d, the steering movement is transmitted from the

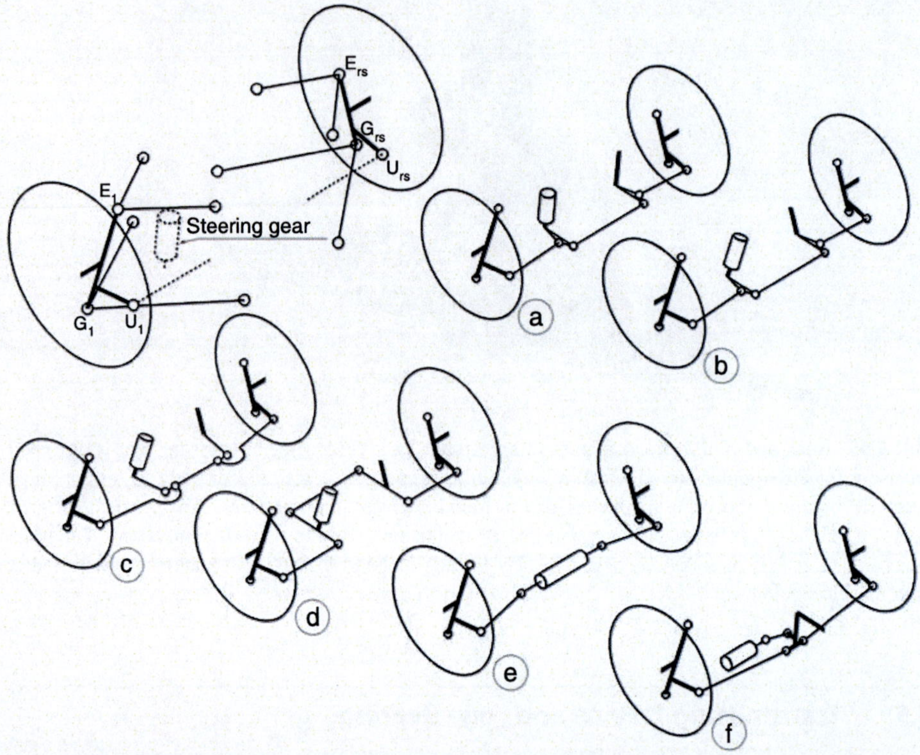

Fig. 5.56 Steering linkage for independent suspension. Starting situation top left: The two joints U_l and U_{rs} of the track levers on the left and right must be suitably connected to the steering gear. (**a**) Steering gear vertical. (**b**) Steering gear parallel to steering axis EG. (**c**) Middle track rod with ball joints. (**d**) Rocker arm as intermediate lever. (**e**) Rack and pinion as track rod. (**f**) Rack and pinion with intermediate lever

steering gear to the track rods via two bell cranks. This results in design free space in the middle of the vehicle (e.g. for the engine), but has the disadvantage of high reaction forces with the correspondingly large elastic deformations. Version e is a rack-and-pinion steering system. The simplicity and the small number of parts become obvious. In version f, the rack does not actuate the track levers directly, but a lever is interposed. The disadvantages result from a comparison with the above considerations. Compared to the simple rack, there are more parts installed, which besides the mass increase backlash and elasticity.

The basic arrangement of levers and push rods (in this case called tie rods) must take into account the transmission angles. These are decisive for the operational safety of the steering system, Fig. 5.57.

The steering linkage ratio results in:

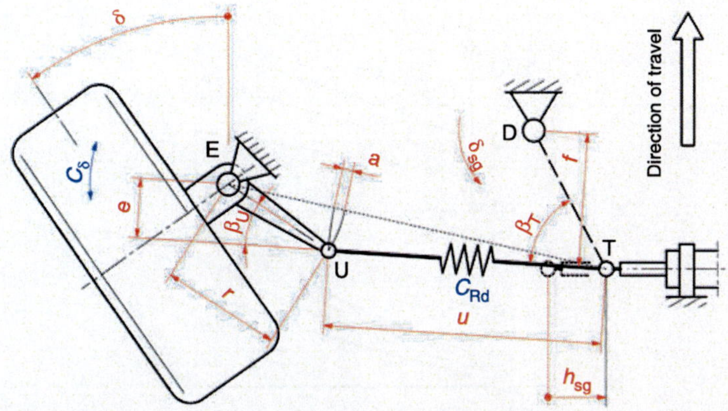

Fig. 5.57 Transmission angle for steering linkages, according to [5]. For a left front wheel, the steering linkage with rack and pinion gear and alternatively with a steering lever (dashed) is shown in point D. β_U, β_T Transmission angle, *u* Track rod length, *a* Length overlap of track lever and track rod, *r* Track lever length, c_{Rd} Stiffness of the tie rod, c_δ Torsional stiffness of a wheel about the steering axis, *e* effective lever arm of the track rod, *f* effective lever arm of the pitman arm, h_{Sg} Stroke of the gear rack

$$i_T = \frac{d\delta_{Sg}}{d\delta} = \frac{e}{f} \tag{5.41}$$

i_T	Steering linkage ratio, –
δ_{Sg}	Angle of rotation of the steering column lever, °
δ	Turning angle of the wheel, °
e, f	Effective lever arms, mm. (see Fig. 5.57)

If the transmission angles β_U or $\beta_T = 0°$, the linkage is unstable. With rack and pinion steering systems, the steering arm is omitted and the pivot point D is a remote point, but the steering arm remains. In particular, the angle β_U of the driven steering arm must not fall below a minimum value to prevent the steering linkage from deflecting. A minimum angle is necessary because play and elasticity must also be taken into account. The transmission angle should not fall below 25°. When assessing the safety reserves of the steering system, the track lever length *r* also comes into play. Figure 5.58 shows the course of important variables over the transmission angle β_U for a numerical example with the track rod length $u = 300$ mm.

From diagram c, which shows the decisive energy absorption up to the point of linkage penetration, it can be seen that larger transmission angles are required for short track levers if a minimum energy absorption is not to be undercut. In fact, the minimum transmission angles for passenger cars are between 20° for long and 30° for short track levers [5].

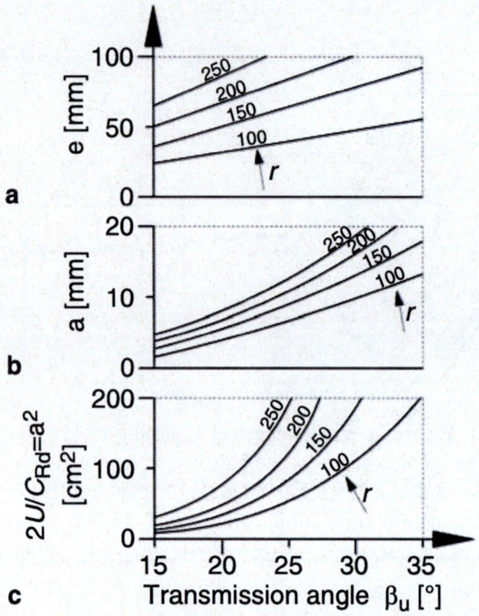

Fig. 5.58 Influence of the transmission angle on the operational safety of the steering system, according to [5]. (See also Fig. 5.57). $u = 300$ mm. (**a**) The effective lever arm e of the track lever around the steering axis E increases with the transmission angle and the track lever length r. (**b**) The length overlap a between the track lever and the track rod is a measure of the distance from the point of intersection. The smaller a becomes, the greater the risk of a breakdown. Large track lever lengths and large transmission angles are also desirable from this point of view. (**c**) Let the tie rod be the only elastic link in the transmission chain with stiffness c_{Rd} . Then the effective torsional stiffness of a wheel about the steering axis is $c_\delta = c_{\mathrm{Rd}}\, e^2$. This means that the energy absorption up to the point of overpressing the linkage $U = c_{\mathrm{Rd}}\, a^2 /2$

Rack-and-Pinion **Steering**

In this case, the steering input is transmitted directly from the stroke h_{Sg} of the steering rack via the tie rods to the steering arms. Like the steering gear ratio, the steering linkage ratio is also dimensionally dependent:

$$i_{T,Rack} = \frac{dh_{Sg}}{d\delta} = e \text{ from } e \cdot d\delta = dh_{Sg} \tag{5.42}$$

$i_{\mathrm{T,Rack}}$	Steering linkage ratio for rack and pinion steering, mm/rad
e	Effective track lever length (normal distance of the pivot point to the track rod), mm. (see Fig. 5.57). $e = r \sin(\beta_U)$, for small steering angles δ $\beta_U \approx 90^\circ - \delta - \lambda$. λ (see Fig. 5.62)

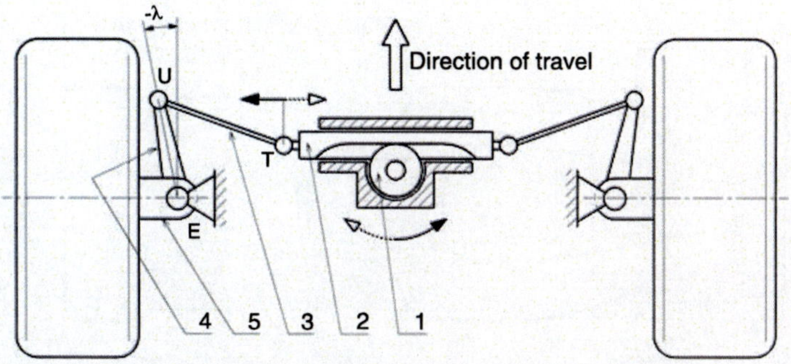

Fig. 5.59 Steering gear in front of the axle
The steering wheel movement is transmitted to the rack (2) via the gear wheel (1), which is located on the steering shaft. This in turn is articulated at its ends to the track levers (4) via the track rods (3). If the rack is shifted, the wheel carrier (5) turns about point E. To ensure that the steering movement is transmitted in the same direction by the steering wheel, the pinion (1) must be located under the steering rack. To fulfil the Ackermann condition, the track levers point outwards (angle λ negative)

The total kinematic steering ratio $i_{S,\mathrm{Rack}}$ corresponds to the ratio of effective track lever length to pinion radius:

$$i_{S,Rack} = i_{Sg,Rack} \cdot i_{T,Rack} = \frac{d\delta_H}{dh_{Sg}} \cdot \frac{dh_{Sg}}{d\delta} = \frac{e}{r_{Pi}} \tag{5.43}$$

r_{Pi}	Rolling circle radius of pinion in rack and pinion steering gear, mm. (see e.g. Figure 5.44a) Pos. 1. $dh_{Sg} = d\delta_H \cdot r_{Pi}$

The position of the steering gear (in front of or behind the axle) determines the arrangement of the pinion to the rack and the alignment of the track levers. The transmission chain must be such that when the steering wheel is turned to the right, the wheels are also turned to the right, Fig. 5.59. The steering movement of the driver is transmitted via the steering shaft to the pinion and thus to the rack.

The track levers can point forwards or backwards regardless of the position of the steering gear. However, to achieve a pure Ackermann steering, the levers must be inclined to the longitudinal plane of the vehicle, Fig. 5.60. The transmission links (tie rods and levers) and the front axle always describe a trapezoid and not a parallelogram in the plan view. That is why this arrangement is also called a steering trapeze. If the track levers point outwards, the track rods become longer with the same steering gear. When the wheels are sprung, there is then generally less relative movement of the track rods and thus less self-steering behaviour.

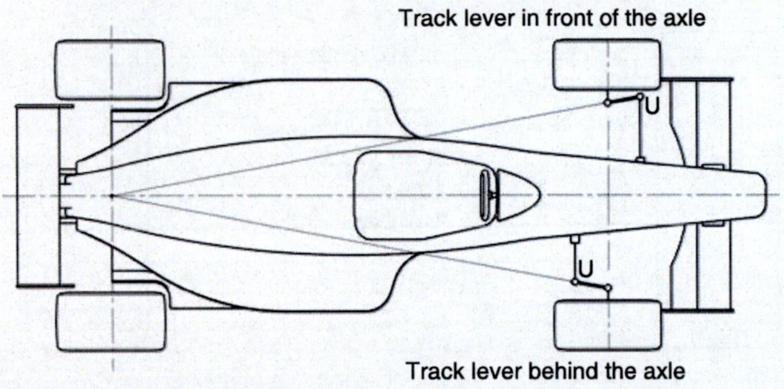

Fig. 5.60 Arrangement of the track levers for an Ackermann steering system. If the tie rod joint *U is* in front of the axle, the track lever must point outwards. In the other case with joint *U* behind it, the lever points towards the body. For ideal Ackermann angles, the lever extension to the rear meets the intersection of the rear axle with the centre of the vehicle

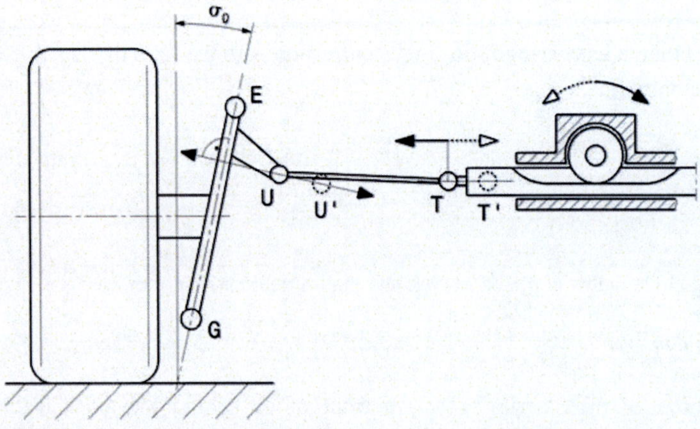

Fig. 5.61 Movement of the track rod during steering. During steering, the point *T* with the rack moves parallel to the ground and becomes *T′*. The second connection *U of* the steering rack rotates around the steering axis *EG* and becomes *U′*. In this example, there is no caster angle, so in the view from behind, the circular path is a normal to the steering axis

However, the basic alignment of the track levers does not guarantee that the desired course of the track difference angle (see Fig. 5.7) is achieved. The movements of the articulated points T and U of the track rod are fundamentally different during steering, Fig. 5.61. The rack joint T moves on a straight line transverse to the direction of travel and the connection U on the track lever rotates around the steering axis$\overline{EG}$, thus describing a circular path in space.

In the design of the steering system, the articulation points determined by the methods described below must be finely adjusted so that the actual curve of the steering angle resulting from steering approaches the desired curve (target curve) as closely as possible.

Basically, there are several possibilities to arrange the steering gear and steering linkage opposite the front axle. The steering gear can be located in front of or behind the axle and independently of this, the track levers can point forwards or backwards. The pinion can be located above or below the steering rack. Figures 5.62, 5.63, and 5.64 show principle arrangements with approximately the same steering geometry. The tie rods provide the articulated connection between the steering rack and the steering levers and transmit push/pull forces.

Ideally, there should be no toe-in change of the wheels, i.e. no steering movement, when the wheels are deflected in and out. In this context, the position of the tie rod pivot points T and U in relation to the frame and chassis is important. If there is a relative movement

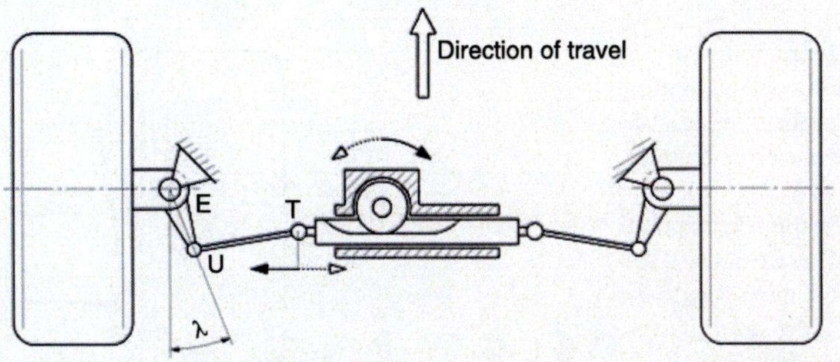

Fig. 5.62 Steering behind the axle. If the steering gear is located behind the front axle, the track levers point inwards. The steering gear is asymmetrical, as is usual for two-seater vehicles (left-hand drive). The pinion sits above the rack. The tie rods are attached to the ends of the steering rack

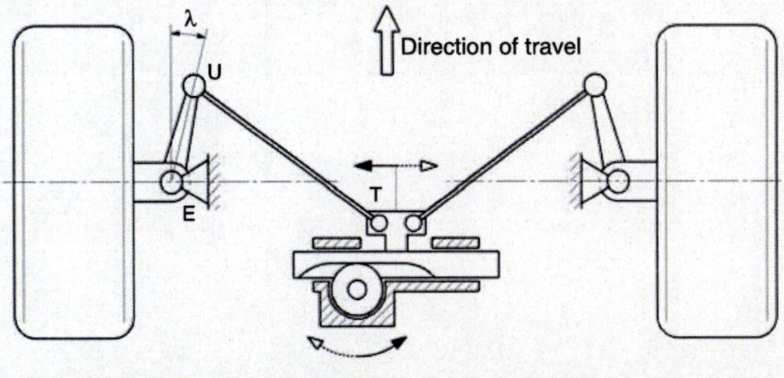

Fig. 5.63 Steering behind the axle. The steering gear is located behind and above the centerline of the front axle. The track levers point forward. To ensure that the track rods can still be long, which is the aim for kinematic reasons, the connection to the steering rack is made in the middle ("centre tapping"). If the tie rods are too short, undesirable self-steering behaviour occurs during compression and extension. The pinion sits under the rack

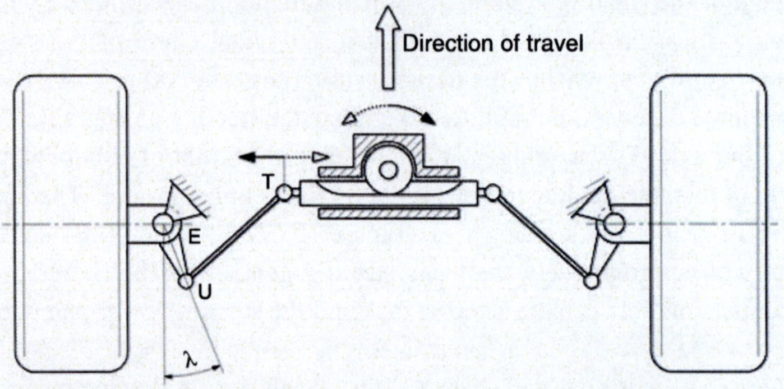

Fig. 5.64 Steering in front of the axle. With this arrangement, the steering gear is in front of the axle and the track levers point backwards and inwards. The pinion sits above the rack

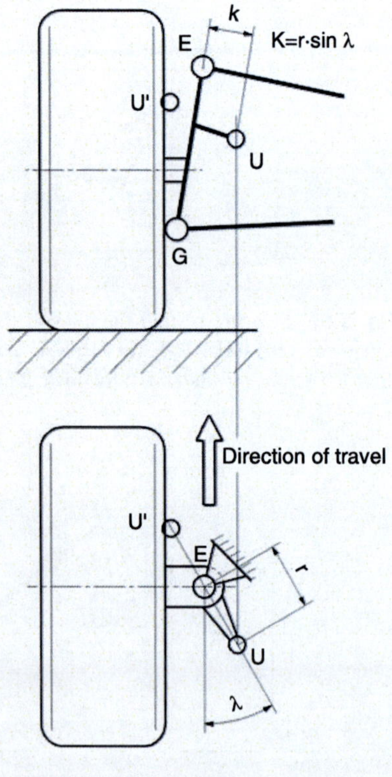

Fig. 5.65 Determining the track lever pivot point U. Before the position of the steering gear and track rod can be determined, the connection to the track lever is needed. The track lever points outwards (U′) or inwards (U), cf. Figure 5.60, thus lying to the left or right of the steering axis EG in the rear view. The distance k can be determined from the specified values of the angle λ and the track lever length r

between the pivot points during springing, a steering movement of the wheels is the inevitable consequence (springing steer, bump steering).

Before kinematic tests are carried out, the toe lever must be determined, Fig. 5.65. Usual toe lever lengths r are roughly in the range around 100 mm.

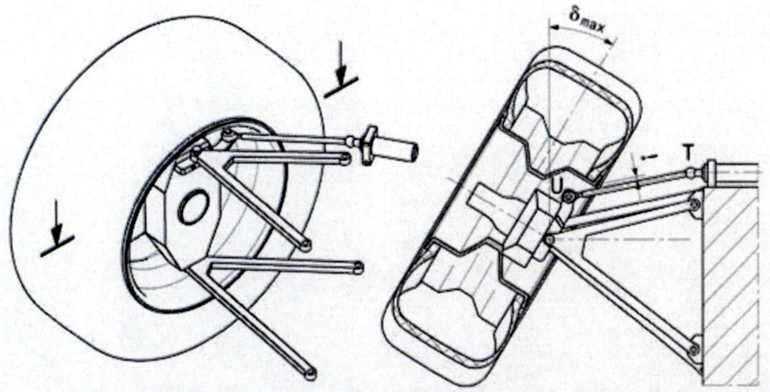

Fig. 5.66 Required free travel of the steering linkage. The left front wheel is shown. Even with the largest steering angle δ_{max}, the tie rod UT must be at a distance from the rim. The available space is thus restricted. It can also be seen that the space for the track lever becomes smaller the higher it is within the rim

Fig. 5.67 Position of the tie rods without self-steering behaviour. Top view (bottom) and rear view (top). In the straight-ahead position, the track rod with the joints T and U lies in the plane of the upper triangular wishbone with the connection points E, C and F. The rack-and-pinion joint T lies on the axis of rotation CF of the wishbone. The connection U between track rod and track lever also lies in the plane of the wishbone

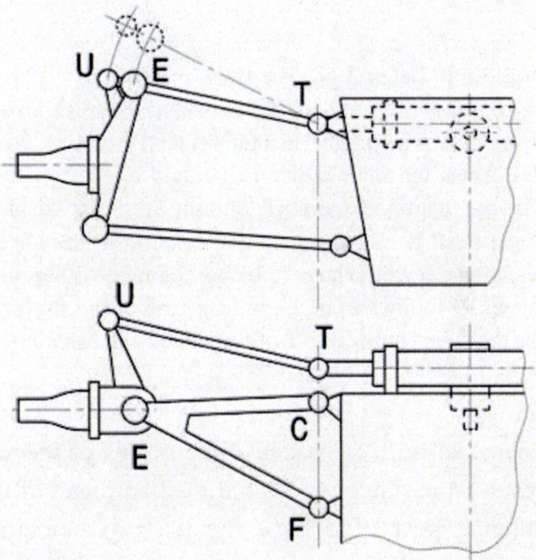

Another consideration for the position of the track lever and the track rod results – with front track levers – from the available installation space within the rim. At the largest turning angle of the wheels, there must be no collision between the track rod and the rim. On the contrary – one will even leave a safety distance, so that despite elasticities in the steering there is always a distance between the parts, Fig. 5.66. If a large steering angle is needed, it may be necessary to move the track lever in vertical direction towards the wheel centre, where the rim gives the largest space.

If the track lever point U is fixed, the second connection of the track rod T must be found. This is possible, among other things, with the aid of instant centers, Figs. 5.67, 5.68, 5.69, 5.70, 5.71, and 5.72. If the connection point T is fixed, the position of the steering

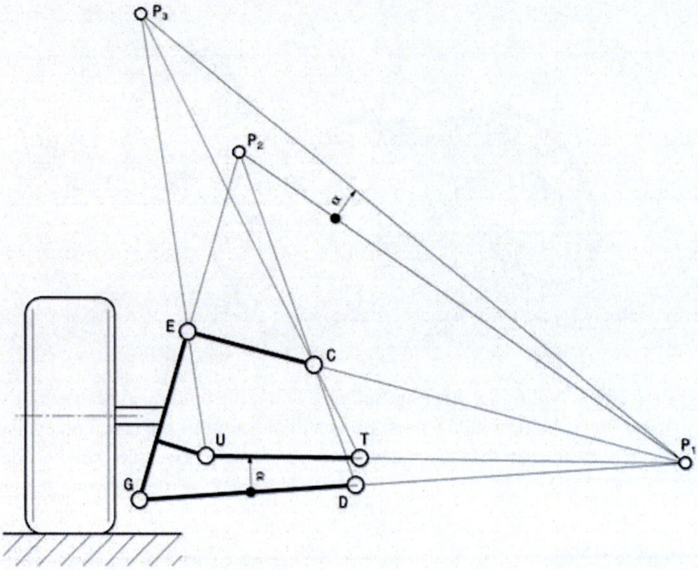

Fig. 5.68 Determining the track rod position. The articulation points of the wishbones are known, namely E and C or G and D, as well as the track lever connection U. This allows the instant center P_1 to be determined. The instant center P_2 results from the intersection of the straight lines GE and DC. Then, the angle α that the straight line UP_1 includes with the lower wishbone GD is determined. The orientation of the angle starting from the wishbone is important, because in the same sense the angle must be plotted from the junction of the instant centers P_1 P_2. That is, if U were below the wishbone, α would have to be plotted in the other direction from the line P_1 P_2. The track rod instant center P_3 follows from the intersection of the straight line UE with the angular arm drawn last. Now the required connection T of the tie rod can be determined. It results from the intersection of P_3 C with P_1 U

gear is also given, because the points of the same name on the other side of the vehicle result from reflection on the median plane of the vehicle. An obvious way to arrange the steering gear is to have the tie rods coincide with the planes of the upper or lower wishbones. If the connection points T of the steering rack are then on the axes of rotation, there is no steering movement during springing when driving straight ahead, Fig. 5.67.

If the steering gear has to be positioned high, the track rod can also be positioned above the upper wishbone, Fig. 5.69.

If the control arms are parallel in design position, the track rod is also parallel, Fig. 5.70.

The position of the track rod can also be used for other wheel suspensions by considering the instant centers. Figures 5.71 and 5.72 show the implementation for the wheel-leading suspension strut axle (McPherson). It can be seen that the higher the linkage point U of the track rod to the track lever, the further the linkage point T to the rack moves towards the centre of the car. This in turn can lead to problems if the track width is small and the steering gear is long. In extreme cases, the only solution is to attach the tie rods not at the ends of the steering rack, but in its centre, see Fig. 5.46c.

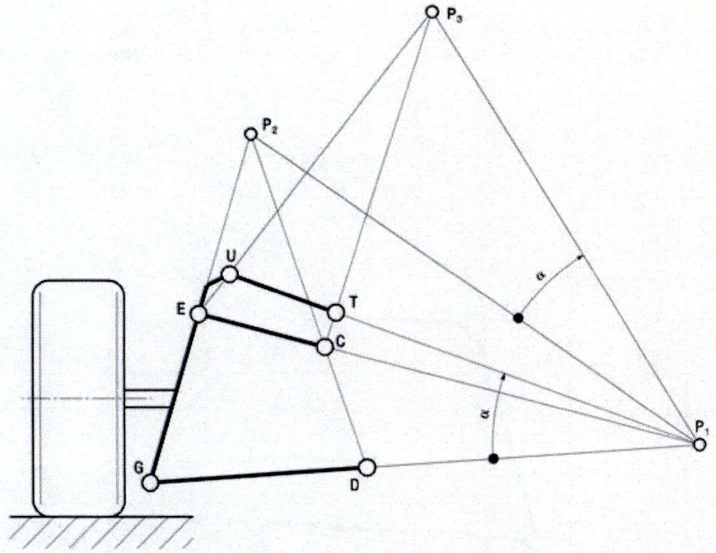

Fig. 5.69 Determining the track rod position. The procedure in the figure is the same as in Fig. 5.68. The only difference is the location of the toe lever UE. It is mounted above the upper wishbone EC and points inwards towards the rear. This also results in the connection T to the toothed rack above the wishbone

The methods described above assume a flat model, which hardly ever occurs in real vehicles. The points of the steering linkage determined in this way can therefore only be used as reference points. The exact determination of the steering linkage is done by the designer by "trying" on drawings or on the computer. This enables him/her to reach his/her goal more quickly than with complex auxiliary constructions that take into account the spatial character of the steering and the chassis [5].

If steering movements occur on the assembled vehicle during springing, the behaviour can be improved by moving the steering gear or by changing the length of the track rods, Table 5.4.

Tie rods must be finely adjustable in length so that toe-in/toe-out and self-steering behavior (springing steer, bump steer) can be adjusted. Similar to the push rods of the suspension, it is also advantageous here if this can be done without disassembly work. For this purpose, the track rod can be constructed in several parts and have an adjustment thread in between, which still needs an appropriate locking device. The design for a typical passenger-car steering system is shown in Fig. 5.73.

The connection to the steering arms is made with ball joints as described in Sect. 2.3.3 *Joints*. The connection to the rack is made by ball joints that can be screwed on or screwed in, which allow the spatial movement between the steering gear and the track lever and are nevertheless stiff enough in the longitudinal direction for a high steering precision. Figure 5.74 shows possible designs.

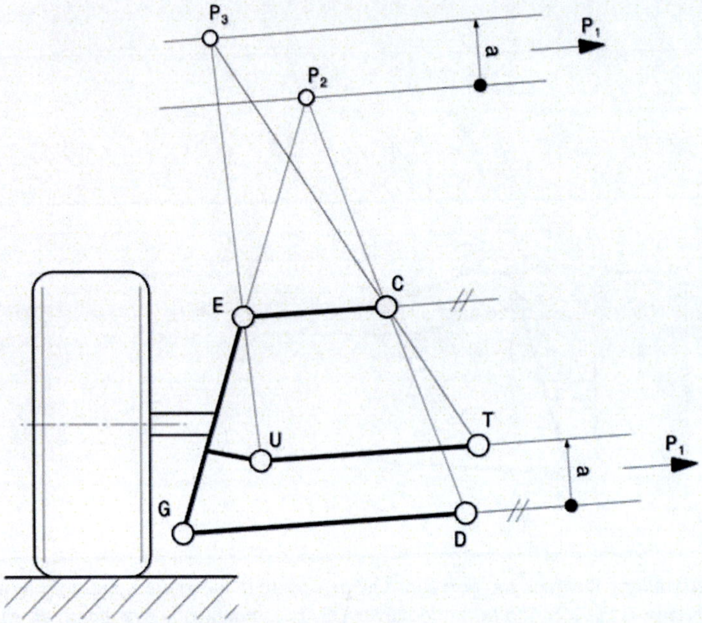

Fig. 5.70 Determining the track rod position. The procedure in the figure corresponds to a large extent to that in Fig. 5.68. The articulation points E, C and G, D are thus known as well as the track lever connection U. Only in the present case, the instant center P_1 is a distant point in the case of parallel wishbones. To determine P_3, the straight line UE is now intersected with a parallel line at a distance a (orientation, i.e. up or down, again important as in Fig. 5.68) from P_2 . With the track rod instant center P_3 the rack linkage point T follows directly from the intersection of the straight line P_3 C with the track rod

For racing vehicles, tie rod connections via rod ends are a good option. The tie rod itself can be designed as a straight tube with a fork or a screw-in thread at the end for a spherical plain bearing. The end of the rack then has a counterpart for the joint connection. The length adjustment is made by a left-right thread combination in the same way as for the push rods of the chassis (see Sect. 4.2.2 *Components of Double Wishbone Axles*), Fig. 5.75.

The axle bearing, which also transmits the wheel loads (corner weight), must be sufficiently dimensioned and have a correspondingly large number of degrees of freedom. It must allow the maximum wheel steering angle and at the same time permit the lifting movement of the wheel. At this point, spherical plain bearings are suitable, which are characterised by a particularly large swivel angle, Fig. 5.76.

An executed steering system of a formula vehicle can be seen in Figs. 5.77 and 5.78.

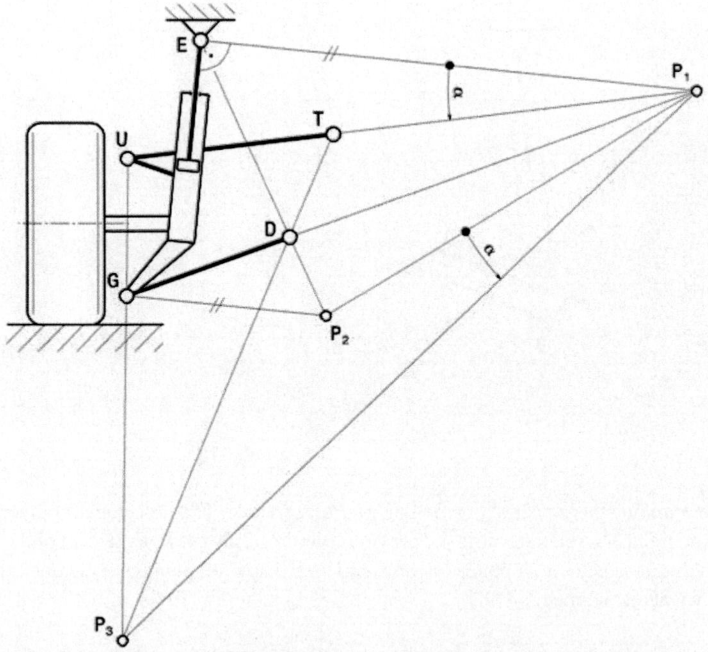

Fig. 5.71 Determining the track rod position on McPherson axle. At this McPherson-axle, the track-arm with its connection U is positioned above and in front of the front-axle. Besides the swivel-knuckle with joints E and G, also the wishbone GD is given. With the body-fixed joint E and the wishbone, the cross-instant center P_1 is constructed. P_1 is the intersection of the normal to the direction of movement of the damper strut with the extended wishbone. The instant center P_2 is now obtained from the intersection of a parallel to the instant center ray P_1 E through G and the straight line ED. The angle α is plotted from the straight line P_1 P_2 in the same sense as the measured angle between P_1 E and P_1 U. The straight line UG can then be used to construct the instant center P_3. With the track rod instant center P_3 the second track rod point T follows directly, because it lies on the straight line P_3 D

5.6 Shimmy (Vibrations of the Steering System)

The steering system consists of several masses that are movably coupled to each other and have at least one degree of freedom in relation to the rest of the vehicle. This system is therefore capable of vibrating. In fact, two types of vibrations are primarily noticeable. Firstly, the steering unsteadiness as a vibration in the range of the natural frequencies of the unsprung masses, i.e. at 10–15 Hz. Secondly, lurching in the range of the rolling frequency of the vehicle. This is around 2 Hz [5].

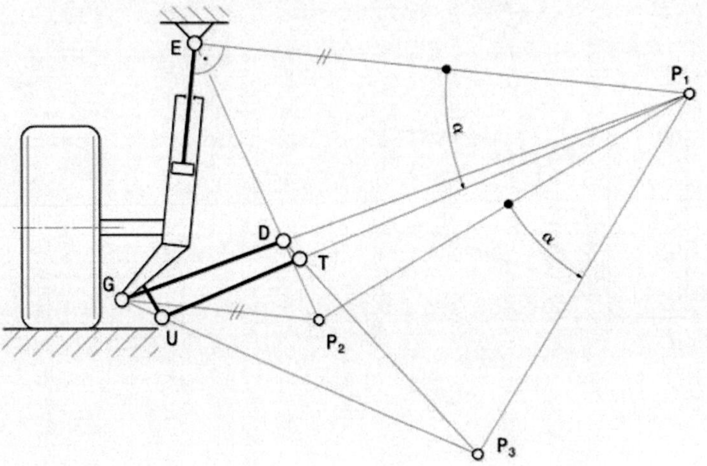

Fig. 5.72 Determining the track rod position on McPherson axle. The procedure for determining the second track rod point T is the same as in the previous picture. Only the arrangement of the track lever is different. In this arrangement, it points inwards and its articulation point U to the track rod can also be below the wishbone connection G

Table 5.4 Remedial measures for self-steering behaviour, according to [11]

Steering movement during		Remedy	
Bump	Rebound	Displacement of the steering gear	Rack length
Toe-in	Toe-out	In front of the axle: lift Behind the axle: lower	–
Toe-out	Toe-in	In front of the axle: lower Behind the axle: lower	–
Toe-out	Toe-out	–	In front of the axle: extend Behind the axle: shorten
Toe-in	Toe-in	–	Before the axle: shorten Behind the axle: extend

Niggle (Steering Unrestiggle)

The excitation is caused by the rotating wheels, i.e. by imbalances of wheel and tire as well as by unequal tire stiffnesses. The tyre stiffness acts on the steering like the wheel load via the wheel load lever arm q. Therefore, the smaller the lever arm q, the smaller the influence of wheel load fluctuations on the steering. Unbalances act in the entire wheel plane and also as deviation moments. They therefore act on the steering axis via the wheel load lever arm and via the wheel-center offset.

Roll

Lurching is basically a torsional vibration of the steering wheel about its axis of rotation. Such torsional vibrations can be triggered during rolling by imbalanced wheels or during

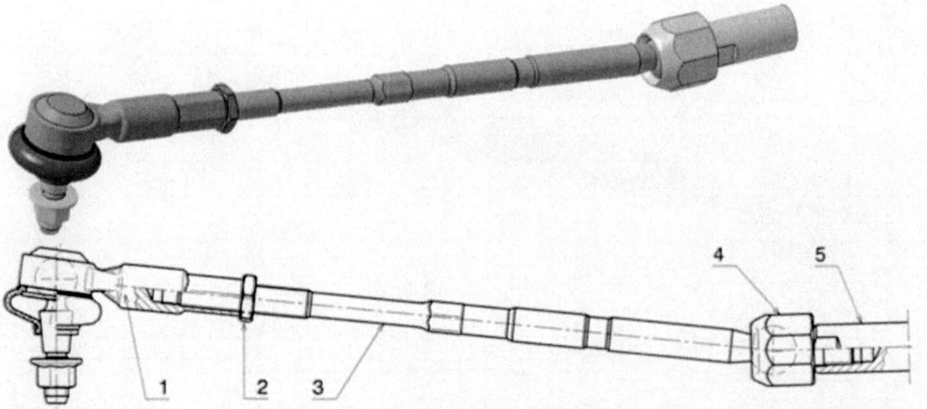

Fig. 5.73 Track rod of a passenger car. Shown is a tie rod connecting a rack end to the track lever. **1** Ball joint to the track lever . **2** Lock nut. **3** Tie rod. **4** Screw-in joint to rack. **5** Rack

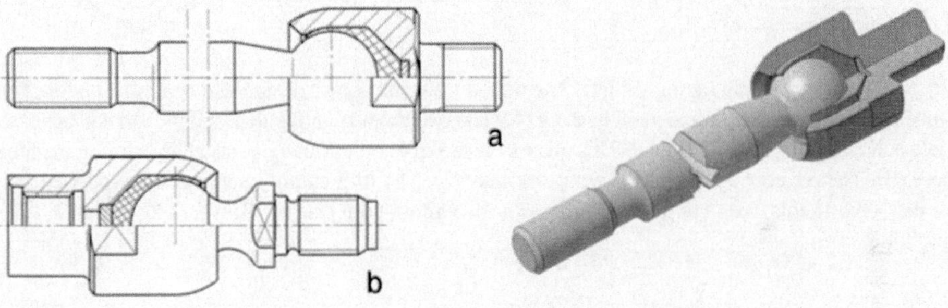

Fig. 5.74 Ball joint connections for racks. These connections are for rack and pinion gear units with side output. (**a**) Housing is screwed into the rack. (**b**) Ball stud is screwed into the rack

braking by brake force fluctuations. Rotational vibrations of the steering wheel typically occur between 10 and 30 Hz [1]. Although the polar moment of inertia of a steering wheel per se is small, in the oscillating steering system it acts via the steering ratio. In the reducing energy consideration ($E = 0.5{\cdot}J_H\ \omega_H^2$) *of* the steering system to a reference angular velocity, the transmission ratio is squared ($E_{red} = 0.5{\cdot}J_H\ i_S^2\ \omega_{reference}^2$). Depending on the steering ratio, influential values of the order of the moment of inertia of the vehicle about the yaw axis can thus become effective. Accordingly, it is obvious to provide a steering wheel with as small a mass moment of inertia as possible. This raises the natural frequency at which roll occurs. A disadvantage of this is that such a steering wheel also "filters" the steering unsteadiness less, i.e. passes it on more perceptibly to the driver.

It is generally important that the excitation and natural frequency are far enough apart in operation. When passing through the resonance outside the operating cycle, sufficient damping must be available. Damping and natural frequency position can already be varied

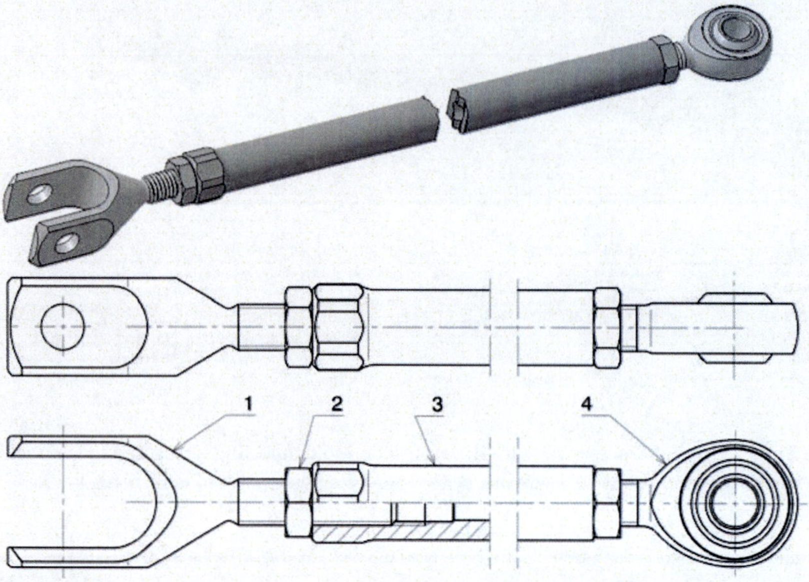

Fig. 5.75 Track rod of a racing vehicle. The tie rod has a multi-part design and consists of a fork (1), a tube piece (3) and a spherical plain bearing (4). The two threads on the fork and the rod end are each a left-hand and a right-hand thread. This allows the toe-in to be infinitely adjusted by turning the tube piece (3). The connection is secured with lock nuts (2). The fork can of course also be screwed into the rack end. In this case, the track rod carries a rod end on both sides. 1 Clevis. 2 Lock nut. 3 Tube piece. 4 Rod end

in some steering systems by using different control arms [1]. Figure 5.79 shows an example of this.

Steering *Damper*

As with any oscillating system, dampers can also be interposed or added to the steering system. In passenger cars, there have been solutions with elastic elements between the track rods and the track levers, but the steering precision suffers noticeably from such installations [24] and these systems also increase the masses involved.

Steering dampers connected in parallel help to inhibit lurching. The operating frequency of the steering unsteadiness is generally too high for conventional steering dampers. They respond much too sluggishly for this application.

Pressureless monotube dampers (see Sect. 3.2.2 *Vibration Dampers*) are used as steering dampers. As a rule, they are installed horizontally and steering dampers must not have any piston rod extension force, as is the case with gas pressure dampers. Otherwise the steering would be hit by the damper towards the pressure side.

Fig. 5.76 Axle bearing on a wishbone with push-rod connection (Reynard D94 F3000). The right front wheel is shown. As supporting bearing (supporting joint, here lower bearing) is installed a spherical plain bearing with extended swivel range, i.e. cylindrical extension at inner ring. The bearing is arranged upright so that it transmits the wheel loads advantageously as radial forces to the push rod (compression strut)

Fig. 5.77 Front suspension of a racing car (Formula Renault 2000), left side of vehicle. The track rod (1) rests against the stop of the steering gear (not visible in the picture), i.e. the maximum steering angle to the right (approx. 16°) can be seen. The pivot point of the steering lever for the track rod forms a steering trapezoid with the pivot point of the upper wishbone

Fig. 5.78 Front view of the vehicle from the above picture (left side of the vehicle). The track rod lies in the plane of the upper wishbone. The collar (1) on the steering rack represents the stop of the steering movement

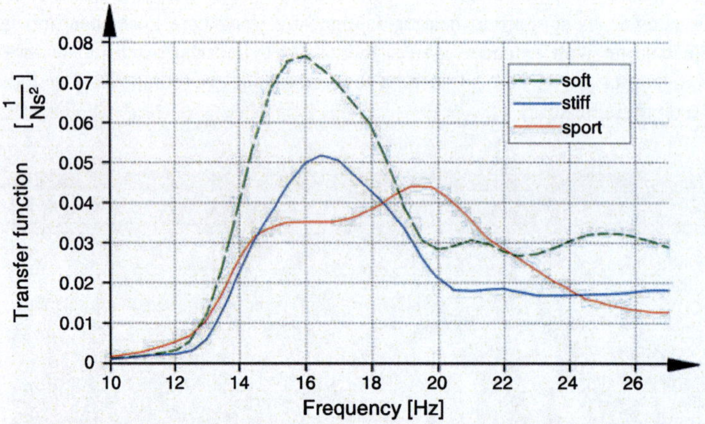

Fig. 5.79 Transfer function of the braking torque variation with different wishbones, according to [1]. The transfer function relates the system response, the amplitude of the steering wheel torsional vibration (a_{SWV} in m/s^2), to the excitation, the braking torque variation (M_B in N m). The investigation was carried out for three different stiffness control arm variants. The softest version is the most permeable in the middle frequency range

5.7 Four-Wheel Steering

Elastokinematic rear suspensions on passenger cars are usually aimed at ensuring understeering handling, e.g. when braking in a corner. Such deliberate toe-in changes of a wheel are, of course, dependent on tire force, only work in the smallest angular ranges and only help at high speeds. The idea is therefore to steer the rear wheels in a targeted manner and to achieve a benefit at high as well as low speeds. However, the effort involved is enormous: the steering angle must be controlled depending on the driving speed, among other things. Additional input variables are the steering wheel angle and the yaw rate of the car. For safety reasons, these values must be recorded several times, preferably using different methods. An electronic control unit determines the driving condition from these input variables and calculates the optimum rear wheel steering angle. This is set by an actuator and checked by a displacement sensor and reported back to the control unit.

The effort seems to be worthwhile because some car manufacturers have already put such systems into series production (Honda, Nissan, Mazda, Toyota in the late 1980s and most recently Nissan in 2009). Such systems are generally referred to as auxiliary steering equipment. In passenger cars, the rear axle stabilizes during rapid lane changes with steering in the same direction as the front wheels and helps reduce steering effort when parking by turning the rear wheels in the opposite direction to the front. Additional steering can also improve the lateral force distribution when cornering, Fig. 5.80.

All-wheel steering can also be worthwhile for racing vehicles. For example, it can solve the problem of power understeer, which occurs in heavily powered vehicles with enormous tire grip on the rear wheels. Thus, the driven rear wheels take over the main steering action for the overtaxed front wheels [11]. The auxiliary steering also helps to solve the trade-off in determining the wheelbase. With long wheelbases, manoeuvrability in tight, slow bends is increased by steering the rear wheels without losing the advantage of a long wheelbase on fast straights by no longer steering the rear wheels.

In Formula 1, the Benetton team developed rear-wheel steering as early as 1993, but it was generally banned in this series the following season [26]. Other regulations, such as GT series, allow this. The following considerations should help in deciding whether four-wheel steering is worthwhile in racing cars. The achievable lateral acceleration is basically the same for vehicles with and without all-wheel steering. The advantage of additional steering is the faster build-up of lateral force at the rear axle and thus the faster lateral acceleration gain. The yaw reaction is considerably less and the (body) sideslip angle (attitude angle) remains smaller – the driving behaviour more stable.[9] This is noticeable to the driver in an advantageously smaller phase delay between steering input and lateral acceleration, Fig. 5.81. At high speeds, the steering ratio (with constant kinematic ratio) can be set more indirectly for the driver by additional steering of the rear wheels and relieve

[9] However, a small sideslip angle is not the only criterion for stability. This also includes, for example, load change behavior and transition behavior in the limit range.

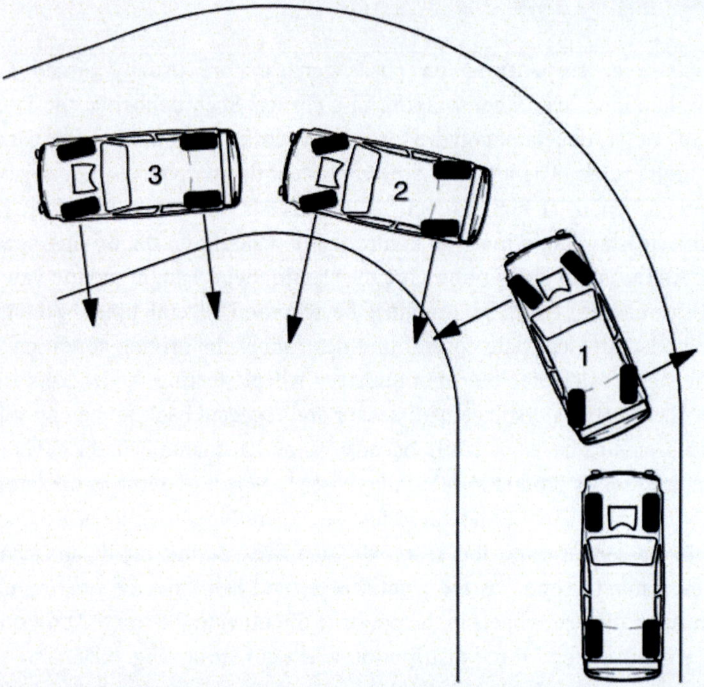

Fig. 5.80 Course of wheel displacement with additional steering, according to Nissan [25]. The position of the rear wheels is not directly linked to the steering angle of the front wheels, but depends on the driving condition. **1** Turning in (corner entry): By turning the rear wheels in opposite directions, the yaw speed required for stable cornering is achieved more quickly. **2** Cornering: Stable cornering with the greater lateral forces on the front wheels. **3** Unwinding steering wheel (corner exit): The lateral forces on the front axle are reduced

him. The load change behaviour can also be avoided with such a measure. The transition to the limit range at the rear axle is smoother with additional steering. A classic drift based on power oversteer can thus be initiated more elegantly.

Setting a sideslip angle of zero, as was the aim of all-wheel steering in passenger cars in the past, is not desirable and leads to understeering and subjectively unfamiliar driving behaviour. A sideslip angle increasing with the lateral acceleration a_y with a maximum value of approx. 1° is better [22], Fig. 5.82. Figure 5.83 shows an example of an executed characteristic curve of an all-wheel steering system of a passenger car.

5.8 Alternative Steering Systems

The Formula 1 team McLaren developed a steering brake (fiddle brake) for the car of the 1997 season, in the manner known from tractors and off-road vehicles. A fourth pedal was placed in the footwell, with which the left or right rear wheel could be braked alone via a

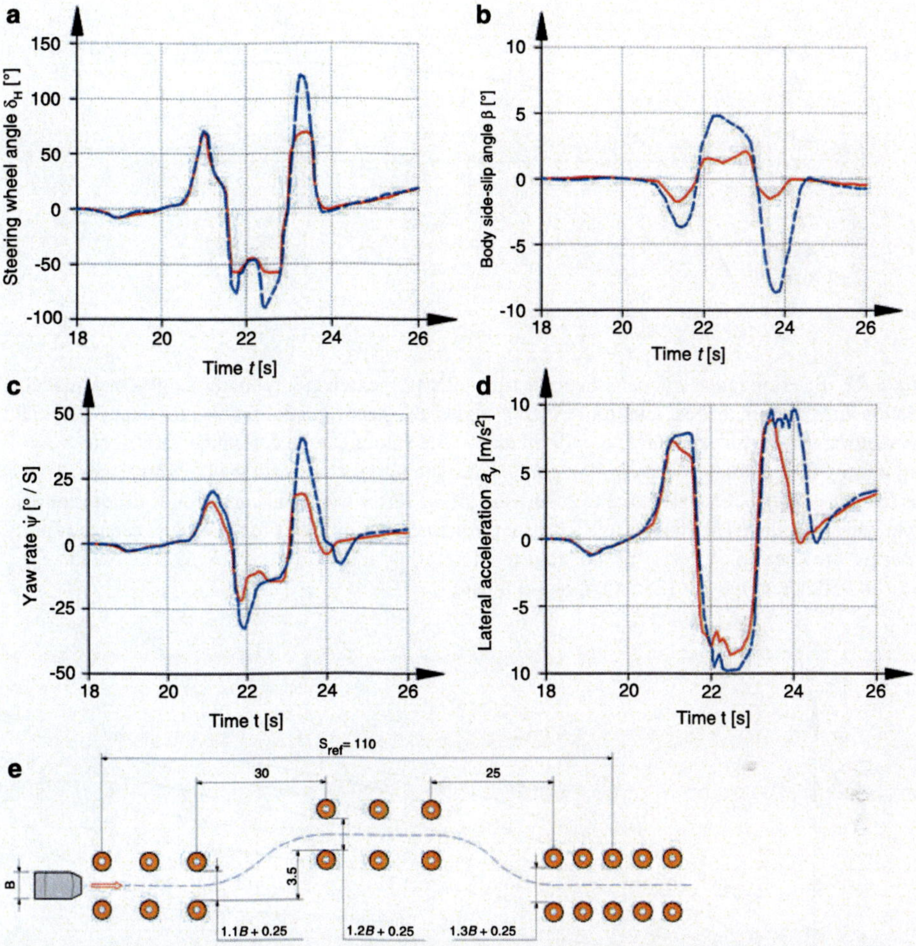

Fig. 5.81 Simulation of ISO lane change, according to [1]. (**a** to **d**) Course of significant driving dynamics values over time with all-wheel steering (red) and conventional front-wheel steering (blue dashed). (**e**) Main dimensions of the driving section of the double lane change according to ISO 3888 in m. *B* Width of the test vehicle, s_{ref} Measuring distance. During the test, the vehicle enters the first lane at a constant speed and follows the path defined by the pylons at a constant accelerator pedal position without touching the marking cones. Among other things, the passage time over the measured distance s_{ref} is measured. Typical speeds for passenger cars are around 80 km/h, sporty vehicles reach around 130 km/h

selector switch on the steering wheel. At the start to midpoint of a tight, slow left-hand turn, for example, the driver applied braking force to the left rear wheel, increasing the car's yaw in the turn while directing drive torque to the outside wheel of the turn, even with an open differential. This effectively combated the troublesome understeer in the cornering phase. The lap time advantage was reported by the developers to be around 0.75 s [26]. It depends

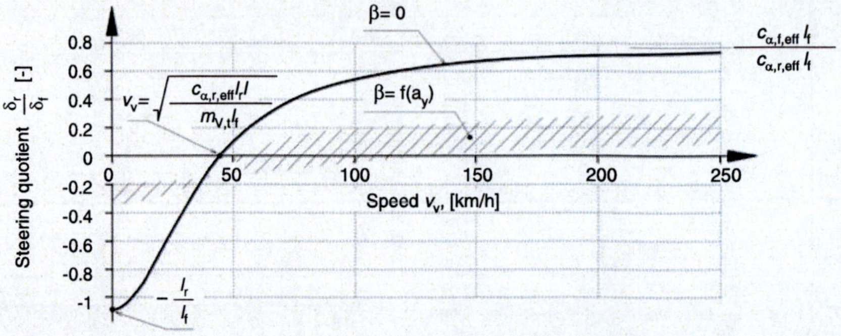

Fig. 5.82 Steering ratios with all-wheel steering, after [22] extended. From the single-track model, a relationship between the rear and front steering angles, the steering ratio, follows for steady states and the requirement for sideslip angle $\beta = 0$. Similarly, the values can be determined at velocity $v_V = 0$ and infinity (asymptote). An interesting point is also the intersection with the horizontal axis. This is the transition from counter-steering to co-steering of the rear wheels and corresponds to conventional front wheel steering. The shaded area is where the actual steering ratios for passenger cars and sporty vehicles are located. Values for the registered steering ratio: $c_{\alpha,\ f,\ eff} = 59{,}330$ N/rad, $c_{\alpha,\ r,\ eff} = 70{,}950$ N/rad, $m_{V,t} = 1356$ kg, $l_f = 1.3$ m and $l_r = 1.4$ m

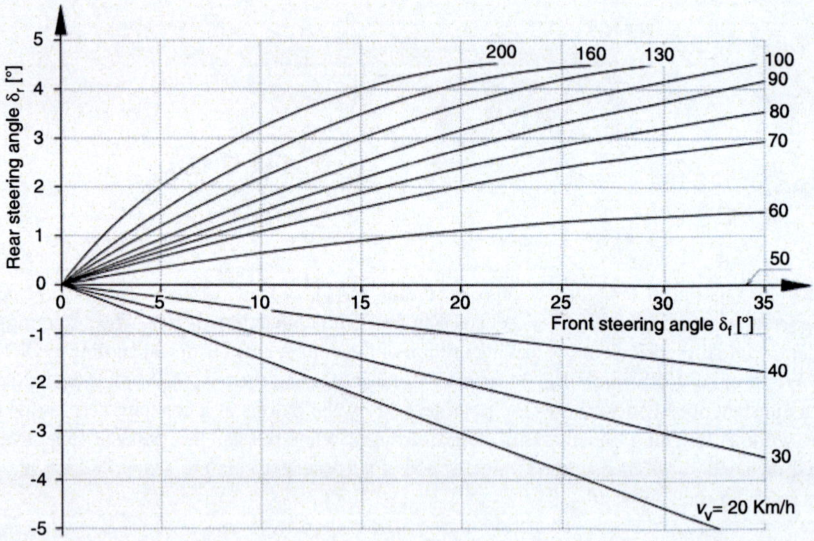

Fig. 5.83 Characteristic diagram of an all-wheel steering system of a passenger car, [27]. The steering ratio was optimised in the road test depending on speed and steering angle. The value zero of the steering ratio (cf. Fig. 5.82) was selected at the driving speed $v_V = 50$ km/h, i.e. below this speed the rear wheels turn in the opposite direction to the front wheels

on the track layout. The greater the number and length of tight corners, the more effective the steering brake proves to be. In any case, it was banned in Formula 1 from 1998 onwards.

References

1. Pfeffer, P., Harrer, M. (Hrsg.): Lenkungshandbuch, 2. Aufl. Springer Vieweg, Wiesbaden (2013)
2. Reimpell, J., Betzler, J.: Fahrwerktechnik: Grundlagen, 4. Aufl. Vogel, Würzburg (2000)
3. Heißing, B.: Moderne Fahrwerksauslegung, Vortrag im Rahmen der ÖVK-Vortragsreihe, Graz 12, Mai (2004)
4. Milliken, W., Milliken, D.: Race Car Vehicle Dynamics, 1. Aufl. SAE International, Warrendale (1995)
5. Matschinsky, W.: Radführungen der Straßenfahrzeuge, 3. Aufl. Springer, Berlin (2007)
6. Voigt, T.: Tourenwagen Story '98. Sportverlag, Hamburg (1998)
7. Arrows A20 Formel 1, Baujahr 2000. Daten ermittelt bei Fa. Leitgeb Motorsport, Wels. Nov. (2003)
8. Tremayne, D.: Formel 1. Technik unter der Lupe. Motorbuch, Stuttgart (2001)
9. BMW Williams: 2001, Ausstellungsfahrzeug Motorfestival in Graz Apr. (2003)
10. Ludvigsen, K.: Mercedes Benz Renn- und Sportwagen, 1. Aufl. Motorbuch, Stuttgart (1999)
11. Staniforth, A.: Competition Car Suspension, 3. Aufl. Haynes, Sparkford (1999)
12. Bosch: Kraftfahrtechnisches Taschenbuch, 19. Aufl. VDI, Düseldorf (1984)
13. Piola, G.: Formel 1. Copress, München (2001)
14. Blundell, M., Harty, D.: The Multibody Systems Approach to Vehicle Dynamics, 1. Aufl. SAE International, Warrendale (2004)
15. Rompe, K., Heißing, B.: Objektive Testverfahren für die Fahreigenschaften von Kraftfahrzeugen. Quer- und Längsdynamik. TÜV Rheinland, Köln (1984)
16. Achleitner, A.: Der neue Porsche 911 Carrera, Vortrag im Rahmen der ÖVK-Vortragsreihe, Wien, April (2005)
17. Heathershaw A.: Variable Ratio Steering Development for Formula 1, in AutoTechnology Feb. 2005, 38. GWV Fachverlage GmbH, Wiesbaden (2005)
18. Hamm, B.: Tabellenbuch Fahrzeugtechnik, 19. Aufl. Holland + Josenhans, Stuttgart (1997)
19. N.N.: Der neue Audi A3, Sonderausgabe von ATZ MTZ, April (2003)
20. Scoltock, J.: World rally championship, racing on a budget. Autom. Eng. **6**(Juli–August), 14–16 (2011)
21. Incandela, S.: The Anatomy & Development of the Formula One Racing Car from 1975, 2. Aufl. Haynes, Sparkford (1984)
22. Mitschke, M., Wallentowitz, H.: Dynamik der Kraftfahrzeuge, 4. Aufl. Springer, Berlin \Heidelberg\New York (2004)
23. N.N.: ZF-Servolectric, Informationsschrift der ZF Lenksysteme GmbH, 2003 Schwäbisch-Gmünd (2003)
24. Milliken, W.F.: Chassis Design: Principles and Analysis, "Based on Previously Unpublished Technical Notes by Maurice Olley". Society of Automotive Engineers, Warrendale (2002)
25. Henker, E.: Fahrwerktechnik. Vieweg, Wiesbaden (1993)
26. Newey, A.: How to Build a Car. HarperCollins Publishers, London (2017)
27. Heißing, B., Ersoy, M., Gies, S. (Hrsg.): Fahrwerkhandbuch, 4. Aufl. Springer Vieweg, Wiesbaden (2013)
28. Braess, H.-H., Seiffert, U.: Vieweg Handbuch Kraftfahrzeugtechnik, 4. Aufl. Vieweg, Wiesbaden (2005)

Braking System

6

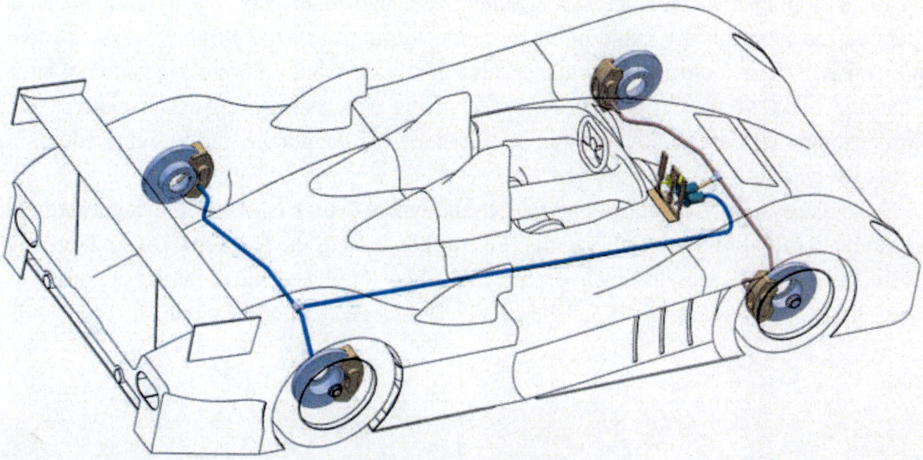

Racing vehicles should be able to accelerate as strongly as possible, reach high speeds and also perform rapid changes of direction. One thinks primarily of the propulsive power of the engine, low masses and resistance, and the grip of the tires. But braking is also acceleration, albeit negative, and at least as important for low (lap) times as positive acceleration.

M. Trzesniowski, *Suspension System*,
https://doi.org/10.1007/978-3-658-39847-7_6

6.1 General

Like wheels, wheel suspension, steering and power units, brake systems are among the most important components of vehicles. Accordingly, the safety requirements and thus automatically the legal requirements must be high.

In any case, two independent brake systems are prescribed for a motor vehicle. A distinction is made between the service brake system, the auxiliary brake system and the parking brake system. The service brake system is designed as a muscle, auxiliary or external power brake. It must be dual-circuit, act on all wheels and be gradable (i.e. meterable), Fig. 6.1. For comparison, Fig. 6.65 shows the brake system of a racing car.

In the event of failure of the service brake system, the auxiliary brake system must perform its task with reduced effect; in the case of commercial vehicles, it must also act on the trailer. In each case, the brake circuit of the service brake that is still intact is used as the auxiliary brake system.

The parking brake system ('handbrake') shall be capable of holding the vehicle stationary on a sloping road with an 18% gradient. On passenger cars, the parking brake is operated mechanically by cable or linkage. On touring cars, the parking brake is often implemented by an intermediate holding valve in a brake circuit. Formula cars do not have a parking brake at all. On production-based racing cars, the locking mechanism of the parking brake is often removed. This so-called fly-off handbrake can be released again immediately after it has been applied.

The brake system is completed by master and wheel cylinders, which are connected via hydraulic lines to form a closed system. The amplification of the foot force is taken over by suction air, in some cases also by hydraulic brake boosters. The tandem master cylinder has two separate pressure chambers for the necessary independent brake circuits.

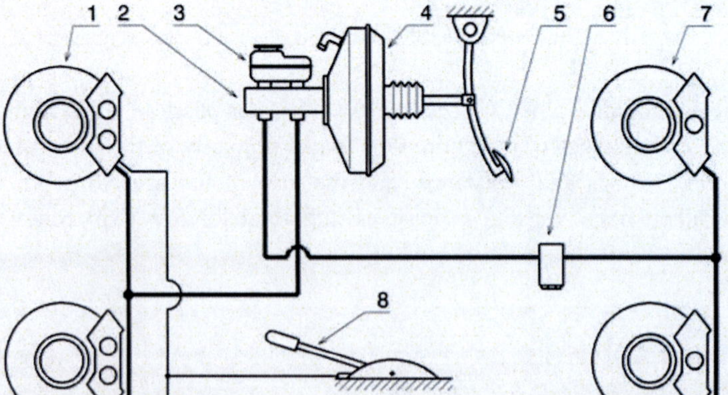

Fig. 6.1 Brake system of a passenger car, diagram. **1** Front disc brake, consisting of disc and caliper, **2** Tandem brake master cylinder, **3** Brake-fluid reservoir, **4** Vacuum brake booster, **5** Brake pedal, **6** Brake-power regulator, **7** Rear disc brake, consisting of disc and caliper, **8** Parking brake

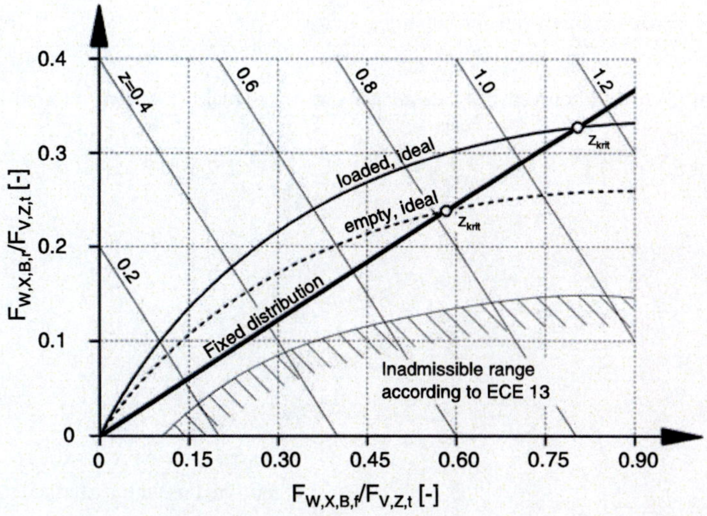

Fig. 6.2 Brake force distribution diagram (brake-balance chart). The sum of the related axle braking forces is the deceleration z. Depending on the braking system, the ratio of the actual braking forces is a fixed value (fixed distribution) or is influenced by a controller. The ideal distribution, at which the greatest braking is achieved, depends on the centre of gravity height

The braking force must be distributed between the axles in accordance with the axle loads. In terms of brake circuit distribution, the diagonal distribution has become established for front-heavy vehicles and the front-axle/rear-axle distribution for vehicles with a greater rear load. In principle, overbraking of the rear axle must be avoided; automatic brake force regulators, which operate load-, pressure- or deceleration-dependently, are used for this purpose on production vehicles. In racing vehicles, a fixed distribution – although adjustable from the driver's seat – is generally chosen, Fig. 6.2.

The braking ratio (normalized deceleration) is the total braking force related to the vehicle weight:

$$z = \frac{F_{V,X,B}}{F_{V,Z,t}} = \frac{a_x}{g} \cdot 100\% \tag{6.1}$$

z	Braking ratio, normalized deceleration, – resp.
$F_{V,X,B}$	Braking force acting at the Centre of gravity of the vehicle, N
$F_{V,Z,t}$	Total weight of the vehicle, N
a_x	Longitudinal acceleration, m/s^2

Table 6.1 shows some minimum braking effects according to relevant regulations.

The braking ratio thus follows from the sum of the related braking forces of the front and rear axles and is limited by the maximum friction between the road surface and the tyres:

Table 6.1 Minimum requirements for braking systems

	Mean deceleration, m/s^2			Braking ratio, %
Brake system	Motorcycle	Passenger Car	Truck	Trailer
Operating mode	5.8	5.8	5.0	45–50
Auxiliary brake	3.1	2.9	2.2 (buses 2.5)	–
Parking brake	–	Must maintain 18% road gradient		

$$z = \frac{F_{W,X,B,f}}{F_{V,Z,t}} + \frac{F_{W,X,B,r}}{F_{V,Z,t}} = \frac{a_x}{g} \leq \mu_{W,X} \tag{6.2}$$

$F_{W,X,B,f}$	Brake force on front axle, N
$F_{W,X,B,r}$	Braking force on rear axle, N
$\mu_{W,X}$	Static friction in longitudinal direction, –

The above relationship can be used to create the brake force distribution diagram, Fig. 6.2.

6.2 Brake System Requirements

The following requirements are placed on a brake:

- Constant braking power under all operating conditions
- No fading tendency at high temperatures
- Maintaining directional stability during braking
- Low weight.

6.3 Physical Basics

In principle, the braking forces must be introduced via the tyre, with the frictional force between the tyre and the road surface determining the force transmission. In general, a frictional force $F_{fr} = \mu\ F_N$, where F_N is the normal force and μ is the static friction coefficient. The normal force depends on the axle or wheel load. Thus, for the braking force of an axle n holds:

$$F_{W,X,B,n} = \mu_{W,X,n} \cdot F_{V,Z,n} \tag{6.3}$$

$F_{W,X,B,n}$	Braking force of the axle n, N
$F_{V,Z,n}$	Axle load of the axle n, N

The axle load of the front axle increases with an initiated braking deceleration a_x by the amount $\Delta F_{V,z}$. The rear axle is relieved by the same amount. If the available adhesion is to be fully utilized for all normalized decelerations z ($z = a_x/g$), then the ratio of the applied braking forces on the front and rear axles must change in accordance with the axle load distribution. In the brake force distribution diagram (Fig. 6.2), the position of the parabola and also the position of the limit curves shift in this case with the change in the vehicle centre of gravity. The braking force distribution remains constant as long as no braking force control intervenes. Above z_{crit} there is a danger of over-braking (locking of the rear wheels) and thus unstable handling, because there is no or very little lateral control force available on the rear axle. The shortest, ideally possible stopping distance (without taking into account an aerodynamic downforce) is achieved when the deceleration becomes equal to the coefficient of friction.

Brake **Force of Vehicle**

The required total braking force of a vehicle $F_{V,X,B}$ can be determined from various approaches, Fig. 6.3.

$$\text{from inertia force:} \quad F_{V,X,B} = m_{V,t} \cdot a_X \tag{6.4}$$

$$\text{from kinetic energy} \quad F_{V,X,B} = \frac{m_{V,t} \cdot v_{V,1}^2}{2 s_B} \tag{6.5}$$

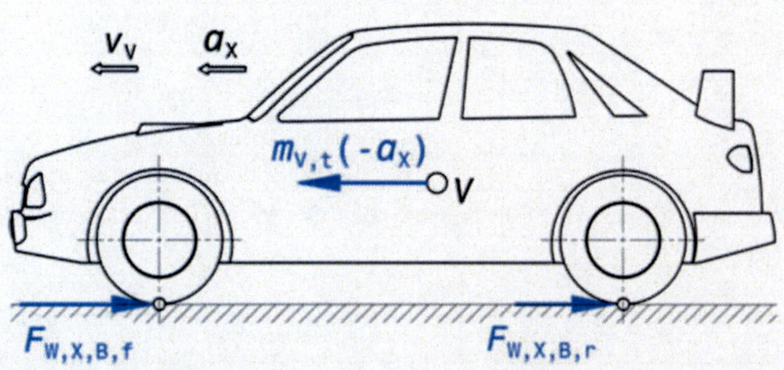

Fig. 6.3 Braking forces on the vehicle
$F_{V,X,B} = F_{W,X,B,f} + F_{W,X,B,r}$
V Centre of gravity of the complete vehicle

$$\text{out of braking } z: \qquad F_{V,X,B} = m_{V,t} \cdot g \cdot z \tag{6.6}$$

$m_{V,t}$	Total vehicle mass, kg
a_X	Mean (constant) braking deceleration, m/s^2
$v_{V,1}$	Vehicle speed before braking, m/s
s_B	Braking distance, m

The maximum braking force acting on the vehicle can be greatly increased in racing cars due to aerodynamic downforce compared to production vehicles.

$$F_{V,X,B} = \mu_{W,X} \cdot \left(m_{V,t} \cdot g + \frac{1}{2} \cdot \rho_L \cdot c_A \cdot A_V \cdot v_v^2 \right) + \frac{1}{2} \cdot \rho_L \cdot c_W \cdot A_V \cdot v_V^2 \tag{6.7}$$

ρ_L	Density of air, kg/m^3. $\rho_L = 1.199$ kg/m^3 at a temperature of 20 °C, an air pressure of 1.013 bar and a relative humidity of 60%
c_A	Aerodynamic downforce coefficient[a]
A_V	Projected cross-sectional area of the vehicle, m^2[b]
v_V	Vehicle speed, m/s
c_W	Drag coefficient, –

[a] See Racing Car Technology Manual, Vol. 2 *Complete Vehicle*, Sect. 5.3 *Downforce*
[b] See Racing Car Technology Manual, Vol. 3 *Powertrain*, Chap. 4, (4.7)

The largest braking force occurring on the tyres of the front axle, which is decisive for the design of the wheel brake, is determined by the axle load displacement in addition to the static axle load and the downforce:

$$F_{W,X,B,f} = \mu_{W,X} \cdot \left(m_{V,t} \cdot g \cdot \frac{l_r}{l} + m_{V,t} \cdot a_X \cdot \frac{h_V}{l} + F_{L,Z,f} \right) \tag{6.8}$$

l	Wheelbase, m
l_r, h_V	Other dimensions (see Fig. 6.8)
$F_{L,Z,f}$	Proportion of downforce acting on front axle, N[a]

[a] This is determined by the position of the pressure point. See Racing Car Technology Manual, Vol. 2 *Complete Vehicle*, Chap. 5

All the forces acting on a Formula 1 car during braking are shown in Fig. 6.4.

Braking Work and *Braking* Power

Roughly, the amount of energy that must be absorbed by the braking system until the vehicle comes to a standstill can be estimated via the conservation of energy, Fig. 6.5.

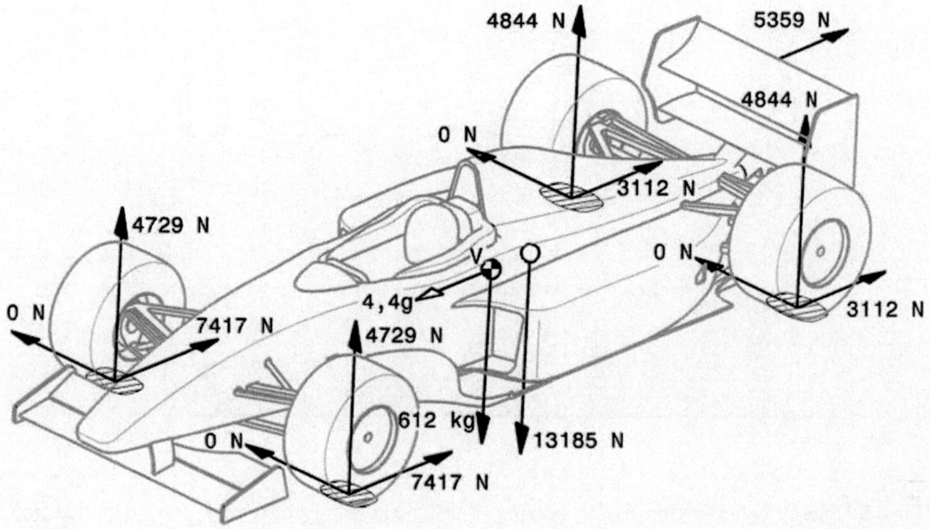

Fig. 6.4 Braking forces on a Formula 1 car, after [8]. All forces during braking at 320 km/h with 4.4 *g* are drawn in the tire contact areas. In addition, the resistance of the rear wing and the sum of the downforce forces at the center of pressure are entered. *V* Vehicle centre of gravity

$$W_B = \int_s F_{V,X,B} ds = \int_t P_B dt \text{ resp.} W_B = F_{V,X,B} \cdot s_B \text{ with } F_{\mathrm{V,X,B}} = \text{const.} \tag{6.9}$$

W_B	Braking work, J. converted to heat by friction

$$\begin{aligned} W_B &= m_{V,t} \cdot a_X \cdot s_B \\ W_B &= \frac{m_{V,t}}{2} \left(v_{V,1}^2 - v_{V,\mathrm{Re}}{}^2\right) \\ P_{B,\max} &= F_{V,X,B} \cdot v_V \end{aligned} \tag{6.10}$$

$v_{V,1}$	Speed before braking, m/s
$v_{V,Re}$	Speed after braking, m/s
$P_{B,max}$	Braking power, largest instantaneous value, W

$$P_{B,m} = \frac{W_B}{t_B} t_B = \frac{\Delta v}{a_X} a_X = \frac{\Delta v^2}{2 s_B}$$

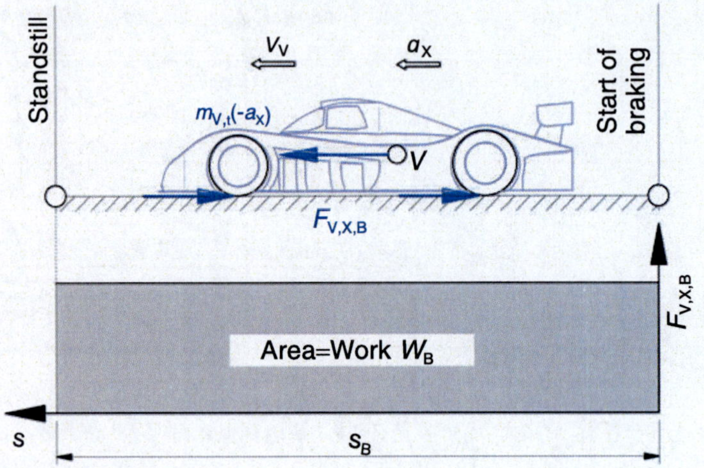

Fig. 6.5 Energy consideration during braking. The kinetic energy of a vehicle is converted into friction work during braking

$$P_{B,m} = \frac{F_{V,X,B} \cdot v_{V,1}}{2}$$

$P_{B,m}$	Mean braking power, W
t_B	Braking time, s
Δv	Velocity difference, m/s with$\Delta v = v_{V,\,\mathrm{Re}} - v_{V,\,1}$

For the general case of non-constant braking deceleration a_x the required quantities can be determined from the following relationships:

$$\int_{s_1}^{s_{\mathrm{Re}}} a_x ds = \int_{v_{V,1}}^{v_{V,\mathrm{Re}}} v dv \text{ resp.} \int_{t_1}^{t_{\mathrm{Re}}} a_x dt = \int_{v_{V,1}}^{v_{V,\mathrm{Re}}} dv$$

with:

$s_B = s_{\mathrm{Re}} - s_1$
$t_B = t_{\mathrm{Re}} - t_1$

Especially for simulations, numerical methods are used for the solution.

Air and rolling resistance help to decelerate a vehicle. These resistances are speed-dependent and represent only a small contribution in the total braking power. Figure 6.6 shows the proportions for a compact class passenger car for a deceleration of 1 g. At 100 km/h the total braking power is about 340 kW. When braking to a standstill, the

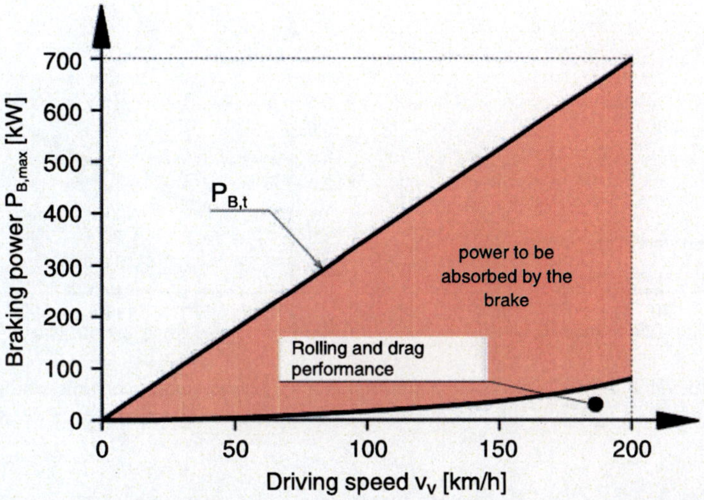

Fig. 6.6 Braking power of a passenger car at 1 *g* deceleration [16]. $P_{B,t}$ total braking power of the vehicle. Rolling resistance and air resistance represent a small contribution to the total braking power. However, the majority must be applied by the braking system. The passenger car has a total mass $m_{V,t}$ of 1250 kg

braking system has to convert most of it into heat and dissipate it to the environment. The contribution of the auxiliary effect of air resistance is also similar in racing vehicles, because downforce-increasing measures inevitably increase air resistance, cf. Fig. 6.7.

While production vehicles achieve maximum values of approx. 1 for braking, the values for racing vehicles with aerodynamic downforce aids are considerably higher, especially at higher speeds, Fig. 6.7. In the case of production vehicles, there are sometimes even lift forces at very high speeds which reduce the wheel load and thus the possible braking force.

Specific Braking Power *N*

To determine this, the maximum braking power, which results from the maximum speed per brake at full deceleration, is related to the friction surface of the brake pad or to the swept area of the brake disc [16]:

$$\begin{aligned} N_{pad} &= P_{B,W,\max} / (2 \cdot A_{pad}) \\ N_{Bd} &= P_{B,W,\max} / A_{Bd} \end{aligned} \tag{6.11}$$

N	Specific braking power, kW/cm^2
$P_{B,W,max}$	Max. Braking power per wheel, kW
A_{pad}	Friction surface brake pad, cm^2
A_{Bd}	Passed over brake disc area, cm^2

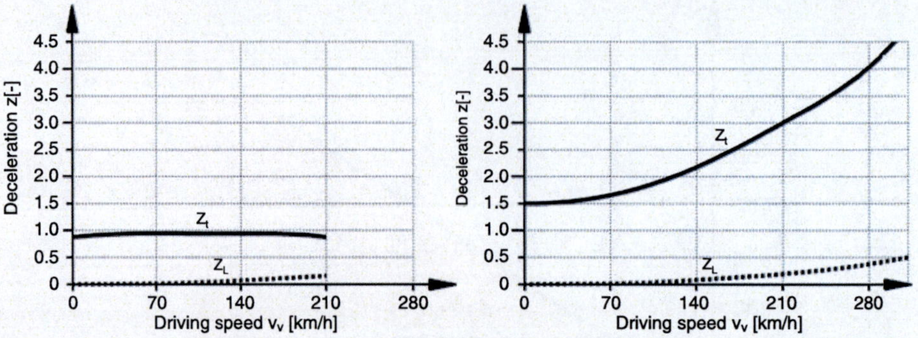

Fig. 6.7 Achievable normalized deceleration z for a mid-size passenger car (left) and a prototype sports car (right) [16]. z_t Actual total deceleration, z_L Proportion of air resistance in deceleration

Brake Force Distribution (Brake *Bias*)

With a fixed distribution of the braking forces between the axles, the following applies to the individual braking forces:

$$F_{W,X,B,f} = \Phi_f \cdot F_{V,X,B} \tag{6.12}$$

$$F_{W,X,B,r} = \Phi_r \cdot F_{V,X,B} \tag{6.13}$$

Φ_f	Brake force ratio front axle, –. $\Phi_f = 0.65$ to 0.75
Φ_r	Brake force proportion rear axle, –. $\Phi_r = 0.25$ to 0.35
it applies: $\Phi_f + \Phi_r = 1$	

Due to the axle-load transfer during braking, a fixed distribution of the braking forces does not allow the maximum possible braking with a given static friction coefficient tyre/road, because one axle blocks while the tyres of the other axle would still have the potential to increase the friction force. The ideal is thus a load-dependent control of the braking force distribution. In this case, the position of the vehicle's centre of gravity h_V together with the current deceleration determines the appropriate ratio of the braking forces of both axles, Fig. 6.8.

$$\frac{F_{V,Z,f}}{F_{V,Z,r}} = \frac{l_r + h_V \frac{a_X}{g}}{l_f - h_V \frac{a_X}{g}} = i_m \tag{6.14}$$

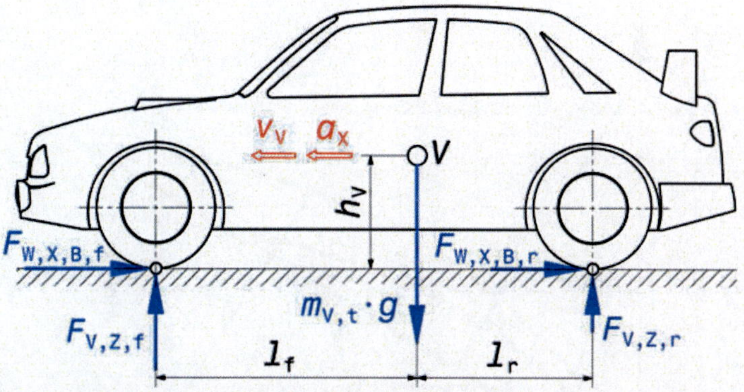

Fig. 6.8 Sketch for the calculation of the brake force distribution

i_m	Axle load ratio front/rear
l_f, l_r	Distance between vehicle Centre of gravity and Centre of front or rear axle, m
h_V	Height of the vehicle Centre of gravity, m

The brake-balance chart, Fig. 6.9, is used to illustrate this relationship. The weight-related braking forces of the front and rear axles $F_{W,X,B,f}$ and $F_{W,X,B,r}$ are plotted on the coordinate axes. The points of intersection of the straight lines of equal static friction coefficients of the front and rear axles form the parabola of the "ideal" brake force distribution (curves 1 and 2). The straight lines of constant deceleration z complete the diagram.

If no brake force distributor is installed, the brake force distribution is fixed and represents a straight line (3). The gradient results as the ratio of the braking forces of front and rear axle determined by the dimensioning of the wheel brakes and brake cylinders. As long as the straight line of the fixed distribution (3) runs below the ideal distribution (1, 2), the front axle always locks first (stable braking behaviour, the cornering force on the rear tyres is maintained). The locking point of the front axle (4) results from the intersection of the "installed distribution" (3) and the straight line of the respective static friction coefficient $\mu_{W,X,f}$.

Some considerations for the choice of a fixed brake force distribution are illustrated in Fig. 6.10 on the basis of three extreme layouts.

Low friction ratios ($\mu_{W,X}$ = const = 0.4) are considered. When the deceleration z is increased from 0 starting at the fixed split Q_1, the rear wheels lock first (point A) because $\mu_{W,X,r} = 0.4$ is fully utilized (front utilized $\mu_{W,X,f} < 0.4$). Further increase of brake application force leads to deceleration at point B. At this point, the front wheels also lock and thus no further increase in deceleration is possible. This design is used for motorcycles.

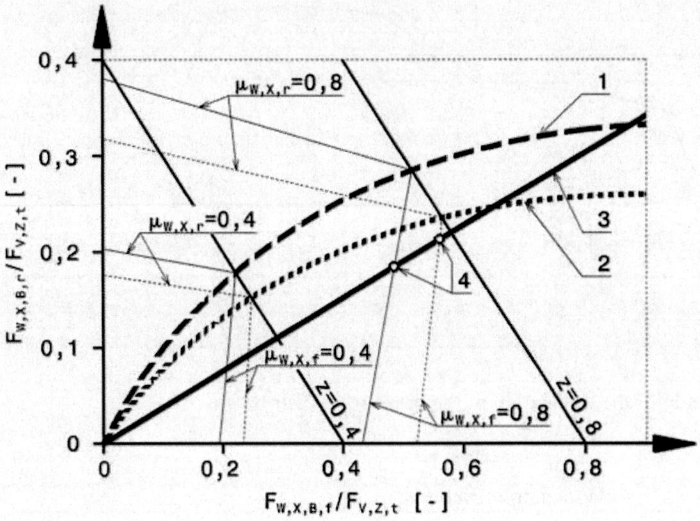

Fig. 6.9 Brake force distribution with fixed distribution. **1** Ideal brake force distribution at permissible total weight, **2** Ideal brake force distribution ready to drive (with driver), **3** Installed, fixed brake force distribution, **4** Front axle locked. $\mu_{W,X,f}$ and $\mu_{W,X,r}$ static friction coefficients in the longitudinal direction at the front and rear respectively

Fig. 6.10 Selection of the brake force distribution [7]. Q_1, Q_2, Q_3 Fixed brake force distributions, Q_{id} ideal braking force distribution, Q_1 is used for single-track vehicles and Q_2 for double-track vehicles

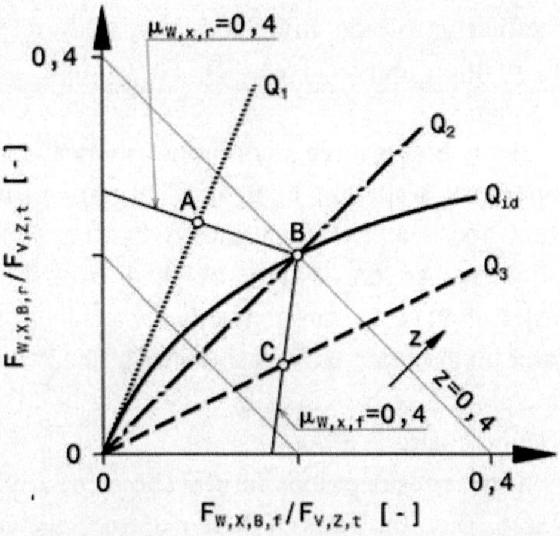

With the braking force distribution Q_2 the front wheels overbrake with an increase of z up to point B. From this point onwards, all wheels lock and thus the maximum deceleration $z = 0.4$ is reached.

In the case of braking force distribution Q_3 the front wheels lock in point C. If the deceleration is increased further, the braking force on the rear axle increases until point B is reached. From this point, the rear wheels also lock and the maximum deceleration for the friction present is reached. This design is used for two-lane road vehicles and for racing

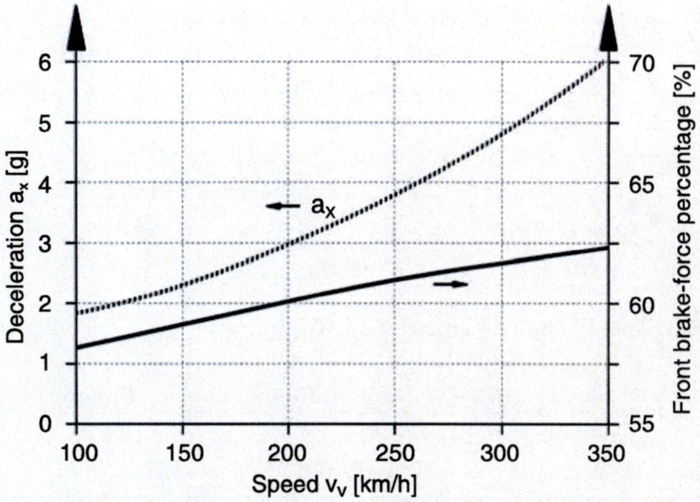

Fig. 6.11 Braking deceleration versus speed and corresponding ideal brake force distribution of a Formula 1 car, according to [1]. With sinking speed the deceleration decreases, because the downforce decreases with the square of the speed and the tires can no longer build up the original braking force

vehicles at high cornering speeds [19]. For braking manoeuvres on the straight, on the other hand, simultaneous locking of both axles is aimed for in racing vehicles, because in this way all tyres exploit their maximum possible braking force [19]. Rally drivers go one step further and set the rear axle to overbrake slightly, because this makes it easier to initiate a drift.

Ideally, the brake force distribution changes during the braking manoeuvre because the deceleration does not remain constant due to aerodynamic downforce aids and speed-dependent friction values, Fig. 6.11. On the contrary, the downforce and thus the tyre forces also decrease. If the aerodynamic balance remains the same during the speed decrease, the braking force distribution must still change because of the axle load shift. In fact, the situation is somewhat different because most cars change downforce distribution due to pitching under braking. The front wing gets closer to the road at high deceleration and front downforce increases disproportionately. If deceleration decreases, the relative downforce share of the front axle decreases. As a result, the front braking force fraction must decrease more at low speeds than in Fig. 6.11.

Drivers make an "adjustment" to the brake force distribution during braking not with the rotary knob in the cockpit, but by braking and accelerating at the same time. This driving style can best be practiced when there is no need to use the foot clutch to shift gears. By accelerating while braking, the driven rear wheels receive a counter-torque to the braking torque. This reduces the braking torque and the braking force is shifted to the (non-driven) front axle.

Creation of a Distribution Diagram of the Braking Forces

1. The curves of the ideal brake force distribution $F_{W,X,B,r} = f\,(F_{W,X,B,f}, h_V)$ follow from (6.3) and (6.14) for constant and equal friction values at both axles:

$$\frac{F_{W,X,B,f}}{F_{W,X,B,r}} = \frac{\mu_{W,X,f} \cdot F_{V,Z,f}}{\mu_{W,X,r} \cdot F_{V,Z,r}} = i_m \quad \rightarrow \frac{F_{W,X,B,r}}{F_{V,Z,t}} = \frac{F_{W,X,B,f}}{F_{V,Z,t}} \cdot \frac{1}{i_m}$$

2. The straight lines of constant deceleration z $F_{W,X,B,r} = \mathrm{f}\,(F_{W,X,B,f}, z)$ follow from (6.2):

$$z = \frac{F_{W,X,B,f}}{F_{V,Z,t}} + \frac{F_{W,X,B,r}}{F_{V,Z,t}} = \frac{a_X}{g} \rightarrow \frac{F_{W,X,B,r}}{F_{V,Z,t}} = \frac{a_X}{g} - \frac{F_{W,X,B,f}}{F_{V,Z,t}} \tag{6.15}$$

3. The straight lines of constant static friction $F_{B,r} = f\,(F_{B,f}, \mu)$ are given by:

$$Mit\, F_{V,Z,f} + F_{V,Z,r} = F_{V,Z,t} \rightarrow F_{V,Z,f} = F_{V,Z,t} - F_{V,Z,r} \tag{6.16}$$

and rearranged (6.14) $F_{V,\,Z,\,f} = F_{V,\,Z,\,r} \cdot i_m$ follows

$$\frac{F_{V,Z,f}}{F_{V,Z,t}} = \frac{i_m}{1 + i_m} \tag{6.17}$$

and

$$\frac{F_{V,Z,r}}{F_{V,Z,t}} = \frac{1}{1 + i_m} \tag{6.18}$$

From (6.2) with (6.3) becomes

$$\frac{\mu_{W,X,f} \cdot F_{V,Z,f}}{F_{V,Z,t}} + \frac{\mu_{W,X,r} \cdot F_{V,Z,r}}{F_{V,Z,t}} = z \tag{6.19}$$

Equations (6.17) and (6.18) substituted into (6.19) and i_m substituted with (6.14) gives an expression for z:

$$z = \frac{\mu_{W,X,f} l_r + \mu_{W,X,r} l_f}{l_f + l_r + h_V \left(\mu_{W,X,r} - \mu_{W,X,f}\right)} \tag{6.20}$$

The related quantities of the braking forces required for the diagram are obtained by substituting (6.16) into (6.19). First of all for the braking force of the front axle:

$$\frac{F_{V,Z,f}}{F_{V,Z,t}} = \frac{z - \mu_{W,X,r}}{\mu_{W,X,f} - \mu_{W,X,r}} \text{ and therefore } \frac{F_{W,X,B,f}}{F_{V,Z,t}} = \mu_{W,X,f} \frac{F_{V,Z,f}}{F_{V,Z,t}}$$

The required braking force of the rear axle as a function of the braking force on the front axle follows from (6.15):

$$\frac{F_{W,X,B,r}}{F_{V,Z,t}} = z - \frac{F_{W,X,B,f}}{F_{V,Z,t}}$$

The points of the lines of constant friction on the coordinate axes of the diagram correspond to the two extreme cases when only one axle brakes alone:

$$\text{Braking only rear}: \quad \frac{F_{W,X,B,r}}{F_{V,Z,t}} = z = \frac{\mu_{W,X,r} l_f}{l_f + l_r + \mu_{W,X,r} h_V}$$

$$\text{Front braking only}: \quad \frac{F_{W,X,B,f}}{F_{V,Z,t}} = z = \frac{\mu_{W,X,f} l_r}{l_f + l_r - \mu_{W,X,f} h_V}$$

4. The straight lines describing a constant brake force Distribution follow to:

$$F_{W,X,B,r} = \frac{\Phi_r}{\Phi_f} F_{W,X,B,f} = \frac{\Phi_r}{1 - \Phi_r} F_{W,X,B,f}$$

Some race cars also use an adjustment of the braking forces left to right. This means that the rear wheel on the inside of the corner is braked more heavily on entry to the corner in order to improve turning in tight corners. In Formula 1, unlike in other racing series, this option is prohibited. The benefit is controversial, not least because of the increased demands on the driver [13].

Brake force control systems to prevent wheel lock-up (ABS, see Appendix)[1] were initially hardly used in racing. The reason was that standard ABS systems were used at the time and these were not designed to meet the needs of a racing vehicle. On undulating surfaces, wheel load fluctuations thus led to an increase in braking distance. In the case of production vehicles, the development goal was also not to optimise the braking distance, but to maintain steerability during braking.

[1] For more details see Racing Car Technology Manual, Vol. 3 *Powertrain*, Chap. 8 *Electronic Driving Aids*.

Table 6.2 Measures to increase stability when braking in corners and with different friction values on the left and right [16]

Measure, characteristic	When braking in corner	With different friction values left and right
Toe-out under front brake force	+	–
Toe-out under rear brake force	–	+
Toe-out on front compression	+	–
Negative srub radius	–	+
Low vehicle center of gravity	+	o
High braking force on the front axle	+	+
High understeer reserve	+	o
ABS with select low on the rear axle	+[a]	+[a]
ABS with yaw moment control	o	+[a]
ABS[b] with CBC/ABS+	+[a]	o
ESP[b]	+	+

Legend:
+: favourable, –: unfavourable, o: no or little influence
[a]causes braking distance extension
[b]see appendix

The optimum brake force distribution also depends on whether the vehicle is driving uphill or downhill due to the centre of gravity height, i.e. the optimum setting on level ground does not suit inclined road surfaces.

An important aspect of braking force distribution is also the maintenance of driving stability. Table 6.2 lists measures for increasing stability during braking. A distinction is made between braking in a corner and different friction values on the left and right. Not all measures prove to be equally favourable. Whereas when braking in a corner, the decisive outer wheels should steer the vehicle outwards, in the case of different friction values, the more important tyres – i.e. those on the side with more grip – should steer to the side with less friction at the front and to the side with more friction at the rear. However, two characteristics show themselves to be “uncompromisingly” favourable: low centre of gravity of the vehicle and a high proportion of braking force on the front axle.

Design Realisation of the Front to Rear Brake Force Distribution

The desired braking force distribution can be realized in different ways if there is a II-division of the braking circuits (cf. Figure 6.15) [13]:

(a) Different wheel brake cylinder diameters produce different braking forces with the same master brake cylinder diameter. This type is preferred for racing vehicles. The slave cylinder diameters are larger on the front axle than on the rear axle, or the brake calipers on the front axle have more pistons. This means that the braking forces are greater at the front than at the rear.
(b) Different master brake cylinders have basically the same effect on the same wheel brake cylinders, but lead to different piston travels in the master cylinders, which must be compensated for by the actuating linkage. The master cylinder for the front axle must have a smaller diameter than the one for the rear axle. For the same pedal force, the pressure at the front brake calipers is therefore greater than at the rear.
(c) Larger disc diameters result in greater braking forces with otherwise unchanged parameters. A larger brake disc also has the advantage of a larger heat dissipating surface. For example, the brake discs on the front axle are larger than those on the rear axle.
(d) A mechanical division of the actuating forces enables different braking forces. This is implemented, for example, by a balance beam system (see below) and used for fine adjustment while driving. If the weather conditions change, the driver can shift the braking force to the rear axle without a pit stop and avoid overbraking the front wheels on a wet track.
(e) The braking force can be limited by a pressure relief valve in the brake line. Such a valve in the supply line to the wheel brake cylinders on the rear axle limits the maximum actuating pressure in this line to a certain value without affecting the actuating pressure on the front axle (see Sect. 6.6). The front to rear actuation pressure ratio can also be varied according to the pressure itself and/or the axle load (see Sect. 6.6).

Balance Beam System

Balance Bar. This system of brake force distribution is widely used and is found in virtually all racing classes. The principle is shown in Fig. 6.12. Figure 6.24 provides an overview of the installation situation with the brake pedal. The brake pedal (1) transmits the foot force via a sleeve to the balance beam (2). At the threaded ends of the balance beam (2) there are pivots which receive the piston rods. One master brake cylinder (3) is actuated by each piston rod. When the balance beam is rotated, it moves along its axis because the trunnions have a nut thread. This changes the leverage ratio between the piston rods. A flexible shaft leads from a receptacle of the balance beam to the driver's place, from where turning can be easily performed.

The dimension *a* between the two main brake cylinders must be identical to the dimension between the two pivots of the pivots. The balance beam must be in a parallel plane to the brake cylinders at all pedal positions.

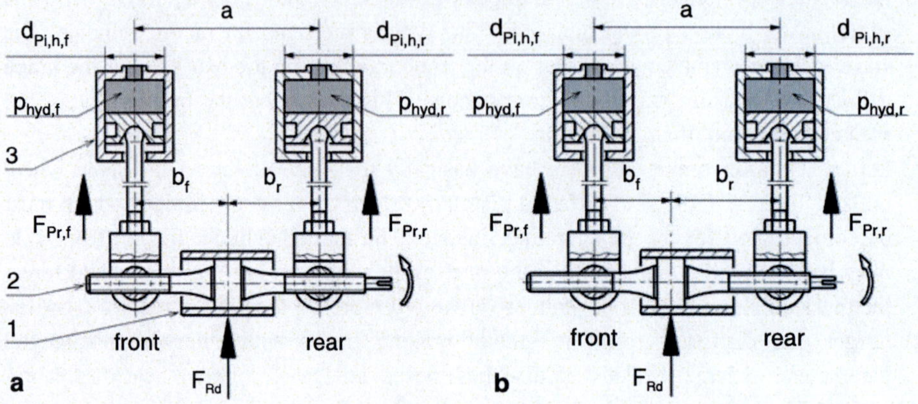

Fig. 6.12 Principle of a balance bar system. (**a**) symmetrical force distribution $b_f/b_r = 1$, (**b**) balance bar displaced, $b_f/b_r < 1$. **1** Brake pedal, **2** Balance bar, **3** Front or rear brake master cylinder

During braking, the foot force is translated via the pedal to the force F_{Rd}, which acts on the balance beam (cf. Fig. 6.18). The beam divides the force F_{Rd} into the piston rod forces via the length ratio b_f/b_r:

$$F_{\mathrm{Pr},f} = F_{Rd}\frac{b_r}{b_f + b_r} \tag{6.21}$$

$$F_{\mathrm{Pr},r} = F_{Rd}\frac{b_f}{b_f + b_r} \tag{6.22}$$

F_{Rd}	Force from brake pedal to balance beam, N
b_f, b_r	Spacings of the spherical plain bearing from the front or rear pivots, mm. (see Fig. 6.12)
	It holds that $b_f + b_r = a =$ const

Table 6.3 shows the effects of a balance beam by means of a numerical example for Fig. 6.12.

Particularly high demands are placed on the brakes of racing vehicles, although different tracks place different demands on the brakes. The Montreal track, for example, places the highest demands on the brakes of Formula 1 cars. On this circuit, a total of six full braking manoeuvres from approx. 300–320 km/h to 60–80 km/h have to be carried out, and this in an interval of only about 4.5 seconds. This sometimes leads to fractures of brake discs. Figure 6.13 shows the corresponding values over time for the fifth braking on this course. Here, the operating temperature of the brake disc rises from about 450 °C to a peak value of over 1000 °C within only 1.5 seconds. After one second, it stabilizes at 700 °C for about five seconds. The deceleration values to which the drivers are subjected are also impressive. In little more than a second, they change from +0.8 to −3 g.

Table 6.3 Effect of a balance beam

Position Balance beam	F_{Rd} N	b_f mm	b_r mm	$F_{Pr,f}$ N	$F_{Pr,r}$ N	$d_{Pi,h,f}$ mm	$d_{Pi,h,r}$ mm	$p_{hydr,f}$ bar	$p_{hydr,r}$ bar
a	1300	30	30	650	650	17.8	19.1	261	22.7
b	1300	25	35	758	542	17.8	19.1	30.5	18.9

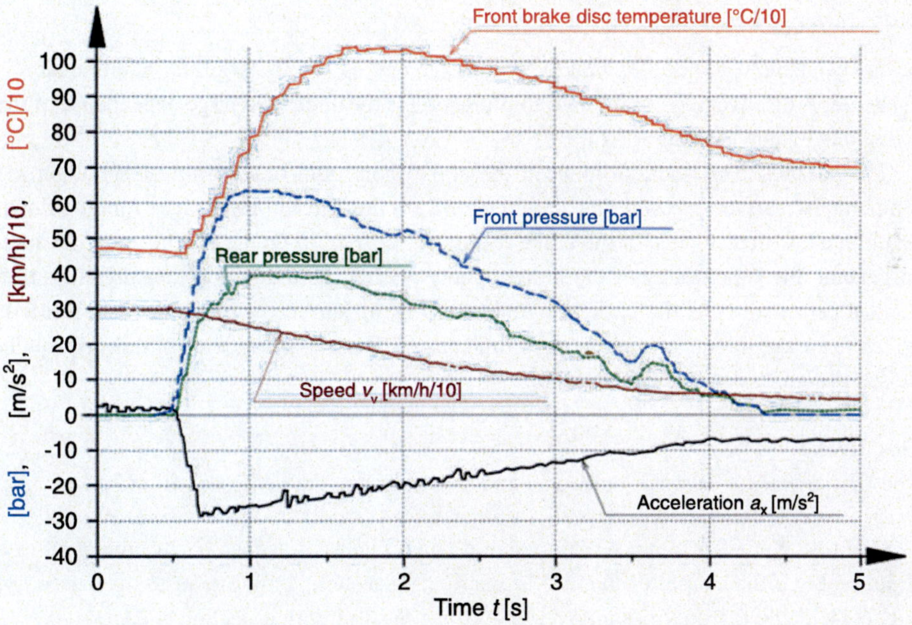

Fig. 6.13 Time course of some variables during emergency braking, according to [14]. Measured values of a braking of a Formula 1 car from 300 to 60 km/h on the circuit in Montreal. This braking manoeuvre is the fifth of a total of six full braking manoeuvres on this circuit

6.4 Brake Constructions and Arrangements

6.4.1 Drum Brake

This design is only important for historical vehicles. Even in passenger cars it is only installed on the rear axle of vehicles in the lower price segment or as a parking brake as additional duo-servo drum brakes in the disc of the service disc brake. Table 6.4 provides some technical data of this type.

The drum brake generates the braking forces on the inner surface of a brake drum. Depending on the design, self-amplification (self-energising) occurs in one accumulating shoe (simplex brake) or in both shoes depending on the direction of rotation (duplex brake)

Table 6.4 Technical data of drum brakes

System pressure	Inlet pressure 0.5–1.2 bar Brake pressure up to 100 bar	Clamping force	Passenger car: approx. 4 kN
Coefficient of friction μ_B	0.3–0.4		
Release clearance	0.3–0.5 mm		

or independently (duo-duplex brake). With servo brakes, a very large self-amplification is achieved. A disadvantage of the self-amplification is the undesired large dependence of the brake rated value (see Fig. 6.16).

The coefficient of friction between the brake lining and brake drum depends on the temperature, surface pressure, air humidity and friction speed. Two shoes running down result in a low dependence of the brake coefficient of friction on the coefficient of friction. In general, the consistency of the lining quality is critical. In addition, heating leads to a conical deformation of the initially cylindrical braking surface of the drum, as a result of which the brake linings are no longer in full contact. The heat dissipation of this principally closed design is also a major problem.

Disc Brake

The friction values between brake pad and disc vary less than with the drum brake.

Due to the smaller characteristic value *C,* high clamping forces occur compared to a drum brake. If the caliper or the disc is axially displaceable, only one piston is necessary, Fig. 6.14. Table 6.5 provides an overview of the most important technical data.

Brake Circuit Configuration

The legal regulations require a dual-circuit transmission system on production vehicles. Also for racing vehicles the regulations generally require more than one brake circuit. The five basic possibilities according to DIN 74 000 are shown in Fig. 6.15.

The II and the X split (configuration) have become established. With a minimum of lines, hoses, detachable connections and static or dynamic seals, they are comparable to a single-circuit brake system in terms of the risk of failure due to leakage. The combination of X-splitting and negative scrub radius on the front axle stabilizes the vehicle in the event of brake circuit failure by "countersteering" caused by the one-sided braking force on the tire. In the event of brake circuit failure due to thermal overload of a hydraulic wheel brake, the HI, LL and HH splits are particularly critical because failure of both wheel brakes on one wheel can lead to total brake failure. In order to comply with the legal regulations regarding the auxiliary braking effect, front-heavy vehicles are equipped with the X-split.

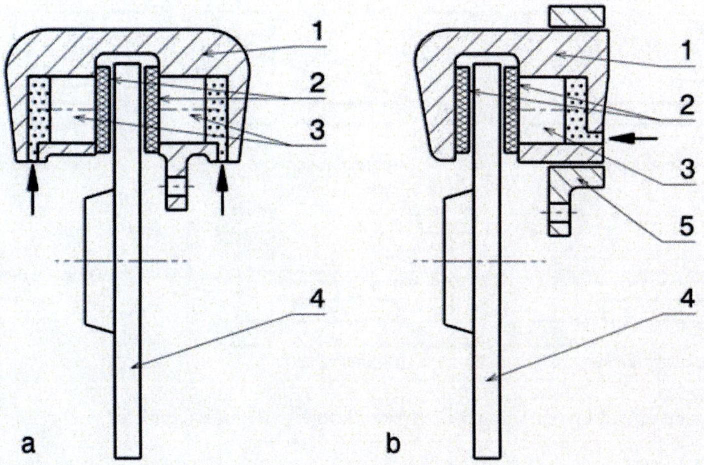

Fig. 6.14 Construction types of disc brakes. (**a**) Fixed-caliper disc brake, (**b**) Floating-caliper disc brake. **1** Brake caliper, **2** Brake pad, **3** Piston, **4** Brake disc, **5** Carrier. The arrows show the hydraulic connections

Table 6.5 Technical data of disc brakes

System pressure	Inlet pressure 0–0.5 bar Brake pressure up to 150 bar	Clamping force	Passenger car: approx. 15 kN
Coefficient of friction μ_B	Passenger car: 0.35–0.5 0.38 on average [4] Racing: 0.42–0.62 [8]	Surface pressure p_{mean}	Up to 600 N/mm^2
Disc runout	Max. 0.1 mm	Specific performance $\overline{P}_{pad}$	Up to 3.3 kW/cm^2
Release clearance[a]	Approx. 0.15 mm per side		

[a]see Fig. 6.66

The II distribution is preferably suitable for rear-heavy vehicles as well as medium and heavy commercial vehicles. For racing vehicles, this type of split is also popular because it allows easy adjustment of front to rear braking forces. In some racing series, this division is even prescribed by the regulations, e.g. in Formula 1.

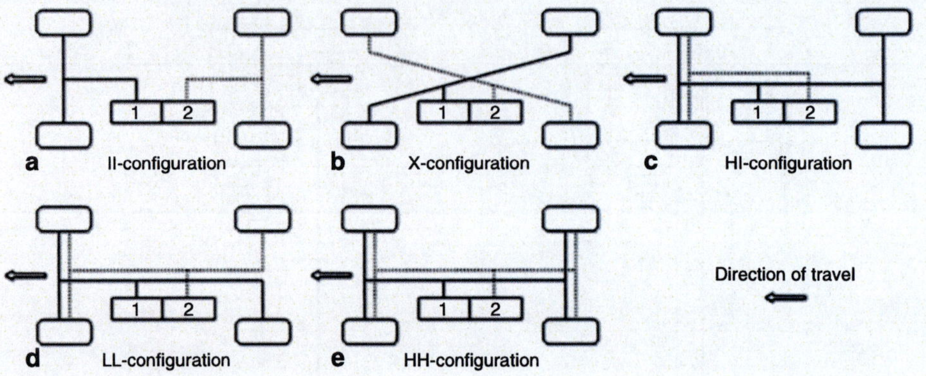

Fig. 6.15 Brake circuit layout variants. **1** Brake circuit 1, **2** Brake circuit 2

6.5 Parameters

6.5.1 Brake Sensitivity and Friction

The braking characteristic value represents the ratio of the braking force achieved at the brake to the applied clamping force:

$$k_B = \frac{F_{Bd,\tan}}{F_{Bd,ax}} \tag{6.23}$$

k_B	Brake sensitivity
$F_{Bd,tan}$	Circumferential force on brake drum radius or disc, N
$F_{Bd,ax}$	Tensioning force of the brake shoes or pads, N

Due to friction and self-amplificatioin influences, the characteristic value is not constant but changes with the friction, Fig. 6.16.

Friction is initially fundamentally dependent on the material pairing of brake disc/lining. However, the coefficient of friction is not constant, but strongly dependent on sliding speed and temperature, cf. Fig. 6.47. In this, the course of the coefficient of friction for an organic and for a carbon brake lining is plotted against speed. A carbon pad produces much more "bite" (braking power at the start of braking) than the organic pad. With organic pads, although the coefficient of friction increases as speed decreases, i.e. as the vehicle slows due to braking, carbon pads remain more consistent in their action, making it easier for the driver to control the braking process. This is crucial because the axle load distribution changes with braking deceleration.

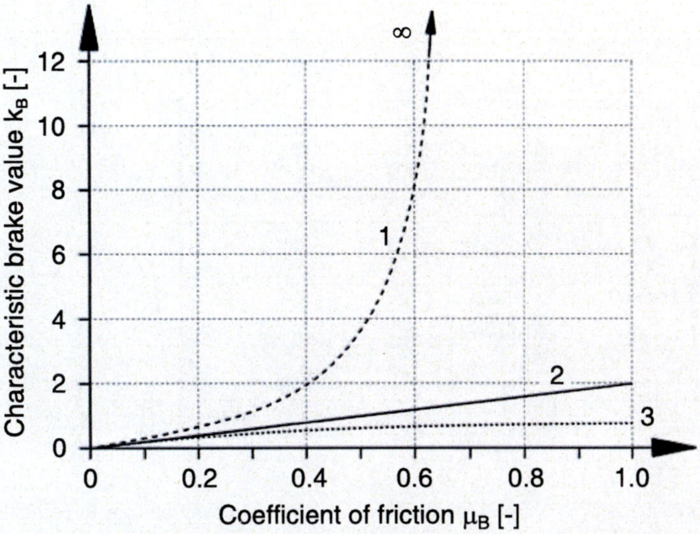

Fig. 6.16 Brake characteristics as a function of the friction coefficient [2]. μ_B Friction between brake pad and mating rotor. **1** Drum brake leading shoe, **2** Disc brake, **3** Drum brake trailing shoe

Table 6.6 Permissible actuating forces in N for series vehicles [3]

	Service brake		Parking brake	
Vehicle class	Manual force	Foot power	Manual force	Foot power
Passenger car	–	≤500	≤400	≤500
Motorcycle	≤200	≤500	–	–

Forces

The actuating forces for braking should not be too high, because otherwise the driver will tire more quickly or not even manage to build up the full braking force. On the other hand, weight can be saved if no additional auxiliary devices are installed to increase the actuating forces. Table 6.6 gives reference values for maximum operating forces.

For single-seater racing cars, average pedal actuation forces during braking are 700–900 N [15]. For full braking in Formula 1 cars, foot forces of about 1500 N are required because no brake boosters are allowed [17].

A maximum actuating force of 2000 N can be assumed for the design of components.

Brake Actuation Leverage **Ratios**

The pedal force and the clamping force of a slave cylinder are related via the geometry as follows.

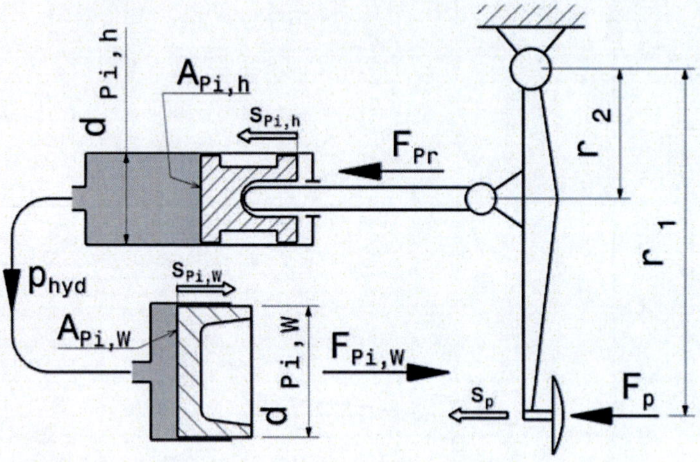

Fig. 6.17 Force ratios of a hydraulic brake actuation system. From the foot force F_p the clamping force of the wheel brake cylinder $F_{Pi,W}$

$$F_{Pr} = \frac{F_p \cdot r_1}{r_2} \tag{6.24}$$

$$p_{hyd} = \frac{F_{Pr}}{A_{Pi,h}} \tag{6.25}$$

$$F_{Pi,W} = p_{hyd} \cdot A_{Pi,W} \tag{6.26}$$

F_{Pr}	Piston rod force, N. (See also Fig. 6.17)
F_p	Pedal force (driver's foot force), N
r_1, r_2	Lever lengths, mm[2]
p_{hyd}	Brake fluid pressure, N/mm^2 (1 N/mm^2 = 10 bar)
$A_{Pi,h}$	Piston area main cylinder, mm^2
$A_{Pi,W}$	Piston area wheel cylinder (brake caliper), mm^2
$F_{Pi,W}$	Clamping force of the piston in the wheel cylinder, N

The displacement of the brake fluid when the brake pads are moved to contact the brake discs results in a travel of the piston in the master cylinder or, via the pedal ratio, the pedal travel.

[2]Typical values can be found in the Racing Car Technology Manual, Vol. 2 *Complete Vehicle*, Chap. 4 *Cockpit*.

$$s_{Pi,h} = j_B \cdot j_{Pi,W} \frac{A_{Pi,W}}{A_{Pi,h}} s_{Pi,W} \tag{6.27}$$

$$s_p = s_{Pi,h} \frac{r_1}{r_2} \tag{6.28}$$

$s_{Pi,h}$	Displacement of the piston in the master brake cylinder, mm
j_B	Number of calipers supplied by the master cylinder, –
$j_{Pi,W}$	Number of pistons in a brake caliper, –
$s_{Pi,W}$	Travel of the piston in the wheel brake cylinder (brake caliper), mm
s_p	Pedal travel, mm

In fact, the travel will be longer than these values, caused by compressibility of the brake fluid and elasticities throughout the transmission system.

Brake Boosting

If a brake booster acts between the pedal and the master brake cylinder, its auxiliary effect is added to the force increased via the brake pedal.

$$F_{\mathrm{Pr}} = F_{Rd} + F_{BG} \tag{6.29}$$

$$F_{BG} = \Delta p \cdot A_{BG} \tag{6.30}$$

$$\Delta p = p_2 - p_1 \tag{6.31}$$

F_{Rd}	Push rod force, N. (See also Fig. 6.18)
F_{BG}	Diaphragm force of the braking device, N
A_{BG}	Diaphragm area of the braking device. Diaphragm diameter up to 250 mm.
Δp	Pressure difference acting on diaphragm, N/mm^2. Δp is max. 0.08 N/mm^2 (0.8 bar)

Forces on the Brake

With the hydraulic pressure generated in the brake line, the forces on the brake disc increase:

$$F_{Bd,\tan} = 2\mu_{Bd,lo} \cdot F_{Pi,W} \tag{6.32}$$

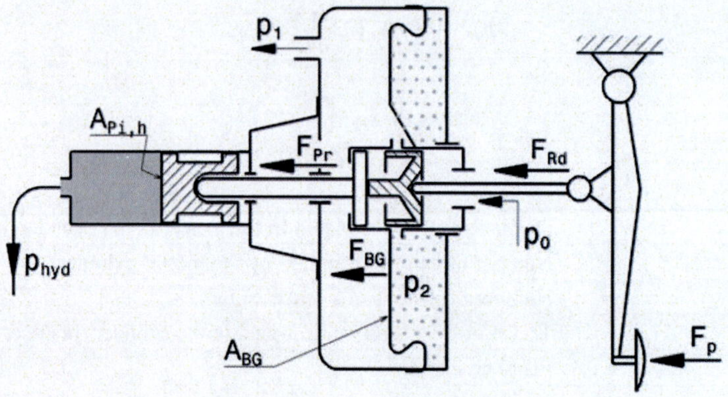

Fig. 6.18 Brake booster, schematic. The pressure difference on the diaphragm acts via the diaphragm area A_{BG} as an auxiliary force F_{BG} to the rod force F_{Rd}

$$M_B = F_{Bd,\tan} r_m = (p_m v_{lo} \mu_{Bd,lo}) \cdot j \cdot A_{pad} \frac{1}{\omega_{Bd}} \tag{6.33}$$

$F_{Bd,\tan}$	Brake force at brake disc, N. (See also Figs. 6.19 and 6.20)
$\mu_{Bd,lo}$	Sliding friction coefficient between brake disc and brake pad, –
M_B	Braking torque, N m
v_{lo}	Sliding speed, m/s $v_{lo} = r_m\, \omega_{Bd}$. With r_m mean friction radius, m
p_m	Average surface pressure, N/mm^2. Permissible values are 5 N/mm^2
j	Number of brake pads, –
A_{pad}	Effective pad area, mm^2
ω_{Bd}	Rotational frequency of the brake disc, s^{-1}

Forces at the Wheel

The forces at the brake disc finally result in the braking force at the wheel:

$$F_{W,X,b} = \frac{F_{Bd,\tan} \cdot r_m}{r_{dyn}} \tag{6.34}$$

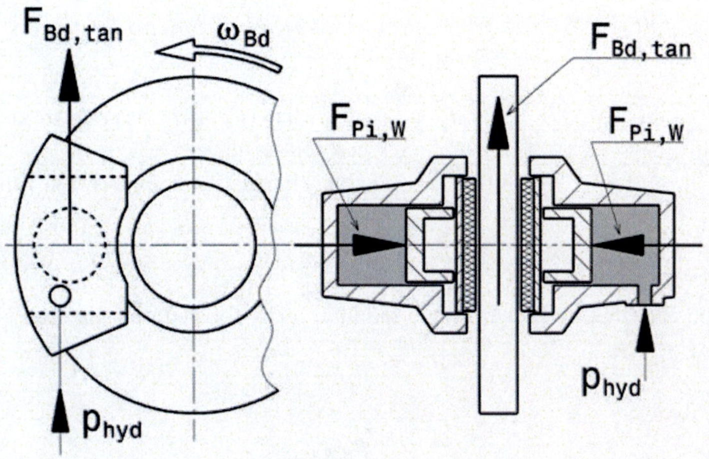

Fig. 6.19 Forces on the brake disc. The hydraulic line pressure p_{hyd} is converted via the pistons in the wheel brake cylinder to the piston force $F_{\text{Pi,W}}$ and via friction to the braking force $_{\text{FBd,tan}}$ at the disc

Fig. 6.20 Forces on the wheel during braking. v_V Vehicle speed. r_m Mean friction radius at which $\text{F}_{\text{Bd,tan}}$ engages

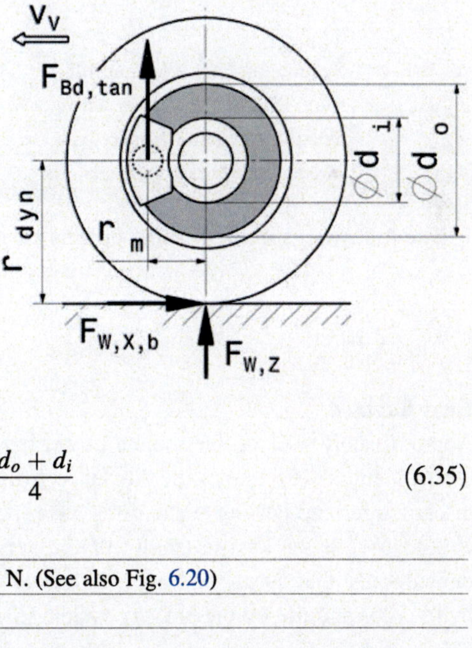

$$r_m = \frac{d_o + d_i}{4} \tag{6.35}$$

$F_{W,X,b}$	Braking force at tyre circumference, N. (See also Fig. 6.20)
r_m	Effective mean friction radius, mm
r_{dyn}	Dynamic Tyre rolling radius, mm
d_o, d_i	Outer or inner diameter of the area on the brake disc covered by the brake lining, mm

The transmittable braking force of the tire still depends on whether it rotates or locks:

$$\text{For the turning wheel :} \qquad F_{W,X,b,\max} = \mu_{W,X} \cdot F_{W,Z} \tag{6.36}$$

$$\text{for the locking wheel}: \quad F_{W,X,b,lo,\max} = \mu_{W,X,lo} \cdot F_{W,Z} \tag{6.37}$$

$\mu_{W,X}$	Static friction coefficient between tyre and road
$F_{W,Z}$	Wheel contact force
$\mu_{W,X,lo}$	Dynamic friction coefficient between Tyre and road

Areas

From the specific braking power *N*, the required areas for pad and disc can be roughly determined.

$$A_{pad} = \frac{P_{B,W,\max}}{2N_{pad}} \tag{6.38}$$

$$A_{Bd} = \frac{P_{B,W,\max}}{N_{Bd}} \tag{6.39}$$

N	Specific braking power at full deceleration from maximum speed per brake, kW/cm^2
$P_{B,W,max}$	Max. Braking power per wheel, kW
A_{pad}	Friction surface of a brake pad, cm^2
A_{Bd}	Passed-over brake disc area, cm^2

The following specific braking powers are achieved by series-production vehicles [16]:

- $N_{\text{pad}} = 2.4$ to 3.3 kW/cm^2
- $N_{\text{Bd}} = 0.45$ to 0.60 kW/cm^2

Heat *Balance*

Approximately 90% of the braking power is absorbed by the brake disc and dissipated to the ambient air [5], i.e. the heat balance is mainly controlled by the design of the brake discs. Larger and heavier brake discs have more heat storage capacity and are thus better able to absorb the kinetic energy of the vehicle [16]. In this context, it should not go unmentioned that during braking with locked wheels, the entire kinetic energy is absorbed by the tires and not by the braking system [5].

The cooling system for the braking system does not have to be designed for the possible maximum amount of heat generated, but a design for the average heat is sufficient because of the storage capacity due to the component masses. Nevertheless, the critical temperatures of discs or linings must not be exceeded during extreme individual braking,

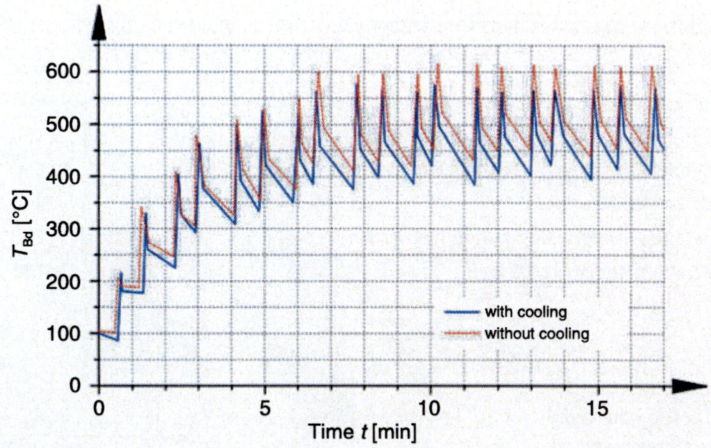

Fig. 6.21 Temperature curve of the brake disc of a passenger car during repetitive braking from high speed with and without active brake cooling [20]. During each braking process, the disc is heated up and cools down again during the following acceleration phase. If the time for this is too short, the temperature of the disc rises until the average temperature difference to the environment is sufficient to release the absorbed energy again by radiation and convection

Fig. 6.21. When designing air ducts to the discs and calipers, consideration must be given to the surrounding air flow so that the desired downforce is not disturbed. Ideally, all cooling air leaves the wheel on its outside. The air velocity across the wheel is in the range 20–30% of the velocity at which the vehicle is flowing [8].

Heat dissipation is increased to improve the performance of braking systems by the following measures:

- Punching, grooving of the brake discs
- Internally ventilated discs
- Supply of cooling air through hoses or ducts.

The temperature increase of the brake disc caused by the heat build-up during braking results in:

$$\Delta T_{Bd} = \frac{0.9 W_{B,Bd}}{m_{Bd} \cdot c_{p,Bd}} \tag{6.40}$$

ΔT_{Bd}	Temperature increase of the brake disc, K
$W_{B,Bd}$	Braking work per disc, J; e.g. from (6.9)
m_{Bd}	Mass of brake disc, kg
$c_{p,Bd}$	Specific heat capacity of the disc material, J/(kgK) (see Table 6.11)

Table 6.7 Relevant reference values for the thermal design of series braking systems [16]

Criterion	Target value, °C
Brake disc surface temperature	<600
Brake fluid temperature pass descent (downhill drive)	<180
Brake fluid temperature high performance stops	<180
Backing plate temperature	<400
Critical temperatures when using aluminium alloys	<180
Critical temperatures for brake hoses	<150

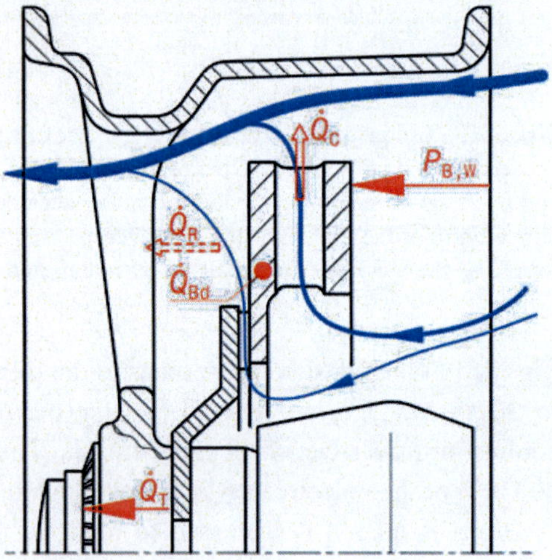

Fig. 6.22 Heat balance of a brake disc. A section through half a wheel with the essential components is shown. In addition to the heat balance, the air flow (blue) through the wheel and the brake disc is also shown

This calculation assumes that 90% of the braking work is stored by the brake disc. In fact, some of the heat is also dissipated via convection and thermal radiation.

Temperatures of conventional (steel) brake discs can reach up to 700 °C. Further guide values for design temperatures are summarised in Table 6.7.

The thermal conductivity of air $\lambda = 0.0242$ W/(K m) is important for heat dissipation [5].

Now a detailed consideration of the conditions should help to specifically adjust the cooling of brakes. The starting point is the heat balance of a brake disc, Fig. 6.22. The heat flow generated by the braking power $P_{\mathrm{B,W}}$ per wheel is absorbed by the brake disc ($\dot{Q}_{Bd}$) and dissipated to the surrounding air by convection ($\dot{Q}_C$) and thermal radiation ($\dot{Q}_R$). The proportion of heat conduction$\dot{Q}_T$ through the mechanical connection to the wheel hub is neglected. The overall balance is thus:

$$|P_{B,W}| = \dot{Q}_{Bd} + \dot{Q}_C + \dot{Q}_R = c_{p,Bd} m_{Bd} \dot{T}_{Bd} + \alpha A_{Bd}(T_{Bd} - T_0) + \varepsilon\sigma A_{Bd}\left(T_{Bd}^4 - T_0^4\right) \quad (6.41)$$

$P_{B,W}$	Braking power per wheel, W
Q_{Bd}	Heat capacity of a brake disc, J. the time derivative describes the heat flow in W
Q_C	Heat dissipated to the environment by convection, J
Q_R	Heat dissipated to the environment by radiation, J
$c_{p,Bd}$	Specific heat capacity of the brake disc material, J/(kg K). (see Table 6.11).
m_{Bd}	Mass of brake disc, kg
T_{Bd}	Absolute temperature of the brake disc, K
T_0	Ambient temperature, K. 20 °C = (20 + 273.15) K = 293.15 K
α	Heat transfer coefficient, W/(m^2 K). For convection in air, $\alpha = 59.7$ W/(m^2 K)
A_{Bd}	Heat-emitting surface area of the brake disc, m^2. For simplicity, the radiation-emitting and heat-emitting surface areas are assumed to be equal in (6.41)
ε	Emissivity (temperature dependent), –. Steel polished $\varepsilon = 0.29$. Steel matt oxidized $\varepsilon = 0.96$
σ	Coefficient of radiation, W/(m^2 K^4). $\sigma = 5.67 \cdot 10^{-8}$ W/(m^2 K^4)

The instantaneous total braking power P_B of the braking system at time t is

$$P_B = (k_R m_{V,t} g + k_m m_{V,t} a_x)(v_{V,1} + a_x t) \tag{6.42}$$

k_R	Rolling resistance coefficient of the tyres, –. (See Chap. 1 *Tyres*, Sect. 2.3)
$m_{V,t}$	Total mass of the vehicle, kg
k_m	Torque mass surcharge factor, –. Takes into account the rotating masses in the drive train. The smaller the gear, the larger k_m. $k_m \approx 1.45$ (first gear) to 1.05 (highest gear)[3]
a_x	Longitudinal acceleration of the vehicle (assumed constant here), m/s^2. Braking deceleration has a negative sign
$v_{V,1}$	Initial velocity, m/s

The total braking power P_B is divided between two brake discs each on the front and rear axle:

$$P_B = 2P_{B,W,f} + 2P_{B,W,r} = \Phi_f P_B + \Phi_r P_B \tag{6.43}$$

with the brake force components Φ_f and Φ_r for front and rear axle, respectively (see (6.12) and (6.13)). This allows the heat load to be represented axle by axle:

$$P_{B,W,f} = \frac{1}{2}\Phi_f P_B \,\text{resp.}\, P_{B,W,r} = \frac{1}{2}\Phi_r P_B \tag{6.44}$$

[3]For more details, see Racing Car Technology Manual, Vol. 3 *Powertrain*, Chap. 4 *Powertrain Design.*

The total heat balance (6.41) represents a differential equation for $T_{Bd}\,(t)$ with the time t as independent variable. If one neglects the thermal radiation term and sets $a_x = \text{const.}$, then an analytical solution for a brake disc can be written down, if (6.44) is still considered:

$$T_{Bd} = e^{-C_1 t}\left[T_0 e^{C_1 t} + \frac{\left|\frac{1}{2}\Phi_j C_3\right| \times \left(\frac{v_{V,1}}{C_1} e^{C_1 t} + \frac{a_x c_{p,Bd}^2 m_{Bd}^2 (C_1 t - 1)}{C_2} e^{C_1 t}\right)}{c_{p,Bd} m_{Bd}}\right] - e^{-C_1 t}\left[T_0 - T_1 + \frac{\left|\frac{1}{2}\Phi_j C_3\right| \times \left(\frac{v_{V,1}}{C_1} - \frac{c_{p,Bd}^2 m_{Bd}^2 a_x}{C_2}\right)}{c_{p,Bd} m_{Bd}}\right] \tag{6.45}$$

The following constants were chosen for a clear presentation:

$C_1 = \frac{\alpha A_{Bd}}{c_{p,Bd} m_{Bd}}$
$C_2 = \alpha^2 A_{Bd}^2$
$C_3 = m_{V,\,t}\,(k_R\, g + k_m\, a_x)$, then (6.42) simplifies to $P_B = C_3\,(v_{V,\,1} + a_x\, t)$

T_1	Initial temperature of the brake disc, K
Φ_j	Braking force component. Index j stands for f or r. in (6.45) the brake disc specific values (index Bd) for C_1, C_2, $c_{p,Bd}$ and m_{Bd} have to be inserted

(6.45) holds as long as the velocity term $(v_{V,\,1} + a_x\, t) \geq 0$ is.

From the heat balance (6.41), the cooling behavior can also be determined, as above, by setting the heat input through the brake pads $P_{B,W}$ to zero. The analytical solution is:

$$T_{Bd} = T_0 + (T_2 - T_0)e^{-C_1 t} \tag{6.46}$$

T_2	Initial temperature of the brake disc during cooling, K

The heat balance (6.41) can be solved numerically and the thermal radiation term can also be taken into account. The time derivative of the temperature is obtained by conversion:

$$\dot{T}_{Bd} = \frac{\left|\frac{1}{2}\Phi_j P_B\right| - \alpha A_{Bd}(T_{Bd} - T_0) - \varepsilon\sigma A_{Bd}\left(T_{Bd}^4 - T_0^4\right)}{c_{p,Bd} m_{Bd}} \tag{6.47}$$

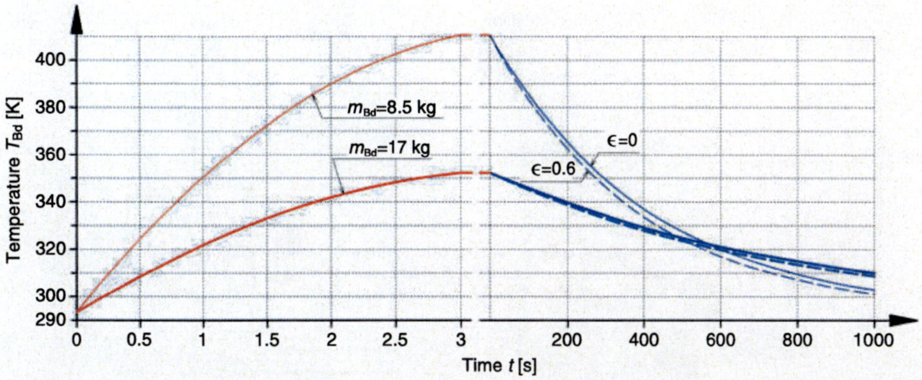

Fig. 6.23 Temperature curve of brake discs (simulation). The temperature curve is plotted over time for front brake discs, whereby the time scale is greatly reduced for the cooling curve (right part, blue). First, braking from 200 to 36 km/h with $a_x = -15$ m/s^2. The disk heats up to more than 350 or 410 K within 3 s. It takes considerably longer to cool down. For comparison, the trace with ($\varepsilon = 0.6$) and without ($\varepsilon = 0$) radiation is entered. The thicker brake disc has a greater heat storage capacity and is therefore more sluggish in its behaviour: it heats up less quickly and also takes longer to cool down. $m_{V,t} = 1000$ kg, $A_{Bd} = 0.18$ m^2, $c_{Bd} = 510$ J/(kg K), $\sigma = 5.77 \cdot 10^{-8}$ W/(m^2 K^4), $\Phi_f = 0.7$

With the aid of numerical integration methods such as Euler's line segment method[4] or Runge-Kutta method, the temperature curve can thus be simulated. During cooling, $P_B = 0$ is set. Figure 6.23 shows the results of simulations for two brake discs with different thicknesses.

6.6 Components of Braking Systems

[4]See e.g. Racing Car Technology Manual, Vol. 5 *Data Analysis, Tuning and Development* Sect. 6.2.2 *Simulation.*

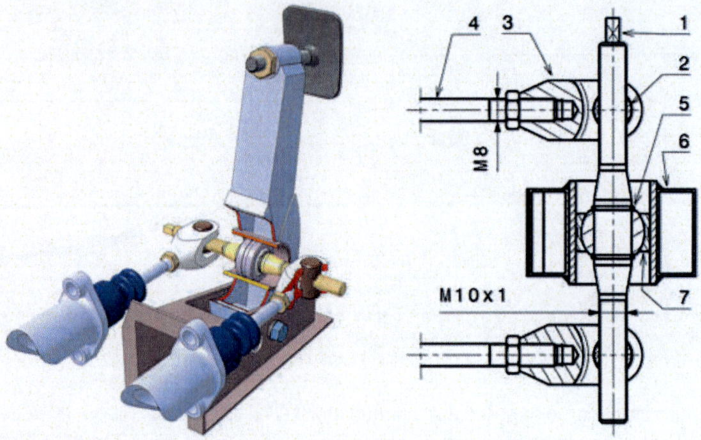

Fig. 6.24 Brake pedal with balance beam. The pedal is shown partly cut open in the left part of the picture. 1 Threaded brake bias bar with square for adjustment, 2 Threaded pivot, 3 Clevis, *4* Piston rod, *5* Round circlips, 6 Brake pedal, 7 Spherical bearing

Pedal

The braking force of the service brake is applied to the pedal[5] with the foot. This translates the force to the piston rod force (Fig. 6.17) of the main brake cylinder or divides the force between two piston rods via a balance beam, Fig. 6.24 (see also Fig. 6.12). In the brake pedal (6), a spherical plain bearing (7) slides in a piece of tubing. The spherical plain bearing is axially secured on a spindle (1) by two round snap rings. The spindle carries forks (3) on both sides which are hinged together with round nuts (2). The forks act directly on the two piston rods (4) of the brake master cylinders. When the spindle is turned, both round nuts move along the spindle thread in the same direction. This changes the lever ratios between the two piston rods and the brake pedal and thus the individual brake pressures with the same foot force.

Brake Fluid

The brake fluid is the medium for transmitting energy between the brake master cylinder, hydraulic control unit if applicable, and the wheel brakes. In addition, it has the task of lubricating moving parts such as seals, pistons and valves and protecting them against corrosion.

Brake fluid must have the lowest possible viscosity, even at the lowest temperatures (down to −40 °C), in order to enable both good brake response and release behaviour and good functioning of the electronic control systems. In addition, the brake fluid must have as high

[5]The design of pedal systems can be found in the Racing Car Technology Manual, Vol. 2 *Complete Vehicle*, Sect. 4.5 *Pedal System and Pedals.*

Table 6.8 Characteristic values of brake fluids (cf. ISO 4925) [3]

Property	Value
Density	Approx. 1.13 kg/l
Low viscosity (at −40 °C)	900–1800 mm^2/s
Dry boiling point	205–260 °C
Wet boiling point (with approx. 3.5% water)	140–180 °C
Flash point	90–140 °C
Rubber swelling	Max. 10%
High aggressiveness against lacquers	
Highly poisonous	

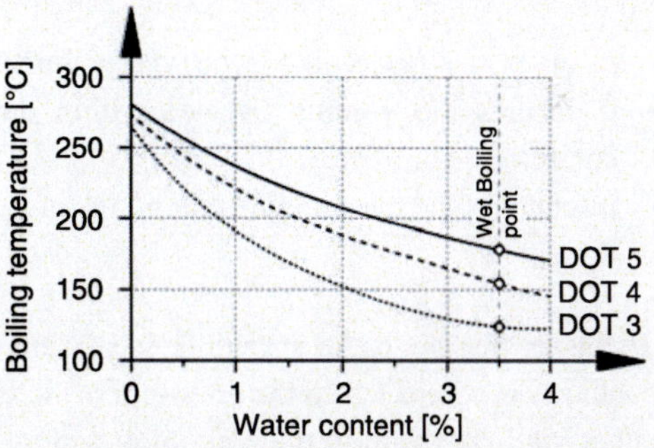

Fig. 6.25 Boiling behaviour of brake fluids

a boiling point as possible so that vapour bubbles do not form even under the most severe thermal load on the brake system. The compressibility of vapour bubbles would result in insufficient pressure being built up due to the limited delivery volume of the tandem master cylinder. Table 6.8 summarises the most important characteristic values.

Conventional Brake Fluids

They are based on polyglycols and polyglycol ethers and are hygroscopic, i.e. they absorb and bind water. This prevents penetrated water from remaining undissolved and forming vapor bubbles during boiling. Numerous international standards, e.g. DOT3, DOT4, DOT5.l, require the highest possible so-called "wet boiling temperature" for the water-enriched brake fluid, Fig. 6.25.

Silicon Based Fluid These are used in racing. Their properties are based on DOT5. They are based on hydrophobic silicone oil, which can only absorb traces of water. Any undissolved water present can boil under certain circumstances (formation of vapour bubbles, see above) or lead to corrosion of components. The compressibility and the amount of dissolved, possibly outgassing air are higher with silicone brake fluids than with conventional brake fluids. For higher brake temperatures (e.g. carbon brake discs), special fluids with a dry boiling point of 310 °C are used [9].

Container (Fluid Reservoir)

The compensation reservoir for the brake fluid is tied into the tandem master cylinder from above by means of so-called "reservoir plugs" and is usually connected to the tandem master cylinder by a further fastening to allow higher pressures during filling on the assembly line and to ensure that the flammable brake fluid cannot escape in the event of an accident. On simple master cylinders, it is either screwed in directly or attached to the frame above the brake cylinder at a suitable point with a clamp.

The expansion tank has the following tasks:

- it serves as a reservoir for the wear volume of the brake pads
- it ensures the volume balance within the braking system under different ambient conditions
- it separates the master cylinder circuits of a tandem master brake cylinder when the level drops.

To ensure that the brake system is depressurised in the release position, the interior of the container is connected to the atmosphere via the container screw connection. This is done either via a labyrinth in the tank lid or a slotted diaphragm integrated in the lid. In racing vehicles, a hose is additionally connected to the lid and guided upwards.

There is usually an insoluble foam in the reservoir so that the movement of the brake fluid due to the extreme driving manoeuvres is kept within limits [15]. Another possibility is to use a bellows, Fig. 6.26. This also prevents the ingress of dirt.

Tank volumes are in the range of approx. 60–300 cm^3. Typical connection threads to the brake cylinder are M12 or 7/16″ UNF.

***Master* Cylinder**

The pistons in the brake master cylinder are subjected to the foot force via the brake pedal and thus build up the hydraulic pressure, which in turn actuates the wheel brake cylinders. The piston diameter (= inner cylinder diameter) is indirectly proportional to the pressure for a given force. A smaller brake cylinder produces a softer pedal feel, extends pedal travel, displaces less volume and generates more pressure. A larger brake cylinder produces a harder pedal feel, shortens pedal travel, displaces more volume and produces less pressure.

In standard passenger cars, two brake cylinders are combined in one housing for the two required brake circuits, Fig. 6.27.

When the pistons are actuated, the small holes ("sniff holes") are overrun by the sealing collars and a pressure build-up is possible because the connection to the compensation/ reservoir is interrupted. In the rest position, these sniff holes must be reconnected to the

Fig. 6.26 Reservoir for single brake cylinder. The reservoir is screwed into the brake master cylinder. It has a diaphragm and a bellows. The screw cap accommodates a hose nozzle

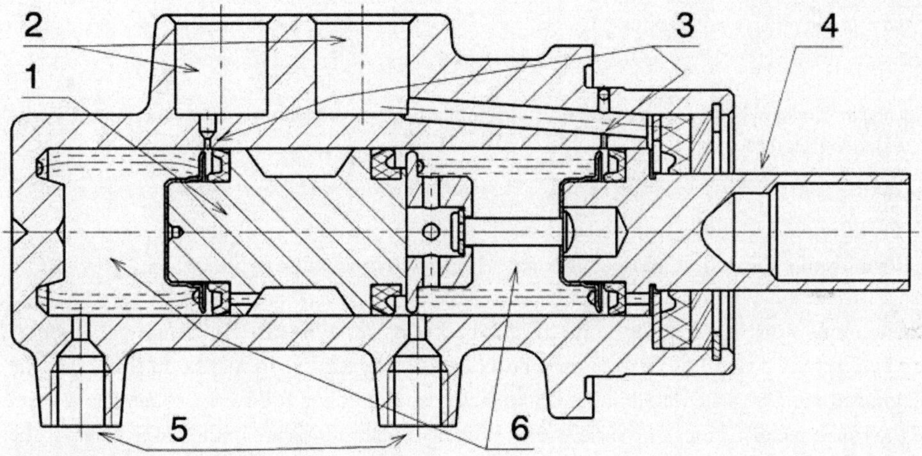

Fig. 6.27 Tandem master cylinder. **1** Floating piston, **2** Connections to compensating reservoir, **3** Compensating holes, **4** Push rod first piston, **5** Outlet for wheel brake cylinders, **6** Pressure chambers. Max. Stroke of the push rod: approx. 25 mm

pressure chamber under all circumstances so that volume compensation and thus brake pad release can take place. This guarantees a clearance between brake pedal and piston in the rest position of the pedal.

In racing, a II brake circuit layout with separate brake cylinders for front and rear is often used. By installing two brake cylinders, a rough tuning of the brake force distribution is possible through different brake cylinder sizes. Fine tuning is then carried out using a balance beam on the brake pedal. The brake cylinder sizes are changed as follows:

- Brake locks rear first and pedal is hard: Reduce front brake cylinder size,
- Rear brake locks first and pedal is soft: Increase rear brake cylinder size,

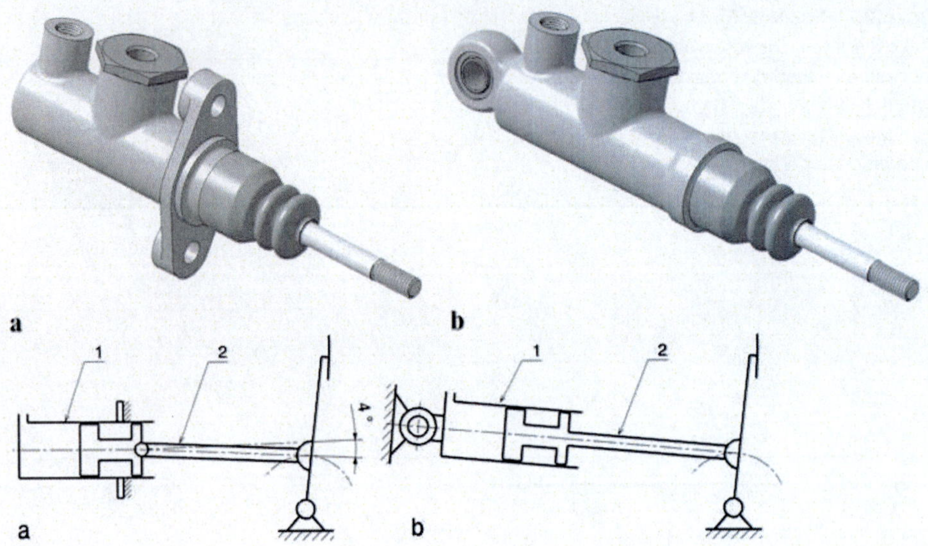

Fig. 6.28 Brake cylinder. (**a**) Version for flange mounting, (**b**) Version for pivot mounting. **1** Brake cylinder, **2** Piston rod

- Front brake locks first and pedal is hard: Reduce rear brake cylinder size,
- Front brake locks first and pedal is soft: Increase front brake cylinder size.

There are two possibilities for mounting the brake master cylinder. The "classical" mounting is done via a two-hole flange on a bulkhead or a bracket, Fig. 6.28a. In this case, the piston rod must be articulated in the brake piston so that it can follow the rotary movement of the brake pedal. The permissible swivel ranges of piston rods are in the range of 4°. If the angles become too large, part of the foot force for forming the hydraulic pressure is lost. This disadvantage is avoided by mounting in a pivot point, Fig. 6.28b. With this arrangement, the full foot force is always transmitted to the piston. For this purpose, the brake cylinder must be pivotable.

Main brake cylinder sizes are in the range of piston diameters 15.9–25.4 mm. In Formula 1 cars, the cylinders measure between 20 and 22.2 mm in diameter [15].

Brake Force Limiter (Pressure Regulating Valve)

A brake force limiter is a valve that is installed in the supply line to the wheel cylinders of the rear axle. Usually in such a way that the adjusting screw can be reached from the driver's seat. If the driver applies the brake pedal with increasing force, the pressure in the brake lines increases accordingly, Fig. 6.29. If the brake pressure of the master brake cylinder exceeds the set value, the valve closes and the pressure in the wheel cylinders of the rear axle remains at this value, even if the foot force is increased further. The valve seat is floating so that after releasing the brake, the rear brake is also released.

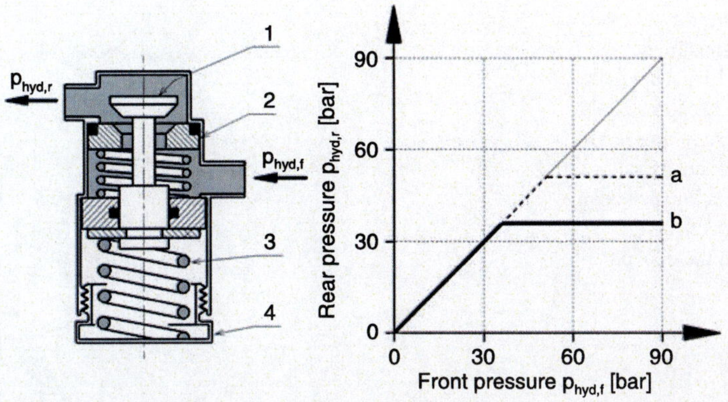

Fig. 6.29 Brake force limiter. **1** Valve piston, **2** Movable valve seat, **3** Piston spring, **4** Adjusting screw. **a** Curve for $p_{hyd,r}$ with high cut-off pressure, **b** Curve for $p_{hyd,r}$ with lower spring preload

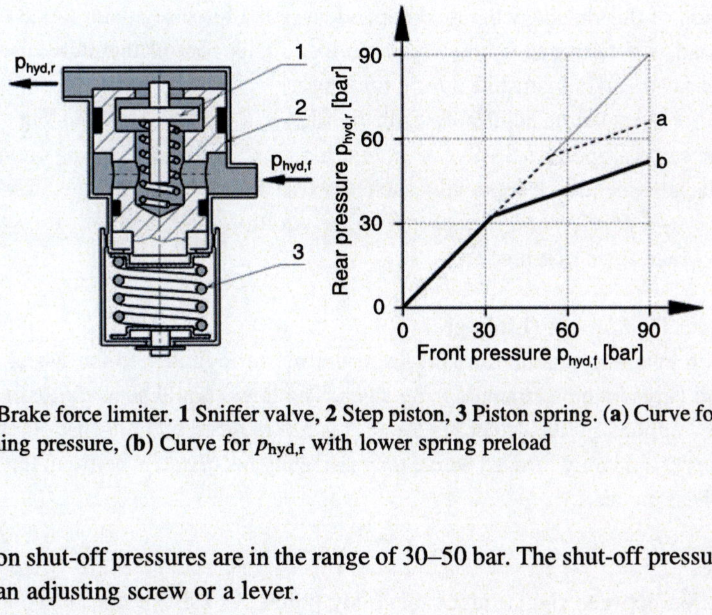

Fig. 6.30 Brake force limiter. **1** Sniffer valve, **2** Step piston, **3** Piston spring. (**a**) Curve for $p_{hyd,r}$ with high switching pressure, (**b**) Curve for $p_{hyd,r}$ with lower spring preload

Common shut-off pressures are in the range of 30–50 bar. The shut-off pressure is set by means of an adjusting screw or a lever.

Brake Proportioning Valve

The operating principle is explained in Fig. 6.30. If the foot force is increased during braking, the same pressure initially prevails in both brake circuits. As soon as the change-over point is reached, which depends on the preload of the piston spring (3), the pressure build-up in the rear axle circuit is reduced in a certain ratio to the front axle. The ratio depends on the area ratio of the stepped piston (2).

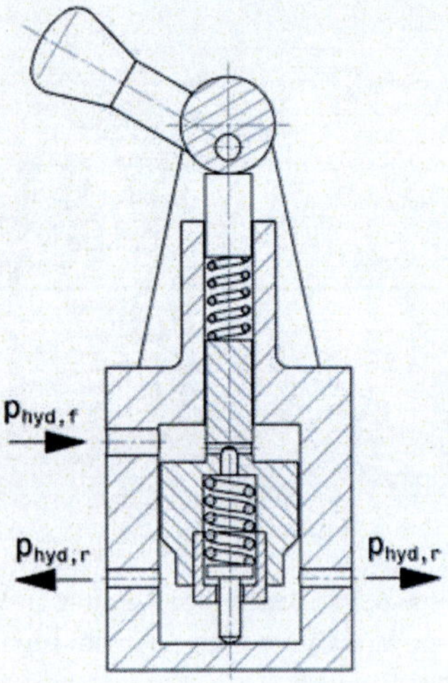

Fig. 6.31 Proportional valve for cockpit adjustment, according to FIA Appendix J Art. 263. This schematically shown valve may be used in super touring cars as an alternative to a balance beam system for brake force adjustment in the cockpit

A variation of this device is the load-dependent brake force regulator (ALB). Here, the spring preload is determined by the rear axle load. This control therefore also advantageously records the axle load shift during braking.

A brake force regulator adjustable from the driver's seat is shown in Fig. 6.31. It is basically the same proportional valve as in Fig. 6.30, except that the spring preload can be adjusted via an eccentric lever. Only such a valve is approved by the FIA for super production cars (WTC-World Touring Car) in addition to balance beam systems for varying the brake force distribution.

Transmission Equipment (Linkages)

Brake lines transmit hydraulic pressure from the master cylinder to the wheel cylinders. T-pieces and valves can be arranged in between. The lines should be as rigid as possible so that the force applied by the driver via the foot is not reduced by deformation of the lines. Flexible lines should only be used where they are really needed, e.g. between the frame and the front wheel brakes.

Brake Pipes

For connection between rigid, immovable body points. Consist of double wound, brazed steel tubes.

To protect against environmental influences, the pipe surface is galvanized and additionally covered with a plastic coating.

Brake Hose Lines
At the transitions to dynamically highly stressed parts such as steering knuckles or brake calipers. Ensure perfect transmission of fluid pressure to the brakes even under extreme conditions. In addition to mechanical load capacity, compressive strength and low volume absorption, chemical resistance, e.g. to oil, fuels and salt water, as well as good thermal resistance are essential requirements [5].

The structure of the brake hose lines is divided into: Inner hose, two-layer braiding as pressure carrier, and outer rubber layer to protect the pressure carrier from external influences.

Flex Lines
(Flexible Line with Steel Braided Outer Hose): Similar to brake hose lines at the transitions to dynamically stressed parts. Due to the lower flexibility resulting from the structure (PTFE (polytetrafluoroethylene) line with stainless steel braiding as pressure carrier and another thermoplastic elastomer as outer protective layer), the range of application is limited to connections with low movement, such as occur at the calipers due to pad wear. This design is used in racing. It enables a brake pressure that is almost independent of temperature.

Connections
In principle, all connection types known from hydraulic systems are used, Fig. 6.32. Rigid lines are screwed together via a cutting ring or ring soldering pieces. Flexible lines are connected to the brake calipers, master brake cylinders or T-pieces (distributors) via pressed-on connectors. Another option is to use a sealing olive for the screw connection.

Table 6.9 compares two common thread sizes and their designations.

Brake Caliper
For road vehicles, durability and reliability are paramount, even with a lack of care over the years. For racing vehicles, light weight is important despite reliability at high temperatures. Brakes on racing vehicles are cleaned and serviced regularly. High stiffness is another goal. This guarantees even as well as even wear of the brake pads and ensures consistent braking behaviour, which is particularly important in endurance races (Fig. 6.34).

The two principal types of construction that have become established for disc brakes are fixed and fist-type calipers (cf. also Fig. 6.14). In racing vehicles, the fixed caliper dominates because there is sufficient lateral space in the wheel and the preferred rear-wheel drive also eliminates the drive shaft on the front axle. Racing calipers are designed and mounted very stiffly. The necessary compensation for temperature expansion and clearance must therefore be ensured by other parts such as the brake disc/disc bell. The brake calipers are mounted by bolting them together using lugs (Fig. 6.35) or radially

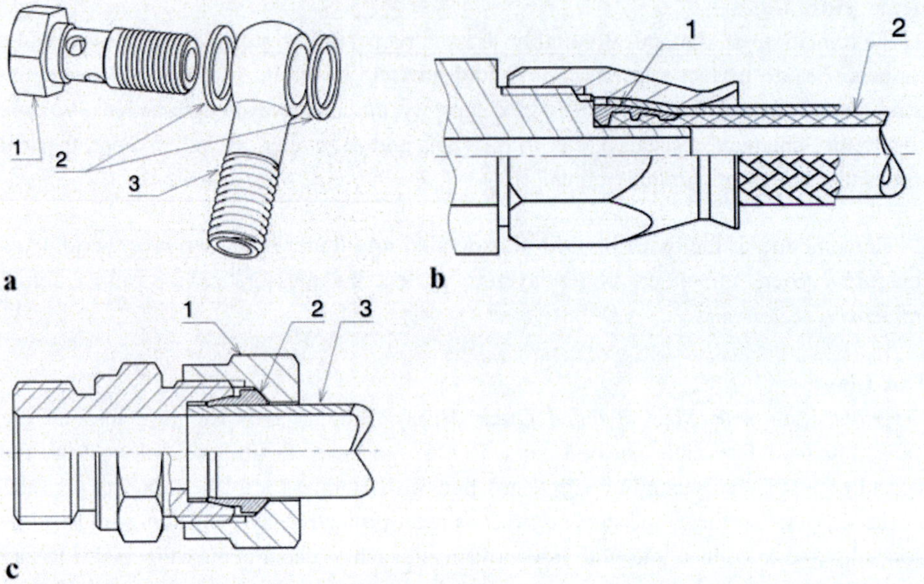

Fig. 6.32 Screw connections of brake lines. (**a**) Ring piece: 1 Banjo bolt, 2 Sealing washer, 3 Banjo ring piece, (**b**) Fitting with compression sleeve: 1 Compression sleeve, 2 Brake line steel braided, (**c**) Cutting ring connection: 1 Coupling nut, 2 Cutter (cutting ring), 3 Steel brake line

Table 6.9 Usual thread sizes and designations of the connections (Designations Fig. 6.33)

Connection thread		External thread dimension d_o, mm	Thread dimension inside d_i, mm
1/8 BSP	D-03	9.52	8.73
¼ BSP	D-04	13.49	11.11
3/8 UNF	D-03	9.52	8.73
7/16 UNF	D-04	11.10	7.93

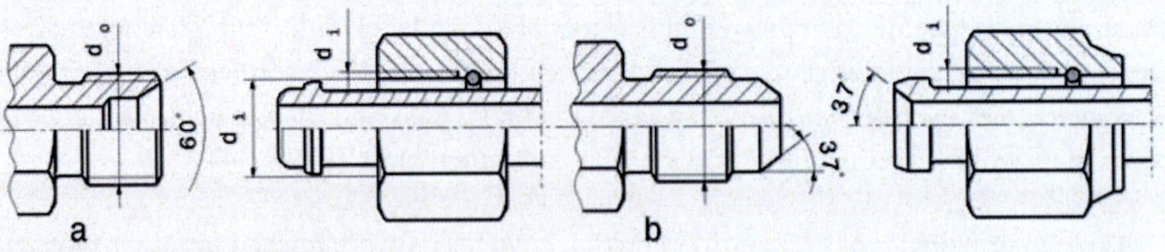

Fig. 6.33 Connectors (fittings) for brake lines (**a**) BSP thread, (**b**) UNF thread

(Fig. 6.36). With the latter type, the position of the holes is fixed and therefore cannot be changed. The running direction of the brake disc must be observed for multi-piston calipers with different brake piston diameters. The smaller piston must be swept by the disc first.

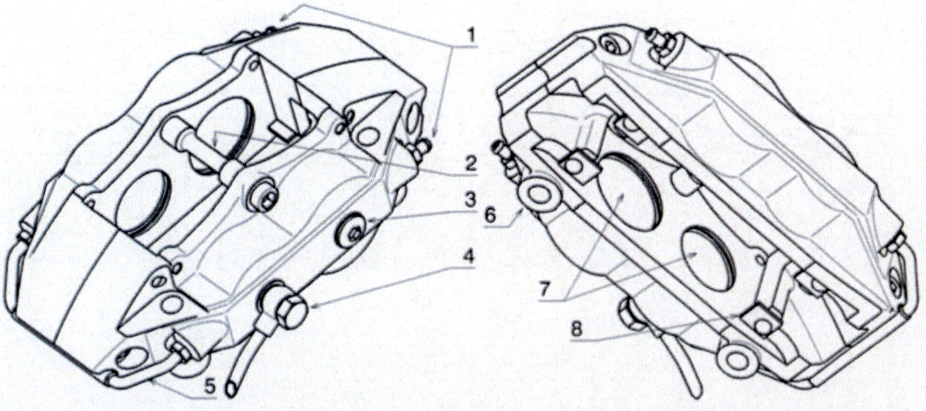

Fig. 6.34 Brake caliper. This two-piece 4-piston brake caliper is radially bolted to the wheel carrier. **1** Bleeding, **2** Securing of brake pads, **3** Screw plug, **4** Connection to master brake cylinder, **5** Connecting tube, **6** Radial screw connection, **7** Holes for brake piston, **8** Holder for brake pads

A disadvantage of the usual fixed caliper design is the connecting hole or pipe (Fig. 6.34, part 5) from one half of the housing to the next. This crosses the brake disc close to its face and therefore the brake fluid is directly heated by the disc. The heat dissipation must therefore also cover this area.

Water-cooled brake calipers also exist. The coolant is pumped through the cooling circuit by an external pump.

Brake callipers are either of two-piece construction, Fig. 6.34, or are manufactured from one piece, Fig. 6.37. One-piece housings offer weight advantages due to integral construction, because there are no boltss and sealing webs. They are cast, forged or machined from solid. Multi-piece calipers are usually cast. In either case, light alloy housings are hard anodized or nickel plated to increase their resistance to environmental influences. During rapid wheel changes, for example, they are completely unprotected. Brake calipers are also called calipers, which directly refers to their main function in braking. Great rigidity, especially in the direction of piston pressure, thus becomes a desirable feature of calipers. Steel bolts on multi-part light alloy brake calipers have an advantageous supporting effect in this context.

Materials
Depending on the manufacturing process and cost specifications, the following materials are used.

Ductile iron EN-GJS-500-7 (was GGG-50), EN-GJS-600-3 (was GGG-60): Low cost, but heavy. Aluminium alloys, aluminium-beryllium alloy (Albemet [17]): More costly, but light; aluminum-lithium. The aluminium material in Formula 1 is restricted by the regulations by its modulus of elasticity ($E < 80{,}000\ \text{N/mm}^2 = 80\ \text{GPa}$).

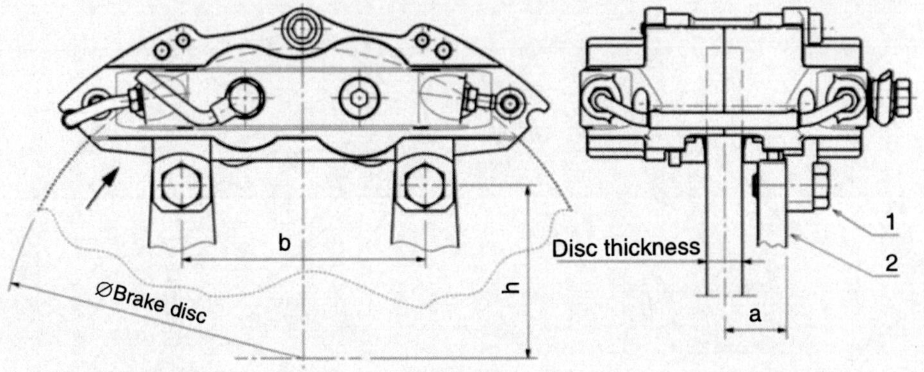

Fig. 6.35 Brake caliper with lug mounting. **1** Fastening bolts, **2** Bar for fastening to wheel carrier. The picture shows important dimensions for mounting the brake caliper. *a* Distance between support and centre of brake disc. *h* Distance from centre of bore to centre of brake disc

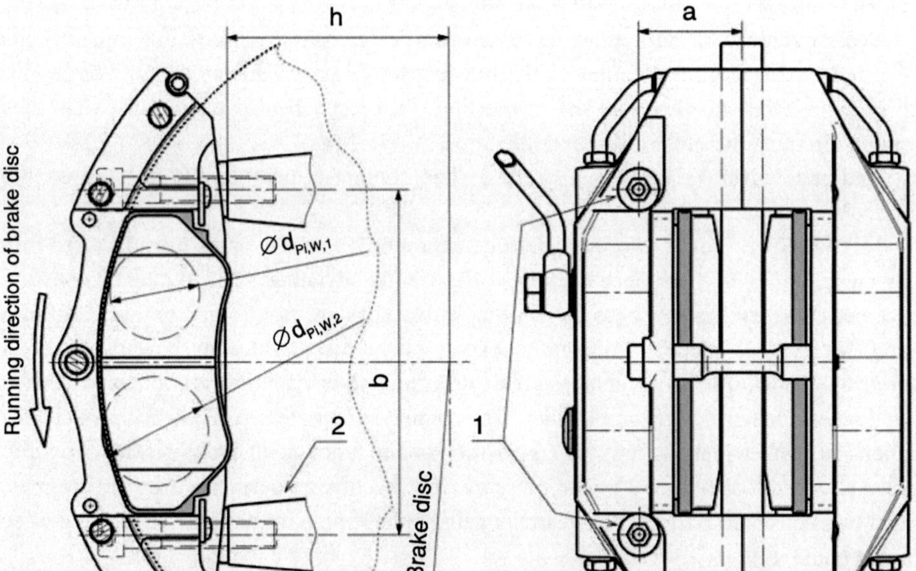

Fig. 6.36 Brake caliper with radial mounting. **1** Fastening bolts, **2** Slugs for fastening to the wheel carrier. The picture shows important dimensions for brake caliper mounting. *a* Distance between screw connection and centre of brake disc, *h* Distance between brake caliper and centre of brake disc, $d_{Pi,W,1\ resp.\ 2}$ Brake piston diameter

Fig. 6.37 Formula 1 brake caliper. The caliper is milled from one piece and is radially bolted to the wheel carrier. It carries six brake pistons. The pistons are radially bored through. This reduces their mass and reduces heat transfer to the brake fluid. The ceramic brake pads, which the caliper holds, are prevented from falling out from below by two screwed plates each

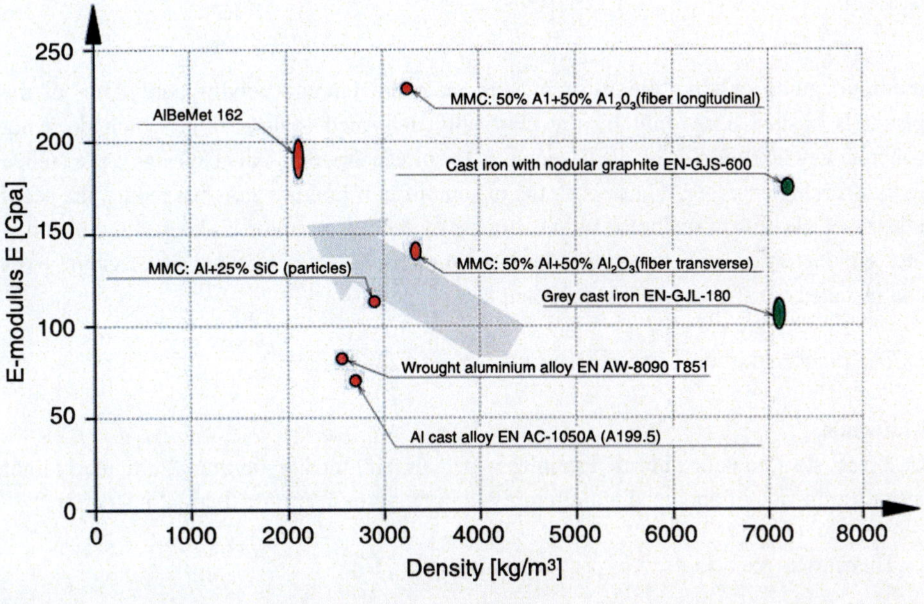

Fig. 6.38 Comparison of materials for brake calipers. MMC are metal-matrix composites, a metal matrix (here aluminium) is reinforced by fibres (longitudinally or transverse). The grey arrow points in the preferred direction, i.e. low density with high stiffness

Figure 6.38 provides an overview of possible materials and their decisive properties.

Brake Piston

The length to diameter ratio should be as large as possible so that the piston does not tend to stick. In addition, during braking with worn linings, i.e. when the piston is extended, a

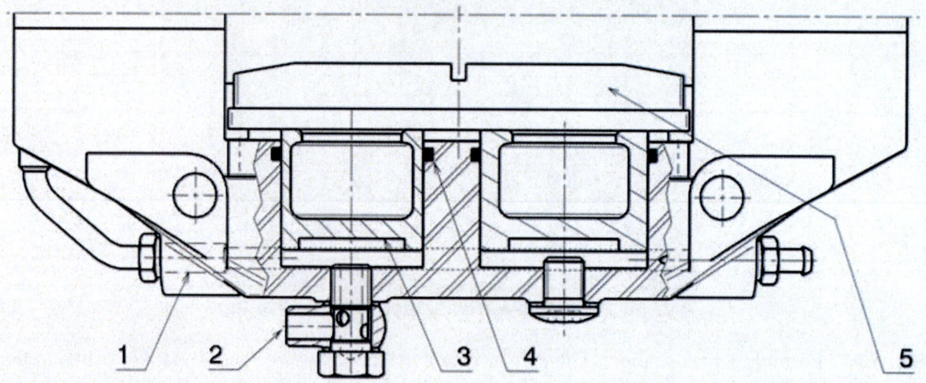

Fig. 6.39 Brake piston in the brake caliper. **1** Connecting holes, **2** Inlet via ring piece, **3** Piston, **4** Sealing ring, **5** Brake lining. The brake fluid transmits the brake pressure via the supply and the connecting bores into the piston chambers. The pistons themselves move the brake lining to the brake disc

minimum guide length of the piston must be maintained in its receiving bore. After all, the piston is pushed back only by the elastically deformed sealing ring, which does not generate any large forces or displacements. In some calipers, conical springs are installed in the working chamber. These push the piston out as if braking and thus reduce the pedal idle travel. Heating of the brake fluid is limited by the design of the brake piston with small heat conducting areas between the fluid and the lining. Brake pistons as individual parts and installed are shown in Figs. 6.39 and 6.40.

The diameters of the pistons are in the range of 25–50 mm.

Materials

Cast iron, steel, titanium-nitrided stainless steel, aluminium and titanium. Particularly light pistons are made of anodised aluminium with a titanium insert for thermal insulation.

Thermosets are also used in production vehicles [16].

Sealing Rings

The sealing rings with a square cross-section are located in a piston groove or in the housing bore. In addition to sealing the piston clearance, which is their name, their function is primarily to return the piston by the clearance (*roll back*) when the brake pressure drops again. If the piston were to remain in the braking position, there would be a residual braking torque that would cause losses. However, the piston should not be moved back too far either. In this case, an extension of the pedal travel would disturb the driver when braking again. The sealing ring also takes over this resetting movement in the direction of the disc (*knock back*).

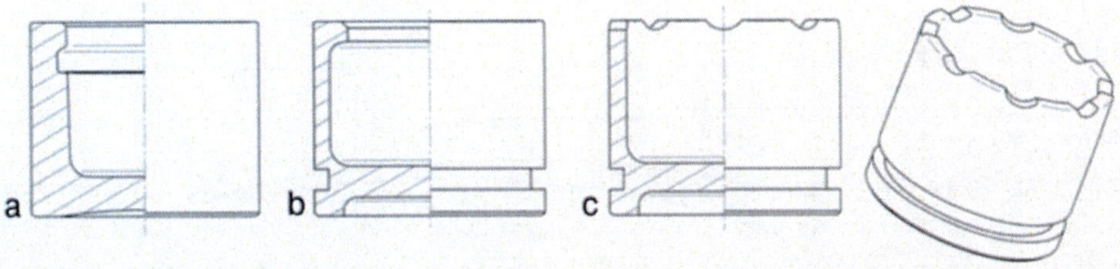

Fig. 6.40 Brake piston. (**a**) Version without groove. The sealing ring is located in the brake calliper bore. Material: aluminium alloy, (**b**) Version with groove. The sealing ring is located on the piston. Material: steel, (**c**) Version with recesses in the contact surface to the brake lining. On the one hand, cooling air can pass through these recesses and, on the other hand, the heat-absorbing surface is reduced

Brake Pad

The two decisive points for the selection of brake pads are the coefficient of friction and the temperature behaviour. The higher the coefficient of friction, the better the braking effect. However, a very high coefficient of friction also leads to increased disc wear. When using low friction linings, on the other hand, the brake control is better.

The brake pads achieve their best deceleration values in the temperature ranges specified by the manufacturer, Fig. 6.41. The temperature ranges are between 100 and 600 °C and 200 and 750 °C. The brake pads are also suitable for use at high temperatures. A flat progression of the coefficient of friction over the temperature makes it easier for the driver to assess the braking behaviour. Pads that show a pronounced "*bite*" when the brake is applied, i.e. a pronounced initial frictional force, also prove to be favourable. Subsequently, the braking force can drop as in Fig. 6.42, type *a*. This also matches the circumferential force potential of the tyres, which can absorb considerably greater forces at the beginning of braking than towards the end.

The temperature dependence of the coefficient of friction is also worth considering in connection with brake balance. At the front and rear axle, without special measures, the temperatures and thus the friction values will generally develop differently. This results in a shift in the distribution of braking force. If you want to keep the brake balance constant, the friction values on both axles must change in the same proportion. This is achieved by using brake discs of different sizes (diameter and thickness), adapted heat dissipation and different brake pads.

Run-in Process

(Running-In, AE: Break-In). During the running-in of brand-new friction linings, the so-called friction layer, which is only a few micrometers thick, forms on their contact surface. It has a decisive influence on the friction and wear properties of the friction lining and disc. It also protects the underlying base material from thermal overload. For this reason, friction pairs should not be subjected to full load until the friction layer is in place.

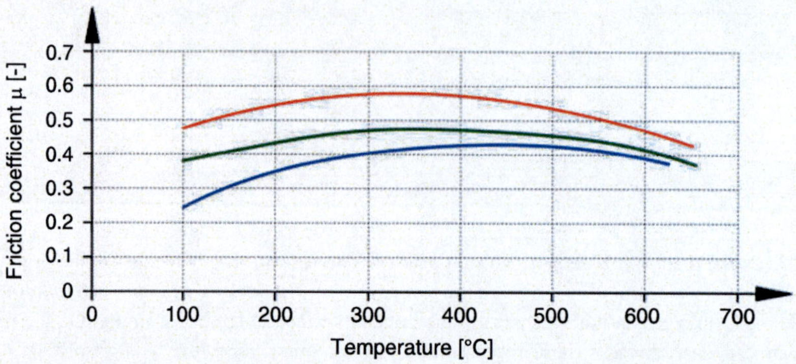

Fig. 6.41 Coefficient of friction of brake pads as a function of temperature, after [22]. The temperature behaviour of three different brake pads shows the same basic curve. The greatest friction is achieved in the middle of the temperature range. Below and above this, the coefficient of friction drops. The pad with the blue curve achieves the lowest friction values, but protects the brake disc. It is used for high performance road cars. Typical racing pads show a curve as shown in green and red. The flatter the curve, the more favourable: If the friction behaviour hardly changes with the temperature, the driver can estimate the behaviour well and vice versa. Of course, friction values depend on the material pairing, so in this case the brake disc is also important. The values shown are mean values from brake tester measurements and are intended to provide an idea of the basic process

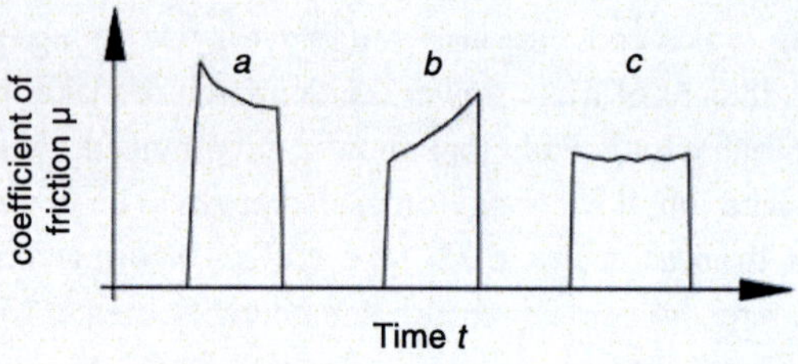

Fig. 6.42 Course of the coefficient of friction of different brake pads during a braking manoeuvre. The curve over time is shown for three different brake pads. Type *a* shows the most favourable behaviour: Strong bite when the brake is applied and decreasing friction value with falling speed. Type *b* is the most difficult for the driver to control. Type *c* lies – as far as dosing is concerned – between the two extremes mentioned

New brake pads must therefore be broken in. New brake discs should not be used in this process.

Figure 6.43 shows how the friction changes in a number of successive tests with the same contact pressure. This initially increases with the measurement time. Only after the adhesion limit has been exceeded does the lining slide on the friction rail. At the beginning of the test series, the smallest friction force and the smoothest transition from the adhesion area to the sliding area are formed. As the number of friction tests increases, the friction

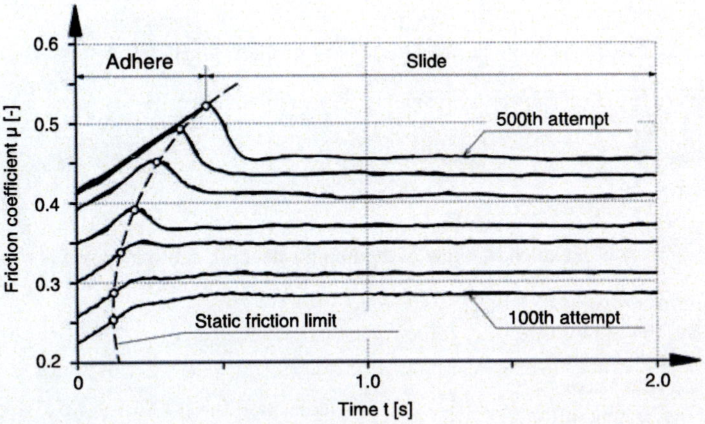

Fig. 6.43 Running-in process [16]: Changes in friction properties in the adhesion/slip transition region in a factory-made friction pairing with organically bonded lining. Measuring system slide on sliding rail. Friction surface = 16 cm^2, pressure $p = 50\ N/cm^2$, sliding speed = 0.04 m/s

coefficient increases. At the same time, an increasingly larger peak is formed in the transition area between adhesion and sliding.

The cause of the change in the friction properties during running-in is, in addition to physical-chemical transformations in the area close to the friction surface, the smallest metal particles which migrate from the metallic friction partner into the friction layer. The running-in process is not complete until equilibrium has been reached between the particles that have migrated in and those that have been removed by wear.

The pad thicknesses are between 10 and 25 mm. Pads must not protrude beyond the brake disc. This can lead to a number of braking problems [12]. Figure 6.44 shows two examples.

The following recommendations can be made for pad selection based on drive type:

- All-wheel drive vehicles and vehicles with rear-wheel drive: Same quality of pads on front and rear axle.
- Vehicles with front-wheel drive: pads with a high coefficient of friction on the front axle and those with a lower coefficient of friction on the rear axle.

The heating of the brake fluid is limited by the design of the brake piston and can be further reduced by a heat shield on the back of the carrier plate, Fig. 6.45.

There are differences between brake pads for production vehicles and those for racing use. Racing pads have the highest deceleration but tend to be noisy, making them unsuitable for everyday use. A performance comparison of different brake pads is illustrated in Fig. 6.46.

In racing, brake pads reach a lifetime of about 300–400 km [15], which means that in some disciplines they are renewed for each race.

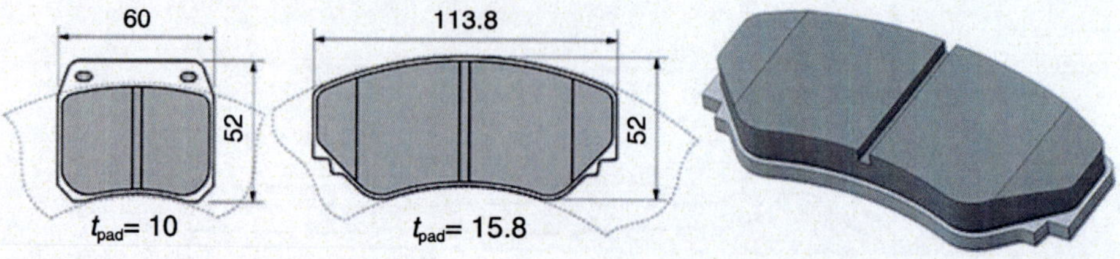

Fig. 6.44 Examples of brake pads. View of the lining side. Disc contour shown in dashed lines

Fig. 6.45 Heat shield for brake lining. A heat shield attached to the carrier plate with a screw rivet reduces the heating of the brake fluid and reduces the formation of bubbles

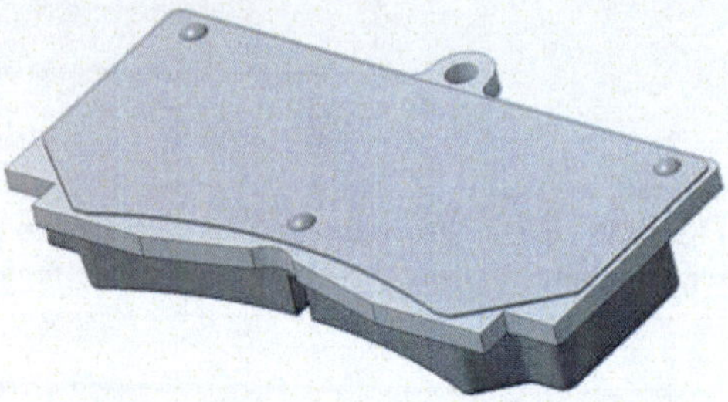

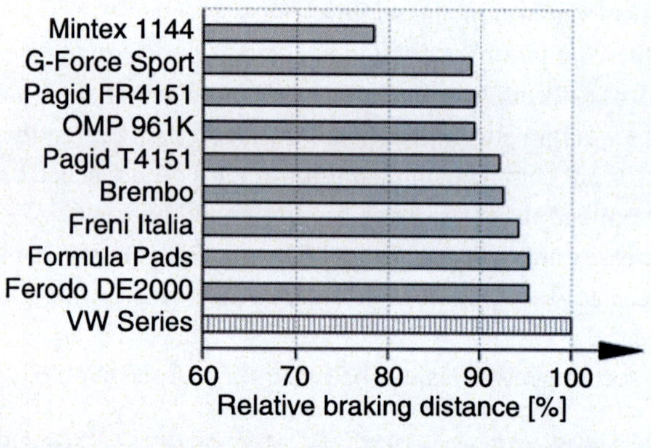

Fig. 6.46 Influence of the brake lining on the braking distance, according to [13]. Different brake pads, including the standard equipment, were tested on a VW Golf GTi (MY '88). Each set of pads was broken in and then the average value was determined from 15 consecutive full braking applications from 88 km/h. Some pads reached their best value at the first, some after a few brakings

Materials

The pad material depends on the friction partner, i.e. the brake disc. Organic and semi-sintered pads are suitable for use with cast brake discs. Ceramic pads (carbon) can only be used exclusively for carbon brake discs.

Depending on the friction coefficient family, the linings themselves exhibit different properties [16]:

- Semimet: Low disc attack, $\mu \leq 0.4$ (unusable for high performance vehicles).
- NAO (Non-Asbestos Organics), organic linings: Low lining wear, $\mu = 0.3–0.4$ (unusable for high-performance vehicles), long service life.
- Low Steel (Low Met): Increased disc wear, $\mu = 0.35–0.5$; high fading stability, lower service life than NAO.
- Hybrid: Combination of the properties of NAO and Low Steel.

Carbon has significant advantages over the conventional friction material [16]:

- Weight: The density of carbon is only one fifth of that of steel or cast iron.
- Thermal conductivity: Since carbon is a composite material, it is possible to influence thermal conductivity by orienting the carbon fibers in predetermined directions.
- Temperature resistance: This results in less fading as the temperature rises, allowing better control and reproducibility of the braking process.
- Bite: Very good response already at the beginning of the braking process.

The only remaining disadvantage is the price. A set of carbon brake disc plus brake pads costs about ten times as much as a set of cast iron brake disc with organic brake pads.

An illustrative comparison of the friction behaviour of both pairings is given in Fig. 6.47. For both pairings, two braking events can be seen starting at 300 and 220 km/h. It can be seen that a carbon brake pad shows much more "bite" (power in the first braking phase) than an organic brake pad. In order to brake the vehicle sharply, a high braking power is necessary at the beginning of a braking phase. The more the vehicle slows down, the higher the coefficient of friction becomes with the organic pads. On the other hand, the carbon pads are much more constant in their action and allow the driver to better modulate

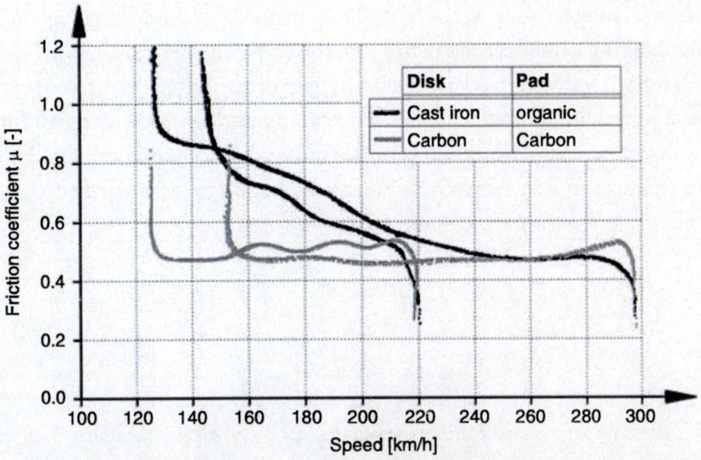

Fig. 6.47 Friction behaviour of different pairings versus velocity, after [16]. For each pairing, two braking operations are shown, starting at 220 and 300 km/h

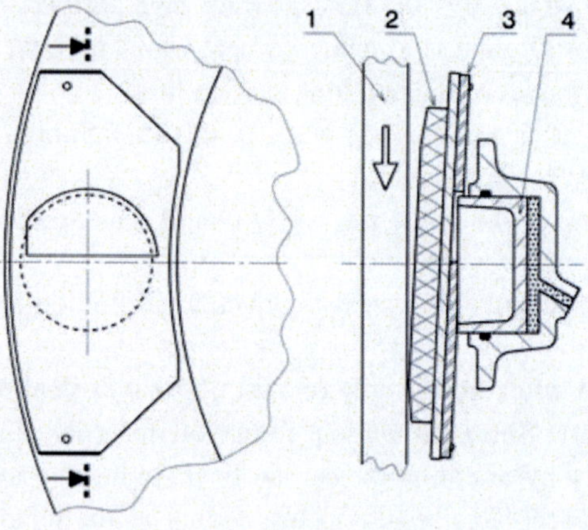

Fig. 6.48 Design of the pad back to prevent brake rubbing (schematic). A plate (3) is riveted onto the back of the brake lining (2). This has an approximately semicircular recess. As a result, when the brake is actuated, the brake piston (4) first presses on the edge of the brake lining facing away from the direction of travel (arrow) of the brake disc (1)

the braking process and control the vehicle as the load distribution changes during braking. This difference is caused by the better temperature stability of carbon.

If the brake pad is at an angle, there is a leading edge and a trailing edge. If the leading edge is running up, there is a self-amplification effect and brake judder due to a stick-slip effect. The leading edge becomes hotter than the trailing edge during operation. In addition, the pad wears out strongly on one side. An asymmetrical pad back ensures that the brake piston presses the rear edge of the pad against the disc first when the brake is moving forward, Fig. 6.48. The same effect is achieved with multi-piston brake calipers with different piston diameters (see Sect. 6.6). This means that there is no self-amplification effect and the braking effect can be better controlled by the driver.

The characteristic value of the specific lining performance (e.g. Sect. 6.4) can be used to estimate the required lining area for a desired braking performance. A small lining surface naturally results in smaller deformations of the lining during operation. On the other hand, longer brake linings are less sensitive to tilting with the effects described above and they reduce the surface pressure for the same actuating force.

Brake Disc

Table 6.10 Reference values of brake disc diameters as a function of rim diameter

	Rim diameter $d_{nominal}$ (inch code)					
	12″	13″	14″	15″	16″	17″
Brake disc outer diameter, mm	221	256	278	308	330	360

The brake disc diameter is decisive for the braking effect to be achieved. Normally, the largest possible disc is fitted, taking into account the rim profile, cf. Table 6.10. However, when determining the disc diameter, the expected braking forces (vehicle weight, load distribution, tyres) must also be taken into account. The disc thickness increases with the diameter of the disc. The vehicle weight and intended use (rally, circuit, long distance, etc.) are also decisive for the thickness of the disc. A thicker disc has a longer service life and stabilises the disc temperature due to its greater mass. These are used for endurance races, for example.

Disc thicknesses are available from 7.10 to 35.50 mm. Outside diameters range from 280 to 378 mm.

The dimensions are partly restricted by the regulations. In Formula 1, for example, the disc thickness may not exceed 32 mm and the diameter 278 mm.[6] For sports prototypes, the diameters are limited to 380 mm (FIA SR1) and 356 mm (FIA SR2). For qualifying or hill climbs many teams use narrower discs. This saves weight with the narrower brake caliper that can be installed for this purpose, and the brake reaches its operating temperature more quickly.

The simplest brake discs are solid and are bolted directly or via a disc bell to the hub, Fig. 6.49. They are installed in vehicles such as Formula Ford or motorcycles. More sophisticated designs have cavities and allow more intensive heat dissipation, which can be forced by turbine-like design of internal ribs. Examples of such internally ventilated brake discs are shown in Figs. 6.50 and 6.54.

In the case of air shafts with scoop action, the direction of rotation must be observed, Fig. 6.51.

In general, when designing brake discs with cavities, recesses and interruptions, care must be taken to ensure that the temperature distribution remains as uniform as possible during operation. Insufficiently connected areas between mass accumulations easily lead to local heat spots (*hot spots*) with selective liquefaction of partial components and formation of an air cushion effect for the brake lining. Conversely, those areas with mass accumulation can become *cold spots*. In this context, shallow grooves prove to be much more favourable than through-holes which completely interrupt the disc.

How the cooling of the brake disc can be improved and the mass reduced at the same time is illustrated by the evolution of Formula 1 brake discs over 10 years, Fig. 6.52. The number of radial cooling holes in the CFC (carbon fibre reinforced carbon) discs has been

[6] Until 2015, only 28 mm was allowed.

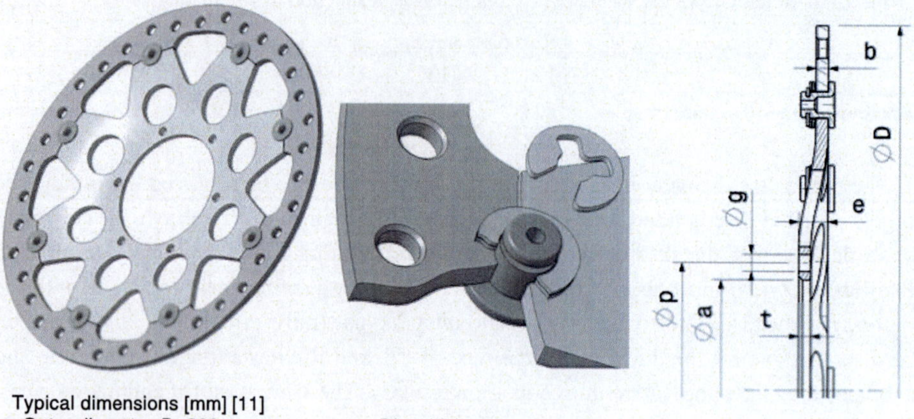

Typical dimensions [mm] [11]
Outer diameter D: 320
Mounting diameter a: 64 to 130
Offset e: 0 to 10
Disc thickness t: 6
Hub width b: 6

Hub screw connection Pitch diameter p: 80 to 110
6 holes with diameter g: 8.35
5 holes with diameter g: 10.2

Fig. 6.49 Brake disc for racing motorcycle made of alloy grey cast iron. The disc is solid and drilled. It is floatingly connected with the hub disk via bolts with two corrugated washers (detail section)

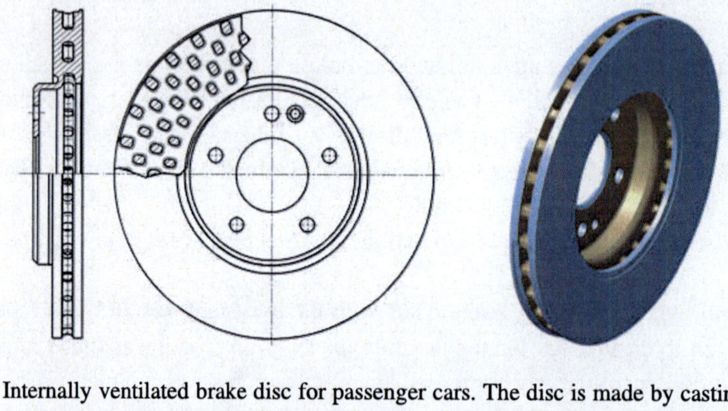

Fig. 6.50 Internally ventilated brake disc for passenger cars. The disc is made by casting

constantly increased. In 2016, a widening of the brake disc was added. Ultimately, 100 holes became over 1400. This not only greatly increased the heat dissipating surface, but also reduced the mass of the brake disc.

Grooved Brake Discs

Brake discs are provided with four or eight tangential grooves which serve to clean the pad surface, Fig. 6.53. This achieves a uniform braking effect and improves wet braking performance. It also lowers the pad temperature. Grooved brake discs have a longer service life than drilled designs. The grooves are about 1.5 mm wide and 0.8 mm deep.

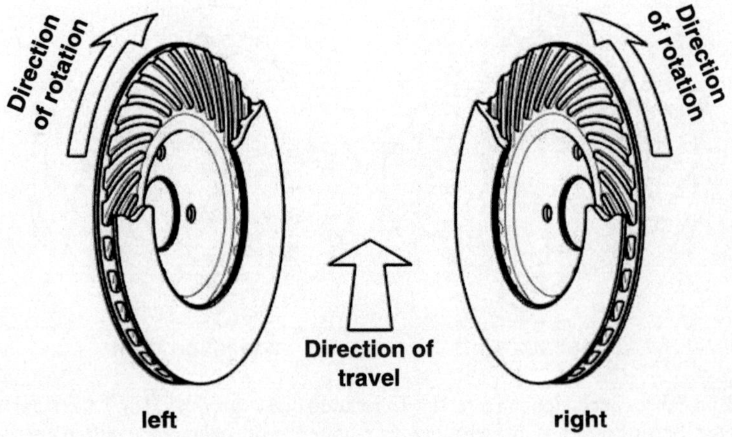

Fig. 6.51 Running direction of ventilated brake discs. Certain brake disc designs are directional. The curved ventilation shafts must point to the rear in the direction of travel because the cooling air is drawn in from the centre of the disc

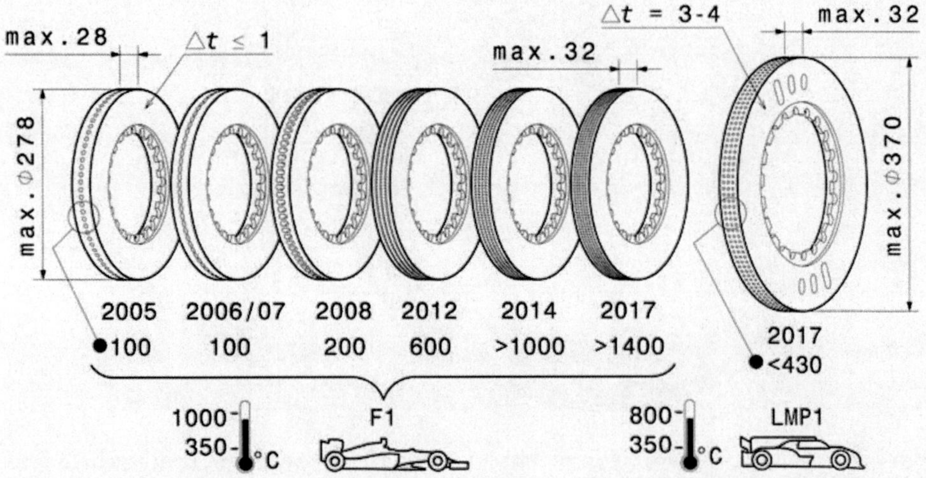

Fig. 6.52 Development of brake disc cooling (Formula 1). The number of radial cooling holes has increased significantly from 100 (2005 season) to over 1400 (2017 season). Although the machining effort has increased just as much as a result, it pays off in terms of increased cooling surface with reduced brake disc mass. A typical Le Mans Prototype Class 1 (LMP1) brake disc is shown for comparison. In addition to the number of cooling holes, the graph also shows the operating temperature range and the permissible thickness reduction Δt in the race. All dimensions in mm

Drilled Brake Discs

These produce a slightly higher braking force (especially at the beginning of a braking = "bite") and achieve a slightly lower brake disc and pad temperature. In addition,

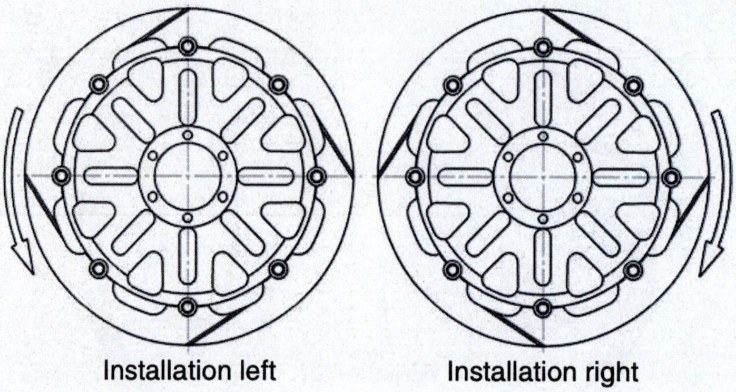

Fig. 6.53 Brake discs dependent on direction of rotation due to grooves [10]. Due to the function of the grooves for abrasion removal, two versions are required for installation on the left and right side of the vehicle. Particles in a groove run from the inside to the outside. The grooves must be kept free of deposits in order to fulfil their function

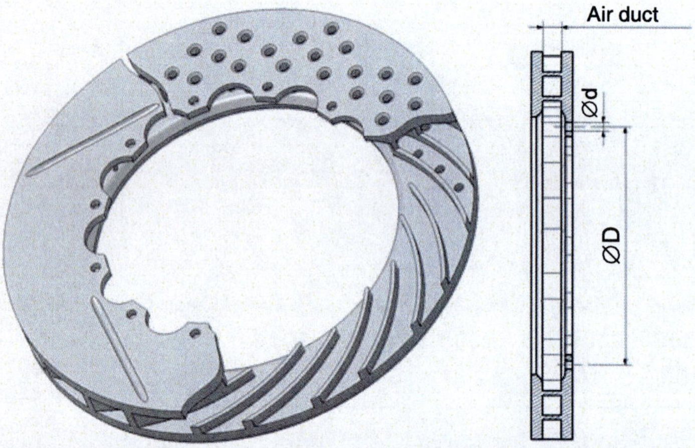

Fig. 6.54 Internally ventilated high performance brake disc for float mounting. Both a grooved and a drilled friction surface are shown. The cooling air is sucked in from the centre of the pane and conveyed outwards through the air shaft. The fastening is carried out via holes (Ød), which are located on the extended side of the disc

the braking performance is better in wet conditions. They are of course lighter, but also more susceptible to cracks in the surface (Fig. 6.54).

Mounting For weight reasons and to compensate for tolerances, most brake discs have a flange mounting. A bell is bolted onto this flange, Fig. 6.55. The advantage of a bell over a flat disc is the larger surface area, which provides better dissipation of heat and thus offers greater resistance to heat cracking. In addition, higher frequency brake noise is better

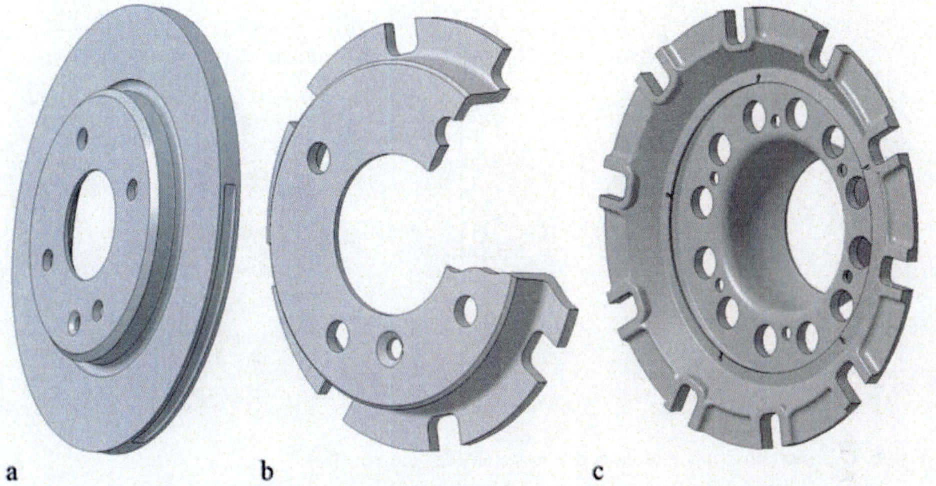

Fig. 6.55 Adapter (disc bell) for mounting brake discs. (**a**) Integrated bell (solid disc), (**b**) Bell for floating mount with bobbin, (**c**) Bell for CFC-disc (Formula 1)

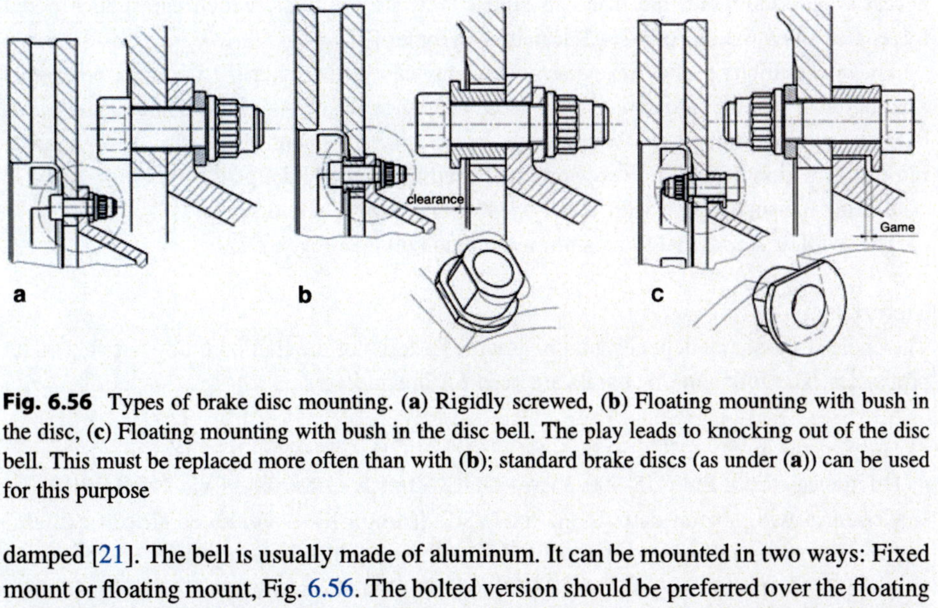

Fig. 6.56 Types of brake disc mounting. (**a**) Rigidly screwed, (**b**) Floating mounting with bush in the disc, (**c**) Floating mounting with bush in the disc bell. The play leads to knocking out of the disc bell. This must be replaced more often than with (**b**); standard brake discs (as under (**a**)) can be used for this purpose

damped [21]. The bell is usually made of aluminum. It can be mounted in two ways: Fixed mount or floating mount, Fig. 6.56. The bolted version should be preferred over the floating version for extremely heavy loads, such as off-road or rally use. Contamination can block the clearance. The brake disc gets a knock and flutters.

Disc bell for internally ventilated discs should not have lightening holes because this allows possible cooling air for the brake disc to escape unused [12].

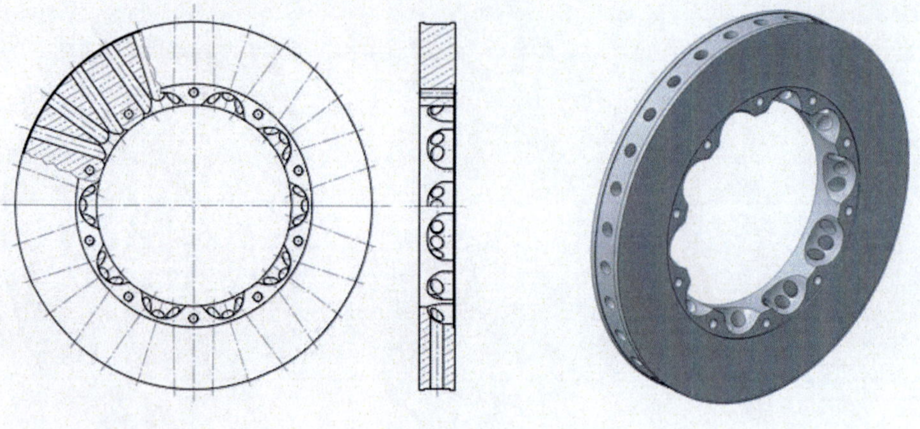

Fig. 6.57 Brake disc made of carbon fibre reinforced carbon (CFC)

With floating brake discs, there is minimal axial and radial play. This allows the disc and the brake disc bell to expand in different directions, which in turn reduces the occurrence of cracks in the disc. For discs over 330 mm in diameter, this type of mounting is definitely necessary. Furthermore, the disc can adjust between the pads, which can reduce pedal travel and allow the use of a smaller master cylinder [9].

The concentricity of the brake disc must in any case be less than 0.15 mm. In the case of floating bearings, the disc should have axial clearance of 0.15–0.2 and radial clearance of 0.05–0.1 mm to the bell [6]. With a fixed bearing arrangement of the disc, a slight axial run-out deviation of 0.05 mm supports the resetting of the brake pads.

Bolting is usually done with 8 to 12 1/4″ UNF screws or M6 screws.

This washer is screwed to an aluminium disc bell (see Fig. 6.57).

Materials

The choice of material depends not only on the material of the friction lining but also on its properties. The following materials are used for brake discs:

- Grey cast iron: EN-GJL-150 (was GG 15), EN-GJL-250 (was GG 25)
- Ductile cast iron: EN-GJS-600-3 (was GGG-60), EN-GJS-700-2 (was GGG-70)
- Ceramics with silicon carbide matrix C/SiC (carbon fibre reinforced silicon carbide) from the group of carbon ceramics (CMC – Ceramic Matrix Composite)
- or CFRC (carbon fibre reinforced carbon), or CC (carbon-carbon): Carbon brake disc, Figs. 6.57 and 6.58.

Table 6.11 compares the properties of some common materials.

The price of a set of carbon disc plus pads is about ten times higher compared to conventional materials.

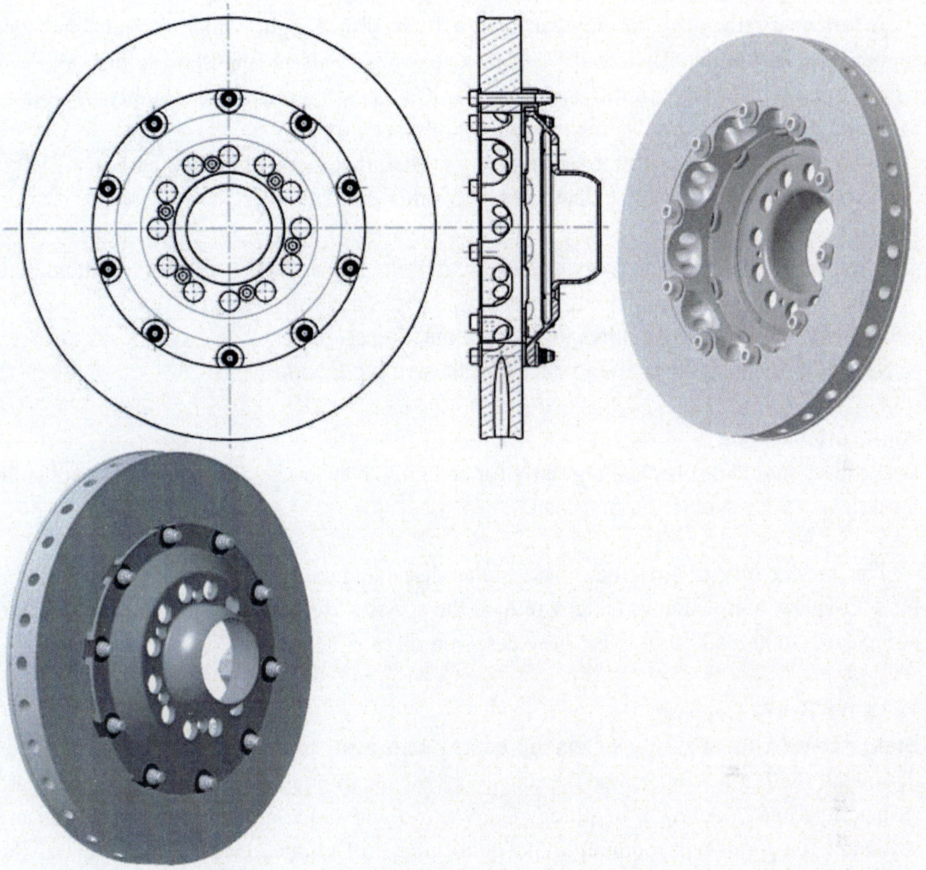

Fig. 6.58 Complete brake disc made of carbon fibre reinforced carbon. Bushings in the disc bell ensure the floating bearing. The disc transmits the braking torque with the holes in the disc bell via driving pins directly to the wheel. Air can pass between the disc bell fastening and the disc to cool this side of the disc

Table 6.11 Comparison of materials for brake discs [16]

	Grey cast iron	Steel	Fiber-reinforced carbon CC	Carbon ceramic CMC
Thermal conductivity, W/(m K)	50	46–50	50–75	25–50
Therm. coefficient of expansion (20–300 °C), 1/K	$18 \cdot 10^{-6}$	$15.1 \cdot 10^{-6}$	$2.6 \cdot 10^{-6}$	–
Specific heat capacity, J/(kg K)	650	540	1200–1300	–
Max. design temperature, °C	600	1150	1000	–
Density, g/cm^3	7.1	7.85	1.7–1.8	2.1–2.3
Mean coefficient of friction μ, –	0.45	0.11–0.35	0.55	$f(t)$

A serious disadvantage of grey cast iron is its specific weight, which is over four times higher, and its high coefficient of thermal expansion. This means that the disc expands when braking from high speeds and warps if it is not designed to be suitably flexible or mounted in a way that allows for expansion, or cracks form.

In contrast to carbon ceramic brake discs (CMC), that made of carbon carbon (CFC) is exclusively suitable for racing. The reasons for this are [18]:

- a temperature-dependent coefficient of friction, which leads to poor cold braking behaviour
- high wear at low temperatures and low contact pressures
- very high material costs due to energy-intensive production.

Application *Notice*
Brake discs should be checked regularly for cracking. New brake discs should be broken in for at least 15 km before the first load.

The service life of grey cast iron brake discs in racing is about 650 to 1000 km [15]. Frequent temperature changes reduce the service life. The rotational speeds should not exceed 3000–3500 min^{-1} for grey cast iron discs.

Parts for Brake *Cooling*
Brakes convert the majority of kinetic energy into heat. In racing vehicles, this heat is dissipated (with a few exceptions) exclusively by air. In general, convection alone is not sufficient; forced cooling is required. Convection and radiation for heat dissipation are sufficient for sprint-type competitions (qualifying, hill climbs, etc.). The design of the cooling air supply depends on the type of brake disc. In the case of solid discs, a shaft leads air to both friction surfaces, Fig. 6.59. The leading edge of the shaft must be temperature-resistant and must be as close as possible to the disc surface (approx. 0.25 mm) so that it removes the majority of the hot boundary layer [6].

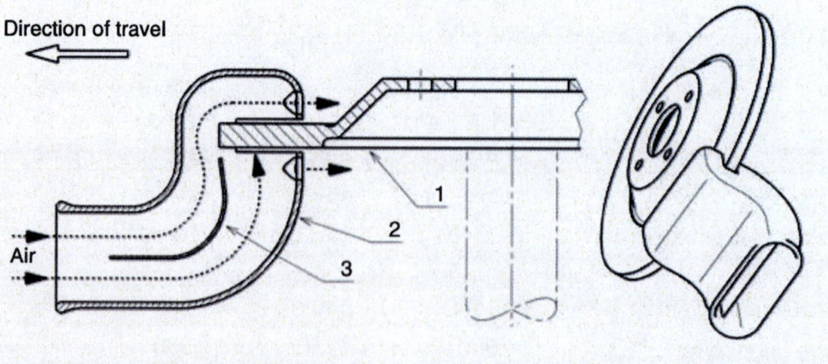

Fig. 6.59 Cooling air feed for solid brake discs. **1** Brake disc, **2** Brake duct scoop, **3** Splitter. An air shaft supplies cooling air to the two friction surfaces of the brake disc. The shaft receives the air between the tire and the car. A possible drive shaft is shown in dashed lines

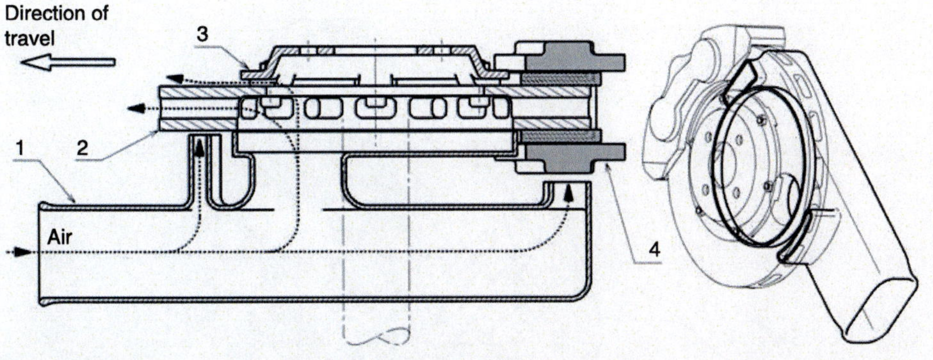

Fig. 6.60 Cooling air supply for internally ventilated brake discs. **1** Air duct, **2** Brake disc, **3** Bell, **4** Brake caliper

In the case of internally ventilated brake discs, the air must be able to flow through the disc from the inside to the outside. Furthermore, it may also be necessary to additionally supply cooling air to their friction surfaces, the brake calliper and the pads, Fig. 6.60.

A shaft (1) located between the tyre and the car guides air specifically to the friction surface of the disc on the inside, the inner area of the disc bell (3) and to the brake calliper (4). The air flows through the disc and a partial flow passes through recesses on the circumference of the disc bell over the friction surface of the brake disc outside. Cooling air for the brake disc can also be supplied through the wheel carrier. Typical diffuser surfaces for racing brakes have 100 cm^2. 80% of the air should pass through the brake disc wells or holes and 10% each through the disc braking surfaces [12].

The overall arrangement of a complete assembly is shown in Fig. 6.61 for a front wheel.

Figure 6.22 shows the total air flow through a wheel. In addition to the design of the brake disc channels, the ventilation effect of the wheel is also important for cooling the brake because it creates the necessary pressure difference for the flow. Additional installation parts such as cover plates, which are standard on passenger cars, in any case impede the air flow and thus worsen the heat transfer. Conversely, air baffles on touring cars can promote air intake from the wheel housing into the wheel.

If the heat dissipation is insufficient, malfunctions will occur:

- The braking effect decreases (sometimes even strongly) (fading)
- Deformation of the brake disc
- Cracks due to thermal stresses in the brake disc
- Loosening of the brake pad from the backing plate (failure of the adhesive bond)
- Thermal damage to surrounding components (wheel bearings, cables, tyres,. . .)
- Boiling and vapour bubble formation of the brake fluid
- Noise generation (however, this is primarily only a problem with series production vehicles)

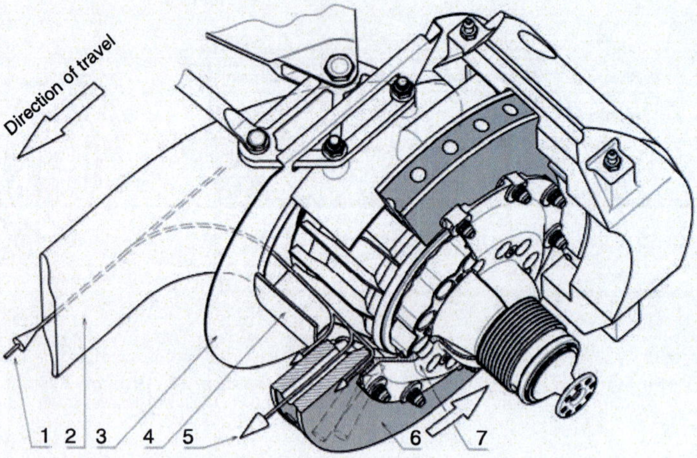

Fig. 6.61 Air supply to Formula 1 front brake (left front). **1** Cool air inlet, **2** Air duct scoop, **3** Backing disc, **4** Upright, **5** Hot air outlet, **6** Brake disc, **7** Disc bell

The noise is caused by substances outgassing from the pads, which diffuse into the disc. So that the outgassing occurs preferentially on the other side, there are pads that are drilled from the backing plate side. These gases can also lead to brake fade by acting as a lubricant between the brake disc and the pad.

In order to achieve an optimum effect of the brake, it is important to achieve the correct disc temperatures. The temperature on the front axle is usually about 100 °C higher than on the rear axle. Different heating behaviour can slightly influence the brake force distribution. On the left and right, the temperature values should be approximately the same, whereby, depending on the cooling system, the outer disc surface can become hotter than the one facing the car. The maximum disc temperature is very dependent on the type of pad. Each race track stresses the brakes differently. Therefore, the temperature values of the brake discs should be recorded. This can be done with the brake disc temperature paint or with a measuring device (pyrometer) during the pit stop. Frequent changes between maximum and minimum temperatures during a race will shorten the life of the discs. Cast iron brake discs should not be driven in temperature ranges above 610 °C.

Recirculating Valve

Recirculation valves can be used to cool and bleed the brake fluid, Fig. 6.62. With each pedal stroke, the brake fluid circulates through the line system and any air bubbles that have occurred are removed, thus preventing local boiling.

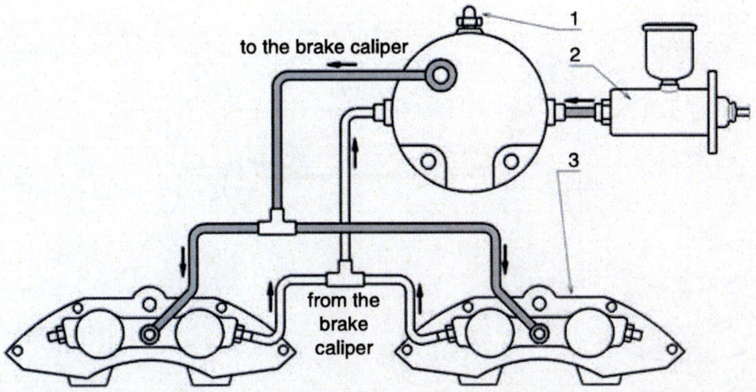

Fig. 6.62 Brake line system with recirculating valve. **1** Bleeding, **2** Brake master cylinder, **3** Brake caliper. The return flow of the brake fluid from the brake calipers takes place via the bore of the bleeder screws

6.7 Brake Installation

Racing brake calipers are designed so that they can be installed on the front and rear axle. If you want to place the masses as close as possible to the overall centre of gravity of the car, arrange the brake callipers as shown in Fig. 6.63. However, the connections of connecting lines and vents must be interchanged for this. Besides this arrangement keeps the loading of the wheel bearings by brake forces low.

The brake disc and the brake calliper must be matched to each other in such a way that a certain clearance is ensured, Fig. 6.64.

The clearance between the brake disc and the brake calliper should be at least 1.8 mm for discs $\leq \varnothing 280$ mm and at least 2.5 mm for larger discs.

Radial bolted calipers allow some adjustment by interposition of discs.

The brake can be housed in the wheel (outboard brake) or on the frame side. The outboard brake has the advantage that the airflow around the car is not affected, which is particularly important in the rear of single-seaters (bottle-neck shaped intake[7]).

The internal brake reduces the so-called unsprung masses, transfers the brake support forces directly to the frame and thus relieves chassis parts. However, a brake shaft is additionally required for non-driven axles. This cancels out at least some of the advantages mentioned and increases the amount of parts required.

Figure 6.65 provides an overall view of a typical racing vehicle braking system.

The limiting environment for the wheel brake is the wheel (inner rim contour) and the wheel carrier. A large wheel and a compact wheel mounting are therefore advantageous for the largest possible brake disc. A large offset of the wheel allows the installation of fixed caliper brakes, Fig. 6.66.

[7] See Racing Car Technology Manual, Vol. 2 *Complete Vehicle*, Chap. 5 *Aerodynamics*, Fig. 5.15.

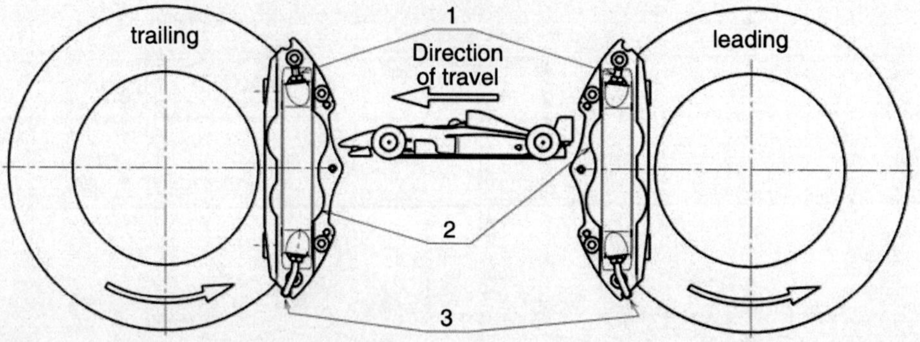

Fig. 6.63 Arrangement of the brake calipers on a racing vehicle trailing/ leading caliper. **1** Arrange bleed screws as high as possible, **2** Brake disc passes small brake pistons first (on multi-piston designs) to avoid stick-slip effects on the brake lining, **3** Arrange connecting lines at the bottom

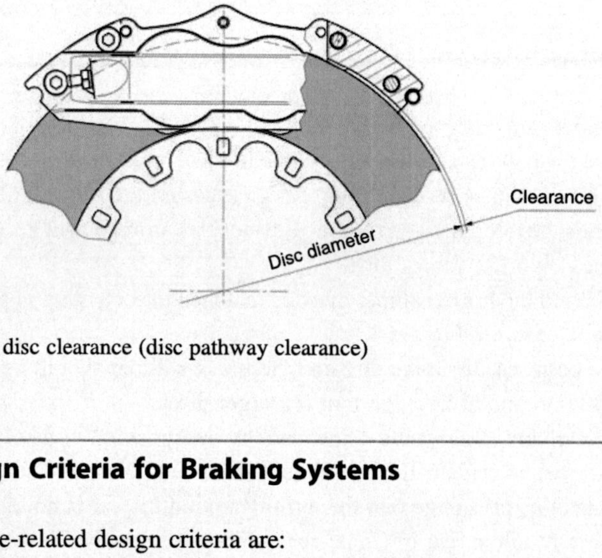

Fig. 6.64 Brake disc clearance (disc pathway clearance)

6.8 Design Criteria for Braking Systems

Essential vehicle-related design criteria are:

- Legal requirements with regard to locking-free minimum braking and locking sequence
- Loading conditions
- Influence of brake fade
- Motor braking torque
- Brake circuit failure
- Brake force distributor (if present)
- Retarder or recuperation (if fitted).

The unit-related design is primarily concerned with the dimensioning of wheel brakes and actuating devices. Design criteria for the wheel brakes are:

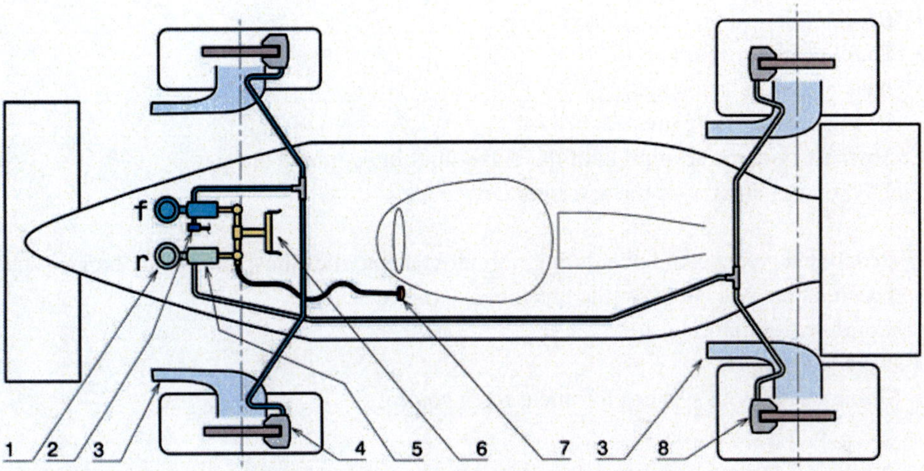

Fig. 6.65 Brake system on a racing car. **1** Brake fluid reservoir, **2** Brake light switch, **3** Front and rear brakes air ducts, **4** Front brake calipers, **5** Front brake master cylinder (f) and rear brake master cylinder (r), **6** Brake pedal, **7** Adjustable brake balance knob, **8** Rear brake calipers

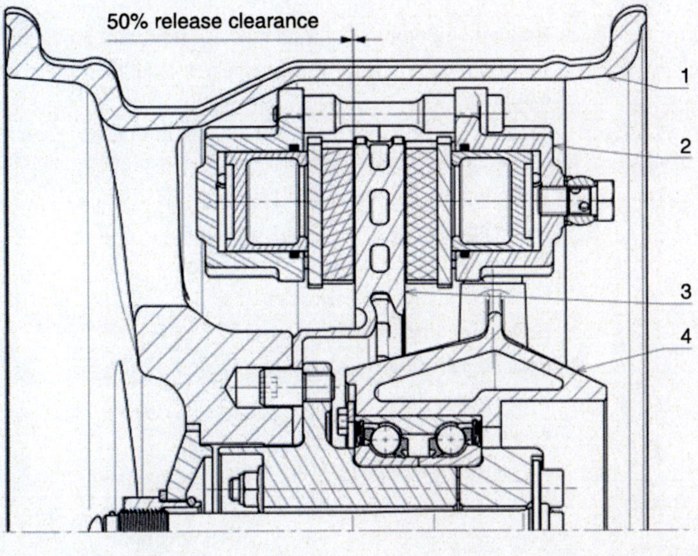

Fig. 6.66 Brake caliper installation in the wheel (Formula Renault 2000 front axle). **1** Wheel, **2** Brake calliper, radially bolted to wheel carrier, **3** Brake disc, internally ventilated, **4** Wheel carrier

- Brake design (disc, drum brake)
- Stability (wear, stress)
- Available installation space
- Permissible pressure level
- Stiffness (volume absorption of the brake fluid in hydraulic brakes).

Finally, design criteria for the actuating device are:

- Pedal travel and pedal force during normal brake applications, emergency braking and failure of a brake circuit or the brake booster.
- Comfort requirements
- Installation space
- Combination with systems for brake force control.

6.9 Standards

Brake systems are relevant to safety. There are therefore numerous standards which are listed below because they represent proven elements and systems.

DIN ISO 611	Braking of motor vehicles and their trailers, terminology
DIN 70024 Part 3	Terms for components of motor vehicles and their trailers, braking equipment
Supplement to DIN 70024	Translations of the terms of DIN 70024 in English, French and Italian
DIN 72571	Fastening clamps, one-sided fastening
DIN 72573	Fastening clamps, two-sided fastening
DIN 74000	Dual circuit brake systems, abbreviations
DIN 74200	Cylinders, hydraulic brake systems
DIN 74225	Brake hose lines, bracket
DIN 74233	Brake pipe fittings, cap screws
DIN 74234	Brake tubes, flare
DIN 74235	Brake system, threaded holes

References

1. Wright, P.: Formula 1 Technology, 1. Aufl. SAE, Warrendale (2001)
2. Dubbel: Taschenbuch für den Maschinenbau, 14. Aufl. Springer, Berlin (1981)
3. Hamm, G., et al.: Tabellenbuch Fahrzeugtechnik. Holland+Josenhans, Stuttgart (1997)
4. Hoepke, E. (ed.): Nutzfahrzeugtechnik. Springer Vieweg, Wiesbaden (2000)
5. Braess, H.-H., Seiffert, U.: Vieweg Handbuch Kraftfahrzeugtechnik, 4. Aufl. Springer Vieweg, Wiesbaden (2005)

6. Smith, C.: Prepare to Win, 1. Aufl. Aero Publishers, Inc, Fallbrook (1975)
7. Breuer, B.: Skriptum zur Vorlesung Motorräder. TH Darmstadt, Darmstadt (1985)
8. Wright, P.: Ferrari Formula 1. Under the Skin of the Championship-Winning F1-2000, 1. Aufl. David Bull Publishing, Phoenix (2003)
9. N.N.: The Competition Catalogue. Coventry: AP Racing, publication ref. P02.002/01
10. N.N.: Motorcycle Catalogue. Coventry: AP Racing, publication ref. P4.005/00
11. N.N.: 2001/2 Motorcycle Catalogue Supplement. Coventry: AP Racing, publication ref. P4.006/01
12. N.N.: Brake Disc Catalogue. Coventry: AP Racing, publication ref. P09.004/03
13. McBeath, S.: Competition Car Preparation, 1. Aufl. Haynes, Sparkford (1999)
14. Piola, G.: Formel 1. Copress, München (2001)
15. Incandela S.: The Anatomy & Development of the Formula One Racing Car from 1975, 2. Aufl. Haynes, Sparkford (1984)
16. Breuer, B., Bill, K.-H. (Hrsg.): Bremsenhandbuch, Grundlagen, Komponenten, Systeme, Fahrdynamik, 4. Aufl. Springer Vieweg, Wiesbaden (2012)
17. Tremayne, D.: Formel 1 Technik unter der Lupe, 1. Aufl. Motorbuch, Stuttgart (2001)
18. Albers, A., Arslan, A., Herbst, D.: Keramik für den Einsatz in Bremsen und Kupplungen. ATZ. **5**, 414 f (2001)
19. Milliken, W., Milliken, D.: Race Car Vehicle Dynamics, 1. Aufl. SAE Inc, Warrendale (1995)
20. Hucho, W.-H. (Hrsg.): Aerodynamik des Automobils, 5. Aufl. Springer Vieweg, Wiesbaden (2005)
21. Füllgrabe, F.: Neue Konzepte für Leichtbau-Bremsscheiben auf Basis metallischer Werkstoffe. VDI, Düsseldorf (2013)
22. N.N.: AP Racing 2017 Product Catalogue, Coventry. www.apracing.com/Aboutus.aspx?cid=16. Accessed on 21.11.2017 (2017)

7 Comparison Series: Racing

Comparisons between production and racing vehicles are made throughout the book to increase clarity, but here the similarities and differences are to be directly contrasted in a compact summary. This says a lot about both industries, their working methods and products in just a few pages.

7.1 Introduction

Developments in motorsport have always gone much faster from idea to implementation than series developments. This is also understandable. Once an idea has been born, it has to be turned into an advantage on the race track as quickly as possible, otherwise the opposition might take care of it. The results of the developments can be described in general terms as increased performance, increased efficiency, weight reduction, increased reliability and simplified handling. In other words, all achievements that are also quite

M. Trzesniowski, *Suspension System*,
https://doi.org/10.1007/978-3-658-39847-7_7

welcome in production vehicles. Numerous innovations in modern passenger cars have their origins in racing. The development results flowed from the sport into the production vehicles (disc brake, four-valve engines, direct injection, light alloy wheels, double clutch transmission, composite materials, . . .). In the meantime, however, the roles have changed to some extent. In the meantime, the race track serves the development departments of large companies as an unbureaucratic test laboratory for preliminary development. For example, when it comes to the practical testing of new materials or extreme designs of components. And if the development is successful, it is duly marketed. From the point of view of the car manufacturers, nothing has changed: Racing success brings sales success.

7.2 Development *Process*

The design and construction of an F1 car takes about 5 months. Depending on the team, around 300 engineers are employed. Thousands work on a production car for about 4–5 years. This may seem astonishing at first glance. But in fact there are significant differences in the goals. The racing car has to comply with the regulations and be ready in time for the start of the season. A passenger car also has a completion date. However, this is self-set and can (and will) be postponed if necessary. There are around 15,000 individual requirements in a car's customer requirements, which not only have to be met, but are also meticulously checked by the legislator or the manufacturer itself. A large part of the development time is spent on optimizing production in high quality despite the large number of units. This is the only way to ensure an affordable purchase price.

The competition car has ONE overriding goal, to be superior to the competition and win races. Cars should find buyers in large numbers so that the immense investments are worthwhile. They must therefore be adapted to the needs of the market and meet the (future) taste of the public.

A direct, reliable comparison between series and motorsport development is provided by companies that carry out both. As an example, consider the chronological sequence of engine development for Ferrari's production cars and Formula 1 cars, Fig. 7.1. While in production an engine matures from concept to producible unit in 42 months, the racing department goes through three full racing seasons. It not only designs, builds and develops new engines for each season, but also constantly improves them during the season in order to keep pace with its opponents or, ideally, to outdo them.

Even if the development processes are therefore different, one circumstance is (meanwhile) the same: the design of the vehicles takes place from the outside in. Based on market analyses, trend research results, strategic product specifications, etc., the design department specifies the size, external shape and appearance of the new car. The size, external shape and appearance of the new car. It is now the task of the design department to accommodate all the necessary and desired assemblies and components. In the case of racing cars, the process is exactly the same in that the aerodynamics department specifies the shape and

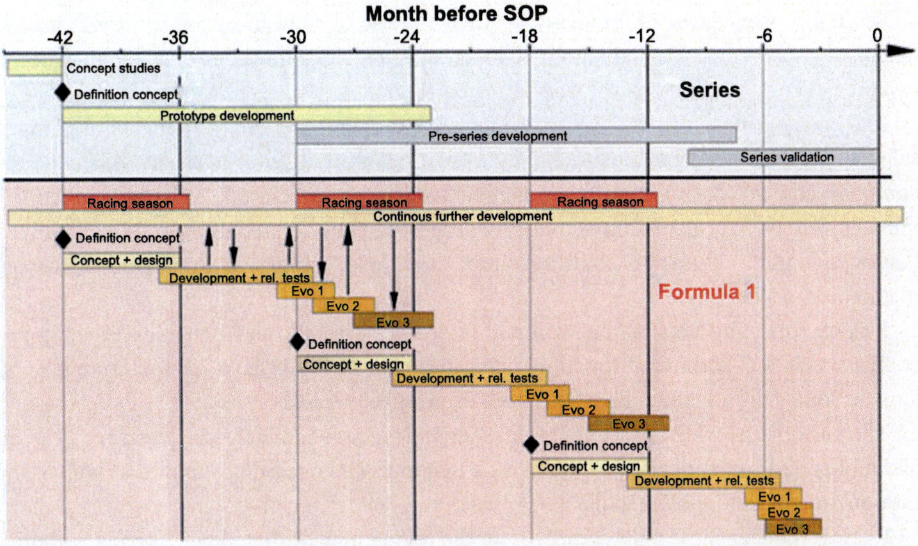

Fig. 7.1 Comparison of the flow charts of Ferrari's engine development, after [1]. Timing of engine development in series production (top) and in the Formula 1 team (bottom). The starting point is the start of production (SOP) of the series engine. Abbreviations: Gen.1: Generation 1, rel. tests: reliability tests, Evo 1: Evolutionary stage 1

thus the desired airflow around and through the car based on CFD calculations[1] and preliminary wind tunnel model tests, and all other development units must subordinate themselves to these specifications.

Often, in the course of a debate about profiting from a racing commitment, the question is asked as to which parts or assemblies have been incorporated from a racing vehicle into a production vehicle. The answer to this question does not have to be about components. Development tools and test methods are also advantageously adopted, as reported in [2], for example.

7.3 Development *Goals*

As mentioned, significant differences can be found in the development goals. Objectives are derived from requirements. In the case of passenger cars, these mainly come from the legislator (sometimes contradictorily from different states), from management, from product planning and from customer service.

Legislation limits emissions (exhaust, evaporation, noise), specifies maximum dimensions and maximum masses, lays down safety standards to prevent accidents and

[1] Computational Fluid Dynamics. Numerical flow simulations on a 2D or 3D computer model.

reduce the consequences of accidents, restricts the use of substances that are hazardous to health, regulates the clear labelling of parts and vehicles and takes care of an orderly, resource-saving procedure at the end of the car's life.

Management positions the car for certain markets with certain attributes and planned sales figures. In addition, cost-reducing synergies with other models available in the company and the use of certain production facilities are required. A target service life must be defined for calculation and testing despite difficult to predict operating conditions (summer/winter, rain/snow, short distance, bad road, motorway, overloaded, garage maintained, driving style, ...).

Product planning acts as the voice of the end customer and demands ergonomics, comfort, practical additional functions, consideration of future legal or social trends and aims to subtly differentiate the new model from its predecessor.

The customer service department thinks about inspection, maintenance and repair of all assemblies and components. It rejects exotic materials and operating supplies as well as the introduction of new special tools.

Racing vehicles must only comply with the regulations of the relevant sports authority and the specifications of the management. In individual cases, the main sponsor or purchaser will also make his demands and wishes known. The choice of material is only limited by the regulations and the time availability.

In general, it can be stated that the goals and methods of series and racing developments converge when the planned production quantity and the period of use of racing vehicles increase. Formula 3 cars are a good example of this. The cars are designed for a specific series but for unknown customers and are built for at least 3 years. Customer taste, cost and spare parts supply thus become significant issues in the design process.

7.4 Research and Development (R&D)

Continuous further developments are necessary – even if not for the same reasons – in both camps under consideration. The financial resources required for this must be generated. 25% of the turnover of a racing team flows back into R&D expenditure. The automotive industry spends only 6% [3]. In this sense, a racing vehicle is never finished, but is subject to constant improvements, which are nourished by own research or ideas "inspired" by the opponent's solutions. Compared to the car industry, a competition vehicle is a permanent prototype.

Traditionally, racing has been extremely stingy when it comes to publishing research results or key findings. On the contrary, they even try to keep solutions they have found secret for as long as possible. After all, they promise an advantage over the opponents in the race. It is not uncommon for ingenious innovations to have decided entire world championships. Lotus, for example, was the lucky discoverer of winged cars using the ground effect. Williams developed an active suspension that made the aerodynamic effects work even more comprehensively. McLaren benefited from the fiber composite monocoque. Ferrari dominated races with a semi-automatic gearbox. Renault won the title with mass dampers on both axles. The double diffuser was the clever exploitation of a loophole in the regulations that brought Brawn GP laurels. And that's just a chronological list of a few cases.

However, history also offers reverse cases, i.e. those in which a team was able to successfully use an alien innovation. The most famous example is the turbocharger, which Renault introduced and developed for years, but in the end it was Honda's turbocharged engines that won multiple world championship titles.

The fear that one's own ideas can be implemented faster or even more successfully by others is therefore not unjustified. Secrecy, protection against industrial espionage and compartmentalization of the development departments are important strategic measures that shape the everyday lives of those involved. There are no major differences in mindset, regulations, contracts and delivery terms between racing and production development. The goals are basically the same: No one should know about groundbreaking creations before they go into the race or are offered in the new model, certainly not the direct opponents or the competition.

This omnipresent fear sometimes produces strange blossoms and makes cooperation with development service providers, suppliers and external manufacturers more difficult. Especially the change of drivers and employees is treated with great suspicion. One of the ways this is solved is that individuals only have access to very specific knowledge and do not have an overview of the big picture. Gone are the days when a handful of engineers designed and developed a complete car, complete with engine. The five-man engineering group of the Formula 1 teams of the 1980s has grown to over 200 engineers today. The developer of the brake caliper has no idea what his colleagues are doing to improve the steering. And drivers aren't privy to the technical details "under the hood" anyway. What's the point? They need to know which switch activates which function and concentrate on the track, literally.

Racing teams have to accept one major disadvantage in this context. Just as successes can be celebrated with great publicity, failures are also not hidden from the public. However, the press has never seen some of the mistakes made by car manufacturers, and for good reason.

7.5 Costs

Even if costs are at the top of the priority list for passenger cars, this does not mean that they do not play a role in motorsport. On the contrary: in numerous racing series, costs became an insurmountable financial problem in the course of the "arms race" between the participants and led to the end of this series of events. Even the top leagues in motorsport are by no means immune to this phenomenon. It is therefore with good reason that rule writers also look at the cost side when regularly adjusting the rule book. Particularly cost-driven racing series are GT3 and GT4, for example. GT3 cars are expected to be able to race for about 6000 km without having to replace parts. In GT4, this figure is as high as 10,000 km or one year. Production cars drive around 250,000 km in 10–15 years (some wear parts such as tyres, brake pads, brake discs, exhaust systems, timing belts, etc. are of course replaced in the process).

However, costs are also becoming a problem in professional racing due to increasing demands, more complex development and internationalisation of the series. History bears witness to several prominent examples where precisely such a development has forced changes in regulations or even meant the end of the series.

Customers' expectations are rising. What was a sensation in the luxury class yesterday is tacitly expected in the small car segment today. Cars must be low-maintenance with long service intervals. Long gone are the days when the valve clearance had to be adjusted after 5000 km or, in winter, a radiator grille with hinged elements was used to partially seal off the cooling air intake so that the engines warmed up (faster). Even punctures are hardly accepted and the average customer wants to be able to drive to the next garage without getting his hands dirty. Sensors that can continuously report the internal pressure of the tyres to the on-board computer while driving have proven their worth both in series production and in racing. Admittedly, pressure from the legislator was required for widespread use in passenger cars. For the race engineer, the current values of tire pressure and temperature are valuable information that can be used to determine the tire behavior and thus the driving performance of the vehicle. Over the course of the race, the development of the tires crystallizes and measures can be derived to influence traction and durability. Sensors located directly on the tire inner liner instead of at the usual rim position turned out to be ideal. They indicate the temperature close to the tread and provide an approximation of the actual internal pressure. The values of the pressure transducers on the rim have to be compensated with its temperature. Because the temperature of the rim rises much faster than that of the air in the tire during hard braking, overcompensation sometimes occurs and as a result annoying false alarms occur as the isochore-compensated cold pressure falls below the warning threshold [4].

In series vehicle production, the aim is to have the same (spare) parts for different versions of a type for reasons of cost and logistics. This can lead to parts being undersized for the strongest variant and oversized for the weakest variant. Rally teams are familiar with a similar dilemma, as they are used to maintaining their vehicles during competitions (under the most adverse conditions) and getting them back on the road again after damage. The fewer parts that have to be taken along, the better. As a concrete example, consider the wheel carriers. They are designed in such a way that one version can be fitted to all four wheels. The same applies to special tools. Incidentally, this is also a request made by customer service to the series designers. It should be possible to adjust and mount spare parts without special tools. During series assembly in the factory, robots carry out many work steps and assembly locations must be correspondingly accessible and assembly parts must be correspondingly clearly designed. This is where racing designs differ. The desired uncompromising fulfilment of function takes precedence. In any case, assembly is done by hand. Parts that work together – such as bearing journals and bearing shells, pistons and bushings, meshing gears, axial washers and housings, inner seal and inner channel contours – are measured, classified and meticulously mated beforehand. Not only dimensional tolerances, but also mass and surface tolerances are taken into account. In Formula 1, even screws with thread pitch matched to the elongation[2] are used.

7.6 Environmental Protection

In one area in particular, which is used by many sides as the main argument against motorsport, fundamental changes have taken place in recent decades. Thanks to environmental management, motorsport, like environmental technology, is regarded as a pacemaker for technical progress. This applies in particular to elementary areas such as energy efficiency, avoidance of pollutants, choice of materials and handling of hazardous substances [5]. Based on the ADAC Environmental Plan 2000+, environmental strategies are successfully implemented in 4 central fields of action: organization, technology, infrastructure, and research and science [5]. Lightweight design, tyre technology, occupant protection, alternative drives, increased efficiency and wear minimisation can be listed under the area of "technology" which is of interest here [5]. "Green racing" or "clean racing" are the buzzwords under which manufacturers, teams and sponsors visibly (and hopefully pioneeringly) pursue sustainable racing.

In the case of series-produced vehicles, the idea of environmental protection has been imposed much earlier. With mass motorization in the middle of the last century, vehicles not only became affordable for everyone, but also became a burden due to the large number of vehicles. Traffic regulations had to be set up, traffic guidance systems became necessary and ultimately exhaust and noise emissions had to be restricted. However, this does not only concern the operation of the vehicles themselves, but also their production and disposal. In the case of racing vehicles, exhaust emissions are not (yet) the focus of the regulation writers. And for good reason. The magnitudes of the effects compared to series production vehicles are completely different: At major DMSB events,[3] around 95% of the total vehicle kilometres travelled are attributable to spectators travelling to and from the event, who thus contribute around 93% of CO_2 emissions [5]. The following estimate illustrates even more clearly the impact of millions of vehicles compared to a few hundred. The total fuel consumption in German motor sports is less than 3% of the evaporative losses during parking and refuelling of passenger car road traffic [5]. To this should be added that the total vehicle evaporative emission of vehicles with gasoline engines at

[2] In common bolted joints, due to the elongation of the bolt and tolerances of the bolt and nut threads during pretensioning, the first thread generally carries approx. 25–35% of the load.

[3] Deutscher *Motorsport* Bund e. V., umbrella organization responsible for motor sports in Germany. Exercises national sporting authority for automobile and motorcycle sport.

standstill is regulated by law and checked in the so-called SHED[4] test. According to the US typing regulations, hydrocarbon emissions are also limited and monitored during the refuelling process.[5]

CO_2 emissions are directly (and linearly) related to fuel consumption with hydrocarbon fuel. A legal restriction on the fuel consumption of production vehicles and thus on climate-damaging CO_2 emissions therefore makes sense. In the case of racing cars, another consideration has led the teams themselves to use liquid energy storage as sparingly as possible. The less fuel a car needs for the targeted distance, the smaller and lighter the fuel tank can be. As a result, the car has lower driving resistance (mass, cross-sectional area) and benefits from higher performance and greater range. Conversely, higher efficiency means lower losses. These become noticeable through the need for heat dissipation and wear. Increased efficiency pays off directly through smaller heat exchangers as well as shrunken cooling air ducts and less coolant required, which in turn reduces drag and helps reduce mass. Reduced wear lowers the need for lubricating oil and its volume, which must neutralize or suspend abrasive particles with special additives (detergents and dispersants). In addition, the component with the wearing surface can be made thinner. It is no longer necessary to maintain so much wear volume so that the residual load-bearing capacity of the affected component wall remains large enough.

7.7 Technology

7.7.1 Frame and Body

Early on in the history of development, the paths of production and racing vehicles diverged when it came to frame construction. A passenger car is supposed to accommodate passengers and luggage, protect them from wind and weather . . . A racing vehicle is, to put it exaggeratedly, an engine on wheels that can be manoeuvred on board by one person, or in exceptional cases two people. Initially, the ladder frame was the standard for all vehicles, but for racing cars it was replaced by tubular space frames, box frames and ultimately spatial shell structures made of fibre-reinforced composites. In passenger cars, these solutions are only found in very small numbers in sports cars, i.e. a preliminary stage to the pure-bred, purpose-oriented racing car. The self-supporting steel body has become established for passenger cars. Strictly speaking also a shell structure. This is no different for apparently production-based vehicles such as rally and touring cars (DTM, NASCAR). They may have a similar shape, which is crucial for recognition value, but under the outer skin there is sometimes completely different technology. Naturally, the difference between the road car and its track counterpart is much smaller for GT3 and GT4 cars. As is so often

[4] Sealed Housing for Evaporative Determination.

[5] ORVR: On-Board Refuelling *Vapour* Recovery, *vapour recovery* system during refuelling.

the case, the degree of specialisation to become a race car is, after all, a question of money. The more cost-effective a racing series is to turn out for the participants, the smaller the extent of the permitted or made conversions may be. This is where the important role of the regulations comes into play, to maintain the orientation of the race series and equal opportunities.

The numerous tasks of the bodywork or body also include the ventilation of vehicle areas (passenger cell or cockpit, engine compartment, brakes, . . .) and thus, in the broadest sense, the aerodynamic behavior. While air resistance, noise generation, soiling of windows and lights are the main issues in passenger cars, downforce dictates the development of high-performance vehicles in particular. The extreme in this respect is Formula 1, which even subordinates chassis designs to this issue. For passenger cars, it is sufficient if the shape does not generate too much lift; sports cars should at least achieve slight downforce. Racing car designers also have an easier time of it; they don't have to consider fields of vision, pedestrian protection, parking pile-ups, or tire covers, but can place wings, chimneys, deflector plates, splinters, spoilers, vortex generators, etc. on the vehicle according to purely technical aspects, subject to the regulations.

7.7.2 *Engine*

The main differences in the internal combustion engine result from the list of requirements. Passenger car engines should be easy to start in all seasons, regardless of previous operation, operate quietly, have low pollutant emissions, be fuel-efficient to run over a wide speed and load range, and function as intended for a long time with long service intervals. A racing engine must also last, but in the crassest ideal case only until the finish line. By then, however, it should have converted maximum power or torque from the energy provided in the fuel. The noise released in the process is perceived favourably, at least by the public, and is not perceived as a noise nuisance.[6]

If there is no need to compromise on suitability for everyday use, component designs can be precisely aligned with the desired objective. In extreme cases, this goes so far that racing engines may only be transported in a certain position and with pressurized valve cups. That they must be preheated before they can be started because the bearing and piston clearances are only suitable at operating temperature. That separate spark plugs are screwed in for the warm-up phase.

The engine is a heavy subsystem and should therefore be installed as low as possible in the vehicle. The flywheel diameter becomes the determining factor in this context. Not least for this reason, racing engines have small flywheels or none at all. High idle speeds and low elasticity only interfere with passenger cars.

[6] A phenomenon that can also be observed at major musical events.

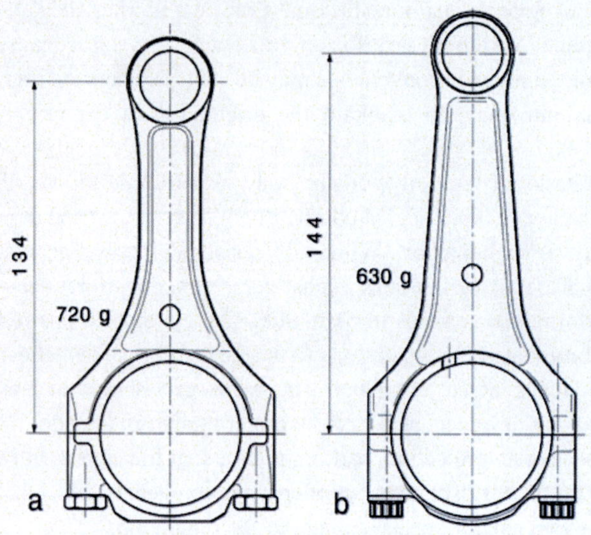

Fig. 7.2 Comparison of connecting rods of a 3.0-l petrol engine, according to [6]. A racing engine was derived from an in-line six-cylinder. With the height of the crankcase remaining the same, the longer connecting rod of the racing engine resulted in a shortened compression height of the piston. Both connecting rods were forged, but the material of the racing connecting rod was of higher quality. This was therefore also lighter in spite of the larger gauge. (**a**) Series connecting rod, (**b**) Racing connecting rod

Again, a direct comparison should make us aware of the differences that characterise this chapter. A DTM engine is derived from a series engine. In the example considered, Fig. 7.2, the racing connecting rod is longer and more flexurally rigid in the transverse direction, yet its total mass is lower. However, it must be mentioned that the material of the racing connecting rod is of higher quality.

Car engines have to cope with different fuel qualities, which, thanks to electronic control and corresponding sensors, involves fewer compromises nowadays than it did a few decades ago. Nevertheless, racing engines have an advantage here because they only have to be designed for a specific fuel and this fuel is also delivered with a much narrower tolerance of its composition.

Exhaust systems have the task of removing the combustion gases in a targeted manner, lowering the noise emission and reducing the proportion of harmful components. At the same time, the design of the pipe/reservoir system enables targeted tuning of the gas-dynamic behaviour for specific speed ranges. The positive scavenging gradient from the intake to the exhaust side is supported and the residual gas quantity (performance!) or the scavenging losses (fuel consumption!) are kept small. In the production vehicle, a compromise is made in the direction of low emissions (noise, pollutants). In the case of a racing vehicle, the focus is on performance. A power-demonstrating soundscape is even enjoyed by the public to a certain extent. Other uses of exhaust gas in racing vehicles are as an energy source in *Energy Recovery Systems* (ERS) and (at least in large-volume, high-revving engines) in supporting aerodynamic elements such as the rear diffuser and rear wing. In production vehicles, ERS will be used in hybrid powertrains. This allows the energy present in the fuel to be used more fully, or at least stored for later use.

Basically, the same applies to electric motors, even if in this case the many years of field experience are still lacking on both sides. Electric vehicles eke out a niche existence in road traffic and on the racetrack. Nevertheless, a lot has already been done in the comparatively short development time for racing vehicles and they often attract attention by setting a new record. The higher electrical currents caused by the greater power output of racing vehicles drive batteries with high internal resistance to the limits of their thermal load capacity and these thus become the bottleneck in the increase in power. And this is true in both directions, i.e. when driving and when (regenerative) braking. Other energy storage devices – such as supercaps – are more advantageous in this context. Improvements have also been made to current-carrying parts in the motor. For example, conductor cross-sections have been optimized (trapezoidal shape, etc.). Cooling types and media and magnetic flux directions are varied with the aim of further increasing the power density. Wheel hub motors or motors close to the wheel make it possible to come remarkably close to the driving dynamics goal of allocating torque to individual wheels (torque vectoring).

7.7.3 *Power Train*

The differences in the powertrain are similar to those in the engines. Service life, operability and comfort characterize the development and results of series-produced vehicles. Low-noise, easy-to-change gears or transmission systems that automatically change gear ratios to save fuel are in demand for passenger cars.

Lightweight transmissions that transmit engine power to the wheels with low losses and are also a load-bearing part of the vehicle are preferred for racing vehicles. Damping elements, synchronising devices and additional masses on the outer gearshift to support the gearshift movement are not to be found on a racing gearbox. These parts increase mass and create losses. The disadvantage of lack of comfort is not an issue on race cars because it is not a competitive factor. Much more important is the ease of adapting gear ratios to engine and track conditions. In production transmissions, oil is generally not even changed in the course of a vehicle's life. Gear geometry in racing transmissions is designed with shaft deflections in mind for the highest transmission efficiencies. Series gearboxes must be able to be manufactured and tested in large numbers using standard tools, and gear meshing must be quiet. This is one of the reasons why hypoid gearing is often used for axle drives, although it has disadvantages in terms of transmission efficiency due to a high degree of longitudinal sliding in the rolling motion.

7.7.4 *Suspension*

The biggest differences between production and racing vehicles are probably in the area of suspension. This is not surprising, since the suspension has the greatest influence on handling. Cars are designed with comfort and safety in mind. Great compromises have to be made in the process. The vehicles are operated without modification – apart from summer and winter tyres – all year round in all weather conditions with different load conditions, with and without roof superstructures, with and without trailers on different road surfaces. Furthermore, very few drivers check the inflation pressure of the tyres or even their settings such as toe-in or camber before setting off. In the case of racing vehicles, these are routine activities that can also make all the difference on the results list. Racing suspensions must therefore be easy to adjust and within the required range. Production tyres should deliver long mileage with low rolling resistance and excellent wet grip (a contradiction that constantly challenges tyre developers). Noise emissions and rolling comfort are also considered and evaluated. Racing tires seem to have an easier time in this regard. They are supposed to have consistently high grip in circumferential and lateral directions and not change their behaviour. The short service life is accepted and sometimes even deliberately used by the organisers of racing series to increase the excitement for the spectators. Compared to their production counterparts, racing tyres are only partially vulcanised and only cure during operation. The rim width of production wheels is made as narrow as possible. Racing tyres, on the other hand, are mounted on wheels whose rim width is 1–2 inches wider than the tyre. This noticeably increases tire volume and lateral stiffness. The offsets of the standard wheels primarily take into account the balanced load on the wheel bearing pairs when driving straight ahead. In the case of racing vehicles, camber stiffness is decisive in this context. Figure 7.3 illustrates how corresponding compromises are made in the suspension designs. If – as in many passenger cars – the focus is on high tyre mileage, the static camber of 0° is selected and the tyres are fully utilised when driving straight ahead (top row). When cornering, the low camber is disadvantageous; the tyre cannot transmit the maximum possible lateral force. For sportier cars and sports cars, the compromise can be shifted in favour of cornering (middle row). For racing cars, the compromise is to the detriment of straight-line driving (bottom row). Tyre mileage is not an issue and for most tracks the importance of lateral dynamics far outweighs longitudinal dynamics (ratio up to 4:1). Tire utilization at maximum lateral acceleration is paramount.

Even though driving safety is the primary concern, there is still a classic conflict of objectives when it comes to spring/damper tuning in production vehicles. On the one hand, ride comfort should be high (soft spring, low damper effect) and on the other hand, wheel load fluctuation should be as small as possible (high damper force), Fig. 7.4. A solution to this conflict of objectives is provided by (technically complex) variable dampers.

To increase driving safety, the suspension joints on passenger cars are deliberately designed to be flexible. When lateral or circumferential forces are applied, the corresponding wheel "steers" in such a way that understeering behaviour results. Not

Fig. 7.3 Types of chassis designs, according to [7]. Three designs each are shown as a view of a wagon from behind when driving straight ahead (left), cornering (centre) and extreme cornering (right). In addition, the tyre contact patch shape is shown under the right-hand wheel or the wheel on the outside of the corner. The ideal contact patch utilization is shown in green. In this case the maximum tyre surface is evenly loaded on the road. In combination with lateral slip and missing or unfavourable camber angle, the worst utilisation of the tyre results, the tyre contact patch is strongly trapezoidal (red). Between these extremes you find the middle contact patch utilization (yellow). Depending on the primary goal (mileage or handling), different camber angles are accepted or aimed for, and thus different utilization or wear of the tire

only can racing drivers not use this elastokinematic wheel suspension, it also no longer functions at all in the saturation range[7] of the tires. Racing suspensions are therefore designed with almost play-free, low-friction joints. The racing driver assumes that the vehicle does what he expects through his steering inputs. The major difference between production and racing suspensions becomes apparent when a sports car is converted to GT3 use. In [9] it is reported that the suspension of the sports car has been simplified for racing use. Not least so that problems on the track could be dealt with more quickly. For similar reasons, the adaptive dampers of the production car were replaced, but the control arms

[7] If the lateral force no longer changes via the slip angle, a steering movement of this tyre also no longer brings about a change in the lateral force.

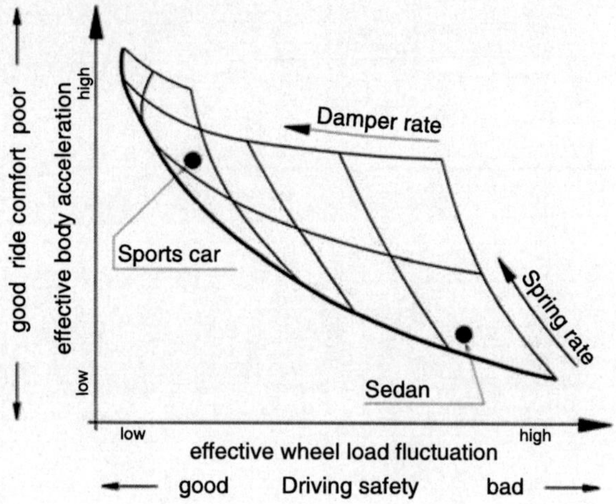

Fig. 7.4 Influence of spring stiffness and damper rate on the driving behaviour, according to [8]. A stiff body spring and a hard damper increase driving safety enormously, but reduce driving comfort due to the high body accelerations caused. For racing vehicles, the choice of spring and damper rate is easier in this respect. For them, only driving safety counts

remained the same. On the GT3 car, the front tires are wider and the weight distribution is more balanced than on the base car. The race car is about 100 kg lighter.

In general, racing vehicles are tuned more neutrally than production vehicles, for which (dynamically stable) understeering behaviour is recommended for safety reasons.

The disturbance force lever arm – the normal distance of the wheel center or the contact point from the steering axis – together with circumferential forces on the front tire (changes in rolling resistance, fluctuations in braking force, . . .) cause disturbance torques around the steering axis that the driver perceives at the steering wheel. In comfort-oriented passenger cars, this disturbance information is kept as small as possible, even if it does contain a certain amount of useful information. In sporty vehicles and racing cars, this information should be more pronounced.

References

1. Jenkins, M., et al.: Performance at the Limit. Business Lessons from Formula 1 Motor Racing. Cambridge University Press, Camebridge (2005)
2. Steiner, M.: Serienmodelle profitieren vom Rennsport-Know-How. ATZ Jubiläumsausgabe 120 Jahre ATZ 03, 138–142 (2018), Serienmodelle profitieren vom Rennsport-Know-how
3. Reuter, B. (ed.): Motorsport-Management. Grundlagen – Prozesse – Visionen. Springer Gabler, Berlin (2018)
4. Kunzmann, S.: Elektronische Reifendrucküberwachung mittels batterieloser Transpondertechnologie. In: Krappel, A. (ed.) Rennsport und Serie – Gemeinsamkeiten und gegenseitige Beeinflussung, pp. 183–197. expert, Renningen (2003)
5. Ziegahn, K.-F.: Umweltschutz und Nachhaltigkeit im Motorsport. In: Reuter, B. (ed.) Motorsport-Management. Grundlagen – Prozesse – Visionen, pp. 311–333. Springer Gabler, Berlin (2018)

6. Indra, F., Tholl, M.: Der 3,0-l-Opel-Rennmotor für die Internationale Deutsche Tourenwagenmeisterschaft. MTZ. **52**(9), 454 ff (1991)
7. Serra, L., Andre, F.: Suspension systems: optimising the Tyre contact patch. AutoTechnol. **4**, 66–68 (2001)
8. Krimmel, H., et al.: Elektronische Vernetzung von Antriebsstrang und Fahrwerk. ATZ. **5**, 368–375 (2006)
9. Scoltock, J.: McLaren MP4-12C GT3. Autom. Eng. Jul.–Aug, 8 f. Caspian Media, London (2011)

Appendix

Glossary

1D simulation:	One-dimensional charge exchange calculation for pre-optimization of pipe lengths, tank volumes and valve timing of an internal combustion engine. Based on one-dimensional unsteady, compressible filament flow (acoustic theory), the engine is simulated as a system of pipes and tanks and wave travel times are determined. The torque curve over the speed can be determined by means of the cylinder filling that occurs. With this method, cam profiles, valve timing, intake manifold lengths, distributor volumes, duct geometries, exhaust pipe lengths and muffler designs can be pre-optimized without an existing test bench engine. Well-known software tools are available from AVL, Gamma Technologies, LMS, Lotus or Ricardo, among others.
ABS:	*Antilock braking system.* A control system in the hydraulic circuit of brake systems reduces the pressure in the brake line applied by the driver via the brake pedal as soon as a wheel threatens to lock. This requires, among other things, sensors that record the wheel speeds and compare them with a setpoint calculated from the deceleration. The main function of an ABS is to maintain the steerability of a vehicle. Locking wheels cannot build up usable lateral forces, which can lead to a loss of stability, especially on the rear axle. If different friction values occur on the left and right (μ-split), the driver must countersteer when braking. ABS can support the driver in this by building up braking force more slowly on the front wheel with more grip (yaw moment control). The yaw moment acting on the vehicle thus also builds up more slowly and there is more time to countersteer. This inevitably increases the braking distance. In

M. Trzesniowski, *Suspension System*,
https://doi.org/10.1007/978-3-658-39847-7

addition, the rear axle is controlled according to the wheel on the low friction value side (select low) [1].

An extension of the ABS control system is ABSplus or CBC (Cornering Brake Control). Here, the system detects the driving situation – in particular cornering – by means of the wheel speeds and regulates the braking forces on the individual wheels accordingly in order to keep the vehicle on track.

Acceleration: Is the rate of change of velocity over time. In purely physical terms, it can be positive or negative, i.e. the speed increases or decreases. In the case of vehicles, however, we generally speak of acceleration and deceleration.

ACO (Automobile Club de L'Ouest): Automobile club that has been organizing the 24 Hours of Le Mans since 1923 and issues the regulations for the vehicles eligible to start. It also sets the regulations for the former European Le Mans Series (ELMS) and the American Le Mans Series (ALMS).

Actuated transmission (*shift by wire*): Manually operated manual transmissions have a mechanical connection (linkage, cables) between the shift lever and the actual actuating device on the transmission housing. If the actual shifting process is carried out via actuators (pneumatic or hydraulic cylinders, electric motors,...), shifting can be initiated by the driver at the push of a button or by the on-board computer (automated transmission).

Air-fuel ratio λ: The air-fuel mixture in the engine ignites and burns satisfactorily only within a certain mixture range. For gasoline, this ratio is about 14.7:1, i.e. 14.7 kg of air are required for complete combustion of 1 kg of fuel (stoichiometric mixture).

The air number λ compares this theoretical demand with the actual mixture present.

$\lambda = \frac{\text{existing mixture}}{\text{stoichiometric mixture}}$. $\lambda = 1$ means that there is a stoichiometric mixture in the combustion chamber. $\lambda < 1$ means there is a lack of air (rich mixture). $\lambda > 1$ means there is excess air (lean mixture).

ALMS: Abbreviation for American Le Mans Series. In this American racing series, the same regulations apply as in the famous 24-hour Le Mans race. However, the races are shorter and last between 2:45 and 12 hours.

Anisotropic: Directional dependence of certain material properties, e.g. modulus of elasticity, strength. The opposite behaviour is called isotropic.

Boundary layer: If air flows around a stationary body with a favourable flow, the air follows the contour of this body the closer to the surface the air layer under consideration is located. Due to friction effects, an air flow slows down the closer it gets to the surface of a stationary body. Thus a static to slow flow is formed at the surface of the body, the thickness of which increases towards the end of the body, the so-called boundary layer. Depending on the shape of the body and pressure conditions, this B. detaches from the surface of the body with increasing thickness and turbulence after a certain distance of flow along it. Outside this B. the friction can be neglected and the velocity of the particles increases with the wall distance.

Buckling: Type of failure of slender, bar-shaped components that transmit compressive forces. Compared to an ideal load that only presses the bar, imperfections occur in reality that lead to additional bending of the component. If the compressive force becomes too great, the bar deflects laterally in the middle and fails due to the excessive bending stress.

CAD: Abbreviation for Computer Aided Design. Components and their assemblies are designed three-dimensionally with the aid of suitable software. Clearances and movement spaces can be controlled more easily than on the drawing board, and numerical simulations (strength, flow investigations, etc.) can also be carried out. Some of the data can be used directly for the production of real components. *See also*: Rapid Prototyping.

CAN: Abbreviation for Controller Area Network. A two-core cable harness used instead of many lines to transmit signals in vehicles. It is a serial bus system in which messages from all participants (ABS control unit, engine control unit, sensors, actuators, ...) can be sent or received one after the other. The CAN controller controls this sequence and sets the priorities if several signals are to be sent simultaneously. The wiring harness in a vehicle with CAN is much shorter than in a conventional system and the number of plug connections is halved.

CART: Abbreviation for Championship Auto Racing Teams. American formula series that was contested in oval stadiums and on road courses. The 2.6-liter V8 engines were powered by methanol and accelerated the single-seaters to 400 km/h. 2003 bankruptcy. Afterwards new start as ChampCar. Champ Car has since (early 2008) initially merged with IRL to form a formula series for financial reasons, and shortly thereafter became officially insolvent.

CFD (*computational fluid dynamics*): Similar to the finite element method (FEM), the geometry to be investigated is broken down into small areas ("grids") for which the equations describing the flow are solved numerically. Depending on the equation used (potential equation, Euler equation or Navier-Stokes equation) and computer performance, even hydrodynamic boundary layers, turbulence and flow separation can be determined.

CFRP: *Carbon-fibre-reinforced* plastic *carbon-fibre-reinforced-plastic*. A composite material in which carbon fibre fabrics are impregnated with reactive resins and processed in several superimposed layers to form moulded parts or with internal honeycomb cores to form sandwich constructions. The targeted arrangement of the aligned fibres makes it possible to influence the mechanical component behaviour in the desired way.

Charging efficiency λ_a: In an internal combustion engine, the C. is the ratio of the fresh charge supplied (this is everything that flows through the air filter) to the charge mass theoretically possible in the cylinder. Thus, the C. is not equal to the degree of delivery. Due to scavenging losses in the charge exchange top dead center, for example, fresh charge can be lost via the exhaust tract. This loss is taken into account in the C., but not in the degree of delivery. In this example, the C. would be greater than the degree of delivery if the mass supplied is greater than the theoretically possible mass. The C. is easier to measure than the degree of delivery.

Chassis: This term is widely used from chassis to body. Probably not least because a clear separation into different assemblies cannot be made in every case. In this book, it is intended to mean the actual, load-bearing structure of a vehicle, to which wheel suspension, powertrain and bodywork components are attached. Accordingly, another term for C. is frame. In most passenger cars, the body is of self-supporting design and thus the bodywork, frame and floor panel are combined to form a structural unit. A clear assignment of the terms to one component each is therefore not possible in this case.

Coefficient of friction μ: Value determined by tests to calculate the frictional force between two bodies. The C. depends, among other things, on the material pairing.

Compression ratio ε: The C. of an engine is the ratio of maximum and minimum cylinder volume. The largest volume results when the piston is at bottom dead center. This volume is therefore the displacement of a cylinder plus the so-called compression volume. The smallest volume is enclosed by the piston at top dead centre. This volume represents the compression volume. The compression volume is made up of the combustion chamber volume and other components that result from the piston crown shape.

Concept: First phase in a design process. In this phase, possible solutions for sub-functions of the overall system are sought and assembled into an effective structure. This phase is followed by the design phase.

Coordinate system: Of the common, vehicle-fixed coordinate systems, the following is used in this book in accordance with DIN ISO 8855 (was DIN 70 000) and ISO 4130: The coordinate origin is the intersection of the vehicle longitudinal center plane with the front axle. The trihedron is aligned to this as follows. The positive x-axis points in the direction of travel, transversely the y-axis points to the left and the z-axis points upwards.

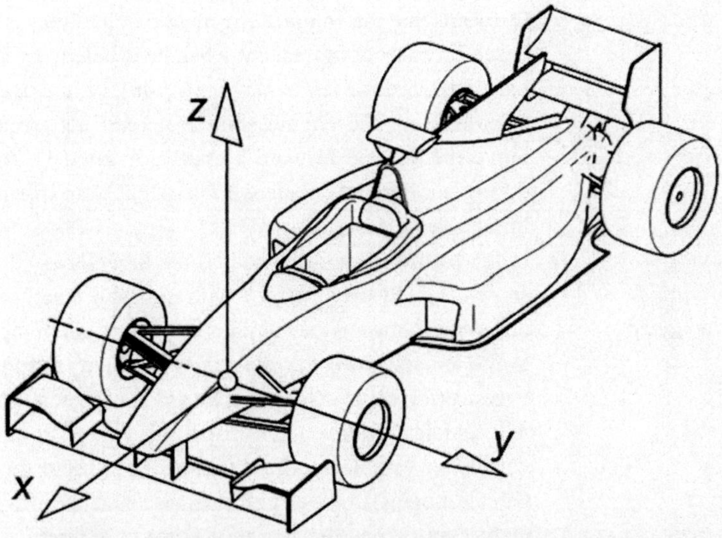

Degree of freedom (DOF): A DOF. is a defined change in position of a rigid body according to a unique and reproducible function. A body has six DOF. in space (three translations and three rotations). The machine elements that enable such DOF. are called joints. A spherical bearing offers three (rotational) DOF. as a ball joint. All displacements (the possible three translations) are locked. The piston rod of a damper leg is a rotational joint. It has two DOF: A translation (compression/deflection) and a rotation (rotation around the piston rod axis).

Design position *(reference standing height)*: Certain position of a vehicle in relation to the roadway, which is used as a basis for the design of chassis. Usually, the car is ready to drive with a half-full tank of fuel and the driver on board. Based on this position, the car can bounce in and out or pitch and roll. All the nominal dimensions, e.g. for ground clearance, ground distance, king-pin inclination, caster, etc., are therefore obtained in D.

Differential construction: Design principle in which a functional carrier (component) is broken down into several parts. Each part can then be optimised for its partial function, e.g. multi-part wheels. The opposite is represented by the integral design.

Draft (*embodiment design*): Phase of design activity in which the proposed solutions literally take shape. The search for solutions before the layout is the concept (conceiving) phase.

Drag: Force acting on moving bodies due to the fact that they displace air and that the air rubs against the surface of the body.

Drivability: For the human driver, a linear, predictable response of a system to his input is best. This is also the case with the accelerator pedal: A good D. means that the engine delivers as much torque as the driver expects based on his foot movement when he accelerates. Special attention is paid to the breakaway behaviour (*tip-in*), i.e. the opening of the closed throttle. Here the engine should increase its torque gently and not move the vehicle forward abruptly. A good D. supports the driver during acceleration maneuvers, especially in overpowered, traction-limited vehicles.

Driving (*operating behaviour*): In the picture (according to [2]) the trajectories of three vehicles are shown, which drive a turn with constant steering angle. The only difference between the vehicles is the position of the centre of gravity. In the understeering car the center of gravity is more forward, in the oversteering car it is more backward compared to the neutral vehicle. All vehicles require a run-in period where slip angles of the front wheels are established first, followed by slip of the rear wheels. The vehicle begins to yaw and deviates from the original straight line. Only then comes the phase of constant cornering. In the case of a neutral vehicle, the slip angles of both axles are the same.

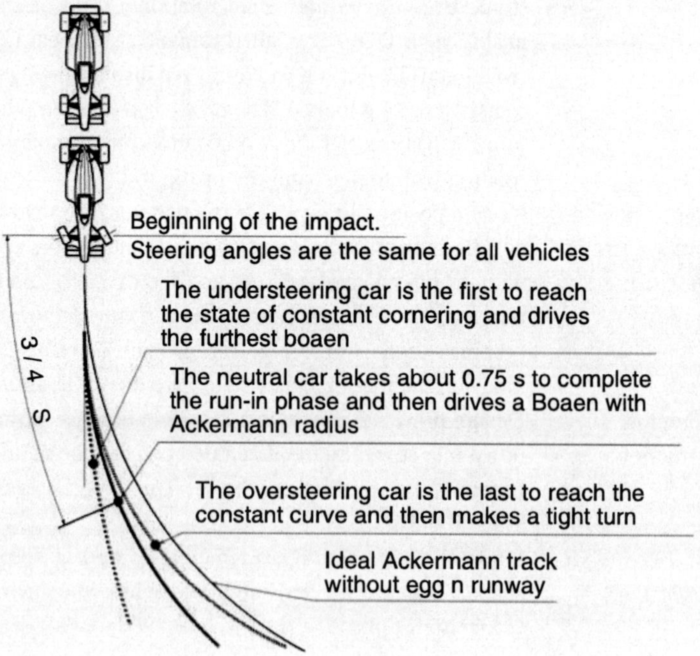

DTM: Abbreviation for Deutsche Tourenwagen Masters (*German Touringcar Championship*). Touring car series whose vehicles must be based on production cars with at least four seats. The engines must be four-stroke gasoline engines with eight cylinders in a V arrangement with 90°. The engine capacity is limited to 4 litres.

Dynamic rolling radius: The distance between the centre of the wheel and the contact patch surface of a stationary wheel is smaller than that of a rolling wheel (static tyre radius). Depending on the tyre design and the wheel speed, the distance increases with increasing speed. The dyn. R. as measured value is calculated from the measured rolling circumference of a tire at 60 km/h.

Elongation at rupture: Relative elongation of a test bar at which fracture occurs. The E. is thus a measure of the toughness of a material. The higher the E., the more favourable is the fracture behaviour of a material, because failure is announced slowly.

ESP: *Electronic stability programme*. Control system that influences driving stability. Sensors record the driving condition of the vehicle, in particular the yaw movement and the steering reaction of the driver. If the state of the car deviates from the calculated target state, the system intervenes by applying brakes to individual wheels or influencing the engine management system. ESP has a stabilising effect, for example, in the event of panic evasive manoeuvres, corners taken too fast or tyre blow-outs.

FIA: Federation Internationale de l'Automobile. World automobile federation with headquarters in Paris. Issues the international sporting code and is thus also the supreme motorsport authority.

Filling pressure (*inflation pressure*): The pressure difference between a tyre and the ambient pressure. The F. is usually measured on a cold tyre. If, for example, the air pressure is 1 bar[1] and the absolute pressure in the tyre is 2.5 bar, the inflation pressure is 1.5 bar. This is also referred to as overpressure.

Finite element method (FEM): Stress calculation of components with numerical methods by a computer. In this process, the component is broken down into (thousands!) finite elements and each element is calculated according to the laws of mechanics. These approximation methods also allow the stress calculation of parts of complex geometry and load, which cannot be calculated with formulas.

[1] 1 bar = 100 kPa. Although the valid SI unit for pressure is Pascal (Pa), the book uses the unit bar, which is more "handy" in practice.

GFRP (*glass-fibre-reinforced plastics*):	Glass fibre reinforced plastic. Plastics reinforced with glass fibres in the form of mats, fabrics and strands of parallel threads to increase strength. GFRP parts are used as bodywork parts, wings, moulded parts.
Glass transition temperature	In plastics, a characteristic change in behaviour occurs when a certain temperature is reached. Below this so-called G., the oscillatory movements of the macromolecules come to a standstill and the materials become brittle. On further cooling, they reach a glassy-hard state. In the case of tires, the greater the difference between the G. of the rubber compound and the operating temperature, the softer the rubber becomes and the more friction it builds up.
Ground clearance:	Distance between the vehicle underbody and the road surface. A distinction must be made between this and the ride height.
Stress:	An external load (force, moment, torque) causes a stress state in the material structure of a component. This stress state is the stress. It is recorded by (technical) stresses (tensile stress, compressive stress, shear stress, ...).
Haptics:	H. is the study of haptic perception. Haptic perception is the active sensing of size, contours, surface texture, weight, etc. of an object through the sense of touch.
Heat exchanger:	Structure in which heat is transferred from one liquid or solid substance at a higher temperature to another at a lower inlet temperature without the two substances being mixed with each other. Depending on the media involved, a distinction is made between, for example, water/air or air/air heat exchangers for the charge air cooling of a turbocharged engine.
High voltage (HV):	Electrical voltages greater than 60 V DC or 25 V AC are referred to as high voltage in the vehicle sector. This concept differentiates this range from "high voltage" in industrial standardisation, which also has completely different numerical values behind it. HV cables and connectors are identified by orange insulation.
IMSA:	International Motor Sports Association. International motorsport authority that runs the American Le Mans races, for example.
IndyCar Series:	Formerly IRL (Indy Racing League): Organizer of the Indianapolis 500 miles (Indy 500 on Memorial Day, May 30) and other races under the same rules on oval tracks. The cars are single-seaters and were powered by methanol-fuelled V8 engines with 3.5-litre displacement. Ethanol is now used as fuel and the engines are 2.2-liter V6 biturbo engines. The cost of the vehicles is limited by the regulations.
Integral construction:	Design principle in which an attempt is made to accommodate all the functions that a component must fulfil in one component. This eliminates the need for weight-increasing and strength-reducing joints. One example of this is sideshafts made from one piece with integrated tripod journals. The opposite is the differential design.

Isotropic: The material properties are the same in all directions. The opposite behaviour is called anisotropic.

Knocking: In a gasoline engine, a limit to the increase in compression results from (partially audible) knock at full load. Knocking is an uncontrolled sequence of a combustion initiated by the spark plug. Especially towards the end of a knocking combustion, high pressure peaks occur which propagate at the speed of sound in the combustion chamber and damage the piston crown, seal surrounds and cylinder head. Therefore, continuous knocking must be avoided at all costs. This is achieved, among other things, by fuel additives, setting a rich fuel-air mixture, reducing the ignition angle, reducing the boost pressure, cooling the intake air, shaping the combustion chamber and targeted cooling of problematic combustion chamber areas (spark plug seat, exhaust valve seat rings).

Labisator, balance spring, Z-bar: Z-shaped connecting rod of the wheels of an axle. In contrast to a stabilizer bar, this arrangement reduces the wheel load differences and thus increases the grip on this axle. The self-steering behavior is thus influenced in exactly the opposite direction compared to the stabilizer.

Laminar flow: The flow runs in superimposed layers that do not mix. This means that there are no cross flows (turbulence).

LMS Le Mans Series: Is a racing series held according to the rules of the famous 24-hour endurance race at Le Mans. The races are usually held over 1000 km. Several drivers are entered per vehicle due to the duration of the race.

Load collective: In general, the load on a component is not constant over time, but changes irregularly. A drive shaft, for example, is subjected to extremely high loads during start-up and after a gear change, but almost no loads at all when braking and driving through a turn. However, simplified representations of loads (forces, moments) are required for the design of components. In test series (e.g. driving through a certain course), loads are therefore recorded and evaluated over time. In such evaluations, among other things, the load heights that occurred and their frequency (temporal proportion, number of load changes) are determined. The figure shows how a load collective is created from a load course.

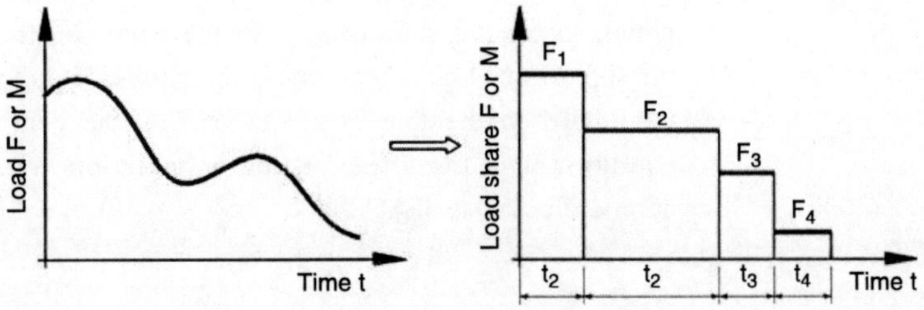

Mass inertia (first Newtonian axiom):	For a body to change its direction of motion or velocity, a force must act on it. This force is proportional to acceleration and mass (Newton's second axiom), $F = m \cdot a$.
Mean pressure, effective $p_{\mathrm{m,e}}$:	During an operating cycle of an internal combustion engine, the pressure in the combustion chamber changes. The mean pressure is a calculated comparative variable. It is an imaginary constant pressure that would perform the same work on the crankshaft as the actual periodically changing pressure in the course of an operating cycle.
Mixture formation:	The task of mixture formation in an engine is to produce an ignitable and combustible air-fuel mixture under all operating conditions. These mixtures only burn satisfactorily in a narrow mixture range. If the air content increases (lean mixture), fuel consumption decreases until combustion misfires increase and the running limit is reached. If the fuel content increases (rich mixture), the engine power increases until the fuel can no longer be completely burned due to a lack of oxygen.
Moment of inertia J_{polar}:	The M. in a rotation is a measure of the resistance to changes in angular velocity and is thus comparable to the mass in a translation. The M. depends on the distribution of the mass in relation to the axis of rotation. The further away mass components are from the axis of rotation, the greater the M.
Momentary pole (*instanteneous centre*):	Every movement between two rigid bodies can be described by a rotation around an instantaneous (=momentary) axis of rotation (= instantaneous pole). The location of the M. is therefore also the location at which no velocity exists between the bodies under consideration. The specification of the M. in coupling gears is done by the combination of the related links. In the picture a four-link gear is shown. If links are mounted in the (fixed) frame 1, the bearing point is taken as the M., i.e. in the example joints 12 and 14 for links 2 and 4. If links under consideration are not directly coupled, the M. can be determined by knowing two velocity vectors belonging to the rigid body. So here the pole for the links 1 and 3. If forces act between two links, the position of the line of action of the force in relation to the M. of these links determines which kinematic state occurs. In the example, the force F_{31} (force on link 3 of link 1) causes a clockwise rotation. If the M. 13 would lie on the line of action of F_{31}, the gear would remain in equilibrium. If the line of action is below the M. 13, link 3 rotates counterclockwise [3].

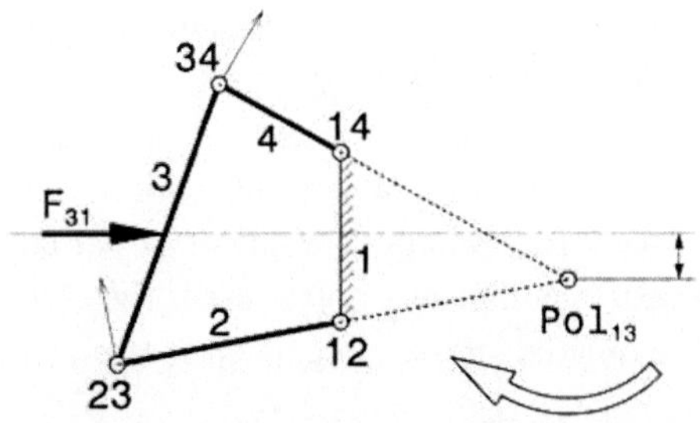

Monoposto (*single-seater*): Single-seater racing car in which the driver's seat is located on the longitudinal centre plane of the vehicle.

Ride height: Is the distance of any point fixed to the vehicle from the road. During set-up, a certain ground clearance is assumed as a reference value and the car is set higher or lower. The ground clearance is therefore only a metrological simplification for determining the ground clearance.

NACA air inlet (*NACA air duct*): (National Advisory Council for Aeronautics). Design of an air shaft according to the recommendations of the NACA.

NASCAR: Abbreviation of National Association for Stock Car Auto Racing Inc. This is the rules authority for the NASCAR Sprint Cup Series (was 2004–2007 Nextel Cup Series, before that Winston Cup), a popular racing series in North America, the majority of which is run on oval tracks in stadiums. It represents the highest level racing series in the United States. The cars look like production cars on the outside, but consist of a tubular space frame and until 2011 were powered by carburetor engines driving a rigid axle on trailing arms via a cardan shaft. Since 2012, gasoline injection (Multipoint Fuel Injection MPFI) has been used.

Natural frequency: An oscillating structure performs an oscillation (= a periodic movement around the rest position) after a single impulse left to itself. The frequency occurring in this process is the natural frequency. If such a structure is excited with a frequency equal or nearly equal to the natural frequency, the oscillation amplitudes become maximum (resonance).

Notch factor: The stress on a component at a point is determined by calculating the mechanical stresses (bending stresses σ, torsion stresses τ etc.). In conventional calculation methods, the so-called nominal stresses are first determined, which result from the cross-section at the notch bottom of the unnotched component and the load. (In contrast, numerical methods exist that allow the approximate calculation of the stress curve, see Finite Element Method). The component is subjected to higher stresses at notch locations. The local stresses at the notch base are considerably greater than the nominal stresses. The notch factor K_f

indicates by how much the maximum stresses become greater than the nominal stresses under dynamic, i.e. time-varying, loading. For bending applies: $\sigma_{b,\ \max} = K_{f,\ b} \cdot \sigma_{b,\ n}$ and for torsion applies: $\tau_{ts,\ \max} = K_{f,\ ts} \cdot \tau_{ts,\ n}$. Where subscript b stands for bending, ts for torsion and n for nominal. A value of $K_f = 1$ therefore means that the component is completely notch insensitive.

Trace of axial stresses in a notched tension bar:

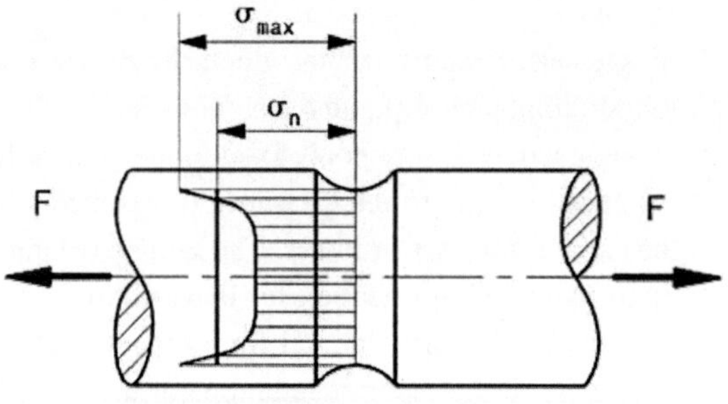

O-arrangement *(back to back a.)*: Two angular contact ball bearings or tapered roller bearings can be arranged in mirror image. If two bearings are fitted in such a way that the pressure lines point outwards (i.e. as in "O"), this is referred to as an O arrangement of bearings. If the pressure lines point towards each other, this is known as an X arrangement.

Octane number: A parameter for the anti-knock properties of a fuel. The higher the octane number, the more resistant the fuel is to knocking. Two different methods are used to determine the octane number: The engine method (engine octane number MON) and the research method (research octane number RON).

Oversteer, AE: loose: See driving behaviour

Planar moment of inertia: Mathematical quantity that follows from the geometry of a cross-section. The MoI. is required in the strength calculation for bending stress on components.

Percentile: Division of a population (normal distribution) into 100 sections. Here statistical division of the dimensions of the human body. This is used to design cockpits and passenger compartments that are suitable for a large proportion of the population. Thus, in car design, the 95% man and the 5% woman cover 90% of the total population. That means only 5% of men are taller and only 5% of women are shorter than the percentiles used.

Pitching: See vehicle movements

Pressure angle: At this angle, the force is transmitted from the outer ring and inner ring in a rolling bearing. The greatest load carrying capacity for a bearing is obtained when the contact angle coincides with the angle of the external bearing force.

Prototype: Racing car of a certain category and group, which is only produced in small numbers or as a one-off.

Quality control: The control of the load (and thus, in the case of constant load, of the speed) takes place in diesel engines by controlling the fuel supply to the combustion chamber. The engine draws in the combustion air unthrottled. This results in the desired air-fuel mass ratio in the combustion chamber solely by changing the fuel quantity.

Quantity control: In gasoline engines with conventional mixture formation (carburetor, intake manifold injection), the load (and thus the speed at constant load) is controlled via throttle elements (throttle valve, slide valve). In the partial load range, the air or mixture quantity supply to the engine is changed by throttling the intake cross-section. At full load, the entire cross-section is released.

Raid, Rally Raid: This generic term covers endurance races that are held cross-country in open terrain, primarily in desert regions. The basic course of the competition is similar to that of a rally, i.e. the vehicles drive from a starting point to a specific destination.

Rally: These are competitions that are held on sections of road that are closed off for the duration of the competition. The road surface can be asphalt or similar, gravel, but also snow and ice. Each vehicle usually drives the route alone. A characteristic feature of R. is that a co-driver announces the course to the driver.

Rapid prototyping (3D printing, additive manufacturing): This includes all processes with the help of which real models can be created directly from 3D CAD information. Some of these processes work like a printer that prints out three-dimensional plastic bodies. Depending on the process and the purpose, these models can be demonstration objects, test parts, casting models or molds. The aim is to quickly arrive at a functional (prototype) part (name!) based on CAD data.

Reynolds number *Re*: Is a dimensionless similarity ratio in fluid mechanics. It compares the inertia forces with the friction forces in a fluid. In a wind tunnel test with a scaled-down vehicle model, the R. values of the model and the original must be the same in order to obtain comparable flow fields and thus useful measurement results.

Rockwell hardness: Indication of the hardness of a material. Is determined by the permanent penetration depth of an indenter (cone, ball) into the workpiece.

Roll: See vehicle movements

Rubber: Collective term for rubber-based elastomers (a subgroup of plastics). The actual rubber is obtained from the thickened sap (latex) of the rubber tree by sulphur treatment (so-called vulcanisation, leading to wide-meshed cross-linking of the molecules). In addition to this natural rubber, there is also synthetically produced rubber. The best known representative is Buna, which is produced by polymerizing butadiene.

For elastomers, the service temperature is higher than the glass transition temperature. For the other plastics (thermoplastics and thermosets) it is exactly the opposite.

Self-steering properties: (See also Driving Behaviour.) In the limit range of the drivable lateral acceleration, the vehicle rotates about its vertical axis in a different way than corresponds to the steering angle during pure rolling of the tyre. The lateral forces increase differently at the front and rear axle (more precisely at each individual wheel). However, lateral forces are only transmitted to the rubber-tyred wheel if it rolls at an angle to its plane (slip). If the slip angle on the front axle of a vehicle increases faster than on the rear axle, the car "pushes" out of the curve via the front wheels. The driver has to turn in more than he would have to if the car was just rolling (understeering S.). The opposite behaviour is called oversteer. The behaviour of a vehicle with (approximately) equally increasing slip angles at all wheels is called neutral. However, a given vehicle need not exhibit the same self-steering behavior over the entire drivable limit range. In addition to vehicles that exhibit constant behavior, there are also those that understeer at low lateral accelerations, but switch to oversteering behavior at higher lateral accelerations, and vice versa. In addition, there is the not inconsiderable influence of circumferential forces on the drive wheels, especially at high engine outputs. For example, a rear-wheel drive vehicle that is neutral when rolling will oversteer when accelerating strongly because the drive forces cause the tires to become laterally softer.

Sequential shifting: A type of gear change in a manual transmission in which the individual gears are only engaged one after the other (sequentially). The driver merely has to make a simple movement. Motorcycle transmissions are an example of this. In contrast, common passenger car manual transmissions have an H-shift, where any gear can be engaged with a compound movement.

Shear modulus (slip modulus) G_{shear}: Material constant determined by shear tests on test specimens. For many materials, the ratio between shear stress and angular distortion remains the same under shear loading. This ratio is the S.

Simulation: Simulations are used to calculate the effects of complex physical relationships, usually over time. For this purpose, the system to be investigated is first represented in simplified form by a model. This model is then described mathematically by a system of equations. With the help of a computer, this system of equations is solved (usually by numerical approximation methods). The results are then displayed (visualized) as graphics or animations. Simulations allow many changes to be made to the system under investigation in a short time, which would either not be possible or too expensive on the real object in isolation. Among other things, the driving behaviour of a car with different tyres, axle loads, centre of gravity heights, downforce, etc. on different routes (which must of course have been recorded three-dimensionally for this purpose) is simulated. Because of the simplifications made, a simulation does not represent reality exactly, but it does provide qualitative information about the factors influencing the system under investigation. By comparison with measured test results, models are tested for their usefulness and subsequently improved.

Spring rate: Specification of the spring stiffness. If the behaviour of a spring is plotted in a force/displacement diagram, the spring characteristic curve is obtained. The slope of the characteristic curve is the spring rate c_{Sp}. The spring rate does not have to be constant, but can change during compression. If the spring becomes stiffer during compression (the line becomes steeper), this is called progressive behaviour. The opposite behaviour is called degressive. The characteristic curve flattens out and the spring becomes increasingly softer when loaded.

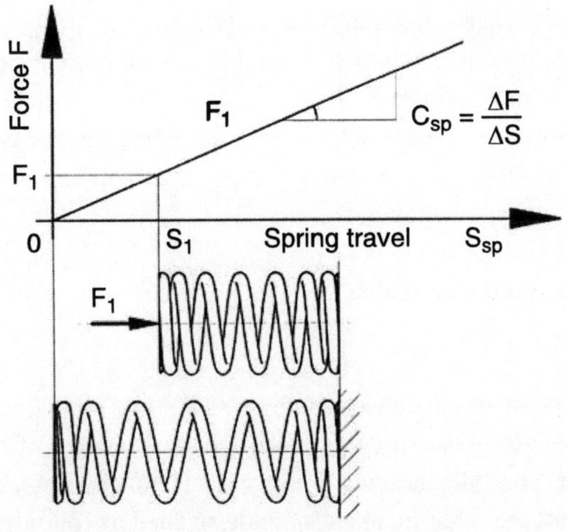

Stroke/bore ratio: The ratio of the piston stroke s to the cylinder bore B in a reciprocating engine. Based on the appearance of a cylinder from the side, a distinction is made between square (stroke = bore), undersquare or long-stroke (stroke > bore) and oversquare or short-stroke (stroke < bore) engine designs. The figure shows schematically a short-stroke (a) and a long-stroke (b) design of a crank mechanism.

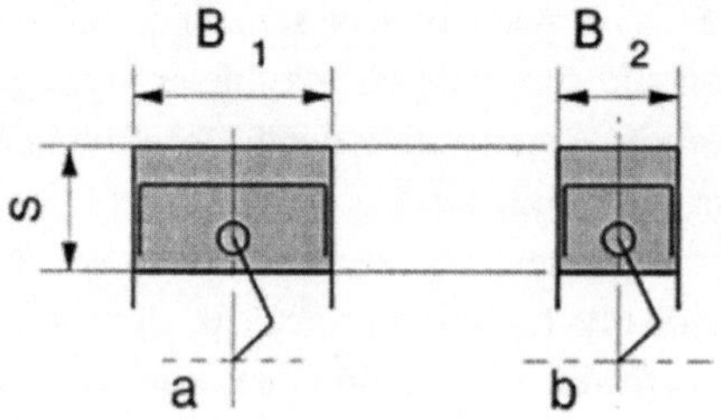

Tensile strength R_m: Material characteristic value determined in a tensile test. It results from the quotient of the maximum force during the test and the cross-section of the test bar before the test. The T. is included in many material abbreviations.

Tension stress: If a component is loaded by external forces and/or moments or if it is hindered in its thermal expansion, a stress occurs in the interior. This stress is recorded mathematically by mechanical stresses, e.g. in N/mm^2 = MPa If the stress at a point in the component exceeds a material-dependent characteristic value, failure (cracking, flow, etc.) occurs at this point.

Tribology: The study of the interaction of friction, lubrication and wear. If relative motion occurs between bodies, this leads to loss of energy (friction) and material removal (wear).

Turbulent flow: Is a flow form in which cross flows and turbulence occur in different sizes and directions.

Tyre contact patch: The contact area of a tyre. All forces between the tyre, and therefore the vehicle and the road, are transmitted via this surface.

Understeer, AE: push: See driving behaviour

Vehicle coordinate system (*axis system*): See coordinate system

Vehicle level: See ground clearance

Vehicle motion: A vehicle – like any rigid body – has six degrees of freedom in space. The possible individual movements (displacements and rotations) about the three main axes are designated as follows:

Displacements (translations): Along the longitudinal axis: twitch (*jerk*).
Along the transverse axis: Push (*drift*).
Along the vertical axis: lifting or lowering (*heave*).
Turns (rotations): Around the longitudinal axis: roll (tilt).
Around the transverse axis: pitch.
Around the vertical axis: yaw.
If a vehicle drives on a roadway, the movements are a combination of the possible individual movements and result from the given movements of the roadway and the driver's influence by steering.

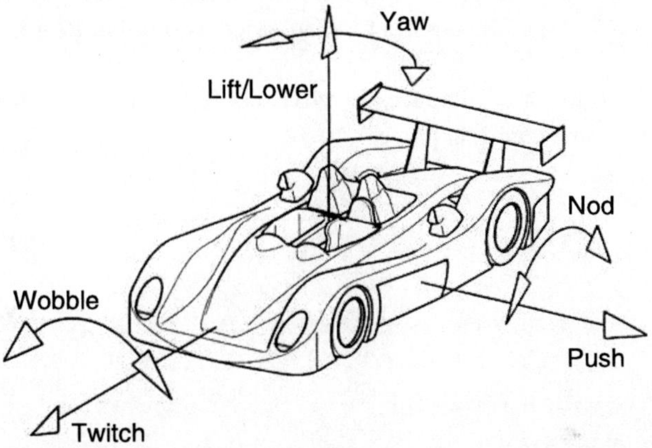

Volumetric efficiency λ_1: In an internal combustion engine, the V. is the ratio of the charge mass actually in the cylinder after completion of the charge exchange compared to the charge mass theoretically possible in the cylinder (= swept volume times air density). In naturally aspirated engines, the V. is less than 1. As the flow velocity (speed) increases, the losses increase due to throttling in the lines and valves. This is partly compensated or even overcompensated by gas dynamic effects at certain speeds.

Wheel frequency: Natural frequency of an oscillating wheel connected to the car body by spring and movable links.

WRC – *World Rally Car*: Rally car based on a generous set of regulations that do not stipulate a minimum number of cars built. The minimum weight is 1230 kg. The number of cylinders in the engines is limited to eight. The displacement depends on the number of valves and the supercharging method. Other rally cars belong to group A and N. For these cars it is required that 2500 basic models are built within one year. To Group A we owe such road cars as the Lancia Delta Integrale, Mitsubishi Lancer Evo and Ford Escort RS-Cosworth.

Yawing: See vehicle movements

Yield strength R_e: Material characteristic value determined in a tensile test. If a bar is pulled with increasing force, it remains elastic until the yield point is reached, i.e. it returns to its original length when the load is removed. For materials without a distinct yield point, a substitute value is determined, the proportional limit $R_{p0,2}$.

Young's modulus: Material constant determined by elongation tests on test specimens. For many materials, the ratio between the stress (load) and the strain obtained (elongation) remains the same. This ratio is the modulus of elasticity. The modulus of elasticity can also be seen as the (of course only theoretical) stress at which the elongation of a bar is 100%, i.e. the bar has reached twice its original length.

Listed below are the differences between corresponding American (AE) and British terms (BE) for some common parts:

Component	American	British
Side shaft	Axle shaft	Half shaft
Drive shaft	Driveshaft	Prop shaft
Wheelhouse	Fender	Wheel arch
(Engine) hood	Hood	Bonnet
Oversteer	Loose	Oversteer
Bevel gearbox	Ring & pinion	Crown wheel & pinion
Understeer	Tight (push)	Understeer
Trunk	Trunk	Boot
Shock absorber	Shock absorber	Damper
Torsion stabilizer	Sway bar	Anti roll bar
Gurney bar	Wicker	Gurney
Windscreen	Windshield	Windscreen

Different racing classes also use different names for what is essentially the same component:

- Wishbone: A-arm/ wishbone, control arm
- Wheel carrier: spindle, knuckle (touring car)/ upright (monoposto)
- tie rod: tie rod/ toe link

References

1. Breuer, B., Bill, K.-H. (eds.): Bremsenhandbuch, 1st edn. GWV Fachverlage/Vieweg, Wiesbaden (2003)

2. Milliken, W.F.: Chassis Design: Principles and Analysis. Society of Automotive Engineers, Warrendale (2002)
3. Neumann, R., Hanke, U.: Elimination of Undesired Motions by Means of Suitable Instantaneous Pole Configuration. Konstruktion, Heft 4, S. 75–77. Springer, Berlin (2005)

Index

M. Trzesniowski, *Suspension System*, https://doi.org/10.1007/978-3-658-39847-7